The Routledge Companion to Urban Imaginaries

The Routledge Companion to Urban Imaginaries delves into examples of urban imaginaries across multiple media and geographies: from new visions of smart, eco, and resilient cities to urban dystopias in popular culture; from architectural renderings of starchitecture and luxury living to performative activism for new spatial justice; and from speculative experiments in urban planning, fiction, and photography to augmented urban realities in crowd-mapping and mobile apps.

The volume brings various global perspectives together and into close dialogue to offer a broad, interdisciplinary, and critical overview of the current state of research on urban imaginaries. Questioning the politics of urban imagination, the companion gives particular attention to the role that urban imaginaries play in shaping the future of urban societies, communities, and built environments. Throughout the companion, issues of power, resistance, and uneven geographical development remain central. Adopting a transnational perspective, the volume challenges research on urban imaginaries from the perspective of globalization and postcolonial studies, inviting critical reconsiderations of urbanism in its diverse current forms and definitions. In the process, the companion explores issues of Western-centrism in urban research and design, and accommodates current attempts to radically rethink urban form and experience.

This is an essential resource for scholars and graduate researchers in the fields of urban planning and architecture; art, media, and cultural studies; film, visual, and literary studies; sociology and political science; geography; and anthropology.

Christoph Lindner is Professor and Dean of the College of Design at the University of Oregon, USA.

Miriam Meissner is Assistant Professor in Urban Studies at Maastricht University, The Netherlands.

The Routledge Companion to Urban Imaginaries

Edited by
Christoph Lindner and Miriam Meissner

Routledge
Taylor & Francis Group
LONDON AND NEW YORK

First published 2019
by Routledge
2 Park Square, Milton Park, Abingdon, Oxon OX14 4RN

and by Routledge
605 Third Avenue, New York, NY 10017

First issued in paperback 2020

Routledge is an imprint of the Taylor & Francis Group, an informa business

British Library Cataloguing-in-Publication Data
A catalogue record for this book is available from the British Library

Library of Congress Cataloging-in-Publication Data
Names: Lindner, Christoph, 1971- editor. | Meissner, Miriam, editor.
Title: The Routledge companion to urban imaginaries / edited by Christoph Lindner and Miriam Meissner.
Description: Abingdon, Oxon : New York, NY : Routledge, 2019. | Includes bibliographical references and index.
Identifiers: LCCN 2018027750| ISBN 9781138058880 (hbk : alk. paper) | ISBN 9781315163956 (ebk)
Subjects: LCSH: Cities and towns. | Sociology, Urban.
Classification: LCC HT119 .R68 2019 | DDC 307.76–dc23
LC record available at https://lccn.loc.gov/2018027750

Typeset in Times New Roman
by Integra Software Services Pvt. Ltd.

ISBN 13: 978-0-367-73304-9 (pbk)
ISBN 13: 978-1-138-05888-0 (hbk)

Contents

Figures

Contributors

Frederick M.C. van Amstel is an Assistant Professor in the School of Architecture and Design and a postdoctoral researcher in the Graduate Program in Informatics at the Pontifícia Universidade Católica do Paraná in Curitiba, Brazil.

Dora Apel is Professor and W. Hawkins Ferry Endowed Chair in Modern and Contemporary Art History at Wayne State University, USA.

Hugh Campbell is Professor of Architecture in the School of Architecture, Planning, and Environmental Policy at University College Dublin, Ireland.

Ben Campkin is Professor of History and Theory of Architecture and Urbanism at the Bartlett School of Architecture, University College London, and Co-Director of UCL Urban Laboratory, UK.

Federico Caprotti is Associate Professor in Human Geography at the University of Exeter, UK.

Federico Cugurullo is Lecturer in Human Geography and Sustainable Urbanism at the University of Manchester, UK.

Ashley Dawson is Professor of English at the Graduate Center, City University of New York (CUNY), and at the College of Staten Island, USA.

Muzi Dong is currently a reporter at The Beijing News Book Review Weekly and holds an MPhil from the Department of Cultural and Religious Studies at the Chinese University of Hong Kong, China.

Nick Dunn is Executive Director of ImaginationLancaster, Professor and Chair of Urban Design, and Associate Director of the Institute for Social Futures, all at Lancaster University, UK.

Barbara Eckstein is Professor of English at the University of Iowa, USA.

Rodrigo Firmino is Professor in Urban Management at the Pontifícia Universidade Católica do Paraná (PUCPR) in Curitiba, Brazil.

Matthew Gandy is Professor of Geography at the University of Cambridge, UK.

Myria Georgiou is Associate Professor in the Department of Media and Communications at the London School of Economics, UK.

Andrea Gibbons is an urban geographer in the Sustainable Housing and Urban Studies Unit, part of the School of Health and Society of the University of Salford, UK.

Rodrigo F. Gonzatto is Adjunct Professor in the School of Architecture and Design at Pontifícia Universidade Católica do Paraná, and PhD candidate in the Graduate Program in Technology and Society at the Universidade Tecnológica Federal do Paraná, both in Curitiba, Brazil.

Tsung-yi Michelle Huang is Professor of Geography at the College of Science of the National Taiwan University, Taiwan.

Sacha Kagan is a Research Associate with the Institute of Sociology and Cultural Organization at the Leuphana University of Lüneburg, Germany.

Nicole Kalms is the Director of XYX Lab and Senior Lecturer in the Department of Architecture at Monash University, Australia.

Jenny Lin is Assistant Professor in the Department of the History of Art and Architecture at the University of Oregon, USA.

Christoph Lindner is Professor and Dean of the College of Design at the University of Oregon, USA, and Honorary Research Professor in Cultural Analysis at the University of Amsterdam, The Netherlands.

Uli Linke is Professor of Anthropology and Department Chair of the Department of Sociology and Anthropology at Rochester Institute of Technology, USA.

Agata Lisiak is Professor of Migration Studies at Bard College Berlin, Germany.

Jason Luger is a Lecturer in City and Regional Planning at the University of California, Berkeley, USA.

Miriam Meissner is Assistant Professor of Urban Studies in the Department of Literature and Art at Maastricht University, The Netherlands.

Gareth Millington is Senior Lecturer in the Department of Sociology at the University of York, UK.

Michele Nastasi is a photographer and researcher of architecture and cities and is currently studying for a doctorate in History of Arts at Ca' Foscari University of Venice, Italy.

Davide Ponzini is Associate Professor of Urban Planning at the Politecnico di Milano, Italy.

Meghan Ashlin Rich is Associate Professor of Sociology and Criminal Justice and Women's Studies at the University of Scranton, USA.

Gillian Rose is Professor of Human Geography at the University of Oxford, UK.

Gerardo Francisco Sandoval is Associate Professor of Planning, Public Policy, and Management at the University of Oregon, USA.

Federico Savini is an Assistant Professor in Urban and Regional Planning at the Faculty of Social and Behavioural Sciences of the University of Amsterdam, The Netherlands.

Debra Benita Shaw is Reader in Cultural Theory in the School of Arts & Digital Industries at the University of East London, UK. She is also co-director of the Centre for Cultural Studies Research.

Tim Simpson is Associate Dean of the Faculty of Social Sciences and an Associate Professor in the Department of Communication at the University of Macau, China.

Richard J. Williams is Professor of Contemporary Visual Cultures at the University of Edinburgh, UK.

Yingjin Zhang is Distinguished Professor of Modern Chinese Literature and Chair of the Department of Literature at University of California, San Diego, USA.

Maite Zubiaurre is Professor in the Department of Germanic Languages and the Department of Spanish and Portuguese at the University of California, Los Angeles, USA.

Acknowledgements

Companions, by their very nature, are highly collaborative book projects. They bring together many authors and involve extensive and extended intellectual exchange. We are tremendously grateful to all of the authors in this companion for joining our shared conversation about urban imaginaries and for contributing their original research to this book. We were fortunate to be able to host a workshop on "City Futures and Urban Imaginaries" at the University of Oregon's historic White Stag Block in Portland, where many of the authors had a chance to meet and discuss their work in person. For generously supporting the workshop, as well as our editorial partnership, we want to thank the University of Oregon's Global Faculty Collaboration program. In Amsterdam, Tijmen Klous was a superhero in providing editorial assistance. At Routledge, Andrew Mould and Egle Zigaite have supported the project with wonderful advice and professionalism throughout. The thinking that went into this book draws on our experiences of teaching and researching at the Universities of Amsterdam, Oregon, Lancaster, and Maastricht. We want to thank our students, colleagues, and friends at these universities for sharing their knowledge, their ideas, and their own, unique urban imaginaries.

We dedicate this book to Hannah and Joseph in Oregon and to Jeannine and Bernd in Troisdorf.

Christoph Lindner and Miriam Meissner

1

Introduction

Urban imaginaries in theory and practice

Christoph Lindner and Miriam Meissner

Why urban imaginaries?

Urban imaginaries are everywhere: from new visions of smart, eco, and resilient cities to postcard images of urban vintage and nostalgia; from architectural renderings of starchitecture and luxury living to performative activism for new spatial justice; and from urban utopias in film, photography, and literature to augmented urban realities in location-based gaming and mobile apps. Urban imaginaries form part of our everyday lives in the city, encompassing tourism, city branding, art and architecture, planning, policymaking, and more. Urban imaginaries travel old and new media, and – importantly – they are studied in disciplines spanning the humanities, social sciences, and art and design fields.

But why study urban imaginaries? Cities tend to be seen as complex sets of social, spatial, and material characteristics. They are understood in terms of design and infrastructure, inhabitation and density, diversity, economic activity, political organization, ecology – the list goes on. Imagination is often posited as secondary to such definitions of the urban. At the same time, however, contemporary urban studies increasingly acknowledges the role that imagination plays in shaping cities, especially in relation to the future. This shift is reflected in a growing body of research on urban visions in social futures and transition studies; in the escalating controversies around smart city futures in urban planning, architectural, and policymaking discourses; and in the scholarly attention given to urban social movements and activism as practices of resistance against authoritarianism, repressive politics, austerity, and global neoliberalism.

The issues tackled in this line of research are wide-ranging. To what extent can eco-city experiments achieve sustainable development? How do urban activist interventions – from Occupy Wall Street and the Hong Kong Umbrella Movement to Istanbul's "Standing Man" and the Rebel Clown Army – revise global ideas of public space and democracy? What are the risks of imagining urban futures predominantly in technological terms? And, under what conditions can smart city plans also become roadmaps to urban social equality and resilience in the face of uncertain futures? As these questions suggest, urban imaginaries are not just "matters of the mind", but also manifest and find expression in lived urban space.

Therefore, to answer the question "why study urban imaginaries?" it is important to understand space as *simultaneously* material, conceptual, experienced, and practiced. Historically, this way of thinking about space is largely attributed to the so-called "spatial turn" in the humanities and social sciences. In particular, it developed from the Chicago School of Urban Sociology as well as the writings of scholars such as Georg Simmel, Walter Benjamin, Kevin Lynch, Henri Lefebvre, Michel Foucault, Michel de Certeau, David Harvey, Doreen Massey, and Edward Soja. In this chapter, we aim to outline the contribution that some of these scholars have made to the concept of the urban imaginary. First, however, we want to highlight that the spatial turn in social and urban thinking has distinct geographical and gender coordinates. While it has propelled an understanding of knowledge as something that is shaped by its locus in social space, the spatial turn itself has been largely ascribed to a predominantly male group of Western thinkers.

The same critique of gender and geographic imbalance applies, of course, to many fields and traditions within urban studies. In *Ordinary Cities: Between Modernity and Development* (2006), for example, Jennifer Robinson shows that current concepts of the "global city" and "modern city" are predominantly based on a Western tradition of urban thought and design that hierarchizes cities according to levels of development and marginalizes a vast array of different potential visions, understandings, and practices of contemporary urbanity (Robinson, 2006).

Much of the recent scholarship on urban imaginaries has developed in an attempt to rethink urbanism beyond this gendered, Western perspective. To this end, it explores urban imaginaries across multiple media and geographies in order to de-centre and diversify current concepts of the city (see Dibazar et al., 2013). This is particularly relevant given the various global challenges that cities face today. These challenges include environmental change, social inequality, transnational migration, and the privatization of public spaces and services, to name just a few. Along with qualities such as solidarity, determination, and ethical capacity, coping with these challenges requires creativity. As creativity draws on imagination, it is paramount within urban studies to map, examine, and critically question the various forms of urban imaginaries that exist in contemporary urban theory, arts, and culture, as well as in everyday life. To that end, this book sets out to provide a broad, interdisciplinary, and critical overview of the current state of research on urban imaginaries. In so doing, the book directs special attention to the role that urban imaginaries play in shaping the future of urban societies, communities, built environments, and ecologies.

In this introduction, we first outline how a number of key thinkers have contributed to the development of the concept of urban imaginaries in critical spatial theory. Building on these insights, we map the current state of urban imaginaries research, which is influenced not only by global political and ecological transformations but also to a large extent by technological and new media innovations. Finally, we reflect on the role that urban imaginaries play in shaping global-urban futures. We conclude by providing an overview of the different sections in this book and their thematic organization and content.

Spatial thinking and the urban imaginary

The twentieth century has produced an extensive range of critical theory on space and urbanism. Exploring the conditions of early-twentieth-century metropolitan life, theorists such as Georg Simmel and Walter Benjamin were amongst the first authors to highlight the entanglements of urbanism and the human mind. While Simmel examined how the myriad stimuli of the modern metropolis affected the "mental life" of its inhabitants, Benjamin analyzed how the urban *flâneur* perceived and made sense of the dazzling shopping arcades of late-nineteenth-century Paris. Theorists such as Simmel and Benjamin were instrumental in establishing urban imaginaries as an object of interest in urban research, even if they did not explicitly employ the concept in their writing. It was not until the second half of the twentieth century, however, that the concept was fleshed out, popularized, and eventually coined.

Urban planner Kevin Lynch's *The Image of the City*, published in 1960, contains one of the first and most explicit explorations of the role of imagination and cognitive processes in shaping the human experience of cities:

> What does the city's form actually mean to the people who live there? What can the city planner do to make the city's image more vivid and memorable to the city dweller? To answer these questions, Mr. Lynch, supported by studies of Los Angeles, Boston, and Jersey City, formulates a new criterion – imageability – and shows its potential value as a guide for the building and rebuilding of cities.
>
> *(MIT Press, 1960)*

As this blurb for *The Image of the City* indicates, Lynch's theory of planning and design makes urban inhabitants' mental images of the city a key field of concern. Exploring the individual impressions that people had of different American cities in the 1950s, Lynch argues that urban inhabitants tend to perceive cities in a highly individualized, partial, and fragmented manner (Lynch, 1960: 2). Beyond that, cities are dynamic places. The city "is the product of many builders who are constantly modifying the structure for reasons of their own" (ibid.). The role of the urban planner, Lynch argues, is to "shape the city for sensuous enjoyment" (ibid.). For him, this means organizing the everyday chaos of urban experience according to principles of imageability. In particular, Lynch insists that cities ought to be "legible" to their inhabitants.

To create this legibility, Lynch proposes organizing the city's "visual quality" (Lynch, 1960: 3) according to coherent structuring elements, such as nodes, districts, pathways, and landmarks. He proposes that this visible organization of the city "is fundamental to the efficiency and to the very survival of free-moving life" (ibid.). Lynch's idea of urban theory favours a totalizing, top-down approach to urban design. The city is to be structured by the planning mastermind, who orders urban chaos into a preconceived grammar and syntax. Moreover, his theory is prescriptive. It aims to provide a manual for modern urban planning and design.

This rationale was later criticized by the urban theorist Michel de Certeau, who regarded modern, top-down visions of the city an "oblivion and a misunderstanding" of everyday urban practices (De Certeau, 1988: 93). De Certeau, writing some 20 years after Lynch, compares different forms of urban imaginary in order to shift the urban focus from plan to practice. In his essay "Walking in the City", De Certeau contrasts the panoramic vision of New York as seen from the top of the former World Trade Center (WTC) with the street experience of the urban pedestrian. De Certeau's main critique is that top-down urban visions exclude a range of relevant urban sensations and activities. Therefore, the urban panorama is wrongly assumed to provide the most knowledgeable perspective on urban life. The urban panorama is a "concept city", a "fiction of knowledge" (ibid.), that neglects crucial details of urban street life, including non-visual qualities such as sound, smell, and touch.

De Certeau's reflection on the urban panorama is not just a critique of top-down planning perspectives but also a critique of vision as such. In urban life, vision tends to be treated as a privileged sense. This tendency is also reflected in Lynch's focus on the "visual quality" of the city. Yet, as De Certeau explains, the ordinary, everyday practice of the city begins "below the thresholds at which visibility begins" (De Certeau, 1988: 93). The city is produced or "written" by urban practitioners "whose bodies follow the urban thicks and thins of an urban 'text' they write without being able to read it" (ibid.). De Certeau demonstrates how urban life is shaped not only by the urban visions of planners, architects, or cartographers, but – just as significantly – through the everyday actions of inhabitants, whose urban imaginaries are neither complete nor fully visual but instead partial and practical.

Regardless of this critique, it is worth noting that Lynch's *The Image of the City* contributed immensely to the recognition of urban imaginaries as key factors in city life. As Lynch suggests, to understand "environments at the urban scale of size, time, and complexity ... we must consider not just the city as a thing in itself, but the city being perceived by its inhabitants" (Lynch, 1960: 3). His work therefore laid an important foundation for the understanding of urban life as simultaneously built form, idea, and experience, combining objective elements ("the city as a thing in itself") with subjective elements.

Social theorist Henri Lefebvre further developed this understanding of urbanity in conceptual terms. In *The Production of Space*, first published in French in 1974, Lefebvre criticizes common notions of space in everyday language as well as in philosophy for failing to recognize that space, much like human ideas, is a social product. Space, he argues, is socially produced. What may, at first glance, appear as a platitude, gains further significance when considering that Lefebvre refers to space not only as a physical "living space" but also as a concept. Lefebvre laments that:

> we are ... confronted by an indefinite multitude of spaces, each one piled upon, or perhaps contained within, the next: geographical, economic, demographic, sociological, ecological, political, commercial, national, continental, global. Not to mention nature's (physical) space, the space of energy flows, and so on. ... [I]t should be pointed out that the multiplicity of these descriptions and sectionings makes them suspect. The fact is that all these efforts exemplify a very strong – perhaps even the dominant – tendency within present day society and its mode of production. Under this mode of production, intellectual labour, like material labour, is subject to endless division.
>
> *(Lefebvre, 1991 [1974]: 8)*

Inspired by Karl Marx's critique of idealism, Lefebvre applies historical materialism to space as both a physical thing and an abstract concept. Both are shaped by the same capitalist system of production, which is based on class division and the exploitation of labour. In this context, hegemonic concepts of space (for instance the distinction between urban and rural space) form part of the so-called superstructure – the ideological system that maintains and reproduces capitalist conditions of production. His project, therefore, is to advance a more critical understanding of space as a "social product", the product of power relations, and to "demonstrate the active – the operational or instrumental – role of space, as knowledge and action, in the existing mode of production" (Lefebvre, 1991 [1974]: 11).

Lefebvre makes an important contribution to the study and critique of urban imaginaries as instruments of power. Interestingly, his writings on social space coincide historically with the elaboration of similar insights in the field of visual studies, including John Berger's book and television series *Ways of Seeing* (1972) as well as Laura Mulvey's (1975) writings on the "male gaze" in narrative cinema. Examining renaissance artworks (Berger) and mainstream Hollywood cinema (Mulvey), Berger and Mulvey show how visual perspectives, as tools of representing three-dimensional space on a two-dimensional canvas/screen, both express and reproduce social power relations, in particular, gender inequalities. Lefebvre himself often resorts to visual art in order to illustrate the ideological nature of space. For example, in his 1956 essay on the painting *L'Ouvrier mort* by Éduard Pignon, Lefebvre interprets the artist's inobservance of geometric rules of visual representation as a break-away from "the mechanism of capitalist technology" (Elden, 2004: 182).

Another noteworthy historical coincidence is that *The Production of Space* was published at almost the same time as Michel Foucault's *Discipline and Punish: The Birth of the Prison* (1975), which is recognized in critical theory for demonstrating a strategic interrelation between space and power in modern Western societies. In this work, Foucault analyzes the spatio-visual workings of Jeremy Bentham's panopticon, a form of prison that facilitates surveillance through architecture. In the panopticon, the prison guard is located in a tower in the middle of the prison, while the cells are positioned in a surrounding circle. While the guard has the ability to observe any and all cells, inmates cannot know if and when they are being observed. Foucault presents the panopticon as indicative of the means by which social control is exerted in modern societies. The intended effect of the panopticon is to "induce in the inmate a state of conscious and permanent visibility that assures the automatic functioning of power" (Foucault, 1995 [1975]: 201).

According to Foucault, architectural structures such as the panopticon constitute technologies for producing "docile bodies". Because they have internalized dominant rules of discipline – both physically and mentally – such bodies integrate seamlessly into a range of other, similar modern institutions, including the school, the factory, and the military. In terms of the history of spatial thinking, Foucault's work is significant because it shows, through empirical examples, how spatial structures produce cognitive-behavioural patterns and vice versa. Additionally, Foucault's theory highlights the way space can acquire an almost technological function, serving as a machine for creating and sustaining power relationships, because "power has its principle not so much in a person as in a certain concerted distribution of bodies, surfaces, lights, gazes" (Foucault, 1995 [1975]: 202).

To return to the topic of urban imaginaries, it is worth noting that Lefebvre and Foucault both emphasize hegemony as a force that conditions the interlocked spaces of architecture, the body and the mind (Lefebvre, 1991 [1974]: 11). Similarly, urban imaginaries – from municipal land use plans to Situationist psycho-geography – do not exist outside of social power structures. For these reasons, it is crucial to examine the relationship between urban imaginaries and power.

To gain a more sophisticated understanding of this relationship, Lefebvre proposes a conceptual triad consisting of what he calls "three moments of social space" (Lefebvre, 1991 [1974]: 40). He calls the first moment "spatial practice", which "under neocapitalism" refers to "a close association . . . between daily reality (daily routine) and urban reality (the routes and networks which link up the place set aside for work private life and leisure)" (Lefebvre, 1991 [1974]: 38). Spatial practice describes space as *perceived* by the people who occupy it. Lefebvre therefore also calls it "perceived space". "Representations of space" or "conceived space" form the second moment of social space. *Conceived* of by "scientists, planners, urbanists, technocratic subdividers, social engineers" and "a certain type of artist with scientific bent", they form the "dominant space in any society" (ibid.). Finally, Lefebvre defines the third moment of social space as "the dominated . . . space which the imagination seeks to change and appropriate" (ibid.). Somewhat counter-intuitively, Lefebvre calls this third moment in space "representational spaces" or "*lived* space".

While most readers would probably associate these attributes – lived and representational – with the first and second moments of social space (spatial practice and spaces of representation), it is not a coincidence that Lefebvre chose a theoretical terminology that makes different concepts of space overlap. Lefebvre sought to develop a theoretical understanding that engenders "space in its totality" rather than "the space of this or the space of that" (Lefebvre, 1991 [1974]: 37). Interesting in this context is that Lefebvre's third moment ("representational spaces" or "lived space") appears to be the one that connects all other moments of social space: the perceived and the conceived, spatial practice, and spaces of representation. In positing "the space which the imagination seeks to change and appropriate" (Lefebvre, 1991 [1974]: 38) as both the *connecting* and *transformative* factor in social space, Lefebvre ascribes considerable significance to the spatial imaginary.

This line of thinking has been taken up and further developed by the geographer and spatial theorist Edward Soja, who seeks to combine modern and postmodern ways of thinking space in his 1996 book *Thirdspace: Journeys to Los Angeles and Other Real-and-Imagined Places*. Taking inspiration from Lefebvre in particular, but also from authors such as Foucault, bell hooks, and Jorge Luis Borges, Soja aims to "expand the scope and critical sensibility of ... already established spatial and geographical imaginations" (Soja, 1996: 1). The concept of "Thirdspace", in this context, is intentionally kept flexible to capture the ever-changing "ideas, events, appearances, and meanings" of space (ibid.). It thus bears similarities with Lefebvre's idea of "representational spaces" but goes further in blurring the conceptual distinctions between different "moments of social space". According to Soja, *Thirdspace* invites a spatial thinking through which

> everything comes together... subjectivity and objectivity, the abstract and the concrete, the real and the imagined, the knowable and the unimaginable, the repetitive and the differential, structure and agency, mind and body, consciousness and the unconscious, the disciplined and the transdisciplinary, everyday life and unending history.
>
> *(Soja, 1996: 56–7)*

What may appear as an arbitrary jumble of space-related concepts is, in a way, the main point of Soja's *Thirdspace*. Space for Soja, as for Lefebvre, needs to be understood in its totality. Yet, at the same time, both theorists insist that the totality of social space is produced in and through a plurality of spaces that are subjective as well as objective, abstract as well as concrete, conceptual as well as embodied, etc. Both theorists assume that the spatial imaginary is central to this plurality in that it reflects, connects, and shapes the co-production of different kinds of spaces in social space.

Drawing on this extended body of spatial thinking, we wish to argue that urban imaginaries play an equally defining role for city space. Urban imaginaries meaningfully interlink the different structures and signs, minds and bodies, facts and subjectivities, actualities and virtualities, economies and ecologies of urban social space. To indicate how this works, Soja describes urban imaginaries as "the mental or cognitive mappings of urban reality and the interpretative grids through which we think about, experience, evaluate, and decide to act in the places, spaces, and communities in which we live" (Soja, 2000: 324). What this definition suggests is that urban imaginaries are always in transition, reflecting transformations in the intersecting fields of culture, politics, economy, nature, and – perhaps most conspicuously – technology.

In the following section, we wish to sketch some of the more recent transformations that have produced what we propose to call "new" urban imaginaries. However, it is important to note that the adjective "new" here is not meant to designate that "new" imaginaries are unaffected by, or radically different from, earlier imaginaries. Nor do we mean to suggest that "new" imaginaries replace previous imaginaries. Instead, we wish to map how urban imaginaries have developed, and continue to develop, in response to the continuously shifting forces of urban change.

Global challenges, "new" imaginaries

To explore urban imaginaries that have appeared in the early twenty-first century, it helps to consider some of the most important transformations that have affected cities throughout the late-twentieth century. In 2000, Edward Soja sought to capture some of the dominant discourses of global-urban change via the concept of the "postmetropolis". Here, the prefix "post" does not suggest a radical break with or replacement of modern urbanism, which is commonly associated with the term "metropolis". Rather, the concept of the postmetropolis brings into focus how, due to the global advent of different forms of "post-isms" – such as postmodernism and post-Fordism – urban social space has diversified into new forms of complexity.

Soja summarizes these developments via what he calls "six discourses on the postmetropolis" (Soja, 2010). While it goes beyond the scope of this introduction to detail all of these six discourses, it is worth listing some of the key social developments that have co-produced postmetropolitan urbanism. Most of these developments are themselves contested in critical scholarship. Yet, it is fair to say that all of them, regardless of their various definitions, have been central to recent social and cultural studies of cities.

The first of these developments is neoliberal globalization. The growing and accelerating worldwide mobility of people, goods, data, and capital, as well as the expansive deregulation of the global market economy, have intensified both diversity and social divisions within and between international cities. Neoliberal globalization has moreover given rise to the process of financialization, broadly understood as "the increasing dominance of financial actors, markets, practices, measurements and narratives, at various scales, resulting in a structural transformation of economies, firms ... states and households" (Aalbers, 2015: 214). Through financialization, cities have experienced new types of urban growth and decline that, as the financial crisis of 2008 has shown, tend to manifest in extreme ways.

The second development, closely related to the first, is the advent of post-Fordism in most industrialized countries. Post-Fordism has caused cities worldwide to restructure and re-profile their means of economic value creation. In the process, cities have re-branded their image under labels such as "global city", "creative city", or "smart city". At the same time, the restructuring of economic activity from the industrial to the post-industrial has brought about stark changes in the socio-economic configuration of cities, creating new forms of struggle for urban justice and equality (Sassen, 2000).

A third, important force that has co-produced the postmetropolis is technological innovation, particularly technologies of mobility and communication. At the time Soja wrote *Postmetropolis: Critical Studies of Cities and Regions*, the Internet was not widely used, and smartphones and mobile apps did not yet exist. These more recent techno-innovations have perpetuated and amplified some of the urban transformations Soja outlines in *Postmetropolis*. This applies, for example, to the notion of the "Simcity", which Soja describes as the "procession of simulacra and the growing hyper-reality of urban life" (Soja, 2010).

Drawing on postmodern philosopher Jean Baudrillard's writings on *Simulacra and Simulation*, Soja uses the Simcity discourse in order to theorize how postmetropolitan urban spaces constitute, to a large extent, a simulated reality, a reality constructed of signs and mediated reproductions. Importantly, however, this does not make those spaces less real. Baudrillard insists that simulations are real – "more real than real" (Baudrillard, 2006 [1981]: 81). With regard to contemporary cities, this means that fiction-inspired urban features, such as the "Harry Potter's London" tours or the futuristic Oriental Pearl Tower in Shanghai, are not any less "real" than other aspects of these cities. They are just as real because, every day, they shape the actions and experience of countless urban tourists and inhabitants.

Technological innovation has further amplified the influence of simulation in contemporary urban life. A striking example of this is the location-based augmented reality game Pokémon Go. Shortly after its release in July 2016, Pokémon Go had a remarkable effect not only on the urban movements and perceptions of its users, but also on the traffic at the sites where users were led by their applications. This is one of many examples of how technologically facilitated urban imaginaries can co-determine urban space and experience. Other examples include the democratization of GPS, mobile apps, geo-tagging, digital photography, and editing software. All of these have advanced new modes and aesthetics of urban street navigation and visualization (Dibazar and Naeff, 2018).

Many of the "new" urban imaginaries that have appeared in the early twenty-first century prolong and intensify tendencies outlined by Soja in *Postmetropolis*. Yet, it is important to note that some of these new urban imaginaries have also developed as symptoms of, and reactions to, crises that have impacted cities over the last decades. Indeed, imaginaries of urban crisis and catastrophe are amongst the most memorable imaginaries of early-twenty-first-century urbanism. The violent destruction of the WTC's Twin Towers on 11 September 2001 is paradigmatic of this tendency and shows how urban catastrophes transmute into globally circulating signifiers and spectacles. As Jenifer Chao argues,

> [t]he sequence of visual images immortalizing 9/11 is firmly fixed: planes soar towards the towers and strike them with precision. Explosions ensue. Trapped workers protrude from broken windows and flail their arms for help. The towers collapse while ash-coated New Yorkers flee the scene. . . . They are extraordinary scenes that defy the banality of prosaic images from everyday city life, and yet, are disturbingly familiar to the artificial aesthetics of Hollywood disaster movies and some popular video games.
>
> *(Chao, 2018: 23)*

The visual imaginary of the falling towers has further inspired a wide range of cultural imagery and has even influenced the ways in which subsequent urban crises have been portrayed in narrative and visual culture. The film *Wall Street: Money Never Sleeps* (2010), for example, visualizes the stock market crash of 29 September 2008, as a fall from a skyscraper on a sunny day in downtown Manhattan (Meissner, 2017: 35). This shows how today's mass-mediated "globalization of fear" (Davis, 2001) establishes specific urban imaginaries as reproducible tropes of crisis.

In addition to terrorism and economic crisis, numerous natural disasters have affected cities in the last few decades. Although natural disasters occur globally and across diverse geographies – both urban and rural – it is notable how disasters affecting cities, such as Hurricane Katrina in New Orleans in 2005 or Hurricane Harvey in Houston in 2017, have attracted particularly intense media attention. Aerial photographs of flooded and desolate urban housing blocks act as recognizable signifiers of New Orleans in the aftermath of Katrina. These urban imaginaries are mirrored and taken to visual and conceptual extremes in natural disaster films such as Roland Emmerich's *The Day After Tomorrow* (2004) or Steven Quale's *Into the Storm* (2014), which demonstrates how "new" urban imaginaries of disaster travel between reality and fantasy.

Yet, our very choice of examples also connects to what may be called a global "geo-politics of the imagination", which tends to be Western-centric. Hurricanes Katrina and Harvey are part of a planetary network of natural disasters that have affected people and places across the globe. These include, amongst many others, the 2008 Cyclone Nargis in Myanmar, the 2008 Afghanistan Blizzard, the 2010 Earthquake in Haiti, and the 2011 East Africa Drought, to name only a few. It would be impossible and also indelicate to compare the severity of these different disasters. However, it is worth emphasizing that, in mainstream Western media, disasters that happen in the Global North receive a different type of media attention than those that happen in the Global South (Joye, 2009).

Postcolonial power relations are a significant part of what shapes global imaginaries of disasters (Carrigan, 2015). In this context, the very question "what is a disaster?" has political implications. Definitions of disasters as "sudden accidents", for example, tend to obscure the fact that many disasters are the outcome – direct or indirect – of so-called "systemic" social, economic, political, and ecological risks created by humans. Beyond that, understanding disasters as "sudden accidents" tends to overlook the slow and chronic disasters that are ongoing without a clear beginning or end, such as poverty or ecological degradation. Imaginaries of disasters as "sudden accidents" thus exclude a broad range of disastrous developments that are unfolding or ongoing around the globe.

What all of this shows is that imaginaries are intrinsically political. They shape and are shaped by socio-spatial relationships of inclusion and exclusion, empathy and apathy, solidarity and segregation, as well as many others. Considering the conditions and processes through which urban imaginaries are produced is therefore crucial. How do socio-spatial power relations *produce* particular urban imaginaries, and how do these imaginaries, at the same time, *reproduce* those relations?

To illustrate this dynamic, it is helpful to envision how imaginaries of an urban "badland" or "no-go area" tend to reproduce the same conditions of urban abandonment that caused them in the first place. For a neighbourhood, the reputation of high crime and violence rates, for example, is likely to create conditions under which residents do not linger in the streets, and non-residents do not visit the neighbourhood for leisure, shopping, tourism, or other activities. The resulting desertion of urban public space is likely to weaken the liveliness and social capital that urbanists such as Jane Jacobs (1961) consider crucial to the safety of urban neighbourhoods and communities. In short, the reputation of being unsafe can cause actual damage to safety. Meanwhile, the reputation of being a "hip and thriving" city district tends to attract locals and tourists alike, thus acting as a self-fulfilling prophecy for neighbourhoods, particularly if it is reiterated in travel guides and other "trendsetting" urban media. Both scenarios exemplify how urban imaginaries – whether positive or negative – can reproduce the socio-spatial conditions that brought them into being.

The question arising from this dynamic is what role urban imaginaries can play in *reconfiguring* – rather than reproducing – the socio-spatial politics of cities. How can urban imaginaries intervene in the interconnected fields of urban class politics, gender politics, geo-politics, and eco-politics? The goal of this companion is to pursue this question through theoretical reflection and international case studies. The chapters in this volume analyze how urban imaginaries co-shape the spatio-politics of cities such as Abu Dhabi, Amsterdam,

Baltimore, Bhopal, Detroit, Dubai, Fenyang, Flint, Hamburg, Hanover, London, Los Angeles, Louisville, Macau, Melbourne, Mexico City, Miami, New York, Paris, San Diego, Shanghai, and Singapore. Before introducing the content of this volume in more detail, we first wish to take the notion of "intervention" as a starting point for discussing a particular type of "new" urban imaginary that has, in our view, proliferated in significant ways in the twenty-first century.

Throughout the last few decades, myriads of urban imaginaries have emerged in connection with efforts to tackle global social, economic, and ecological challenges. Urban imaginaries, in this context, emerged largely as "tools", in the sense of being instrumental to the interventions of both top-down and bottom-up urban initiatives. In urban architecture, design, planning, and policymaking, for example, visions of the "smart", "green", and "creative" city have gained increasing popularity. To varying degrees, these visions are currently put into practice in projects such as the London DataStore project, the hyper-planned, renewable-energy Masdar City project in Abu Dhabi, the circular economy initiative Circular Buiksloterham in Amsterdam North, or the government of India's Smart Cities Mission for sustainable urban renewal and retrofitting.

All of these projects are intended to solve acute challenges of urban energy consumption, food provision, traffic and mobility, waste and recycling, and health and safety. Yet, critical scholarship has shown that current smart, green, and creative plans/policies frequently tackle urban problems in either a partial or shortsighted manner. Smart city discourses, for example, have been extensively criticized for envisioning urban futures predominantly in technological terms, neglecting issues of urban social justice as well as the skills and wishes of cities' inhabitants (Hollands, 2008; Thomas et al., 2016). Robert Hollands argues that:

> Debates about the future of urban development in many Western countries have been increasingly influenced by discussions of smart cities. Yet despite numerous examples of this "urban labelling" phenomenon, we know surprisingly little about so-called smart cities, particularly in terms of what the label ideologically reveals as well as hides.
> *(Hollands, 2008: 303)*

Part of Holland's critique is that current smart city discourses tend to be motivated by entrepreneurial, rather than social, agendas. Optimistic connotations that are attached to the "smart city" brand (because who wants to live in a "dumb city"?) tend to divert attention from this basic misconception. Similar critiques have also been directed towards the "creative" and "green" city labels (Campbell, 1996; Peck, 2005). Fashionable city visions – as imaginaries echoing Lefebvre's notion of "conceived cities" – require careful scrutiny.

In contrast to top-down planning visions, bottom-up urban interventions of recent decades often emerged under the labels of "self made", "grassroots", "guerilla", "insurgent", and "tactical" urbanism; as "the city from below"; and as urban "social movements", "protests", "uprisings", and "revolutions". For example, urban imaginaries of bottom-up intervention emerged in and through the Iranian Green Movement (2009), the Arab Spring (2010–2012), the Gezi Park Protests (2013), and the Hong Kong Umbrella Movement (2014). Moreover, they were created during the international Occupy Wall Street and Spanish Anti-Austerity/Indignados Movements, amongst many others.

Part of what links these distinct interventions is that they all made strategic use of urban space, creating imaginaries of urban upheaval that subsequently spread via global news media. It has been shown that social media are central to the mobilization, organization, and global communication of contemporary urban activism (Gerbaudo, 2012: 3). As Paulo Gerbaudo notes in *Tweets and the Streets* (2012), social media are a means to "*choreograph* collective action" (Gerbaudo, 2012: 4; emphasis in the original). In so doing, they facilitate new methods of performing urban activism, as in the case of flash mobs or "cloud activism".

Creativity and aesthetics are crucial to these new forms of urban activism. The art-activist John Jordan has argued that "art and activism together can make incredible new forms" (Jordan, 2016). Jordan is a member of the artist-activist collective Liberate Tate, and of the Laboratory of Insurrectionary Imagination. The former received much public attention in the period between 2010 and 2013 for organizing a series of interventions that combined art with civil disobedience. Its activism is directed against the fossil-fuel industry's practices of "greenwashing" through art sponsorship, and its initial goal was to "liberate" the Tate from its sponsor British Petroleum (BP), which eventually happened in 2017 after 26 years of sponsorship. An example of Liberate Tate's art activism is the unauthorized performance "Human Cost" (Figure 1.1), which was carried out at Tate Britain in 2011, on the anniversary of the 2010 BP oil spill in the Gulf of Mexico. For this performance, a member of the collective posed naked on the floor of the exhibition "Single Form", which was part of BP British Art Displays. Lying in the fetal position with an oil-like fluid poured over the body, the performance evoked familiar images of animals washed ashore and covered in petroleum in the aftermath of the Gulf oil spill.

What was important about the performance, Jordan underlines, was not that it represented the oil spill but rather that it intervened in a problematic coalition between the cultural and fossil-fuel industries (between Tate Britain and BP). For Jordan, aesthetics are crucial to this goal-oriented endeavour. He argues that in order to reach an activist goal – in order to "win" – it is necessary to "apply aesthetic discipline" (Jordan,

Figure 1.1 "Human Cost" (2011) activist performance by Liberate Tate. Photo by Amy Scaife.

2016). Liberate Tate's art activism experiments with how aesthetics can be used in such an applied way, in this case to serve an eco-political goal.

Urban imaginaries have a particularly central role in the convergence of aesthetics and activism in cities. In recent years, social movements, protests, and activists have made strategic use of the centrality, visibility, and symbolism of urban places and landmarks (such as Tahrir Square or Zuccotti Park), as well as urban institutions (such as Tate Britain). While this is particularly relevant for urban activism, it also applies to less extreme forms of bottom-up urban intervention. Urban farming, for example, is not only about communal food production, it is also about transforming the image of a given neighbourhood or space and boosting a "sense of community" (White, 2011). Taking an optimistic view, some scholars argue that urban farming provides a vehicle to envisioning "alternative and more sustainable socio-ecological futures" (Follmann and Viehoff, 2015: 1148). Others, however, contend that urban farming is co-opted by neoliberal strategies of state "de-responsibilization" (Rosol, 2012) and that it leads to gentrification (Quastel, 2009).

This introduction is not the place to take a stand in the urban farming debate. What we wish to emphasize instead is the importance of examining processes of urban imagination and exploring the potential effects they may have on the future development of urban built environments, communities, and ecologies. In particular, we think it is crucial to question the *politics of urban imagination*. Politics, in this context, needs to be understood in broad terms as a process of "making sense" of the socio-spatial conditions that surround us (Rancière, 2009 [2000]). In the following section, we further flesh out this notion of a politics of urban imagination. In so doing, we pay particular attention to the role urban imaginaries can play in the construction/creation of urban futures. Our argument is that, in order to consider urban futures critically, it is vital to map and examine the various subjects and subjectivities of urban imagination that coalesce in the process of urban future-ing.

Urban futures and the politics of imagination

Throughout the last decade, the concept of "post-politics" has gained increasing significance in critical urban studies (see Lahiji, 2014; Metzger et al., 2014; Wilson and Swyngedouw, 2014). Building on the political philosophy of thinkers such as Jacques Rancière, Slavoj Žižek, Alain Badiou and Chantal Mouffe, post-political theory questions and reconsiders the very definition of the term "politics". Following the French distinction between *le politique* and *la politique*, it distinguishes between "politics" and "the political". "The political" (*le politique*) refers to the fundamental constitution of the polis as a political sphere. It concerns "a broadly shared public space, a rational idea of living together" (Swyngedouw, 2011: 373). As such, "the political" is distinct from the notion of institutionalized "politics" (*la politique*), which "refers to the power plays between political actors and the everyday choreographies of policy making within a given institutional and procedural configuration" (ibid.). In simplified terms, "the political" refers to the process of establishing (and re-establishing) a political sphere, while "politics" is what happens within an already established political system.

Post-political theory's main critique is that Western post-Cold-War societies tend to debate matters of public concern in ways that foreclose questions of "the political". This dynamic is called "post-politics". To illustrate this critique, it is helpful to turn to the example of climate change politics. Post-political theorists contend, for example, that climate change politics today is limited to the techno-managerial administration of environmental risks while leaving unquestioned the systemic social and economic conditions that produce global warming in the first place. Post-political governance regulates carbon emissions through the global trading of "carbon permits" – that is, via pre-established market institutions. In so doing, it generates the impression that the risk of global emissions is "managed" (although emissions have not been reduced so far), which forecloses urgent debates about how contemporary industrialized capitalism is generating ever-increasing greenhouse gas emissions. The risk appears contained, while its source remains unaddressed.

What we wish to underline here and adopt from post-political theory is its sensitivity towards the interrelation between politics and the imagination. When applied to the city, post-political theory insists on questioning urban visions, policies, and practices in relation to the socio-spatial questions and debates that they either open up or foreclose. The questioning also extends to the ways in which certain subjects and subjectivities (i.e. topics and people) are either included or excluded. This implies a need to continuously interrogate urban imaginaries – from green city plans to acts of guerilla-gardening – according to how they envision the relationship between urban space and the interrelated fields of social justice, political economy, and the environment.

To interrogate urban imaginaries in this way, it is helpful to turn to Jacques Rancière's *The Politics of Aesthetics*. Addressing political debates within critical theory extending from Marxian ideology critique to Foucauldian discourse analysis (and beyond), Rancière advocates an understanding of aesthetics that is intrinsically political. Aesthetics, for him, describes the process of "making sense of a sense given" (Rancière, 2009 [2000]: 1). Making sense of a sense given is intrinsically political because it involves questions of "what is seen and what can be said about it, around who has the ability to see and the talent to speak, around the properties of spaces and the possibilities of time" (Rancière, 2009 [2000]: 13). Aesthetics thus includes certain topics and perspectives while excluding others. It is a power dynamic that privileges certain "ways of making sense of a sense" over others.

Rancière's "politics of aesthetics" is therefore close to how critical spatial theorists have conceptualized space as a dynamic that is simultaneously conceptual, material, experienced, and practiced; and as the product of social power relations. In this companion, we use the term "urban imaginaries" to bring these perspectives on space, aesthetics, and politics together. Specifically, we advance an understanding of urban imaginaries as socially conditioned. Urban imaginaries thus involve intrinsically political dynamics of making sense of urban space in its full, multiple dimensions: conceptual, material, experienced, and practiced. This dynamic is what we mean by the "politics of urban imagination".

Pursuing this line of thinking a little further, we want to highlight two challenges that require additional consideration. Both concern the relationship between the urban and what might be called the "non-urban" (if such a condition still exists), as well as the relationship between the urban present and its future. The theory of "planetary urbanization" specifically addresses the challenge of rethinking the relationship between urban and non-urban spaces. Building on Lefebvre's *The Urban Revolution* (2003 [1970]), it contends that, in the late-twentieth century, "non-urban" life ceased – in a way – to exist. Societies across the globe had become irreversibly intertwined with, and dependent upon, urban modes of living, which are shaped by industrialized forms of production and consumption. For Lefebvre, "urban society" describes the "society that is born of industrialization and succeeds it" (Lefebvre, 2003 [1970]: 2). Thus the concept of "urban society" does not refer to every space becoming a city, but rather to the expansion of a particular way of life. Echoing the term "Industrial Revolution", Lefebvre's term "Urban Revolution" refers to a situation in which "industrialization dominates and absorbs agricultural production" (Lefebvre, ibid.), thus reshaping rural lifestyles.

The theory of planetary urbanization takes up and further develops Lefebvre's thinking, and argues that "urbanism", as a specific form and organization of life, has massively expanded over time. Yet what does "urbanism" really mean in this context? In 1938, Louis Wirth coined the term "urbanism as a way of life" to describe the human "concentration into gigantic aggregations around which cluster lesser centres and from which radiate the ideas and practices that we call civilization" (Wirth, 1938: 2). Wirth's theory tends to hierarchize the globe into urban centres and their peripheries. It further marginalizes "peripheries" by understanding them as being the "receiving end" of civilized innovation emanating from urban centres. While it is important to resist this form of cultural marginalization, it is worth pointing out that Wirth's vision of "urbanism as a way of life" captures and reproduces an uneven spatial power dynamic that continues to operate in the present era of globalization.

Researching the spatial power dynamics of globalization, Saskia Sassen has shown how governmental institutions, multinational corporations, and technological infrastructures tend to agglomerate in "global cities". Because of this urban centralization of "control, ownership, and profit appropriation", cities emerge, in Sassen's words, as "central places where the work of globalization gets done" (Sassen, 2000: 81). This not only creates an unequal distribution of political power and economic resources globally, it also means that non-urban places all over the world are – to a certain extent – dependent on specific urban centres. Thus, urbanization does not necessarily mean that urban centres "radiate the ideas and practices that we call civilization" into the periphery, but rather that they dominate peripheries through practices of "control, ownership, and profit appropriation" (Sassen, 2000).

The theory of planetary urbanization problematizes such global power relations but goes a step further in seeking to overcome the distinction between urban and non-urban. It considers worldwide socio-spatial transformations that intertwine city centres and their "peripheries" in multiple new ways, including regional urbanization, the relocation of formally "central" urban functions (such as shopping facilities) into suburban regions, the appropriation of city hinterlands for urban functions (e.g. recreation, resource extraction, or waste disposal), and the lasting ecological impact of urbanization on the planet (Brenner and Schmid, 2012: 11–12). Because of this blurring of boundaries, and because of the inextricable interdependencies between urban "centres" and "peripheries", planetary urbanization postulates that the very "category of the 'city' has today become obsolete as an analytical social science tool … Today the urban represents an increasingly worldwide condition in which political-economic relations are enmeshed" (Brenner and Schmid, 2012: 13).

This poses a challenge to the urban imagination. If the planet is becoming urbanized, then to which distinctive feature does the "urban" in "urban imaginary" actually refer? In an article on "city becoming world" David Madden has sought to address this challenge and capture in critical terms the "global-urban imagination" that results from planetary urbanization:

> To speak of an urban planet is not to imagine that highways and skyscrapers cover the entire earth … What most statements of the global urban imaginary do tend to share, despite their differences, is the notion that the planet no longer hosts discrete urban islands. Instead, there is a sprawling worldwide urbia, massively uneven and unequal and ranging across radically different social spaces, which has covered the globe…
>
> *(Madden, 2012: 773)*

This provides a productive starting point to envision the new imaginaries that result from planetary urbanization processes, and – returning to our argument that urban imaginaries are intrinsically political – we wish to highlight the value of critically interrogating such new forms of "global-urban imagination". Global-urban imaginaries interpret global challenges, including poverty, environmental pollution, health hazards, and the list goes on. The way in which they interpret these challenges, however, always includes and excludes certain subjects and subjectivities.

This argument can be illustrated by Maria Kaika and Erik Swyngedouw's critique of contemporary scholarship and governance surrounding the idea of "sustainable urbanism" (Kaika and Swyngedouw, 2014). For them, planetary urbanization is at the root of contemporary environmental problems. Planetary urbanization implies the "urbanization of nature" – that is, "the process through which all manner of natures are socially mobilized, economically incorporated (commodified), and physically metabolized/transformed in order to support the urbanization process" (ibid.). Kaika and Swyngedouw's main argument, however, is *not*

> to show the urban roots of the environmental conditions, but rather why and how these urban roots are customarily ignored in much of urban theory and practice, and how the feeble techno-managerial attempts to produce more *sustainable* forms of urban living ... actually continue to sharpen the combined and uneven socio-ecological apocalypse that marks the contemporary dynamics of planetary urbanization.
>
> *(ibid., emphasis in original)*

In seeking to answer these questions, Kaika and Swyngedouw identify problems that are, in our view, directly related to issues of urban imagination. They contend that urban scholarship of the twentieth century has, to a large extent, considered urbanization and nature as separate entities: "Nature became relegated to the material and discursive domains outside the city and practically monopolized by technocratic engineering professions" (ibid.). Because of this conceptual distinction between city and nature, urban imaginaries of the twentieth century tended to disregard the ecological and environmental entanglements of cities. In short, cities were imagined as separate from their supposedly "natural environments".

This imaginary is misleading because, as Kaika suggests in an interview on urban political ecology, "cities are made of transformed nature, nature that has been processed and produced through the input of human labour and capital investment" (Kaika, 2017). Pointing to a brick wall, she argues that the wall is "fired earth",

> but for this earth to be fired it takes human labor, it takes capital investment, it takes technology. So if we think about everything around us – from this brick to our garments to everything as transformed nature – I started talking not about cities anymore ... I prefer to talk about the process of urbanization of nature.
>
> *(ibid.)*

The political implication of this change in conceptual imaginary – from "city" to "the urbanization of nature" – is that ecology becomes included *within* the urban imaginary.

One practical consequence of understanding cities as an urbanization of nature *could be* that ecological issues receive significantly increased attention in future planning and decision-making. Any type of urban social, economic, or cultural development – from social inequality, to economic growth, to consumer behaviour – has ecological preconditions and repercussions that need to be accounted for. Clearly, this is not yet happening in any kind of systematic way. Kaika and Swyngedouw's criticism is that urban governance acts *as if* nature were distinct from urbanization processes. While growing attention is being given to urban "sustainable development", most sustainability policies are still designed as if the socio-cultural and economic development of cities were independent from the environment.

For example, most urban governments pursue an agenda of economic growth. This growth agenda, as environmental economists argue (Demaria et al., 2013; Jackson, 2009), is often irreconcilable with global sustainability targets, such as the minimization of resource usage, packaging waste, and carbon footprints. However, current urban sustainability policies tend to ignore this paradox and instead attempt to tackle environmental pollution through techno-managerial measures, such as recycling and carbon emissions trading. While these efforts should not be dismissed, critical environmental scholarship shows that meeting sustainability targets requires going beyond such "eco-modern" approaches and implementing post-growth economic systems (D'Alisa et al., 2014; Demaria et al., 2013; Jackson, 2009; Rockström et al., 2009).

Why is this an issue of urban imagination? In mainstream political discourses, economic growth and environmental sustainability are generally not imagined as inter-related. Instead, the environment is treated as an object external to the urban political economy. What is more, the urban economic system *as such* (including its emphasis on growth) tends to remain undisputed. This supports post-political theory's critique in which questions about the basic *constitution* of the common socio-economic and political systems tend to be excluded from mainstream political debates, contributing to a false consensus on these matters of public concern.

It is against the backdrop of these blind spots within the political discourses and imaginaries of sustainable urbanism that Kaika and Swyngedouw call for "radical urban political-ecological imaginaries":

> Ultimately, the intellectual challenge posed by the socio-environmental conditions shaped by planetary urbanization must be to extend the intellectual imaginary and the powers of thought and practice to overcome the contemporary cultural injunction identified by Jameson that "it is easier to imagine the end of the world than changes in the eco-capitalist order and its inequities" (Jameson, 2003: 76). This is the courage of the intellect that is now required more than ever, a courage that takes us beyond the impotent confines of a sustainability discourse that leaves the existing combined and uneven, but decidedly urbanized, socio-ecological dynamics fundamentally intact, and charts new politicized avenues for producing a new common urbanity.
>
> *(Kaika and Swyngedouw, 2014)*

This call for a radical re-imagination of urbanity, however, leads to another challenge for contemporary urban research and theory: understanding the ways in which urban imaginaries actually intervene in the relationship – or the transition – between the urban present and its future. This challenge is addressed in a growing body of research on social futures and transition management. The fields of sociology, behavioural economics, and science and technology studies are contributing very actively to this research. A recurring concern in this work is how to *predict*, as well as purposefully and democratically *intervene*, in urban socio-economic, environmental, and technological transformations.

This naturally raises questions about the accuracy of future imaginaries and about the ways in which future imaginaries are socially conditioned, and therefore shaped by power relations. As John Urry argues in *What is the Future?* social science analysis has shown "how futures are often the opposite of what is planned and imagined" (Urry, 2016: 10). As a result, there is a need to examine the extent to which future-ing practices, such as scientific forecasting or municipal planning, can capture and account for the complexity as well as the contingency of urban futures.

In addition, we need to question "who or what owns the future" – that is "the power to make futures" (Urry, 2016: 11). This is particularly relevant given that the future is nowadays also an object of trading. In *Financial Derivatives and the Globalization of Risk*, Benjamin Lee and Edward LiPuma show how financial derivatives are used to commodify global futures. Derivatives, as they define it, are "financial instruments that derive their monetary value from other assets, such as stocks, bonds, commodities, or currencies. The peculiarity of all derivatives is that they give individuals the right to buy or sell certain assets by a specified date" (Lee and LiPuma, 2002: 204). The profitability of trading with derivatives thus depends on the speculator's capacity to adequately predict – or influence – a specific future. The complexity and contingency of socio-economic futures is instrumental to this dynamic of derivative trading. Derivatives "speculate on risks associated with ... the forms of connectivity brought about by globalization" (Lee and LiPuma, 2004: 18). Such risks may include environmental risks (such as weather effects on agricultural production), political risks (such as the effect of economic policy changes on currency values), and many others. Making profits with derivatives depends on the speculator's capacity to assess such risks. In addition, the financial speculators' collective risk assessments actually co-shape the underlying risk. For example, if speculators collectively expect a currency such as the U.S. dollar to increase in value, then this expectation is likely to actualize, because it raises demand for the currency, leading in this scenario to an appreciation of the dollar. The example shows how the question of "who owns the future" can sometimes be taken quite literally.

However, setting aside the fact that global futures constitute an object of financial speculation – with extensive real economic consequences, as the 2008 global financial crisis has shown – the activity of "making" futures itself is not evenly distributed throughout social space. Certain actors adopt particularly powerful roles in anticipating social futures. As examples, Urry mentions

> companies like Google and Shell, environmental organizations, such as the Intergovernmental Panel on Climate Change (IPCC) or Forum for the Future, government bodies like Foresight in the UK or the European Strategy and Policy Analysis System (ESPAS) in the EU, military organizations such as the Pentagon, academic bodies such as the Oxford Martin School or the Tyndall Centre, and very many others.
>
> *(2016: 2)*

The power of these organizations, however, is not limited to the activity of future forecasting because "some of those futures anticipated by these organizations have performative consequences" (ibid.). We therefore need to take seriously the idea that future-imagining practices have future-making consequences.

One implication is that, when it comes to urban space and its imagination, there is a need to map and question who participates in the process of urban future-ing. Earlier in this chapter, we drew on post-political theory to argue that the politics of urban imagination revolve around questions of who, and what, is included in the urban imagination process. This is particularly relevant given that critical scholarship sees a democratic deficit within contemporary future-making processes (Crouch, 2004; Rancière, 1999; Urry, 2016; Wilson and Swyngedouw, 2014). Not only do select corporate, governmental, and technological elites have a disproportionately powerful role in making futures, but also "those yet to be born generally possess no voice in what we can call the 'parliament of generations'" (Urry, 2016: 4). Though they will be affected by future-ing conducted in the present, future generations cannot participate in present future-ing activities. In similar ways, the voices of cyborgs, animals, and other species remain largely unexpressed or unheard within current future-ing discourses – a point repeatedly made in transhumanist, posthumanist, and animal rights scholarship (Garner, 2017; Haraway, 1991 [1983]; Hughes, 2004; Meijer, 2016).

All of these subjects will be affected by the ways in which planetary urbanization progresses, and so there exists a strong case to include them within the politics of urban imagination. But how can this inclusion be achieved? According to Urry, "thinking and democratizing futures involves what we might call 'post-modern planning' in the contemporary era of civil society, global change, wicked problems, the limits of markets, multiple 'unknown unknowns', and so on" (2016: 13). To analyze and facilitate this process, Urry advocates for "future-forming interdisciplinary research" (2016: 14). In this companion, we seek to produce precisely that kind of experimental, far-reaching research.

Overview of the companion

Our collective endeavour across the following chapters is to engage multiple disciplines – spanning architecture, design and art history, geography and planning, anthropology, history, and sociology, as well as literary, media, and cultural studies – in mapping, examining, and critiquing the politics of urban imagination and their impact on shaping urban futures. Reflecting this aim, the book is organized into five closely related and interlocking sections, each of which emphasizes different aspects of the creation, study, and critique of urban imaginaries.

Section 1 – *Eco and Resilient* – focuses on urban imaginaries that address both actual and potential situations of urban crisis and catastrophe, including threats of climate change, natural disaster, pollution, and toxicity. In particular, this section discusses imaginaries that emerge from attempts to tackle pervasive ecological crises facing cities around the globe. It presents urban development ideas, designs, and interventions that attempt either to prevent/reduce environmental pollution in cities, or to mitigate the effects of environmental degradation on cities. Crucially, Section 1 questions the very concepts of urban ecology and resilience. What is the relationship between nature and the city? To what extent does the ecological resilience of cities – defined as the capacity to adapt to and recover from environmental shocks – depend on questions of urban social justice? How do visual and literary forms work together in fostering new understandings of urban ecological damage? Can artistic intervention, performance, and experimentation in cities foster the creation of new sustainable development strategies? And what is the role of urban, bottom-up initiatives – such as protest movements or climate action plans – in rethinking environmental social futures?

Section 2 – *Smart and Digital* – is concerned with urban imaginaries that emerge from or focus on the use of new media and digital technologies in urban contexts. Smart city designs and applications that promise to tackle urban challenges of energy consumption, food provision, traffic and mobility, or waste and recycling are extremely popular in contemporary urban design, research, and policymaking. Recently, however, critical scholarship has also drawn attention to the risks of envisioning urban futures predominantly in "smart" technological terms. This section presents critical debates surrounding different types of smart city imaginaries. In so doing, the section compares case studies of smart urban visions and implementations for various urban environments, ranging from the virtual to the actual, and from the fictional to the real. At the same time, it addresses the role of new media and digital technologies in *producing* urban imaginaries. How do digital modelling and camera techniques picture complex processes of urban renewal? What roles can smart devices and software such as crowd-mapping applications play in promoting social inclusion in cities? What happens when strategies of oppression and tactics of resistance in cities transition into cyberspace? Moreover, what are the political risks of envisioning urban futures in terms of smart "transparency"? To tackle these and related questions, this section also reflects on the limits of framing the concepts of digital smartness and human imagination in oppositional terms.

Section 3 – *Connected and Consuming* – examines how urban imaginaries affect the mobilities of people, goods, capital, signs, and data characterizing twenty-first-century globalization. In particular, the section explores how urban governments, planners, and developers strategically deploy urban branding in order to sell cities in various ways: notably to tourists and investors. How are photographic images of iconic architectural projects used to make cities globally recognizable? What are the risks of focusing urban planning and regeneration on iconic "starchitecture", rather than on less spectacular forms of urban design and development? Is a city that brands itself as "open" really open to cosmopolitan diversity, inclusion, and interaction? And, how are cosmopolitan urban imaginaries instrumental to economic agendas of promoting new consumerist outlets and subjectivities? To address these issues, the section critically investigates the various ways in which urban imaginaries that celebrate global status and connectivity are shaping communities, economic development, and social justice in cities. At the same time, the section addresses changing imaginaries of mobility and connection that emerge in contemporary urban culture. How do urban imaginaries re-configure metropolitan forms of subjectivity and solidarity in the era of globalization? What are their effects on experiences of freedom and practices of everyday life?

Section 4 – *Uneven and Divided* – focuses on imaginaries produced by the conflicts, ruptures, and imbalances of urban social, economic, and political life. A special concern in this section are imaginaries connected to struggles over urban housing and public space. These conflicts manifest in the contrasts between urban sprawl versus urban density, formal versus informal, slum versus luxury (and gated), as well as private versus public living. In critical urban theory, such conflicts are associated with the concept of the "right to the city", while in urban design, planning, and policymaking they tend to be framed as issues of urban wellbeing, community, participation, and strategic development. This section aims to bring into focus the ways in which issues of urban justice and inclusion connect to different modes of imagining and living in cities. The section also critically questions established ideals of urban progress, such as the logic of art-led urban development. Whose interests are reflected in culture-led urban revitalization efforts? What is the impact of urban visual culture and aesthetics on the accessibility of urban housing and infrastructure? What role do urban trends – such as the popular re-discovery of brutalist architecture or the celebration of "global city lifestyles" in mass fiction – play in processes of gentrification and urban exclusion? To what extent do conflicts over urban resources epitomize much broader problems of the global political economy? To address these questions, this section explores how urban imaginaries are instrumental to the creation of – but also the resistance against – the unevenness and divisions that exist within and between contemporary cities.

Section 5 – *Speculative and Transformative* – explores imaginaries that attempt to go beyond or to rethink existing/established ideas of the city. The section zooms in on "urban future-ing" and considers the dynamics through which different actors – such as urban planners, policymakers, architects, designers, and think tanks, as well as urban movements, activists, and civil society organizations – co-determine the socio-political and environmental futures of cities. The section begins by providing a historical perspective on the practice of urban future-ing. How were city futures imagined in past urban discourses, designs, and master-plans? Are contemporary urban visions radically different from those of the past? What new types of urban "futurologists" have emerged in the context of postmodern, post-democratic, and post-political urban governance? What roles do urban experiments and "living labs" play within the institutional practices of crafting urban planning for the future? Are contemporary urban visions fit to tackle today's pressing social, political, and environmental challenges, or is there a need to invent radically different forms of urban imagination? To address these questions, this section critically questions the processes and actors involved in urban future-ing and identifies what limits and potentials are latent in contemporary urban design and activism, planning and policy, and even everyday life.

What connects the companion's chapters across the different sections is a shared attentiveness to the diverse ways in which urban imagination is socially produced, responsive to global challenges, and enmeshed in the politics of urban future-ing. The result, we hope, is not only a compelling case for the value of studying urban imaginaries but also a series of new insights into the ongoing transformation of cities, including our co-evolving lives within them.

References

Aalbers, M.B. (2015) The Potential for Financialization. *Dialogues in Human Geography* 5(2): 214–19.

Baudrillard, J. (2006 [1981]) *Simulacra and Simulation*. Trans. S.F. Glaser. Ann Arbor: University of Michigan Press.

Berger, J. (1972) *Ways of Seeing*. London: BBC.

Berger, J. (1973) *Ways of Seeing; A Book Made by John Berger*. New York: Viking Press.

Brenner, N. and Schmid, C. (2012) Planetary Urbanization. In: Gandy, M. (ed.), *Urban Constellations*. Berlin: Jovis, 10–13.

Campbell, S. (1996) Green Cities, Growing Cities, Just Cities? Urban Planning and the Contradictions of Sustainable Development. *Journal of the American Planning Association* 62(3): 296–312.

Carrigan, A. (2015). Towards a Postcolonial Disaster Studies. In: DeLoughrey, E., Didur, J., and Calligan, A. (eds.), *Global Ecologies and the Environmental Humanities: Postcolonial Approaches*. London: Routledge, 117–40.

Chao, J. (2018) *Cultural Resistance, 9/11, and the War on Terror: Sensible Interventions*. New York: Routledge.

Crouch, C. (2004) *Post-Democracy*. Cambridge: Polity.

D'Alisa, G., Demaria, F., and Kallis, G. (eds.) (2014). *Degrowth: A Vocabulary for a New Era*. New York: Routledge.

Davis, M. (2001) The Flames of New York. *New Left Review* 12: 34–50.

De Certeau, M. (1988) *The Practice of Everyday Life*. Trans. S. Rendall. Berkeley: University of California Press.

Demaria, F., Schneider, F., Sekulova, F., and Martinez-Alier, J. (2013) What Is Degrowth? From an Activist Slogan to a Social Movement. *Environmental Values* 22(2): 191–215.

Dibazar, P., Lindner, C., Meissner, M., and Naeff, J. (2013) Questioning Urban Modernity. *European Journal of Cultural Studies* 16(6): 643–58.

Dibazar, P. and Naeff, J. (2018) *Visualizing the Street: New Practices of Documenting, Navigating and Imagining the City*. Amsterdam: Amsterdam University Press.

Elden, S. (2004) *Understanding Henri Lefebvre*. London: Continuum.

Emmerich, R. Dir. (2004). *The Day after Tomorrow*.

Follmann, A. and Viehoff, V. (2015) A Green Garden on Red Clay: Creating a New Urban Common as a Form of Political Gardening in Cologne, Germany. *Local Environment* 20(10): 1148–74.

Foucault, M. (1995 [1975]) *Discipline and Punish*. Trans. A. Sheridan. New York: Vintage.

Garner, R. (2017) Animals and Democratic Theory: Beyond an Anthropocentric Account. *Contemporary Political Theory* 16(4): 459–77.

Gerbaudo, P. (2012) *Tweets and the Streets: Social Media and Contemporary Activism*. London: Pluto Press.

Haraway, D. (1991 [1983]) A Cyborg Manifesto: Science, Technology, and Socialist Feminism in the Twentieth Century. In: Stryker, S., and White, S. (eds.), *The Transgender Studies Reader*. New York: Routledge, 103–18.

Hollands, R.G. (2008) Will the Real Smart City Please Stand Up? Intelligent, Progressive or Entrepreneurial? *City* 12(3): 303–20.

Hughes, J. (2004) *Citizen Cyborg: Why Democratic Societies must Respond to the Redesigned Human of the Future*. Cambridge, MA: Westview Press.

Jackson, T. (2009) *Prosperity without Growth: Economics for a Finite Planet*. New York: Earthscan.

Jacobs, J. (1961) *The Death and Life of Great American Cities*. New York: Random House.

Jameson, F. (2003) Future City. *New Left Review* 21: 65.

Jordan, J. (2016) On Creative Activism. Presentation at the Fossil Free Culture NL launch event, Amsterdam, 6 April. Available at: https://vimeo.com/162248834.

Joye, S. (2009) The Hierarchy of Global Suffering: A Critical Discourse Analysis of Television News Reporting on Foreign Natural Disasters. *Journal of International Communication* 15(2): 45–61.

Kaika, M. (2017) Urban Political Ecology, Interview by Ibai Rigby. Available at: https://vimeo.com/180669461.

Kaika, M. and Swyngedouw, E. (2014) Radical Urban Political-Ecological Imaginaries: Planetary Urbanization and Politicizing Nature. *Derive Journal* 15. Available at: https://www.eurozine.com/radical-urban-political-ecological-imaginaries/.

Lahiji, N. (ed.) (2014). *Architecture Against the Post-Political: Essays in Reclaiming the Critical Project*. London: Routledge.

Lee, B. and LiPuma, E. (2002) Cultures of Circulation: The Imaginations of Modernity. *Public Culture* 14(1): 191–213.

Lee, B. and LiPuma, E. (2004) *Financial Derivatives and the Globalization of Risk*. Durham: Duke University Press.

Lefebvre, H. (1956) Pignon: Douze Hors-Texte. In: Fall, G. (ed.), *Musée de Poche*. Paris: Éditions Fall.

Lefebvre, H. (1991 [1974]) *The Production of Space*. Trans. D. Nicholson Smith. Malden, MA: Blackwell.

Lefebvre, H. (2003 [1970]) *The Urban Revolution*. Trans. R. Bononno. Minneapolis, MN: University of Minnesota Press.

Lynch, K. (1960) *The Image of the City*. Cambridge, MA: MIT Press.

Madden, D.J. (2012) City Becoming World: Nancy, Lefebvre, and the Global-Urban Imagination. *Environment and Planning D: Society and Space* 30(5): 772–87.

Meijer, E. (2016) Interspecies Democracies. In: Bovenkerk, B. and Keulartz, J. (eds.), *Animal Ethics in the Age of Humans: Blurring Boundaries in Human-Animal Relationships*. Cham: Springer, 53–72.

Meissner, M. (2017) *Narrating the Global Financial Crisis: Urban Imaginaries and the Politics of Myth*. London: Palgrave.

Metzger, J., Allmendinger, P., and Oosterlynck, S. (eds.) (2014). *Planning against the Political: Democratic Deficits in European Territorial Governance*. London: Routledge.
Mulvey, L. (1975) Visual Pleasure and Narrative Cinema. *Screen* 16(3): 6–18.
Peck, J. (2005) Struggling with the Creative Class. *International Journal of Urban and Regional Research* 29(4): 740–70.
Quale, S. Dir. (2014). *Into the Storm*.
Quastel, N. (2009) Political Ecologies of Gentrification. *Urban Geography* 30(7): 694–725.
Rancière, J. (1999) *Disagreement: Politics and Philosophy*. Trans. J. Rose. Minneapolis, MN: University of Minnesota Press.
Rancière, J. (2009 [2000]) *The Politics of Aesthetics*. Trans. G. Rockhill. New York: Continuum.
Robinson, J. (2006) *Ordinary Cities: Between Modernity and Development*. London: Routledge.
Rockström, J., Steffen, W., Noone, K., Persson, Å., Chapin, F.S., III, Lambin, E., Lenton, T.M., Scheffer, M., Folke, C., Schellnhuber, H., Nykvist, B., De Wit, C.A., Hughes, T., Van Der Leeuw, S., Rodhe, H., Sörlin, S., Snyder, P.K., Costanza, R., Svedin, U., Falkenmark, M., Karlberg, L., Corell, R.W., Fabry, V.J., Hansen, J., Walker, B., Liverman, D., Richardson, K., Crutzen, P., and Foley, J.A. (2009) A Safe Operating Space for Humanity. *Nature* 461(7263): 472–75.
Rosol, M. (2012) Community Volunteering as Neoliberal Strategy? Green Space Production in Berlin. *Antipode* 44(1): 239–57.
Sassen, S. (2000) The Global City: Strategic Site/New Frontier. *American Studies* 41(2–3): 79–95.
Soja, E.W. (1996) *Thirdspace: Journeys to Los Angeles and Other Real-and-Imagined Places*. Malden, MA: Blackwell.
Soja, E.W. (2000) *Postmetropolis: Critical Studies of Cities and Regions*. Malden, MA: Blackwell.
Soja, E.W. (2010) Six Discourses on the Postmetropolis. In: Bridge, G. and Watson, S. (eds.), *The Blackwell City Reader*. Malden, MA: Blackwell, 374–82.
Swyngedouw, E. (2009) The Antinomies of the Postpolitical City: In Search of a Democratic Politics of Environmental Production. *International Journal of Urban and Regional Research* 33(3): 601–20.
Swyngedouw, E. (2010) Apocalypse Forever? Post-Political Populism and the Spectre of Climate Change. *Theory, Culture & Society* 27(2–3): 213–32.
Swyngedouw, E. (2011) Interrogating Post-Democratization: Reclaiming Egalitarian Political Spaces. *Political Geography* 30(7): 370–80.
Thomas, V., Wang, D., Mullagh, L., and Dunn, N. (2016) Where's Wally? In Search of Citizen Perspectives on the Smart City. *Sustainability* 8(3): 207. (Refer: https://doi.org/10.3390/su8030207.)
Urry, J. (2016) *What is the Future?* Cambridge: Polity.
White, M.M. (2011) Sisters of the Soil: Urban Gardening as Resistance in Detroit. *Race/Ethnicity: Multidisciplinary Global Contexts* 5(1): 13–28.
Wilson, J. and Swyngedouw, E. (2014) *The Post-Political and Its Discontents: Spaces of Depoliticization, Spectres of Radical Politics*. Edinburgh: Edinburgh University Press.
Wirth, L. (1938) Urbanism as a Way of Life. *American Journal of Sociology* 44(1): 1–24.

Part I

Eco and resilient

2

Thirsty cities

Who owns the right to water?

Dora Apel

Human right or corporate right?

Is access to clean water a basic human right or is water a commodity to be bought and sold by corporations? Should water be part of a democratic public commons or should water distribution be run like a business that values the bottom line over public health and human lives, denying water to those who cannot afford it? As clean water becomes increasingly scarce, water access and control become more urgent. Globally, the scale of ecological dispossession through corporate control of water, combined with chemical agriculture, the destructive diversion of water through large dam projects, and the climate chaos produced by greenhouse gas emissions, has led to the creation of deserts, the depletion of rivers and lakes, widespread disease, battles and wars (often masked as ethnic or religious conflicts), and a massive refugee crisis. Environmental activist Maude Barlow asserts that "The global water crisis has become the most powerful symbol of the growing inequality in our world", based on the three fundamental water crises that pose a threat to the survival of the world's populations: dwindling freshwater supplies, inequitable access to water, and the corporate control of water (Barlow, 2007: 1, 142). Yet to many, the global water crisis remains largely invisible.

As case studies, Flint and Detroit provide object lessons in how we use water, distribute it, contaminate it, and waste it, while also demonstrating that water crises are not only local but have wide-ranging implications. They provide object lessons about problems whose solutions hinge on whether we view water as a free right and human commons or as a corporate right. Although it is the poor and marginalized in declining cities and regions who suffer most at this point, ultimately, the effects of water scarcity will cross race and class lines, countries and continents, and once water has been depleted, it cannot be replenished.

Approximately an hour apart in the state of Michigan, Detroit and Flint are impoverished majority black cities that have become emblematic of urban centers seen by the neoliberal state as profit-making opportunities with expendable populations. The state's Republican Governor Rick Snyder suspended democracy in both cities when he voided the powers of the elected mayors and city councils and instead appointed emergency managers to run the cities, who were accountable only to himself. In Detroit, the emergency manager (EM) turned off water for 20,000 residents before giving control back to the mayor (who then continued the water shutoffs). In Flint, the EM recommended switching the city's water source to the polluted and highly corrosive Flint River, which caused massive damage to an already crumbling infrastructure, an outbreak of Legionnaire's disease, and the lead poisoning of the water, potentially exposing the city's 100,000 residents, with especially devastating effects on its children.

Flint: public health and the bottom line

Shortly after the EM switched Flint's drinking water source from the Detroit Water and Sewerage Department (DWSD) to the Flint River in March 2014, residents began complaining that the dirty brown and foul smelling water coming out of their taps caused chemical burns and skin rashes, caused their hair to fall out, and made them ill in other ways. The Michigan Department of Environmental Quality (DEQ) and other officials repeatedly and falsely reassured city residents for eighteen months that the water was safe for drinking, cooking, and bathing. How could this happen?

Flint officials decided to draw water from the Flint River rather than the DWSD while the new Karegnondi pipeline connecting Genesee County to Lake Huron was built, calculating that they would save about five million dollars. The cost for the Karegnondi (from the Petan Indian language meaning "lake" and an early name for Lake Huron) pipeline was calculated at $285 million and the entity in charge of building it was the quasi-public Karegnondi Water Authority (KWA). The cost of Flint's share of the new pipeline was $85 million, a price the bankrupt city could not pay.

Jeff Wright, the Genesee County drain commissioner and KWA CEO, spearheaded the drive to build the new pipeline. Wright argued that it would save Flint money because the DSWD was price-gouging the city: "Detroit charged whatever Detroit felt like charging for drinking water, and our community was forced to pay whatever it cost" (Guyette, 2017). As Curt Guyette (2017), an investigative reporter for the ACLU of Michigan, points out, Wright's complaint ignored the fact of Michigan law, which requires public water departments to charge municipalities only what it costs to treat and deliver water to its customers. Thus, like most public utilities, the DWSD is highly regulated and prohibited from making a profit. Where there are disputes over costs and reasonable rates, the DWSD may be challenged in court. The mayor of Flint at the time, Dayne Walling, was chair of the KWA board, and the governor appointed a former interim mayor as Flint's EM. Both strongly supported the KWA pipeline. Were they acting as watchdogs for the city or were they, along with KWA CEO Jeff Wright, seeking ways to secure political influence?

The KWA did not need Flint's participation to build the new pipeline, but it did need the city's approval to increase the diameter of the pipeline. In this one instance, the EM gave power back to the mayor and city council in a ceremonial vote for widening the pipeline, hoping this would prevent any future court challenges. Although they expected a quick decision, the process took more than a year. During this time, the governor asked the DWSD to come up with a new cost-saving proposal. "That proposal, according to the DWSD", writes Guyette (2017), "would save Flint and the Genesee County more than $900 million as compared to the KWA alternative over the 30-year contract period".

Yet the new Flint EM flatly rejected the offer. Despite the prevailing narrative that Flint was bankrupt and switching water sources to save money, the EM dismissed the cost-saving proposal by the DWSD that undermined the rationale for the switch. In addition, no one addressed the problem of Flint not having the money to pay its $85 million share of the KWA pipeline. It appears, then, that the switch was not about saving money at all;

instead, it seems intended to fill the campaign coffers of the politicians who supported the new pipeline by the private companies that would build it.

The decision about where Flint would get its water ultimately rested with the Michigan Department of Treasury. Despite the fact that no plan existed for financing the KWA option, or that the DWSD plan was cheaper, the Treasurer accepted the EM's recommendation to proceed with the KWA. When officials finalized the KWA agreement, the contract stipulated that Flint use the river as a temporary water source until the KWA water became available.

The first problem following the switch to the Flint River was that the water was infested with bacteria such as *e-coli* derived from fecal matter, and other toxins. Chlorine dumped into the water to deal with this problem then caused the production of cancer-causing chemicals called trihalomethanes (TTHMs), which affect the liver and kidneys (Democracy Now!, 2016). A deadly outbreak of Legionnaires' disease, which is caused by a water-borne bacteria, spread throughout the city, killing at least twelve people and affecting seventy-eight more (DeVito, 2017). While this was occurring, the river water, to which the city had added no anti-corrosive, proceeded to corrode Flint's aging water pipes and to poison the drinking water with high levels of lead. In one home, an EPA manager detected lead levels seven times greater than the EPA's acceptable limit of 15 parts per billion. A Virginia Tech research team found that 40% of Flint homes had elevated levels of lead, with some homes at 13,200 parts per billion. But the DEQ expressed skepticism about the study to the *Flint Journal.*

Lead is a powerful neurotoxin that affects every system in the body and is especially harmful to pregnant women and children, where it can cause irreversible developmental delays because of damage to developing brains and nervous systems. In 2015, a research team at a local health center found that the proportion of children with elevated lead levels in their blood had nearly doubled since the water switchover, affecting around 9000 children under the age of six (Moore, 2016; NOVA, 2017).

Finally, in October 2015, Snyder signed a bill appropriating $9.35 million, almost double the alleged "savings" of switching to river water in the first place, to reconnect the city to the Detroit water system. In the following months, both the city mayor and Snyder declared a state of emergency, the National Guard was mobilized to distribute bottled water, and a federal investigation began.

The corrosion of the city's lead water pipes would not have occurred if an anti-corrosive had been added to the water, as required by federal law for large municipal systems. Why did the DEQ fail to add this crucial and inexpensive corrosion control chemical? At first, officials claimed there *was* an anti-corrosive plan, then that they had misunderstood the federal law, and finally that they were concerned an anti-corrosive would foster bacteria (Kaffer, 2016; NOVA, 2017). The DEQ knew there was a problem with lead contamination because they consistently tried to manipulate water safety tests by avoiding homes that had already tested high for lead and instructing city residents to flush their lines before collecting water samples, which would dilute the lead readings (NOVA, 2017).

The reason for the failure to add corrosion control finally became clear in testimony at a legislative hearing in March 2016. A reporter explained, "When Flint began to pump drinking water from the Flint River, the city's water treatment plant wasn't capable of adding corrosion control treatment, not without equipment upgrades the broke city couldn't afford" (Kaffer, 2016). Thus, the city switched its water source to the highly corrosive river knowing it could not afford to update the water treatment plant to provide corrosion control. In order to claim that the switch would save money, officials omitted the cost of updating the water treatment plant from the financial calculations. That cost has been estimated at $108 million, in addition to the $85 million (Fonger, 2017). This deliberate evasion and the collusion of the DEQ clearly suggests motives for the switch other than saving money or safeguarding public health.

In February 2015, Flint officials hired Veolia North America, a division of the French conglomerate that is the largest private water company in the world, to conduct a study of the city's water for $40,000. While making a series of minor recommendations, Veolia announced in its report the following month that the water quality was "in compliance with State and Federal regulations", including nothing about corrosion or elevated lead levels, and thereby reinforcing the position of the DEQ and city officials (Veolia, 2015). Veolia afterwards proposed more extensive work for an estimated $1.8 million, which the city declined (Smith and Bosman, 2016).

Cities cannot be run like corporations. To thrive, they must sustain the life and wellbeing of the community through access to essential resources like clean water and services such as education and health care. To run the city as a corporation and focus exclusively on the bottom line means sacrificing residents seen as "surplus" and "expendable" – people who may therefore be allowed to suffer slower or faster deaths from the harmful effects of poisoned water. Not just in Flint, but also in other cities across the nation, state and local officials have compromised the welfare of citizens. According to a report by the Natural Resources Defense Council (Khan, 2016), more than 18 million Americans received their drinking water from systems with lead violations in 2015. This is a conservative estimate and the number is actually thought to be higher, because many cities that are gaming the system and avoiding corrosion control, including Flint, are not entered into the EPA's database. Moreover, enforcement is lax, further helping to create unsafe drinking water conditions for millions of people. Nonetheless, the 18 million who drank lead-contaminated water are part of a larger estimate of 77 million people served by drinking water systems with other contaminants, including arsenic and dangerous bacteria (WJBD, 2017). About 70% of these people live in smaller cities and rural communities, leading to a two-tiered system of inequality that has only worsened under the Trump administration as cuts to the EPA cause cuts to drinking water programs and further encourage privatization and other cost-cutting measures.

The buying of political influence that helped trigger the Flint fiasco was revealed in an ACLU analysis of contributions to the 2016 political campaign by KWA CEO Jeff Wright for the position of Genesee County drain commissioner. The study shows that of the nearly $270,000 he raised, at least $188,000, or roughly 70%, came from political action committees and employees of companies doing business directly with KWA or working on the pipeline in some capacity. One of Wright's biggest contributors was a Houston-based engineering and project management firm that was one of KWA's primary contractors and helped conduct a 2009 study asserting the value of leaving the Detroit system and building its own pipeline. This company, along with other firms involved in the study, helped build the KWA pipeline, which was completed in 2017 (Guyette, 2017).

The tragedy of Flint is that it was an avoidable and knowing failure of government, produced by a combination of greed, political power grabbing, and indifference to the public health and safety of the poor and majority black city. Only a few months after the switch, when General Motors factory executives complained to Governor Snyder that the river water was corroding their car parts when washed on the assembly line, Snyder quietly spent $440,000 to switch the plant back to Lake Huron (DWSD) water, while keeping Flint's residents on the corrosive river water (Moore, 2016).

Appointed Flint city officials, the DEQ, and the State of Michigan are all responsible for the fiasco, along with the EPA's Michigan regional administrator, who colluded in the disaster and has since resigned, plus the federal EPA, which also was complicit and responded too slowly. Ultimately, however, responsibility lies at the feet of Governor Snyder and the neoliberal ideology he upholds. After taking office in 2011, Snyder got the Republican legislature to award the wealthy and corporations enormous tax breaks, and then slashed public services such as schools, welfare, pensions, and safe drinking water to compensate for the lost revenue. It was Snyder who suspended democracy for four years and put Flint's fate in the hands of a series of appointed emergency managers whose job was to do what seemed cheapest, most expedient, and corporate-friendly. And it was Snyder who prioritized removing environmental safeguards: a 2012 statement entitled "Energy and the Environment" announced the elimination of more than 100 environmental regulations and cuts to staff levels at the DEQ and other "quality of life" agencies, in advance of similar appalling policies under the Trump administration.

Structural racism is also clearly at work (see Hammer, 2016). The Michigan Civil Rights Commission, which spent nearly a year investigating the crisis to see if racism played any role in the fiasco, issued a report determining that in a black-majority city with a poverty rate of 40%, where democracy was curtailed and replaced by state-appointed emergency managers, racism and political influence-buying took precedence over serving the public good. The Commission wrote that such a crisis could not have occurred in affluent white Michigan cities, and asserted, "The lack of political clout left the residents with nowhere to turn, no way to have their voices heard" (Guyette, 2017). The role of structural racism in placing profit above the welfare of citizens is increasingly visible across the nation as it becomes increasingly destructive.

Flint residents fought back, however, with protests, citizen journalism, a new mayor, and a massive resident testing project, which turned the Flint disaster into a national scandal and brought renewed attention to the problems of clean water, official malfeasance, and lead poisoning.

In June 2016, new Flint Mayor Karen Weaver announced that the city was legally bound to the KWA contract despite the fact that a 2016 analysis of Flint water rates predicted that water bills for the city's residents, already among the highest in the nation, would double by 2022 under KWA. However, ten months later Weaver announced support for a plan to stay permanently with the Great Lakes Water Authority (GLWA), who had served as Flint's temporary water source after the city abandoned river water. The GLWA took over the DWSD regional water structure during the Detroit bankruptcy and became the provider of wholesale water and sewer services to member communities in Southeast Michigan in 2017. This new plan would reportedly save $1.8 million annually over the life of a new thirty-year contract compared to staying with KWA (WXYZ, 2017). It took the city council another seven months to approve the plan because members were concerned that it would keep high water rates in place for residents for years. The truth is that, no matter which system is in place, water rates far exceed the national average.

The Flint water crisis is a warning about what is at stake when cities lose democratic control over their resources. By some estimates, it will cost $1.5 billion to upgrade the entire system and provide services to those affected by lead poisoning (NOVA, 2017). Far from saving money for the city of Flint, the real costs of this catastrophe in financial terms are exponentially higher than any alleged savings ever could have been, and in human terms the cost is incalculable. Although Flint returned to the regional system that provided clean water for nearly fifty years in 2015, city water still must be filtered for it to be safe to drink while the corroded lead service lines are replaced, and many are still afraid to use the water. There are also major problems with the city water mains, which leak 40 to 50% of the treated water that is pumped into them, and there are 20,000 lead service lines corroded by the river that need to be replaced, a process that will take years. Many residents stopped paying for their poisoned water, but controversy arose again when the city warned 8000 homeowners that tax liens would be placed on their homes if they were more than six months behind on their water accounts, making them vulnerable to home foreclosure (Delaney, 2017). Despite the continued need for water filters, the water was officially declared safe in 2017 (although no blood level of lead is safe), four years after the switch to river water.

Four government officials, one from the city of Flint, two from the DEQ, and one from the EPA, resigned over the mishandling of the crisis, and one more DEQ staff member was fired. Investigations continue at the federal, state, and local levels, and so far fifteen current and former state and local officials have been criminally indicted in connection to the disaster. Five of them have been charged with involuntary manslaughter in connection with the deaths from Legionnaire's disease, including the head of the state's health department, who did not warn the public about increases in the disease for a year after he knew about it and said, "Everyone has to die of something" (Atkinson and Davey, 2017). Four officials, including two of the city's former emergency managers, have been criminally charged with using false pretenses to obtain the loan needed by the city to finance its portion of the KWA option (Guyette, 2017). The Michigan attorney general is suing Veolia, as well as the company Lockwood, Andrews, and Newnam, which had been hired to advise the city about the switch in water sources. Nationally, environmental activists and the Massachusetts attorney general have accused Veolia of violations that include polluting waterways and overbilling customers, leading to lawsuits in courts around the country (Smith and Bosman, 2016).

In a powerful visual statement about race and second-class citizenship, New Orleans-based artist Ti-Rock Moore's artwork *Flint* depicts a water fountain with a constant stream of brown running water under a plaque that says "Colored". It was displayed at the international art competition *ArtPrize* in Grand Rapids, Michigan (20 September – 8 October 2017). Recalling the segregated water fountains of the Jim Crow era and the brutal inequality that segregation signified, Moore's work poignantly evokes the long legacy of American racist oppression in the poisoned water of the majority black city of Flint.

Detroit shutoffs: prelude to privatization?

The DWSD has not been immune to corruption. Suburban leaders have long criticized the DWSD for its annual rate increases and what many considered questionable contracting practices. In the 1980s, the head of the water department was imprisoned for taking bribes from a sludge-hauling company. More recently, department head Victor Mercado, who was appointed by former Mayor Kwame Kilpatrick, pleaded guilty to illegally steering contracts to a crony of Kilpatrick's in one of the largest corruption scandals in the city's history.

In March 2014, when Flint left the Detroit water system, the DWSD asserted that half of its 323,000 accounts in Detroit were delinquent and began turning off water for up to 3000 residents every week while EM Kevyn Orr requested proposals from private water companies interested in the DWSD (Gaist, 2014). The following September, the EM and federal bankruptcy mediators constructed a deal in which the DWSD was forced to lease its regional water infrastructure to the GLWA as part of a restructuring plan that also undermined unions, eliminated jobs, and reduced city pensions. The EM used the

Figure 2.1 Ti-Rock Moore, *Flint*, 2016. Photo courtesy of the artist.

bankruptcy crisis to engineer the transfer of control of Detroit's water out of the city's hands in an action that may be read as a move toward the privatization of water, a policy supported by the Obama administration through the Water Resources Reform and Development Act of June 2014. This act dismantles restrictions on private investment in public utilities contained in the Clean Water Act and other federal legislation. The new act encourages financially strapped cities to enter "public-private partnerships" as federal and state funding for infrastructure declines (White, 2014). Public-private partnerships, while suggesting accountability, actually disguise and promote the privatization of public services. Obama's appointee at the EPA, who was complicit in covering up the lead poisoning of Flint residents, pushed for the privatization of essential social services, from public education to water systems (Brewer and White, 2016).

Under the regional authority agreement signed by Detroit Mayor Mike Duggan and three area counties, the GLWA has a forty-year lease for pipelines servicing suburban communities, at a cost of $50 million per year. A six-member board of appointed officials makes decisions on water rates, contracts, labor agreements, and the privatization of services. The agreement overrides Detroit City Council's right to approve privatization of any portion of the water system, which, according to the City Charter, also must be approved by a majority of city voters in a regular or special election (White, 2014). By superseding these provisions, the new arrangement allows the future privatization of the GLWA at the mandate of an unelected board.

Like Flint's EM, Orr then contracted Veolia North America to recommend cost-savings for Detroit's water system. This raises more critical questions. Is it a coincidence that pulling Flint out of the Detroit water system and depriving the DWSD of a large share of revenue helped cause the crisis that justified creating the GLWA and calling in Veolia? Is it a coincidence that Veolia is a major client of the Jones Day global law firm for which Orr works? Is it a coincidence that when Veolia was called in, water already had been shut off for nearly 20,000 households? Was this an attempt to eliminate "bad debt" and clear the way for refinancing and privatization, as many observers have suggested? Under Victor Mercado, who had previously worked for the private water company Thames Water North America, the DWSD shut off water for 40,000 people.

Many of the water cutoffs impacted children, allowing child welfare authorities to remove them from their homes, since working utilities in all homes with children are legally required. The water shutoffs left everyone, including the sick, without water to drink, without toilets, and without the ability to cook or bathe. No such cutoffs were implemented for large corporations and institutions, such as the Red Wings' hockey arena and the Ford football stadium, which were also delinquent on their bills, and thousands of vacant and abandoned buildings with water often gushing through broken pipes (Apel, 2015: 32).

A recent study conducted by researchers from Henry Ford Health System's Global Health Initiative (Plum et al., 2017) found a significant relationship between water shutoffs imposed by the City of Detroit and water-related illnesses experienced by a sample of Henry Ford Hospital patients, including skin and soft tissue infections and water-borne bacterial infections. Researchers analyzed data from more than 37,000 cases of water-borne illness for over a year starting in January 2015, which was then compared with a list of Detroit addresses whose water had been shut off. Even accounting for the effects of poverty, the study shows a clear connection between water-related illnesses and shutoffs.

United Nations investigators denounced the water shutoffs as a human rights violation, asserting, "Disconnections due to non-payment are only permissible if it can be shown that the resident is able to pay but is not paying. In other words, when there is genuine inability to pay, human rights simply forbids disconnections" (OHCHR, 2014). The city maintained it was responding to unpaid bills, but this turned out not to be entirely true, as even residents with up-to-date bills had their water shut off. Matthew Kovac (2015) notes that Detroit officials did not know exactly whose water they were shutting off, but, like banks that engaged in wrongful home foreclosures when they saw an opportunity to seize people's homes, they did not really care. Instead, they saw an opportunity to drive out low-income African Americans in order to clear the way for urban "development" and an influx of young white professionals, in a form of "hyper-gentrification" (see Apel, 2015: 31–3). As environmentalist Martin Lukacs observes,

> ... more than half of the city's commercial and industrial users are also owing – a sum totaling $30 million. But no contractors have showed up on their doorstep. The targeting of Detroit families is about something else. It is a ruthless case of the shock doctrine – the exploitation of natural or unnatural shocks of crisis to push through pro-corporate policies that couldn't happen in any other circumstance.
>
> *(Nichols, 2014)*

In response to growing national pressure over the mounting Detroit water crisis, full-scale media coverage, and a series of protest actions in Detroit, including a mass protest organized by National Nurses United, the largest labor union of registered nurses in the country, EM Orr finally announced a moratorium on shutoffs and relinquished control over the DWSD to the city's Mayor Duggan. Duggan himself, however, also has a history of supporting privatization, having overseen the sale of the nonprofit Detroit Medical Center to a for-profit hospital conglomerate in Virginia. Not surprisingly, water cutoffs eventually continued under Duggan.

Many who have bought inexpensive homes in Detroit have unwittingly inherited responsibility for water bills amounting to thousands of dollars. Despite pledges by city officials that water shutoffs would decline, they instead rose by 18% in 2016, denying water to an additional 27,000 Detroit homes, a ratio of one in five. By the end of 2017, Detroit had shut off water to more than 100,000 homes (Kurth, 2017). Moreover, residents who cannot afford increases in their water bills have been further threatened with foreclosure through liens on their homes, like many Flint residents. Critics also note the intensification of the crisis under the city's "10/30/50" payment plan:

> [T]hose delinquent on their water bills are required to pay 10 percent of the overdue bill up front, with the remainder spread over a 24-month period on top of the regular already-unaffordable monthly bill. [I]n a city with an official poverty rate of 40 percent, avoiding a shutoff means increasing one's monthly payments for two years. City residents already pay water bills that are nearly twice the national average and this year [2015] rates went up another 3.4 percent for water and 16.7 percent for sewerage! This plan is a mathematical algorithm designed to suck the maximum amount of money from poor residents before they are shut off completely. Fifty cents of every dollar collected by DWSD goes directly to finance debt principally owned by wealthy bondholders, big banks and hedge funds.
>
> *(Martin, 2017)*

Without a progressive water rate in place to insure that all families can afford water, the crisis will continue. The Detroit City Council in 2006 adopted such a plan, the Water Affordability Program, to create a safety net allowing low-income residents to pay on the basis of their income, but the plan was significantly downgraded by DWSD head Victor Mercado, and cut altogether during bankruptcy proceedings. Detroiters, who have been fighting for the right to water for years, need a water affordability program enacted and current assistance programs expanded.

The problem of water affordability is by no means unique to Detroit or Flint. In 2015, water was cut off to 25,000 households in Baltimore in a similar assault on low-income people and African Americans; in Cleveland, 44,000 homes were sent shutoff notices in 2016 and over 8000 people lost their homes through foreclosure because of tax liens for non-payment of water bills. Data analysis by *Bridge Magazine* (Martin, 2017) reports that "some studies have forecasted that water prices could quadruple in the next 20 years", and notes that cities such as Austin, Charlotte, Chicago, San Francisco, and Tucson "have all experienced water rate hikes of over 50% in the past five years". This suggests that one-third of households in the US may not be able to pay for water by 2020.

The bigger picture: planetary urbanization

In the face of water shutoffs, water contamination, and lack of access to clean water, neoliberal narratives seek to marginalize and isolate low-income people as responsible for their own problems or as unfortunate victims of "innocent errors" or decisions for which the corporations and the state take no responsibility. Such narratives serve to legitimize the treatment of poor, black, and immigrant populations in deindustrialized cities as "disposable" and "surplus populations", and to justify the transfer of infrastructural resources out of their control with a goal of future privatization.

Food & Water Watch, which has helped stop almost forty privatization efforts and fights for local, public, and democratic control of water, conducted a comprehensive survey of the 500 largest US water systems. They found that, on average, private systems charge 59% more than public systems (Grant and Miles, 2016). Privatized water services not only have higher costs, but also worse services, profit-motivated decision making, and less accountability. This means they cut corners, respond slowly to service requests, and

Figure 2.2 National protest against Detroit water shutoffs in downtown Detroit, 18 July 2014.

let infrastructure deteriorate in order to improve short-term profits. They expand into affluent areas and cut back services in poor areas, where services are needed most, and when problems occur, people have few options.

Moreover, these pushes for privatization are occurring at a time when crumbling water infrastructure is pervasive. Most pipes were laid out in the thirty-year period between 1920 and 1950, including those in cities such as Detroit, Los Angeles, and Chicago. These water infrastructures are all likely to break down at about the same time in the coming years (Cooper, 2009). The plastic and iron industries are currently waging a $300 billion lobbying battle to replace the country's deteriorating water and sewer pipes in the next decade, driven by burst pipes, leaks, and public health scares such as occurred in Flint (Tabuchi, 2017). A massive New Deal-style jobs program is needed to get America's old and leaky water infrastructure updated and the nation's drinking water systems able to deliver clean water to everyone. Instead, the Trump administration announced plans in June 2017 to curtail the federal government's role in funding the nation's infrastructure, calling on cities, states, and corporations to take on the cost.

The selloff of public water systems, supported by both major political parties, represents a form of social regression. Privately owned water systems became municipal entities during the nineteenth century in response to the failure of private water companies to adequately maintain infrastructure, meet sanitation needs, and provide services for the masses of workers and residents gathering in cities. It was only through public ownership that sanitary urban conditions and dependable access to water became possible. The turn to neoliberal globalization and a "free market" agenda in the late twentieth century has fostered the return to privately owned water systems and the growing water wars and crises. In 1989, Margaret Thatcher's government privatized all public water and sewerage operations in England; Paris and Berlin privatized their water infrastructure in the 1990s. The International Monetary Fund and World Bank have aggressively insisted on water privatization across Africa, India, South America, Eastern Europe, China, and Saudi Arabia (Barlow, 2007: 36–8; Gaist, 2014; Shiva, 2016), causing ecological havoc and making water inaccessible or unaffordable to millions of mostly poor people. The sale of water systems to powerful water corporations such as Veolia and Suez provides huge profits for a tiny elite that claims the right to own and control water for the world, as if water were a renewable resource to be bought and sold like t-shirts. As scientist and activist Vandana Shiva (2016: 87) argues,

> While privatization is generally couched in rhetoric about the disappearing role of the state, what we actually see is increased state intervention in water policy, subverting community control over water resources. Policies imposed by the World Bank, and trade liberalization rules crafted by the World Trade Organization, are creating a sweeping culture of corporate-states all over the world.

Further complicating the struggle of Michigan residents for control over their water, the Swiss-based corporate giant Nestlé Waters has become the largest owner of private water sources in the state. Nestlé has been embroiled for years in lawsuits over its water bottling activities, which include pumping more than 200 gallons of fresh water per minute out of wells in northern Michigan and selling it as Ice Mountain in the Upper Midwest and as Pure Life nationally. In 2017, Nestlé applied to increase its pumping allowance at its well in Evart, Michigan, by 60%.

Although Evart's city manager supports the increased pumping allowance, claiming that it provides economic development and has minimal ecological impact, thousands of citizens vehemently oppose it, in part because Nestlé pays exactly nothing for the water and only a $200 annual permit fee to pump from the nine wells it owns or leases. In a state where cities like Flint and Detroit are in water crisis, many are outraged that a giant bottling company can make enormous profits for so little while depleting the Michigan water table, which damages rivers, lakes, and streams, and impacts wetlands, fish populations, and other wildlife. Operating at an immense scale, Nestlé packages an average of 4.8 million bottles of water *a day* and plans a $36-million expansion for its Michigan Ice Mountain bottling operations in Stanwood (Friess, 2017).

Not surprisingly, Nestlé has ties to the Snyder administration: Deb Muchmore, the spokesperson in Michigan for Nestlé, is the wife of Dennis Muchmore, Snyder's former chief of staff, who was intimately involved in the Flint water switch. As celebrity filmmaker and Flint activist Michael Moore notes (2016), "The Muchmores have a personal interest in seeing to it that Nestlé grabs as much of Michigan's clean water as possible – especially when cities like Flint in the future are going to need that Ice Mountain". Indeed, the *Detroit Free Press* reported that in March 2015, Dennis Muchmore proposed spending $250,000 to buy bottled water from Nestlé or Absopure for the citizens of Flint (Matheny and Egan, 2016).

As previous ecological victories of the 1980s continue to be rolled back by environmental deregulation at the state and national levels, public health and the environment continue to deteriorate. One of the first acts of the Trump administration was to overturn a late Obama-era regulation prohibiting surface-mining operations from dumping waste into nearby waterways. Trump further proposed cutting funding to the Great Lakes Restoration Initiative. The Great Lakes represent 21% of the world's supply of natural fresh water, especially critical at a time when the world is running out of this precious resource. The Great Lakes Restoration Initiative was intended to clean up giant harmful algae blooms, toxic shipping discharge, and invasive species. In 2014, toxic algae blooms in Lake Erie contaminated water for 400,000 people in Michigan and Ohio and left them without drinking water for three days, while some residents on the Canadian side of Lake Erie had no drinking water for almost two weeks. Because of climate change, such algae blooms are expected to become larger, more frequent, and more toxic. To make matters worse, since 2000, Nestlé has been allowed to package and sell water from the Great Lakes Basin (aquifers feeding the Great Lakes) for billions of dollars in profits. Given the growing dangers to the Great Lakes, the commercialization and export of Great Lakes water is yet another threat to the 26 million people who depend on the Great Lakes Basin for their water supply.

The pathways of capitalist development create dramatic transformations of the spaces around cities, often at a considerable distance from them, including processing plants and subterranean industrial infrastructure. The progressive industrialization and ruination of the hinterland to support the extraction and bottling of water is a major destructive effect of water privatization. Such colonization of the environment moots boundaries between urban settlements and their allegedly non-urban exteriors and impacts millions of people across state and national boundaries. This suggests the need for a new understanding of urbanization that replaces the traditional concept of the city as circumscribed and autonomous. As critical urban theorist Neil Brenner argues, the nineteenth-century concept of the city as bounded and self-sufficient has become obsolete as capitalism produces "intensifying degradation of surrounding landscapes, ecosystems, watersheds, rivers, seas and oceans through their intensifying role in supplying cities with fuel, materials, water and food, and in absorbing their waste products" (2017: 217). Brenner urges us to realize that we are witnessing "nothing less than the intensification and extension of capitalist urbanization at all spatial scales, across the planetary space as a whole, including not only the earth's terrestrial surfaces, but the underground, the oceans, and even the atmosphere itself" (2017: 36). To conceptualize "planetary urbanization", in Brenner's terms, is to recognize the way capitalist urbanization has already massively transformed and degraded the environment, causing widespread and ongoing suffering among the world's populations.

Critical urban theory is a way of thinking about urbanization and its future founded on a Marxist critique of capitalism in both spatial and political terms. Rethinking of the urban imaginary began at least as far back as the 1970s with radical scholars such as Henri Lefebvre and David Harvey, among others, who shared a concern over the way that cities under capitalism became key sites for commodification processes. Such processes focus on expanding profits for a tiny percentage of the wealthy upper class even as they lead to global economic crises such as that of 2007–2009, in which, among other effects, millions of people lost their homes. Planetary urbanization points to the need for a vision of urban life beyond capitalism as a structuring principle.

The linked crises in Flint, Detroit, the Great Lakes and many other cities and regions pose the urgent question of a worldwide urban imaginary that takes into account the importance of collective rather than private ownership of resources, responding to social needs rather than capitalist profit-making. Instead of trying to make capitalism "work better" by regulating itself and reining in its own "excessive" greed, it must be recognized that a liberal response such as this fails to see the problem as systemic rather than individual. Capitalism is a system that depends on aggressive commodification and an expanding rate of profit, and regards cities as vehicles of profit for the few in a steadily expanding network of exploitation. Planetary urbanization makes it clear that borders are relatively meaningless, not only between hinterland and urban center, but among regions, nations, and continents.

The continued pillaging and privatizing of water and other public utilities, as part of the dispossession of common assets and publicly owned property for the purpose of capitalist accumulation, will only lead to future catastrophes. Will the fierce resistance movements that have arisen around the globe to the corporate takeover of water be sufficient to stop water privatization and to establish water justice and water democracy? Popular outrage generated by crises such as those produced in Detroit and Flint must ultimately reckon with the rule of profit over the economy as a whole and champion the idea of cities for the benefit of people, where the hopes and dreams of a productive and creative life may be fulfilled. Access to clean water – like the fundamental need for housing, jobs, public transportation, and access to education and healthcare – is a basic human demand, yet its realization ultimately requires the economic and political reorganization of society.

Conclusion

We cannot afford to let private interests seize and maintain global control of water for their own profit as the world runs out of freshwater and ever-greater numbers of people live without access to clean water or simply cannot afford to pay for it. The potentially irreversible ecological disaster caused by corporate control of the

Figure 2.3 Exterior view of the Nestlé Waters Ice Mountain bottling plant near Stanwood, Michigan.

Figure 2.4 Exterior view of the Nestlé Waters Ice Mountain bottling plant near Stanwood, Michigan.

non-replaceable resource of water has positioned the global water crisis at the center of the most pressing challenges we face today: diminishing fresh water, social inequality, and the privatization of public resources and services. The appropriation of water resources dispossesses people of their just share of natural resources and will eventually destroy the planet. In response to planetary urbanization, an emancipatory urban imaginary must be based on common public ownership of resources, managed according to need and the public good rather than profit, including the resource that is the basis of all life: water.

References

Apel, D. (2015) *Beautiful Terrible Ruins: Detroit and the Anxiety of Decline*. New Brunswick: Rutgers University Press.

Atkinson, S. and Davey, M. (2017) 5 Charged with Involuntary Manslaughter in Flint Water Crisis. *The New York Times*, 14 June.

Barlow, M. (2007) *Blue Covenant: The Global Water Crisis and the Coming Battle for the Right to Water*. New York: The New Press.

Brenner, N. (2009) What Is Critical Urban Theory? *City* 1(2–3): 198–207.

Brenner, N. (2017) *Critique of Urbanization: Selected Essays*. Basel: Birkhäuser Verlag.

Brewer, J. and White, J. (2016) Questions Emerge over Motives for Ending Flint's Ties to Detroit Water System. *World Socialist Web Site*, 29 January. Available at: https://www9.wsws.org/en/articles/2016/01/29/dwsd-j29.html.

Cooper, M. (2009) Aging of Water Mains Is Becoming Hard to Ignore. *The New York Times*, 17 April.

Delaney, A. (2017) Some Flint Residents Could Face Foreclosure Over Unpaid Water Bills. *Huffington Post*, 18 May. Available at: http://www.huffingtonpost.com/entry/flint-residents-unpaid-water-bills-foreclosure_us_5909e494e4b05c397684e4ec.

Democracy Now! (2014) Water Is a Human Right: Detroit Residents Seek U.N. Intervention as City Shuts off Taps to Thousands, 24 June. Available at: https://www.democracynow.org/2014/6/24/water_is_a_human_right_detroit.

Democracy Now! (2016) Thirsty for Democracy: The Poisoning of an American City: Special Report on Flint's Water Crisis, 17 February. Available at: https://www.democracynow.org/2016/2/17/thirsty_for_democracy_the_poisoning_of.

DeVito, L. (2017) Centers for Disease Control Find Link between Flint Water and Legionnaire's Disease. *Detroit Metro Times*, 16 February. Available at: http://www.metrotimes.com/table-and-bar/archives/2017/02/16/centers-for-disease-control-finds-link-between-flint-water-and-legion naires-disease.

Fonger, R. (2017) Consultant Puts Cost of Flint Water Plant Fixes at $108 Million. *MLive*, 7 February. Available at: http://www.mlive.com/news/flint/index.ssf/2017/02/consultants_final_report_puts.html.

Friess, S. (2017) Where Nestlé Guzzles Water, Locals Doubt Claims of Negligible Impact. *The New York Times*, 24 May.

Gaist, T. (2014) Privatization of Detroit's Water Moving Forward at "Lightning Pace." *World Socialist Web Site*, 8 April. Available at: https://www.wsws.org/en/articles/2014/04/08/dwsd-a08.html.

Grant, M. and Miles, J. (2016) Lessons from Flint and the Price of Water Privatization. *Food and Water Watch*, 16 February. Available at: https://www.foodandwaterwatch.org/news/lessons-flint-and-price-water-privatization.

Guyette, C. (2017) A Deep Dive into the Source of Flint's Water Crisis: Tunnel Vision. *Detroit Metro Times*, 19 April.

Hammer, P.J. (2016) The Flint Water Crisis, KWA and Strategic-Structural Racism. Written Testimony Submitted to the Michigan Civil Rights Commission Hearings on the Flint Water Crisis, 18 July. Available at: https://peopleswaterboard.org/wp-content/uploads/2016/07/Hammer-MCRC-Testimony.pdf.

Hauter, W. (2015) Detroiters Need an Income-Based Approach to Water Bills. *Huffington Post*, 29 May. Available at: http://www.huffingtonpost.com/wenonah-hauter/detroiters-need-an-income_b_6956744.html.

Kaffer, N. (2016) Why Didn't Flint Treat Its Water? An Answer, At Last. *Detroit Free Press*, 31 March. Available at: http://www.freep.com/story/opinion/2016/03/30/flint-water-crisis/82421546/.

Khan, K. (2016) 18 Million People Served by Water Systems with Lead Violations in 2015, Report Says. *PBS*, 29 June. Available at: http://www.pbs.org/newshour/rundown/18-million-people-served-by-water-systems-with-lead-violations-in-2015-report-says/.

Kovac, M. (2015) The Truth about the Detroit Water Shutoffs. *Progress Michigan*, 9 April. Available at: http://www.truth-out.org/news/item/30118-the-truth-about-the-detroit-water-shutoffs.

Kurth, J., Bridge Magazine, and Detroit Journalism Cooperative. (2017) Detroit Shut Water to 1 in 10 Homes This Year. Yes, That's Progress. *Michigan Radio*, 5 December. Available at: http://michiganradio.org/post/detroit-shut-water-1-10-homes-year-yes-s-progress.

Martin, K. (2017) Detroit Water Shutoff Crisis Intensifies under Bill Payment Scheme. *World Socialist Web Site*, 10 May. Available at: https://www.wsws.org/en/articles/2017/05/10/wate-m10.html.

Matheny, K. and Egan, P. (2016) Nestlé Bottled Water Company Seeks to Take More Michigan Water. *Detroit Free Press*, 20 November. Available at: http://www.freep.com/story/news/local/michigan/2016/11/20/nestl-bottled-water-company-seeks-take-more-michigan-water/93175144/.

Moore, M. (2016) 10 Things They Won't Tell You about the Flint Water Tragedy. But I Will. *Huffington Post*, 1 February. Available at: http://www.huffingtonpost.com/michael-moore/10-things-about-flint-water-tragedy_b_9132150.html.

Nichols, J. (2014) Against Austerity in Detroit: 'Water Is a Human Right'. *The Nation*, 11 July. Available at: https://www.thenation.com/article/against-austerity-detroit-water-human-right/.

NOVA. (2017) Poisoned Water. *PBS*, 31 May.

OHCHR. (2014) Detroit: Disconnecting Water from People Who Cannot Pay – An Affront to Human Rights, Say UN Experts. *OHCHR*, 25 June. Available at: http://www.ohchr.org/EN/NewsEvents/Pages/DisplayNews.aspx?NewsID=14777&LangID=E.

Plum, A., Moxley, K., and Zervos, M. (2017) Water Shutoffs Impact Public Health: A Collaborative Study with Henry Ford Health System. *We the People of Detroit*, 11 April. Available at: https://wethepeopleofdetroit.com/2017/04/11/water-shutoffs-impact-public-health-a-collaborative-study-with-henry-ford-health-system/.

Shiva, V. (2016) *Water Wars: Privatization, Pollution, and Profit*. 2nd edn. Berkeley: North Atlantic Books.

Smith, M. and Bosman, J. (2016) Michigan Attorney General Sues 2 Companies Over Flint Water Crisis. *The New York Times*, 22 June.

Tabuchi, H. (2017) A $300 Billion Ground War. *The New York Times*, 11 November.

Veolia. (2015) Flint Michigan: Water Quality Report, 12 March. Available at: http://www.cityofflint.com/wp-content/uploads/Veolia-REPORT-Flint-Water-Quality-201503121.pdf.

White, J. (2014) Regional Authority a Prelude for Privatization of Detroit Water System. *World Socialist Web Site*, 12 September. Available at: https://www.wsws.org/en/articles/2014/09/12/dwsd-s12.html.

WJBD. (2017) Nearly 77 Million Served Water from Contaminated Systems, Report Says. *WJBD*, 2 May. Available at: http://www.wjbdradio.com/health-news/2017/05/02/nearly-77-million-served-water-from-contaminated-systems-report-says.

WXYZ. (2017) City of Flint Intends to Remain on Detroit's Water Supply. *WXYZ*, 18 April. Available at: http://www.wxyz.com/news/flint-mayor-karen-weaver-to-give-recommendations-for-water-sources.

3

Rapid adaptation and mitigation planning

Ashley Dawson

Introduction

We are losing. Losing beloved people and places to anthropogenic climate disasters, and losing hope that global elites will take the action urgently needed to mitigate carbon emissions. Some victories have been won, it is true. The steadfast resistance of the Water Protectors and their allies at Standing Rock in the face of the heavily armed police who pepper-sprayed, tear-gassed, beat and arrested nonviolent protestors eventually prodded the procrastinating Obama administration to block construction of a section of the Dakota Access pipeline. But this eleventh-hour decision was promptly reversed with the arrival of the Trumpocene. More recently, the World Bank announced that it would cease funding oil and gas extraction by 2019, a signal victory for the growing fossil fuel divestment movement (Elliot, 2017). Yet notwithstanding such important victories, the overall trajectory is towards planetary ecocide, as underlined by the announcement at the 2018 United Nations Climate Change Conference in Bonn that none of the world's industrialized countries are living up to the relatively tepid, voluntary pledges to cut carbon emissions they made two years before in Paris (Plumer and Popovich, 2017). And even if governments were to take further steps to meet their individual pledges, the world will still be on pace to warm-up well in excess of 2° C. This level of warming means irreversible sea-level rises of at least two meters (six and a half feet), which means that the entire bottom third of Florida will be covered in water by 2100, and that Miami and New Orleans will have to be abandoned. Many other major global cities will face catastrophic flooding, with Asian megacities like Shanghai, Shenzhen, Bangkok, Mumbai, Dhaka and Osaka worst affected (Holder et al., 2017). We cannot and will not give up the fight, but we must confront the fact that we are losing the struggle to forestall climate catastrophe.

Cities consequently need to adapt to increasingly destructive heat waves, droughts, hurricanes, and other "natural" disasters. Various adaptations will happen – indeed, they are already unfolding in fits and starts. The danger is that these steps to adapt to climate chaos will intensify the already appalling inequalities that characterize the world's extreme cities. The absence of concerted social action, adaptation and even post-disaster "recovery" efforts tends to wear even deeper the grooves of quotidian economic and social marginalization that bear down on urban communities the world over. As the climate emergency intensifies, adaptation threatens to exacerbate already nascent forms of climate apartheid (Dawson, 2017). For the global climate justice movement, it is imperative to imagine alternatives to such dystopian urban futures. How might climate adaptation be conceived of as an opportunity to heal the suppurating wounds of the extreme city, rather than to widen these gashes? How can we frame (and implement) redemptive urban imaginaries in order to shore up social solidarity and heal the divisions that imperil urban sustainability?

Climate action plans are emerging as an increasingly essential tool in envisioning an equitable urban future during the age of climate chaos. If the immediate cessation of all new fossil fuel infrastructure projects is the key demand of the global movement for climate protection, climate action plans offer a positive path beyond our current fossil-fueled social order. Such plans must offer a credible pathway towards zero carbon emissions no later than 2050. Climate action plans should be rolled out in verifiable stages, and the really heavy lifting of emissions reductions should not be left until the final stages. The climate justice movement is calling for the development of such plans by all organizations and at all levels of the state, but cities will clearly play a key role in articulating these, both because they are on the frontlines of climate chaos and because they often offer the greatest purchase for the democratic governance that is necessary to move societies beyond fossil fuels. Climate action plans certainly need to include sophisticated assessment of the technological challenges of moving to renewable energy, including the difficulties associated with transforming today's fragmented (and increasingly privatized) energy grids. But climate action plans are not simply technical documents. They are also – and above all – schemes for social transformation. If we accept the basic premise that urban resiliency is crucially linked to the strength of urban denizens' social networks, then successful climate action plans must hinge on imagining urban futures that capitalize on and strengthen communities. Building urban solidarities will be an essential task in order to help cities weather climate chaos. Such solidarities will also be imperative to fight against efforts by reactionary social forces to spark and exploit xenophobic and racist reactions to those displaced by the climate crisis.

In what follows, I examine two exemplary climate action plans: *The Upper Manhattan Climate Action Manual* (*UMCAM*) created by the Harlem-based environmental justice group WE ACT, and the Urban Land Institute's proposal for strategic retreat from a low-lying neighborhood in Miami. With their diverse strengths and weaknesses, these two plans offer not just roadmaps for imagined urban futures but also paradigms for the more general global effort to draft climate action plans. The process by which these plans were drafted is in many ways as important as the actual content of the plans. Also of cardinal significance is the conceptualization each organization produces for how such plans may be implemented, since it is one thing to draft a climate action plan, and quite another to see that plan actually being realized. Finally, such plans should ideally mix adaptation to climate chaos with mitigation of carbon emissions – with efforts to bolster the former also boosting the latter. After all, if adaptation is not an integral part of efforts to move us beyond fossil fuels, it may become nothing more than a glossy supplement to the current trajectory toward planetary ecocide.

Toward a just transition: the Northern Manhattan climate action manual

WE ACT has a long history of fighting against environmental injustice. The organization was founded during the 1980s in response to plans by New York City authorities to open a sewage treatment plant on the Hudson River along the western fringe of Harlem

(Dawson, 2010). In addition to challenging such environmental injustices in New York, the group has also played a role in originating paradigms for an alternative social and environmental order. The drafting of the *Upper Manhattan Climate Action Manual* (*UMCAM*) by WE ACT and a broad swath of partner organizations carries forward this struggle for climate justice, and is informed by some of the climate justice movement's cardinal principles. Among these is a framework which insists that the vulnerabilities of a city like New York are not simply a function of rising seas or aging infrastructure, but also of social inequality and exploitation:

> Without addressing fundamental issues such as poverty and racism, millions of New Yorkers will be displaced, regardless of climate. Therefore, methods of policymaking, infrastructure developments, financial systems, and more should work at the intersection of ending inequality and preparing for the specter of climate change.

This emphasis on the conjoined vulnerabilities that affect urban populations informs every aspect of the climate action plan, meaning that efforts to create resilient infrastructure and landscapes are always embedded in discussions of ameliorating the conditions for the communities that inhabit and use them. This approach contrasts significantly with dominant practices of even the most progressive architects, whose focus falls almost exclusively on shoring up physical infrastructures in order to climate-proof vulnerable cities, with consideration of the denizens of such cities coming in most cases as an afterthought. *UMCAM* deploys its highly holistic approach to climate action across a wide variety of sectors of urban society, including energy production, green infrastructure, waterfronts and coastal flooding – mitigating the urban heat island effect, affordable housing provision, and civic governance. In what follows I will touch on a few of the most indicative plans for these various sectors.

The intersectional lens of climate justice is perhaps most powerfully exemplified by *UMCAM*'s approach to urban energy. This is a key issue because the climate crisis is, above all, an energy emergency. The energy sector is responsible for at least two-thirds of all greenhouse gas emissions (International Energy Association, 2016). The world is growing increasingly hungry for power, and so, despite every effort, carbon emissions continue to rise. According to the International Energy Agency (IEA, 2016), overall global energy demand is projected to accelerate significantly in the coming decades, expanding by roughly one-third between now and 2040. This inconvenient truth has generated a sense of urgency around the need for an energy revolution, a wholesale shift from dirty fossil fuels to clean renewable energy. *UMCAM* registers this urgency and intervenes in the campaign for renewable energy, but situates this campaign within a much broader social struggle:

> Over the next several decades, billions of dollars will be invested in designing, building, and maintaining new energy systems. These systems can double-down on the centralized grid, gas, oil, and nuclear systems that New York State is already dependent on, or they can be transitioned to sources of renewable energy that are not managed by large bureaucracies, but rather by community-based institutions that can reinvest resources back in the community, including in the form of access to financial capital, jobs, educational opportunities, and more.
>
> *(Khawarzad, 2017: 30)*

Making the shift to solar, wind, and other renewable energies is thus not simply about introducing novel technologies for the generation of power. It is, equally crucially, about establishing the forms of community control and benefit that are increasingly known as *energy democracy* (Angel, 2016).

Energy democracy is predicated on the fundamental principle that energy production must not harm the environment or people. This seemingly obvious proposition comes with a radical corollary: planet-killing fossil fuels must be left in the ground. But it is not enough simply to switch to wind-, solar-, and tide-based renewable energy sources. This is not, in other words, simply an issue of technological innovation – as the shift to renewable energy is so often represented by mainstream commentators. Advocates of energy democracy insist that everyone should have sufficient access to renewable energy sources. Since the free market is notoriously unconcerned with such questions of egalitarian access, energy democracy thus also implies that the means of power production must be socialized and democratized. This entails conceptualizing energy as a form of commonwealth rather than, as at present, a commodity produced by for-profit entities. This shift makes sense since the new forms of renewable energy are predominantly produced in decentralized forms, making collective ownership of the means of production the most sensible way to regulate them, and to shift power to community and grassroots organizations.

UMCAM grasps and builds on the positive implications of energy democracy, treating the deployment of renewable energy as providing potential income and jobs to economically marginalized communities in Northern Manhattan. The *Manual* states that "organizing residential and commercial tenants into consumer and producer cooperatives can increase investments in energy, reduce costs, and provide needed ownership and employment within energy industries" (Khawarzad, 2017: 30). Collective investment in renewable energy would help speed the transition away from fossil fuels, and, in tandem, would allow communities to own the means of power production – allowing them to benefit when the energy they generate is fed back into the grid. Using progressive legislation such as New York City's *Community Shared Solar Act*, public housing developments as well as other institutions such as universities and hospitals could all develop shared solar generation capacity. Using this act, residents of public housing could also form co-ops that could receive solar power from remote installations, meaning that they need not own their rooftops and could still use solar power even when their needs exceed the capacity of their roofs (Khawarzad, 2017: 34). In addition, if done correctly, Climate Action would generate skilled and well-paying jobs for local communities that have been economically marginalized for many decades. Finally, the establishment of energy democracy would also counter strong trends toward energy poverty. As *UMCAM* notes, "low-income New Yorkers pay up to 13 percent of their income on energy. The average family in the US pays 1.5 percent" (Khawarzad, 2017: 32). This striking disparity in the economic burden of energy is of course added to other disproportionate burdens shouldered by low-income communities in New York for basics amenities such as housing. Fighting such injustices will require dedication to a vision of energy democracy that sees the city not as a homogenous entity but as the site of extreme social and environmental inequalities. Thus, although New York has fairly progressive legislation on the books, such as the *80 x 50* plan to cut carbon emissions by 80 percent by 2050, very little of the renewable energy systems that will be essential in reaching that goal have actually been deployed in northern Manhattan. The city needs to recognize that poor communities have borne far more than their fair share of environmental injustices, and that true urban resilience can only come through policies that lift up historically marginalized and disenfranchised communities.

As these examples suggest, the social component of urban resiliency is always primary for *UMCAM*. Although this theme is threaded through all of the *Manual*'s proposals, it is most apparent in the initiatives the *Manual* advocates for coping with disasters. In the section on emergencies, the *Manual* discusses the need for community education and the broader fostering of social networks as a key component in disaster harm-reduction. These proposals draw on insights about the centrality of social connections – rather than any particular technological system in isolation – in determining who survives and who does not in emergencies such as heat waves (Klinenberg, 2015). The *Manual* states:

> Connecting tenants with each other, and with organizations that can provide support, is critical to surviving an emergency. In areas with high social cohesion, vulnerable populations can be identified and receive evacuation and medical support. In places with low social cohesion people risk not receiving the services they need because they can't communicate and service providers don't know where/how to find them.
>
> *(Khawarzad, 2017: 44)*

Building on such insights, *UMCAM* advocates for the construction of *social centers* as spaces to weather storms. In the short term, such spaces, provisioned with microgrids, can offer safe harbor from blackouts caused by either severe storms or the overloading of the energy grid during heat waves (Khawarzad, 2017: 45). But, beyond such emergencies, social centers – as their name suggests – offer venues dedicated to fostering urban sociality. For example, a proposal of particular note in the *Manual* is the diversity of uses that such social centers may have:

> These places support programs ranging from hosting community meetings, providing facilities for meetings of tenant organizations, housing a library, showing film screenings, providing public health programming like yoga and self-defense, provide incubator space for community organizations, and access to technology and tools for art-making, among other things. The goal is to have a flexible space that can be programed and managed by the community and therefore caters to local needs.
>
> *(Khawarzad, 2017: 76)*

Underlying these diverse uses of the social center is an emphasis on generating social cohesion, on providing access to key services and skill sharing, and on a myriad of other forms of informal community building.

Perhaps the most developed contemporary form of the social centers proposed by *UMCAM* can be found in the city of Jackson, Mississippi, where Black activists have been fighting back against the legacy of centuries of oppression by organizing a striking variety of vibrant cooperative institutions (Akuno and Nangwaya, 2017). Happily, there are many other examples of such cooperative social spaces, both around the US and the rest of the world, from the many collective institutions that have grown up in post-industrial Detroit to the profusion of autonomous social centers across Europe (Lee Boggs and Kurashige, 2012; Pusey, 2010). But all of these alternative, fundamentally anti-capitalist social spaces face not simply pushback from reactionary social forces – including, most obviously, the police – but also the fundamental challenge of countering the power of capital on an urban scale. While there is much talk of greening cities today, all too often the transformations that result from such discourse are *relatively* superficial. One need only visit one of the booming cities in the US (a profusion of craft breweries is often a good guide to finding such cities) to find examples of skin-deep environmentalism such as the introduction of a lightweight light rail system in a city that is simultaneously throwing up superhighways helter-skelter to cope with the traffic congestion caused by galloping suburban sprawl. Urban greening is in this case part of a neoliberal ethos of inter-city competition, in which progressive cities must compete to lure businesses and affluent consumers without fundamentally challenging the growth of carbon emissions that comes with US-style urban sprawl (Fox Gotham and Greenberg, 2014). In addition, efforts at ameliorating environmental conditions in cities can also generate green gentrification, as more attractive living spaces draw affluent settlers (Checker, 2011). All too often, successful struggles against environmental injustice by grassroots social movements can paradoxically create the conditions for the ultimate displacement of the very low-income residents who have engaged in this battle. In the absence of significant, systematic efforts to provide good jobs and affordable housing, the dynamics of capital on the urban scale can displace communities, perpetuating relatively invisible but nonetheless deep forms of injustice.

Many of the proposals for green infrastructure in the *Climate Action Manual* address the problem of urban capital and inequality in a relatively indirect manner, but *UMCAM* also tackles these challenges head-on in its discussions of housing and governance. Housing is a particularly difficult issue in New York. Despite the city's much-ballyhooed program of constructing affordable housing under Mayor Bloomberg, New York still suffers from a massive dearth of housing that is truly affordable – and currently has all-time high numbers of homeless people as a result. Why is this? The Bloomberg affordable housing program worked by incentivizing developers to build putatively affordable apartments within luxury developments; in return for building these apartments, developers were given tax breaks under the city's fifty-year old 421-a housing scheme. The program generated around 2,500 units of affordable housing each year, but cost tax payers $1.4 billion annually (Bagli, 2017). In addition to this large tax giveaway to real estate, the housing program was flawed by the fact that "affordable" was defined according to median income averages. As a result of the massive amounts of money garnered by the city's 1 percent, median income for the city as a whole is totally unreflective of the income of the city's working class. The affordable housing program effectively became purely symbolic, benefitting only a small slice of the city's middle-class rather than the working poor it was truly intended to aid.

The situation may be particularly dire in New York, but it is part of a much broader, global crisis of housing. According to recent research, residential displacement due to development, extraction, and construction has reached a scale rivaling that caused by natural disasters and armed conflict (Madden and Marcuse, 2016). From the bulldozed townships of South Africa, to the demolished favelas of Brazil, the foreclosed suburbs of the American southwest and the gentrifying outer boroughs of New York, real estate is attacking housing, subordinating social belonging to private profit (Aalbers, 2016). Today more than a billion people are unable to find a decent or affordable home. The global housing crisis is a product of the underlying crisis of contemporary capitalism, as a tide of over-accumulated capital sloshes around the globe, commodifying housing, setting up urban "regeneration" schemes that generate mass displacement, and generally spreading insecurity and dispossession.

UMCAM confronts the housing crisis and the festering problem of gentrification in New York neighborhoods like Harlem and Washington Heights by highlighting the problem of impermanent and illusory "affordable" housing, and by calling for permanent (i.e. non-market-based) affordability as an alternative paradigm to the present capitalist one. One of the primary mechanisms the *Manual* suggests for realizing such ambitions is the Housing Development Finance Corporation (HDFC) co-op. Such co-ops are "social purpose corporations committed to the conservation of affordable housing" (Khawarzad, 2017: 114). Since, as the *Manual* notes, there are over 3,000 HFDC co-ops in New York, the idea of leveraging this cooperative institution is not simply a pipe dream. But the HFDC co-op is not a genuine solution for the city's deep housing crisis: these co-ops require a tricky combination of fixed income caps *as well as* a significant down payment – a combination that very few urban residents are able to satisfy. New Yorkers with low incomes tend to spend a significant amount of their incomes on housing, precluding the amassment of the types of financial assets required by HFDC co-ops (Higgins, 2014). While social lending institutions might be able to help poor urban residents by lending them some of the money required for HFDC co-op down payments, the concept of encouraging additional debt for the poor is a rather dubious one.

UMCAM's ideal of permanent affordable housing needs to be radicalized around the demand for a universal right to housing. This demand, although unrealizable under present political conditions, can be a rallying cry to mobilize radical coalitions around transformative demands that can be won, which might include a fight to defend and expand public housing, a secure system of rent controls, public ownership of land, public financing, limits on speculation, and the adoption or reintroduction of regulations on home finance mechanisms. These and other specific struggles could go some way toward the broader goal of making housing a right. That said, *UMCAM*'s intuition to link struggles over the right to housing to environmental struggles is a vital contribution. As Raquel Rolnick, former UN Special Rapporteur on Housing, argues, "the notion of the human right to adequate housing is not restricted to the access of the house itself . . . the right to housing has to be apprehended in a much broader context" (Madden and Marcuse, 2016: 330). The struggle for social justice must, in other words, be closely tied to the fights for environmental and climate justice.

Solving the global housing crisis will clearly take more than piecemeal, reformist solutions. A climate action plan that addresses fundamental aspects of urban sustainability must consider the basic conflict between capital and the good of the vast majority of those who inhabit cities. Resolving this crisis will take radical action by urban movements dedicated to rethinking the neoliberal social order. In the section of the manual dedicated to urban governance, *UMCAM* discusses the issue of how such radical transformation might take place, and how new commons-based ideas and movements are fighting to reappropriate urban spaces and cities more broadly:

> The urban commons framework raises the question of how best to produce and consume shared resources like energy, land, and water. Collaborative governance strategies ensure that resource allocation is not only based on who has power but that resources are made available to everyone.
>
> *(Khawarzad, 2017: 92)*

The urban commons framework is predicated on moving political struggles beyond the market-state dichotomy that characterizes the great majority of the political struggles of the twentieth century. While it does not eschew the state-based gains of social democracy during the past century, it strives to move beyond the top-down orientation that often characterized such modes of governance through the empowerment of citizen-led collective initiatives and institutions (Ostrom, 2009).

The idea of an urban commons provides leverage for urban environmental justice movements to challenge the remorseless commodification of cities by real estate interests and neoliberal capitalism more broadly. But, as David Harvey notes in his thoughtful critique of autonomist ideas of popular self-determination, commons movements can all too easily become the vehicle for producing greater inequality when they are appropriated by elite, neoliberal interests (Harvey, 2013: 83). As Harvey notes, urban commons exist alongside public spaces and institutions (such as public housing) that have been actively created by the state through direct conflict with the interests of capital. Both of these collective goods are being dismantled and privatized by capitalism. Effective commoning is predicated on taking collective goods out of the realm of private property relations, treating the commons as a good that is managed collectively and collaboratively. Recent proposals to achieve this include stipulating that those who use urban land exclusively should pay a "community land contribution" as compensation, creating a fund that would be administered either by local government or a community land trust to take land off the market or provide funding to those wanting to buy homes (Adams, 2016). In fact, New York already has such a program, known as the NYC Community Land Initiative, an organization formed in response to the housing crisis and the inadequacy of the city's responses. This movement is doing important work, but it operates within a political context in which city governance largely remains tied to the prerogatives of real estate. It will take far more political mobilization and a significant diffusion of the radical ideas behind community land trusts for these movements to make significant inroads in a city ruled by the likes of David Rockefeller, Michael Bloomberg, and Donald Trump (Moody, 2007). But this is the terrain on which the struggle for a truly sustainable New York must be waged. The *Upper Manhattan Climate Manual* gives us one of the clearest and most comprehensive imaginaries of this terrain and some of the means through which the battle for urban resiliency must be waged.

Planning retreat: the Urban Land Institute's adaptation action area framework

People will fight to hold onto and remake their cities, but some of these cities – many, in fact – will have to be abandoned, in part or in whole. The danger is that retreat from these drowning cities will take place in piecemeal fashion. The wealthy will eventually heed their financial advisors' assessments about declining property values in repeatedly inundated zones in cities like Miami and Shanghai. They will jettison their coastal land holdings, moving to higher ground within cities or away from the coast altogether. The poor will be stranded behind, left underwater in both an economic and material sense. If we do not talk forthrightly about the ultimate need for retreat, this will necessarily be the fate of the world's extreme cities.

But it is not easy to talk about retreat, and even harder to actually make it happen. Although the word has multiple meanings, including *a quiet and secluded place in which one can rest and relax*, the term has an overwhelmingly military connotation today. Military imagery and terminology has bled into virtually every aspect of life in this superpower that wages perpetual warfare. Even the environment has come to be represented using a military lexicon: for example, during the disastrous oil spill caused by the explosion of the Deepwater Horizon explosion, the struggle to stop the spill was described as a military campaign and, even more symptomatically, as an effort to "kill the leak" (McClintock, 2010). In this kind of martial culture, one thoroughly imbued with settler-colonial values of regeneration through violence, the notion of retreat is anathema (Slotkin, 2000).

The name of the post-Sandy recovery program in New York City says it all: Build It Back. This commitment to rebuilding at all costs is not purely rhetorical. Although coast-shaping policies such as the National Flood Insurance Program are nominally committed to supporting retreat from repeatedly flooded properties, out of every $100 of federal flood insurance money spent to rebuild homes damaged by floods, less than $2 goes to help people relocate themselves away from flood-prone areas (Geiling, 2017). As a nation, the US is committed to systematically ignoring the rising tides that imperil the cities that line its coasts, not to mention the increasingly searing temperatures that are prostrating the improvidently sprawling cities of the southwest. Yet, as well-funded and muscular an ideology as climate change-denial may be in the contemporary US, the wrathful force of climate chaos will ultimately prove far more potent. How may we begin to imagine retreat from climate change-imperiled cities while we still have the means to make such plans real? We need to plan now, to begin imagining alternatives, when we have both the capital and the state capacity left to make bring such schemes to fruition, and to do so in a manner that respects the ethical imperative of an egalitarian transition.

The Urban Land Institute's (ULI) scheme for the community of Arch Creek Basin in Northern Miami tackles this challenge of imagining a just urban retreat. The *Arch Creek Basin* report begins with a frank admission of the stark inequality that afflicts Miami. The Miami-Dade County metropolitan area is "one of the nation's least-affordable housing markets and faces severe income inequality" (ULI, 2016: 9). As a result of this, "day-to-day realities for low-income families are extremely difficult, without even considering long-term vulnerabilities and exposure to sea-level rise, flooding, and storm risk". In addition to this combination of quotidian and long-term precariousness, residents of Arch Creek Basin and other low-income neighborhoods in Miami are faced with what ULI calls *climate gentrification*: "fears that low-income, high-ground neighborhoods may become unaffordable as residents of the city's coastal and low-lying neighborhoods seek to relocate to avoid their areas' inherent risk" (ULI, 2016: 9). Concerns about climate gentrification are being articulated by low-income communities across the US, but they are particularly urgent in Miami and surrounding areas as a result of Florida's uniquely dysfunctional racialized and classed geographies. Wealthy people typically live as close as possible to the state's beaches, while poorer people usually inhabit less desirable (and heretofore less valuable) inland areas. But as sea-level rises begin to imperil coastlines, developers and wealthy individuals have begun acquiring property in neighborhoods on higher ground. The result is climate gentrification, a particularly vicious form of retreat in which poor communities of color are displaced from their land by wealthy people, who are ironically disproportionately responsible for the carbon emissions driving climate change.

ULI proposes to address this emerging climate change-related social injustice through a project of planned retreat in the zone of north Miami designated as a site for adaptation action. The *Arch Creek Basin* report sets out four interrelated areas of intervention: building social resilience in the area; managing water by restoring some of Arch Creek Basin's historical, natural systems; encouraging resilient and connected development patterns in a future transportation hub located on high ground; and using the Adaptation Action Area framework to address governance, financing, and implementation (ULI, 2016: 14–15).

The first of these interventions – building social resilience – hinges on expanding the definition of this fashionable buzzword, which has been adopted across an astonishing variety of contemporary institutions and endeavors, from architecture and urban design, to high finance, to counter-terrorism. These days everyone wants to be resilient, to be able to cope in a flexible and nimble way with the risky and unstable environments we increasingly inhabit. Yet most definitions of resilience hinge on almost exclusively technical definitions of the term. While material infrastructures clearly need to be shored up to cope with climate chaos, these (often extremely expensive) efforts will make little difference if communities are not educated about the threats they face and the adaptation options available to them. Resilience consequently must include a social component, one that crucially acknowledges the deleterious impact of extreme economic inequalities. The ULI team lays out a number of recommendations for how such social resilience may be built, and foremost is an effort to engage local communities through the creation of a social center very much along the lines proposed in WE ACT's *Climate Action Manual*, called, in ULI's case, a Resilience Resource Center (ULI, 2016: 15). This site would become a resource for recovery assistance during natural disasters as well as a venue for year-round outreach efforts to community organizations. Importantly, the ULI report recognizes that Miami in general and Arch Creek in particular is already home to many civic groups working around the intertwined issues of social and environmental resilience, and suggests that these groups must play an integral role in planning efforts through participation in the Adaptation Action Area Steering Committee. Although ULI is undeniably an external actor, and, as we shall see, might be viewed with suspicion by working class communities of color, it nonetheless sounds the right note through this insistence not just on community inclusion but on giving affected people a directing role in the adaptation solutions under development.

The extent of ULI's dedication to social justice only becomes truly apparent, however, in the section of the report dedicated to restoring Arch Creek's natural flood-absorption capacities. Some of these proposals flow naturally out of efforts at community education, including, for instance, the encouragement of household-level efforts to capture roof runoff using rain barrels. But other proposals are more sweeping, including the signature suggestion that the city construct a "slough" or wetland park to absorb flood waters at appropriate times, and provide much-needed recreational space for the community in drier times. This proposal builds on examples of flood-absorbing public space pioneered elsewhere, such as Rotterdam's Benthemplein Water Square (Keeton, 2015). ULI innovates, however, by marrying this proposal to an effort to provide safer, affordable housing to flood-affected residents of Arch Creek Basin. As the report puts it, "The slough would be implemented over time in concert with the alternative safe housing strategy, through the acquisition and assembly of persistent flood-prone properties" (ULI, 2016: 15–16). This scheme recognizes the imperative to retreat from certain flood-prone portions of a city like Miami, but in doing so seeks to ensure communities have a right to reside in a more attractive location rather than simply leaving them to fend for themselves as the waters rise. It is an immensely important intervention, one that acknowledges low-income communities' concerns about climate gentrification as well as the realities that must prompt retreat for imperiled coastlines.

ULI has come to this social-justice oriented framework around urban retreat as a result of the firestorm of criticism elicited by previous proposals by the organization for shrinking New Orleans after Hurricane Katrina decimated the city. Shortly after Katrina, Bring New Orleans Back (BNOB), an organization led by mega-developers such as Joseph Canizaro, persuaded Mayor Ray Nagin to invite ULI to propose a redevelopment plan for New Orleans (Davis, 2014: xii). As Mike Davis argues in his stinging review of the resulting plan,

> ULI's recommendations reframed the historical elite desire to shrink New Orleans's socioeconomic footprint of Black poverty as a crusade to reduce the city's physical footprint to contours commensurate with flood safety and a fiscally viable urban infrastructure. The outside "experts" proposed an unprecedented triage of a US city, in which low-lying neighborhoods would be targeted for mass buyouts and future conversion into a greenbelt to protect New Orleans from storm surges. ULI proposed setting up the Crescent City Rebuilding Corporation, armed with eminent domain and able to bypass city council. For veterans of the 1960s civil rights movement, it reeked of a return to the paternalism of plantation days.
>
> *(Davis, 2014: xiii)*

Under a barrage of criticism for embracing this plan to appropriate largely working class, historically Black portions of the city by legal fiat, Mayor Nagin reversed course and condemned the ULI/BNOB plan. Community mobilization against forceful displacement scored an important victory against a top-down, developer-led initiative with a genealogy extending back to the destructive "urban renewal" programs of previous decades. But the upshot is that many perennially flood-threatened neighborhoods in New Orleans remain at risk, their residents given no recourse as communities throughout Louisiana begin contemplating – and, increasingly, engaging in – managed retreat from imperiled zones (Gass, 2017). The solution cannot be to obstinately remain in place: this is a recipe for an eventual repeat of the mass devastation caused by the largely anthropogenic disaster that resulted from Hurricane Katrina (Smith, 2006). If current trajectories of carbon emissions are not radically altered, hundreds of millions of city-dwellers will have to be resettled in the coming decades. It is becoming increasingly apparent that we need policies of planned, just retreat from threatened coastlines (as well as drought-prone areas) if we wish to avert civilization-destroying social chaos (Watts, 2017). We need, in short, to imagine what it would mean to move portions of and even entire cities out of harm's way.

To what extent has ULI learned from the debacle in New Orleans? In its proposals for Arch Creek Basin in Miami, ULI lays out a plan for a new high-density, transit-oriented housing development located on higher ground along the area's Coastal Ridge (ULI, 2016: 15). The proposal for the new development hews to now well-established New Urbanist criteria: according to ULI, it should be "mixed use, walkable, and mixed income, creating a unique sense of place and economic development opportunities for downtown north Miami". Yet as critics such as Andrew Ross have argued, New Urbanist ideals often perpetuate auto-centric, sprawl-based urbanization through segregated zoning of commercial and residential areas. It is not clear that the ULI proposal has learned from the failures of New Urbanist design in recent decades. Indeed, ULI's discussion of the proposed Coastal Ridge complex as a "hub for economic development for a larger area" raises suspicion that the organization has retained the role of a "self-interested voice of corporate land developers", as Mike Davis characterizes it in his discussion of planning for post-Katrina New Orleans (2014: xii). And yet immediately after describing this economic potential, the ULI report states that "with this development potential comes the opportunity to serve nearby vulnerable communities" (ULI, 2016: 33). In addition to integrating mass transit and energy-efficient green buildings and systems into the new neighborhood, that is, ULI proposes a mixed-income housing strategy within a half mile of the new transit station. Crucially, ULI suggests that residents of Arch Creek Basin who choose to relocate from their repeatedly flooded properties would be given "first right of refusal" to the affordable housing units integrated into the new Coastal Ridge development. On first glance, this seems like a highly progressive proposal, one that overturns the policies of careless displacement that marred ULI's proposals for shrinking New Orleans and that potentially offers a paradigm for policies of just retreat from endangered coastal zones around the world.

Yet there are significant reasons to be skeptical of ULI's proposals. For one thing, there are problems, to put it mildly, with urban policies based on leveling public housing complexes – which are routinely perceived by urban elites in nakedly racist terms as hives of criminality and social dysfunction – and decanting residents into "mixed-use housing". Mixed-income developments billed as liberating the poor from the projects are, in fact, usually characterized by a wall of security measures, including lease compliance checks, housekeeping checks, criminal background checks, credit checks, employment verification, and drug testing (Dukmasova, 2014). If they are lucky enough to make it through this gauntlet and successfully resettle in the new developments, the residents of set-aside affordable housing find themselves subjected to a bevy of disciplinary mechanisms designed to maintain putative middle-class social standards. They find themselves living in a "sort of well-outfitted prison", where they are "pacified through rigid discipline privileging the perceptions and priorities of their wealthier neighbors" (Dukmasova, 2014). Many residents of such panoptical mixed-use developments unsurprisingly find themselves feeling isolated, disempowered, and not-too-subtly ostracized. But it is in relation to the notion of "affordable housing" that the pitfalls of ULI's proposal is most clear. As we have seen in relation to New York, affordable housing schemes pegged to "area median income" all too easily become boondoggles that benefit a small number of better-off urbanites while doing little to ameliorate the grave housing crisis afflicting the truly needy. True to the ersatz affordable housing protocols in rapidly gentrifying cities such as New York, ULI's proposal for Miami specifies that housing assistance should be directed to "low-income families (meaning families with 80 percent or less of area median income)" (ULI, 2016: 35). And yet, in an earlier section of the report, the authors note the city of Miami's gaping economic polarization: the "highest-income residents earn $40 for every $1 earned by Miami-Dade's low-income residents" (ULI, 2016: 25). Given this disparity, affordable housing based on area median income is inevitably destined, by simple mathematical necessity, to be completely out of reach for the vast majority of low-income residents of the city.

In summary, the ULI plan for Arch Creek Basin has the great virtue of taking the necessity of coastal retreat seriously, and attempts to offer a detailed plan for facilitating that retreat in a manner that cleaves to principles of economic and social justice. The plan's insistence on providing a social context for resiliency-building efforts, and its consequent effort to marry the revival of natural flood-protection features in the urban fabric to a voluntary relocation program is highly laudable, at least in principle. But these admirable elements are marred by a failure to be sufficiently serious in confronting the fundamentally unjust, economically exclusionary aspects of the extreme city. As we have seen, contemporary capitalism is quite remorseless in its development of socially and environmentally unsustainable cities. Without plans to fight back against the destructive aspects of the commodification of land and the natural world, vulnerable people and communities will continue to find themselves in harm's way as they struggle to survive in the inhospitable niches of the planet's increasingly extreme cities.

The shortcomings of ULI's work should perhaps not be so surprising given the organization's relatively elite background and the technocratic process through which the report on Arch Creek Basin was generated. How, we might well imagine, might a report on this topic have looked had a Miami-based organization similar to Harlem's WE ACT taken up the task of reimagining the area? We can only hope that grassroots climate justice organizations in Miami and other cities will respond to ULI's efforts to plan for retreat with their own overtly anti-capitalist imaginings of the sustainable, just cities of the future. Such radical urban imaginaries, which hinge above all on social solidarity, hold the key to winning the struggle for climate justice.

References

Aalbers, M. (2016) *The Financialization of Housing: A Political Economy Approach*. New York: Routledge.

Adams, M. (2016) *Land: A New Paradigm for A Thriving World*. Berkeley: North Atlantic Books.

Akuno, K. and Nangwaya, A. (eds) (2017) *Jackson Rising: The Struggle for Economic Democracy and Black Self-Determination in Jackson, Mississippi*. Montreal: Daraja Press.

Angel, J. (2016) *Strategies of Energy Democracy*. Brussels: Rosa Luxemburg Stiftung.

Bagli, C. (2017) Affordable Housing Program Gives City Tax Breaks to Developers. *New York Times*, 17 April.

Checker, M. (2011). Wiped Out by the 'Greenwave': Environmental Gentrification and the Paradoxical Politics of Urban Sustainability. *City & Society* 23(2): 210–29.

Davis, M. (2014) Foreword: Sittin' on the Porch with a Shotgun. In: Fontenot, A., McMichael Reese, C., and Sorkin, M. (eds), *New Orleans under Reconstruction: The Crisis of Planning*. New York: Verso.

Dawson, A. (2010). Climate Justice: The Emerging Movement against Green Capitalism. *South Atlantic Quarterly* 109(2): 313–38.

Dawson, A. (2017) *Extreme Cities: The Peril and Promise of Urban Life in the Age of Climate Change*. New York: Verso.

Dukmasova, M. (2014) The Problem with Mixed-Income Housing. *Jacobin*, 21 May.

Elliot, L. (2017) World Bank to End Financial Support for Oil and Gas Extraction. *The Guardian*, 12 December.

Fox Gotham, K. and Greenberg, M. (2014) *Crisis Cities: Disaster and Redevelopment in New York and New Orleans*. Oxford: Oxford University Press.

Gass, H. (2017) Tactical Retreat? As Seas Rise, Louisiana Faces Hard Choices. *CSI Monitor*, 2 August.

Geiling, N. (2017) Harvey Flooding Is Going to Break the Failing National Flood Insurance Program. *Think Progress*, 30 August.

Harvey, D. (2013) *Rebel Cities: From the Right to the City to the Urban Revolution*. New York: Verso.

Higgins, M. (2014) Bargains with a 'But': Affordable New York Apartments with a Catch. *New York Times*, 27 June.

Holder, J., Kommenda, N. and Watts, J. (2017) The Three-Degree World: The Cities that Will Be Drowned by Global Warming. *The Guardian*, 3 November.

International Energy Association (IEA) (2016) World Energy Outlook 2016, 1.

Keeton, R. (2015) A Storm-Water Drainage System Cleverly Disguised as a Park. *Next City*, 27 January.

Khawarzad, A. (2017) *Upper Manhattan Climate Action Manual*. AurashKhawarzad.com.

Klinenberg, E. (2015) *Heat Wave: A Social Autopsy of Disaster in Chicago*. Chicago: University of Chicago Press.

Lee Boggs, G. and Kurashige, S. (2012) *The Next American Revolution: Sustainable Activism for the Twenty-First Century*. Berkeley: University of California Press.

Madden, D. and Marcuse, P. (2016) *In Defense of Housing: The Politics of Crisis*. New York: Verso.

McClintock, A. (2010) Militarizing the Gulf Oil Crisis. *Counterpunch*, 24 June.

Moody, K. (2007) *From Welfare State to Real Estate: Regime Change in New York City, 1974 to the Present*. New York: New Press.

New York City Community Land Initiative. https://nyccli.org/about/background/.

Ostrom, E. (2009) *Beyond Markets and States: Polycentric Governance of Complex Economic Systems*. Nobel Prize Lecture. www.nobelprize.org

Plumer, B. and Popovich, N. (2017) Here's How Far the World Is from Meeting Its Climate Goals. *The New York Times*, 6 November.

Pusey, A. (2010). Social Centres and the New Cooperativism of the Common. *Affinities: A Journal of Radical Theory, Culture, and Action* 4(1): 176–98.

Ross, A. (2000) *The Celebration Chronicles: Life, Liberty, and the Pursuit of Property Value in Disney's New Town*. New York: Ballantine.

Slotkin, R. (2000) *Regeneration through Violence: The Mythology of the American Frontier, 1600–1860*. Norman: University of Oklahoma Press.

Smith, N. (2006) There Is No Such Thing as a Natural Disaster. *Understanding Katrina: Social Sciences Research Council*. 11 June.

Urban Land Institute (2016) Arch Creek Basin, Miami-Dade County Florida: Addressing Climate Vulnerabilities and Social Equity with an Adaptation Action Area Framework.

Watts, J. (2017) From Miami to Shanghai: 3C of Warming Will Leave World Cities below Sea Level. *The Guardian*, 3 November.

4

Urban nature and the ecological imaginary

Matthew Gandy

Introduction

The artist Lucian Freud is perhaps best known for his striking figurative representations of the human body. In the recent retrospective of his work held at the Tate Gallery in London, however, we find an intriguing exception to these studio portraits, represented by a painting entitled *Wasteground with Houses, Paddington* (1970–1972). This intricate tableau, which reveals a remarkable glimpse of London from the window of his studio, is framed by the rear elevation of a typical Victorian terrace. The drab greyish-brown brickwork and stained cornices are enlivened by ranks of chimney stacks with their jumble of fulvous earthenware chimney pots. Cutting through the middle of the scene is a mews of former stables now appropriately converted into a row of smart garages, and in the foreground is an expanse of rubble-strewn waste ground. Despite the twisted remains of abandoned furniture and rusted metal this former bombsite is now brimming with botanical interest: the faded spikes of the ubiquitous *Buddleia davidii* are interspersed with other characteristic colonizers of London's post-war landscape such as Ground-elder *Aegopodium podagraria* and Rosebay Willowherb *Chamaenerion angustifolium*. This, then, is an urban landscape, a seemingly unremarkable fragment of urban nature yet a critical reminder of the intricate combination of nature and human artifice that has produced urban spaces. An "urban ecology" is by definition a human ecology and is no more or less "natural" than any other type of modern landscape, whether it be a managed fragment of wild nature in a national park or those accidental pockets of nature of the type that Freud observed from the window of his studio in West London.

The interaction between nature and the modern city raises a series of conceptual complexities. If we understand the city to be a special kind of nodal point within an extending hyphal mesh of urbanization this still leaves the idea of urban nature as a somewhat ill-defined entity. The urbanization of nature, a transformation that has gained accelerated momentum over the last few decades, is clearly much more than a gradual process of appropriation until the last vestiges of "first nature" have disappeared. The production of urban nature is a simultaneous process of social and bio-physical change in which new spaces are created and destroyed, ranging from the technological networks that give sustenance to the modern city to new appropriations of nature within the urban landscape. The word "nature" is used here to encompass two somewhat different clusters of ideas: on the one hand, nature is used to denote a menagerie of concrete forms ranging from the human body to parks, gardens or complete ecosystems; and on the other hand, nature is evoked as an ideological and metaphorical schema for the interpretation of reality. In practice, however, these abstract and concrete elements are often interwoven to produce a densely packed urban discourse within which the origins and implications of different conceptions of nature are often afforded only cursory reflection.

The rise of the modern industrial city necessitated a refashioning of relations between nature and culture. Yet to refer to this transformation simply as the production of "urban nature" does not fully capture the complexity of this transition. The term "metropolitan nature" is probably more apposite since it can be deployed to signal recognition of the specific ways in which cultures of nature evolved in response to the socio-economic development and technological complexity of the modern city (see, in particular, Green, 1990). The urbanization of nature – and the concomitant rise of a metropolitan sensibility towards nature – encompasses not just new approaches to the technical management of urban space, such as improved housing and sanitation, but also extends to different types of cultural interactions with nature as a source of leisure. The transformation of nature in the modern city thus extends from new modes of urban governance or "governmentality" to use Foucault's term to changing modes of cultural perception, so that both the strategies and techniques of negotiating urban space become inseparable (Gordon, 1991).

Conceptions of the modern city have often been framed in terms of degrees of deviation from a supposed "natural" mode of living or in terms of analogies made with the body of a living organism. Ideas drawn from nature have played a significant role in developing an "ecological imaginary" in which ideas or metaphors drawn from the bio-physical and medical sciences have been used to understand the form and function of the modern city. The dynamics of urban change have been conceived, for example, in terms of processes such as ecological succession, the metabolic transmutation of nature, or even the post-industrial impetus towards putrescence and decay. Underlying many formulations of the ecological imaginary, however, there is an implicit naturalization of urban processes so that urbanization is no longer conceived as the outcome of historical change but rather as a cyclical dynamic alterable through technological modifications rather than by political contestation. By developing a concept of urban nature as a medley of different elements, we can begin to critically dissect some of the nature-based metaphors that have played such an influential role in the development of critical urban discourse. This essay seeks, therefore, to explore a hiatus between the conceptual stasis emanating from organicist conceptions of urban form and an alternative set of readings of urban space that place greater emphasis on the malleable, indeterminate and historically specific dimensions to the urban experience.

Aberrations and Utopias

From the middle decades of the nineteenth century onwards, the urban experience became increasingly synonymous with the experience of modernity itself. In Britain, for example, the urban population increased from just 25 percent in 1800 to over 75 percent in 1900, so urban life had changed from a minority experience to the majority experience. London, for instance, had a population of around 1 million in 1800, rising to 4.5 million by 1881. Similarly, Berlin saw its population of 200,000 in 1800 rise to 1.5 million by 1890 and Paris experienced a five-fold increase in population over the course of the nineteenth century to reach over 2.5 million by 1900. Other cities grew dramatically during the nineteenth century, including Chicago, Glasgow, Manchester, New York, Naples, Rome, St Petersburg, Vienna and Moscow. These industrial cities necessitated a new synthesis between nature and culture extending from the construction of urban technological networks to the establishment of new modes of municipal governance. The modern metropolis that emerged out of the chaos of the nineteenth-century city was driven by a combination of factors: advances in the science of epidemiology and later microbiology which gradually dispelled miasmic conceptions of disease; the emergence of new forms of technical and managerial expertise in urban governance; the innovative use of financial instruments such as municipal bonds to enable the completion of ambitious engineering projects; the establishment of new policy instruments such as the power of eminent domain and other planning mechanisms that enabled the imposition of a strategic urban vision in the face of multifarious private interests; and the political marginalization of agrarian and landed elites so that an industrial bourgeoisie, public health advocates and other voices could exert greater influence on urban affairs (Gandy, 2004).

From the nineteenth century onwards, the urban experience began to take an increasingly dominant place in modern culture. It is paradoxical that although modern cities were frequently evoked in organicist terms as bodies or organisms in their own right, cities were at the same time widely perceived to reside outside nature or the "natural order" as parasites or monsters. Thomas Hardy, for example, described London as "a monster whose body had four million heads and eight million eyes" and a spate of nineteenth-century novels such as James Greenwood's *The Wilds of London* (1874) dwelled on the poverty, darkness and danger associated with the industrial metropolis. Yet developing in parallel with these eschatological responses to urbanization, we can also detect a different set of discourses focused on the implications of urbanization for modern consciousness. By the early twentieth century, we find a proliferation of interpretations of the jarring and disorientating quality of urban life: Georg Simmel, for example, explores the "blasé outlook" of the city dweller in Berlin as a means to handle the "rapidly shifting stimulations of the nerves"; Virginia Woolf describes the atomism of urban life in the seemingly vast and alienating expanse of metropolitan London; and James Joyce conceives of early twentieth-century Dublin as a series of fragmentary encounters between different characters struggling to make sense of their lives. In these literary and sociological evocations of the modern city the urban experience enables the development of new forms of social, political and sexual awareness. As the city played a role in the enlargement of identity we find that modern consciousness finds expression through intensified pleasures of nature within the industrial metropolis. With the gradual distancing of the body from the fatigue, illness and malnourishment of the past, new cultures of metropolitan nature developed, including excursions into semi-wild fragments of nature at the urban fringe and the development of new aesthetic sensibilities towards landscape (Green, 1990). As the social and political possibilities engendered by modern metropolis unfolded, however, the progressive potential of the city began to increasingly conflict with traditional conceptions of urban order. The idea of "nature" was to play a defining role in this emerging tension between modernity and tradition as an ambiguous motif capable of underpinning both radical approaches to urban design and at the same time questioning the very foundations for urbanism itself.

The increasing association of the industrial city with the destruction of rural life sharpened the perceived antinomy between "city" and "country". The modern city was widely characterized as an aberrant spatial form that threatened to undermine existing ties of social and communal solidarity (Williams, 1973). The perceived superiority of rural life forms part of a powerful anti-urban sentiment connecting between the Jeffersonian ideals of small town America and a succession of later writers on cities – from Ferdinand Tönnies in the 1870s to Louis Wirth in the 1930s and Jane Jacobs in the 1960s. In a contemporary context, anti-urban views have resurfaced as part of an ecological critique of modernity that has a wide ranging influence on architecture and urban design. The so-called "New Urbanism", for example, owes much to the perceived superiority of small town life within which ideological motifs of stability and sustainability draw heavily on nature-based conceptions of urban design. A determinist conception of spatial form as a dominating influence over human behaviour is combined with a form of ecological nostalgia for an imagined past.

In contrast with reactionary visions of the modern city as an aberration, we can find an alternative lineage of urban thought originating within the Renaissance ideals of the city-state, which drew inspiration from the designs of Hippodamus, Vitruvius and other early advocates of symmetrical urban form. Renaissance scholars such as Leon Battista Alberti asserted that beauty in architecture was derived from the mimesis of nature but Alberti went beyond neo-Platonic conceptions of creativity to emphasize the critical role of human skill in the full realization of aesthetic perfection (Bacon, 2000; Forty, 2000). Urban design becomes an extension of beauty in nature whether reflected in the geometric arrangement of space or the embellishment of urban life through gardens, fountains and other meticulous appropriations of nature within the fabric of the city. The cultural utilization of nature is thus both an aesthetic quest to change the urban landscape but also an attempt to foster greater degrees of social and spatial order. The emergence of the nineteenth-century urban beautification movements, for example, sought to reintroduce nature into cities in order to prevent urban space from becoming an uninterrupted vista of development. A myriad of new organic spaces began to appear as a means to re-establish contact between nature and urban society. With the introduction of features such as public parks, botanical gardens and tree-lined boulevards, we find the explicit inclusion of a designed nature within the heart of the modern city.

The city beautiful movement evolved into the garden city movement of the early twentieth century and the search for a more ambitious synthesis between nature and urban form. The garden city brought together what was at best an inchoate mix of different ideas ranging from the utopian planning ideals of Ebenezer Howard to the naturalistic landscape designs of Frederick Law Olmsted, who drew a stark contrast between the cultural vibrancy of industrial America and the "rustic vice" of plantation agriculture. The various approaches to the garden city as it diffused through Europe and North America combined an eclectic mix of influences, including romantic and Beaux Arts traditions, but transcended the earlier ad hoc interventions through the articulation of a comprehensive approach to urban planning and design. Yet this apparent reconciliation between "city" and "nature" masked the actual transformation of nature under the impetus of capitalist urbanization. The principal legacy of the city beautiful and garden city movements was not the creation of utopian fragments in the urban landscape, important though these were, but the linking of landscape design and city planning ideals with burgeoning middle-class aspirations. In subsequent decades this earlier attempt to find a synthesis between nature and culture would, in fact, lead towards ever greater degrees of spatial polarization through the growth of suburbs, peripheral housing estates and other twentieth-century efforts to dismantle the inner core of modern cities.

From design to function

Developments such as the urban beautification and garden city movements were largely tangential to the underlying dynamics of capitalist urbanization, yet they remain one of the most influential dimensions to urban design. If we shift our attention, however, towards the function rather than the design of urban space, we find that the transformation of nature is far more pervasive and complex than it might first appear. The production of urban nature is inseparable, for example, from the development of urban technological networks that served to bind the modern city into a more integrated spatial form. The central cores of older cities were modernized to make way for roads, railways and speculative land development, forcing the working classes into ghettos and industrial districts of intense poverty. The "Haussmann approach" of comprehensive reconstruction pioneered in Second Empire Paris was also extended to Amsterdam, Barcelona, Cologne and many other cities (Centre de Cultura Contemporània de Barcelona, 1994). Yet this impetus towards spatial rationalization was not without its critics. In Vienna, for example, Otto Wagner attempted to create a modern city based around light, space and ease of movement, but his vision was challenged by Camillo Sitte with his rejection of utilitarian rationalism (Harvey, 1989; Schorske, 1981). Wagner's disavowal of nature as an organizational impetus for urban design was not shared, however, by all of the leading figures within the modernist movement: Le Corbusier and Frank Lloyd Wright, for example, continued to place nature at the centre of their work even if their conception of the modern city conflicted with the more vernacular urbanisms of the past.

From the late nineteenth century onwards, urban planning emerged as a clearly defined discipline accompanied by the growing role of technical elites in the institutions of modern governance. From the 1880s, for example, innovations such as land use zoning and regional planning gathered momentum. Technical and administrative expertise played an increasingly significant role in the modern repertoire of "governmentality" and the management of complex urban societies. By the early twentieth century, we find increasing emphasis on the "scientific management" of cities in the segregated and hierarchical ordering of urban space. Earlier attempts to create a utopian synthesis of nature and culture were gradually supplanted by a more radical technologically inspired vision. Progressively greater emphasis was placed on the radical separation of land uses in the "hygienic city" so that light, air and movement took precedence over the congested mingling of land uses in the nineteenth-century city. The idea of "speed" became the focal point for a new urban imaginary rooted in the creative destruction of the past. These technological fantasies reached their apogee in the designs of Italian futurists such as Antonio Sant'Elia, with his emphasis on multi-level roadways as a means to perfecting the circulatory dynamics of urban space. The sketches and plans for these technological utopias depict towering new buildings and virtually empty roads in an era before post-war congestion and the grassroots political challenge to the excesses of technological modernism.

The ideology of "hygienism" in twentieth-century urban planning belied the persistence of environmental and miasmic conceptions of the healthy city in combination with the post-bacteriological revolution in the scientific management of space: free circulation of both air and people was both a utopian gesture towards the horrors of the nineteenth-century city but also an attempt to create a new type of organic unity within the modern metropolis. The early decades of the twentieth century saw attempts to forge closer links between urban nature and the public realm so that the earlier innovations of municipal parks, improved sanitation and pedagogic displays of nature in zoos and museums could be extended to encompass a more ambitious conception of the role of nature in the modern city. Under the American New Deal, for example, we can discern a shift towards an expanded concept of urban nature to encompass a wider programme of social reform, including improvements in health, housing and urban infrastructure. The reshaping of nature on behalf of the modern city also encompassed vast engineering projects to provide water and power; for example, the new landscapes of dams and aqueducts in the American West cannot be conceived independently from the vast urban agglomerations with which they are connected. Yet the earlier associations of water engineering projects with a progressive political agenda, whether in Roosevelt's America or Nehru's India, have now waned to the extent that many of these engineering projects have become a leitmotif for the rapacious impact of modern cities on impoverished and politically marginalized rural communities (Cutler, 1985; Roy, 2002).

In the twentieth century, the changing relationship between nature, technology and urban space was driven to a significant degree by the spread of car ownership. This technological dynamic transcended national differences to the extent that we can discern striking similarities between the landscaped highways of Germany, Italy and the United States. In Martin Wagner's plans for 1920s Berlin, for example, the need for regional mobility was combined with the development of new peripheral housing estates. Wagner attempted to re-organize urban space in order to promote the greatest possible human happiness so that the rationalization of social and economic life and the rationalization of space became inseparable facets of the same process (Scarpa, 1986). Similarly, in Fritz Schumacher's plans for Hamburg (1909) and Cologne (1920) the centres of these cities were to be opened out with parks and public spaces to foster a new leisure-oriented metropolitan culture (Centre de Cultura Contemporània de Barcelona, 1994). And in the United States, Robert Moses brought a distinctively car-dominated vision to the modernization of the New York metropolitan region, within which middle-class consumer aspirations would play a decisive role. During the late 1950s and early 1960s, however, the construction of urban highways began to open up a conflict between the centralized engineering dominated ethos behind infrastructure development and growing demands for greater public participation in urban planning. The ideal metropolis conceived by technical experts and urban managers was increasingly in conflict with the lived reality of the modern city. Urban planning faced the disintegration of the kind of putative "public interest" that had sustained the ideal of comprehensive urban renewal. Planners themselves increasingly recognized that the ideal of "master planning" was illusory and began to explore ways of bolstering their legitimacy through wider public consultation. Patterns of infrastructure investment that had previously been conceived as integral to urban revitalization had now become directly implicated in post-war urban decline and the destruction of city life (Gandy, 2002). In the US, for example, the collapse of the consensus over highway construction in the 1960s mirrors the broader dissolution of the New Deal bipartisan consensus in public policy. The close interrelation between discourses of urban planning and the progressive impulses of modernist thought gradually began to unravel in the face of combined fiscal and political challenges. As urban planning became increasingly dominated by massive state subventions for corporate sectors such as cars and real estate, we find an increasing polarization in space between grim housing projects for the working classes and the burgeoning suburbia of middle-class consumer aspirations. With the rise of the fragmentary metropolis, the designed landscape was increasingly an adjunct to corporate atria, speculative housing developments, and other market-led responses to the urban crisis of 1960s and 1970s. In a sense, therefore, the ideological resonance of nature had come full circle to emulate the ad hoc interventions of the past: the role that metropolitan nature had played in the building of a functional public realm had been gradually supplanted by a more piecemeal emphasis on the decorative contributions of nature to the design of urban space.

Ecology, modernity and the post-industrial metropolis

Urbanization has now become synonymous with the globalization of economic and cultural life. In 1900 there were no more than a dozen cities in the world with more than a million people, and agriculture remained the dominant economic activity except for a relatively small number of industrialized nations. By the end of the twentieth century, however, over 500 cities had populations exceeding 1 million people and over half of the world's population were urban. The contemporary "urbanization revolution" dwarfs the experience of nineteenth-century Europe and North America yet is distinctive from this earlier transition in a number of critical respects. The so-called "brown agenda", which dominated the rancorous UN environmental summit held in Johannesburg in 2002, reflects the scale of the public health challenge facing contemporary cities, but the current sanitation crisis has originated in a fundamentally different context to that of the nineteenth-century city. These rapidly growing cities in the global South reflect an urban dynamic unrelated to the classic paradigms of city governance and planning, whether in the sprawling slums of São Paulo or the construction frenzy underway in China's Pearl Delta. It is increasingly difficult to talk in terms of any general or identifiable model for urban development, as each element takes shape within its specific context and parameters. The place of technical expertise has been superseded by a new entrepreneurial vista ranging from the most precarious slum settlements to the latest generation of immense skyscrapers that dwarf those of twentieth-century Europe or North America. The scale and complexity of this global urban transformation militates against any teleological extension of past experience and necessitates new insights into the urban process. The need to connect policy deliberation with the establishment of effective and legitimate forms of urban governance remains as important now as it was in the past, but such arguments can no longer rely on either the scientific logic of public health advocacy or rationalist conceptions of urban space promoted by a coterie of technical experts.

The gathering critique of modernist planning and design from the 1970s onwards has fostered new intersections between urban design and the bio-physical sciences. In the place of a cogent critique of the inequities engendered by capitalist urbanization we find a growing engagement with socio-biological ideas such as "defensible space", which were eagerly incorporated into critiques of public architecture and urban design. Increasing emphasis on individual property rights and demands for fiscal independence from the urban poor gradually coalesced around a new type of urban agenda exemplified by new surveillance strategies and the rise of gated communities. These developments have intensified the ambiguity of urban nature as both an inherent element within a functional public realm but also as a means to enhance property values as the management of hitherto public spaces has been increasingly taken over by quasi-public agencies or private foundations dependent on the whim of individual or corporate benefactors.

The post-war crisis in the rationale and impact of urban planning has been a central element in the ecological critique of modernity, yet a closer inspection of urban environmental discourse reveals the innate ambiguity of "ecological politics" as the basis for any progressive response to urban problems. The emergence of anti-nuclear movements in Europe and the environmental justice movements in the USA reflects a very different appropriation of ecological and environmental discourses to the reactionary anti-urbanism of the past. The impact of urban environmental disasters such as Seveso (1976) and Bhopal (1984), as well as the chronic ill health experienced in the poisoned cities along the US-Mexican border and other toxic locales, has spurred a new synergy between the politics of social and environmental justice (Hofrichter, 1993; Hurley, 1995). Although these radical political challenges to militarism, industrial negligence and the productivist logic of consumer capitalism share important elements with the ecological critique of modernity, they nonetheless embrace a more dialectical, inclusive and culturally determined conception of nature.

The return to nature in the post-industrial metropolis also denotes a conscious rejection of the kind of aridity engendered by the concrete landscapes associated with technological modernism. The understanding and utilization of urban ecosystems has become increasingly sophisticated, to embrace a more holistic conception of the interaction between biophysical processes and urban society. The development of new approaches to "ecological restoration", for example, marks a self-conscious attempt to recreate the bio-diversity of ecosystems that preceded the growth of the industrial metropolis in order to foster a different kind of synthesis between nature and culture. For instance, in the case of river channelswe can find examples of ecological restoration efforts that not only add aesthetic interest to the landscape but also contribute towards improvements in flood control and waste water treatment to produce a post-industrial or late modern synthesis between advances in ecological science and new approaches to landscape design (Gauzin-Müller, 2002; Gumprecht, 1999).

These developments have in part been fostered by the return of nature to post-industrial cities, so that the inner areas of some formerly industrial cities such as Baltimore, Detroit or Pittsburgh have taken on an increasingly Arcadian feel. In the photographic essays of Camilo José Vergara, for example, we can observe how inner urban areas have been reclaimed by nature through a mix of abandonment, neglect and structural change to produce "green ghettos". "In many sections of these ghettos", notes Vergara (1995: 16), "pheasants and rabbits have regained the space once occupied by humans, yet these are not wilderness retreats in the heart of the city". The growing presence of nature within former industrial landscapes can be conceived as a kind of urban entropy whereby the distinction between human artifice and ecological succession becomes progressively blurred. In the literature of J. G. Ballard, for instance, the fragility of the modern city is repeatedly portrayed through a tendency towards dilapidation and decay. In the post-industrial landscapes of Ballard, Iain Sinclair and other authors, we find that elaborate highway interchanges, high-rise apartments and other characteristic features of the twentieth-century city take on the form of urban ruins set amidst a complex palimpsest of new social and technological structures (see also Davis, 2002; Picon, 2000). A similar topographical trope of urban decay is also reflected in cinematic representations of the post-industrial metropolis. In Terry Gilliam's *12 Monkeys* (1995), for example, we encounter a post-apocalyptic Baltimore that has been taken over by elephants, lions, spiders and other organisms. This eerie spectacle is hardly an example of ecological restoration but rather a futuristic zoöpolis where urban space is controlled by animals rather than by human beings. The post-industrial metropolis and its cultural representations are suggestive of a very different kind of city to that of the nineteenth-century metropolis but this is a city for which we are still seeking an appropriate conceptual vocabulary. The characterization of urban segregation in terms of "ecological zones" by the Chicago School of urban sociology, for example, has more recently been reworked, albeit somewhat ironically, in Mike Davis's exploration of the "ecology of fear" in contemporary Los Angeles (Davis, 1998). The nineteenth-century metabolic insights of Karl Marx and Justus von Liebig have been reprised in order to provide a counterfoil to the functionalist emphases of "industrial metabolism", "ecological footprints" and other static conceptions of the modern city (Swyngedouw, 2004a, 2004b). And the dystopian genres of monstrous urbanism originating in romanticist reactions towards the nineteenth-century industrial city have been widely appropriated within more savvy examples of science fiction cinema and literature as a means to provide allegorical critiques of contemporary social and political developments.

Conclusions

The politics of urban nature are characterized by a range of "political ecologies" that can be differentiated from one another on the basis of their contrasting approaches to the conceptualization of nature. Any critical engagement with urban environmental change must contend with problems of terminology and historicity; although many aspects of contemporary urban discourse derive from the nineteenth-century city, we can nonetheless identify a critical change since the 1960s in which the "ecological imaginary" has played an enhanced yet deeply problematic role. The ecological imaginary, which comprises a cluster of dichotomous, ethological and neo-romantic readings of nature, remains rooted in organicist conceptions of urban space. The dynamics of urban change are widely conceived in terms of an adjustment towards a notional "equilibrium state" or as a set of processes that must be forcibly realigned towards a putative set of "natural" parameters. Yet this appeal to nature as something that resides outside of social relations is a corollary of fragmentary conceptions of cities as discrete entities that remain unconnected with wider processes of social and political change.

Ranged against the organicist lineage of the "ecological imaginary" we can identify alternative approaches to the understanding of urban nature that recognize the cultural and historical specificities of capitalist urbanization. The urban ecology of the contemporary city remains in a state of flux and awaits a new type of environmental politics that can respond to the co-evolutionary dynamics of social and bio-physical systems without resort to the reactionary discourses of the past. By moving away from the idea of the city as the antithesis of an imagined bucolic ideal, we can begin to explore the production of urban space as a synthesis between nature and culture in which long standing ideological antinomies lose their analytical utility and political resonance. Thus far, however, the development of more fluid and mutually constitutive conceptions of urban nature have had relatively little impact on popular discourses of "ecological urbanism", where the emphasis has tended towards the functional dynamics of metabolic pathways or the promotion of new forms of bio-diversity as a corollary of social and cultural complexity. It is perhaps only through an ecologically enriched public realm that new types of urban environmental discourse may emerge that can begin to leave the conceptual lexicon of the nineteenth-century city behind.

Acknowledgements

This essay was originally published in Erik Swyngedouw, Nik Heynen, and Maria Kaïka (eds.) *In the nature of cities: urban political ecology and the politics of urban metabolism* (London: Routledge, 2006).

References

Bacon, E.N. (2000) *Design of Cities*. London: Thames and Hudson.

Centre de Cultura Contemporània de Barcelona. (1994) *La visiones urbaines*. Barcelona: Centre de Cultura Contemporània de Barcelona.

Cutler, C. (1985) *The Public Landscape of the New Deal*. New Haven: Yale University Press.

Davis, M. (1998) *Ecology of Fear*. New York: Metropolitan Books.

Davis, M. (2002) *Dead Cities and Other Tales*. New York: The New Press.
Forty, A. (2000) Nature. In: *Words and Buildings: A Vocabulary of Modern Architecture*. London: Thames and Hudson, 220–39.
Gandy, M. (2002) *Concrete and Clay: Reworking Nature in New York City*. Cambridge, MA: The MIT Press.
Gandy, M. (2004) Rethinking Urban Metabolism: Water, Space and the Modern City. *City* 8: 371–87.
Gauzin-Müller, D. (2002) *Sustainable Architecture and Urbanism: Concepts, Technologies, Examples*. Basel: Birkhäuser.
Gordon, C. (1991) Governmental Rationality: An Introduction. In: Burchell, G., Gordon, C. and Miller, P. (eds.) *The Foucault Effect: Studies in Governmentality*. Hemel Hempstead: Harvester Wheatsheaf, 1–52.
Green, N. (1990) *The Spectacle of Nature: Landscape and Bourgeois Culture in Nineteenth-Century France*. Manchester: Manchester University Press.
Gumprecht, B. (1999) *The Los Angeles River: Its Life, Death and Possible Rebirth*. Baltimore: The Johns Hopkins University Press.
Harvey, D. (1989) *The Condition of Postmodernity: An Inquiry into the Origins of Cultural Change*. Oxford: Blackwell.
Hofrichter, R. (ed.) (1993) *Toxic Struggles: The Theory and Practice of Environmental Justice*. Philadelphia: New Society.
Hurley, A. (1995) *Environmental Inequalities: Race, Class, and Pollution in Gary, Indiana, 1945–1980*. Chapel Hill, NC: University of North Carolina Press.
José Vergara, C. (1995) *The New American Ghetto*. New Brunswick, NJ: Rutgers University Press.
Picon, A. (2000) Anxious Landscapes: From the Ruin to Rust. *Grey Room* 1: 64–83.
Roy, A. (2002) *The Algebra of Infinite Justice*. London: Flamingo.
Scarpa, L. (1986) Martin Wagner Oder Die Rationalisierung des Glücks. In: *Martin Wagner 1885–1957. Wohnungsbau und Weltstadtplanung: Die Rationalisierung des Glücks*. Berlin: Akademie der Künste, 8–24.
Schorske, C.E. (1981) *Fin-de-siècle Vienna: Politics and Culture*. Cambridge: Cambridge University Press.
Swyngedouw, E. (2004a) Circulations and Metabolisms: Hybrid Natures and Cyborg Cities. Paper presented to the research colloquium *Re-naturing Urbanization*, held at the University of Oxford, 30 June 2004.
Swyngedouw, E. (2004b) *Social Power and the Urbanization of Water: Flows of Power*. Oxford: Oxford University Press.
Williams, R. (1973) *The Country and the City*. Oxford: Oxford University Press.

5

Litter and the urban imaginary

On chewing gum and street art

Maite Zubiaurre

Introduction

Waste often attracts artists because of its threatening monumentality. Trash easily turns into a monster (one only needs to think of the infamous "Whitechapel fatberg," a gigantic "tumor" primordially made out of cooking oil, diapers, and sanitary products that blocked the sewage system of the British capital), or creates its own, expansive topographies. *Fresh Kills*, the massive landfill in Staten Island that was reopened in 2001 to receive the debris from the Twin Towers after the 9/11 terrorist attack, is said to be the "one-time largest man-made structure in the world" ("Fresh Kills Landfill...") and is clearly distinguishable from outer space.

This chapter, however, is not about the threatening monumentality of trash as depicted, for example, in the works of a number of renowned artists such as Andreas Gursky, Daniel Canogar, Chris Jordan, Vik Muniz, and HA Schulte. It is not about the dystopian urban imaginaries that mega-trash immediately inspires, but about imaginaries of a more amiable kind, suggestive of cities that embrace humans instead of overwhelming and even crushing them. It is a reflection ultimately about what art does with, and to, urban refuse at its early stages when it is still tiny, a newborn really that will only later accumulate and grow into gigantic proportions, and about what British street artist Ben Wilson and Mexican artists Ilana Boltvinik, Rodrigo Viñas, and Mariana Mañón, the members of art collective TRES, are able to accomplish with the help of one particular type of litter: the spat-out chewing gum one routinely finds on the streets.

The artists mentioned above work with "little" trash, and figure among the urban artists who favor "looking down" over the sweeping panoramic gaze. What interests them is not what is in front of them, or above them, but what humbly lies on the floor, discarded and ignored. This does not mean, however, that they are not interested in the rather monumental goal of creating new urban imaginaries. Artists sitting on the pavement and bending over discarded chewing gum are in fact imagining new possibilities for the city they inhabit. When Ben Wilson creates a quaint micro-landscape on a chewing gum blob, he is not only "decorating" it, but suggesting that urban life could be quite different from what it is (Figure 5.1). For example, it could be calmer and less rushed (with time to pause and create and enjoy art), and, in fact, what Wilson's paintings accomplish is precisely a new-found respite in the midst of urban chaos. Passers-by stop and bend to take a closer look at Wilson's mini-portraits of urban life, and ponder about this "other" urban existence Wilson paints – greener, with minuscule urbanites that have a name and tiny neighborhoods that irradiate a sense of belonging. Similarly, TRES's archeological study and lab work around spat-out chewing gum amasses biological and environmental data not only to determine the nature of human behavior and urban reality, but, more importantly, to imagine new possibilities for the contemporary megalopolis (Figures 5.2 and 5.3). What if we would not spit out chewing gum, what if we would not soil the streets thoughtlessly, what if the stress that comes with big city frenzy would not make us frantically chew gum, what if.... These are all questions that are born naturally from TRES's scientific-artistic interventions and give the Mexican art collective its characteristically speculative edge.

Boltvinik, Viñas, and Mañón define TRES as a collective that "explores public space with the help of artistic practices, and puts emphasis on garbage both as conceptual residue and political stance" (Boltvinik and Viñas, 2009). Wilson is equally fixed on the discarded, and when his eyes concentrate on the floor, they identify chewing gum blobs (sometimes still pink or mint green in color, but more often than not already soiled and thus turned grayish or even black) as ideal canvasses for his art. Wilson shares with TRES the fixation with all things "small," namely, with all that is tiny but also "inconsequential" and not desirable, such as litter: the British artist states, "I love the idea of taking something that is thrown away and transforming it" (Renaud, 2017). More importantly, TRES is as intrigued as Wilson with chewing gum soiling the streets, although its *modus operandi* and the nature of its artistic intervention, "Chicle y Pega" (It is Chewing Gum, and it Sticks, 2012) are quite different from Wilson's. Whereas the Mexican art collective is keen on integrating archeological and ethnographic methods into its artistic exploration of spat-out chewing gum in Mexico City's historical downtown, the British street artist "simply" paints tiny landscapes on the chewing gum stuck to the London pavement (Figure 5.1).

Wilson's approach is certainly part of an easily identifiable trend, where artists resort to floor-level miniatures to attract the attention of the urbanite. Instead of focusing on the mega-features of a city, artists such as Slinkachu and Isaac Cordal highlight the beauty of tiny topographies and their ability to replicate both nature and the urban landscape in small: in Cordal's "mini'art" and his series of tiny cement men, for example, a street becomes a grey river; a rain pool becomes a lake; a discarded TV becomes a cinema screen; and a tiny, dried-out splash of red paint on the pavement becomes a crime scene (Cordal, 2011). Slinkachu's equally minuscule replicas of urban dwellers similarly succeed in aggrandizing and giving relevance to the small. A tiny sculptural ensemble, no bigger than a thumb, shows an old lady pushing a trash can, and all this happens in the shadows of a "real" (and thus gigantic in comparison) trash bin, identical to its small replica; a minuscule woman wearing only a red blouse and a straw hat, her legs spread apart and her genitalia exposed, sits on a tossed cigarette butt, as if it were a bench; two men carry an "enormous" *Cheeto* found near a dumpster, among other *Cheetos* scattered around a McDonalds Plastic bag; and two micro-children, a boy and a girl, venture into an empty Marlboro cigarette pack (Slinkachu, 2008).

In the case of Cordal and Slinkachu, the "too small" has a double and contradictory effect on the viewer: it is reassuringly familiar and welcoming (more often than not, one is compelled to smile tenderly at the tiny urban creatures), but can also be disturbing, if not downright sinister: Slinkachu does not refrain from representing starkly erotic scenes,

Figure 5.1 Ben Wilson (aka Chewing Gum Man), 2008.

as the one described above, and both Slinkachu and Cordal are eager to recreate scenes of violence, such as a sparrow eagerly feeding on a dismembered body covered in blood (Slinkachu), or a man hanging from a barbed wire, the sharp metal digging into his stomach (Cordal).

Wilson's art, however, is quite different: although not unlike Slinkachu's and Cordal's tiny creations, it also centers around the urban miniature, it shuns violence or harshness nonetheless, and caters only to amiable scenes. For the British artist, chewing gum blobs are the perfect canvasses to his paintings, meant to be a reflection – often an utopian and intimate version – of the neighborhood where the spat-out chewing gum is found, or even of the street art that adorns its walls. As Renaud recounts, while

> taking a walk through the neighborhood of Shoreditch, a hotspot for street artists in London, Ben Wilson insisted on showing us his latest creations. He has just spent several days paying tribute to his peers by reproducing graffiti on chewing gums placed at the foot of the corresponding tagged wall. Each of these miniatures that have passed through the transmuting filter of his brush is absolutely remarkable. It's a way of recognizing his fellow artists and expressing his respect. If Ben Wilson is willing to spend a part or all of his day transforming a chewing gum into a work of art for you, it means he deems you worthy.
>
> *(Renaud, 2017: 7)*

One of Wilson's main purposes is to highlight the importance of community and a sense of solidarity among urban dwellers, artists and their work included. Always fixed on "small," Wilson is not interested in the grandeur of London as a big city and European capital, but in the small and humble neighborhood as the repository of a more human and less alienated (and alienating) way of life: "I'm doing work that's for the people," the artist says. "It's about social cohesion. Every time I do a picture for a different person, it's making links between people. It's vital we think more deeply about our human connection both to nature and one another" ("Meet the Artist: Ben Wilson"). As part of his effort at humanizing urban existence, Wilson makes sure to include "real" people and/or their names as part of his tiny urban landscapes, and to dedicate his paintings to his friends and neighbors, even to tourists strolling along the Millennium Bridge. Frequently, passers-by and locals are integrated into the micro-paintings; in fact, many of Wilson's creations are "commissioned." As the artist points out, "people are always coming up to me asking me for pictures, for different reasons" (Lyall, 2011).

Despite their naïve flair, Wilson's over 10,000 chewing gum paintings, found predominantly in London but also in other European cities, are heavily infused with social activism. One of the goals of his chewing gum paintings is to reclaim the city for its inhabitants and to remind citizens that they have an inalienable right to public space. Wilson's art certainly reclaims the city for himself by leaving his artistic imprint on the pavement, a regular performance – routine is important here, since it is through routine that one takes possession of space – that more often than not ignites the curiosity and admiration of the passers-by, but also provokes resistance and hostility. On occasions, people willfully destroy his art (Renaud mentions how some of the chewing gums Wilson used as canvass to reproduce graffiti art, "have been covered up with opaque paint" (2017: 7)), and quite frequently the artist has to respond to queries from police officers eager to declare his activity on the pavement unlawful. However, in this case recurrent incidents were met with a legal solution: a judge determined that as long as Wilson's creations remain within the confines of the chewing gum blob and the paint does not soil property, his art is indeed legitimate and does not contravene city ordinances.

Street art is quite effective at signaling the lack of urban spaces of freedom. Muralists and graffiti artists are routinely banned from walls, the pavement, and urban furniture, and as a consequence street artists searching for paintable surfaces are often forced to resort to all things discarded. Some find meaning in chewing gum blobs, as Wilson does, or leave their imprint on garbage bags, as is the case with Spanish artist Francisco de Pájaro. Ultimately, true freedom is unavoidably linked to the abject: trash accumulating on curbsides and spat-out detritus are among the very few sites of artistic expression able to escape dehumanizing regulation.

Wilson's art is a keen reflection of dehumanized spaces, policies, and behaviors, and one of his strategies to restore humanity to neighborhoods and urban life is to play with eloquent oppositions. For example, his delicate miniature paintings have a domestic quality (in *Talking Trash. Cultural Uses of Waste*, I compare them with little home adornments or bibelots, reminiscent of Victorian or even *Biedermeier* aesthetics), and yet they are street roamers and provocateurs, quick at igniting the contempt of public authorities. In addition, Wilson's pictorial miniatures heavily rely on naïve art (and thus evoke "cuteness" and innocence), but then again, it is difficult to forget that they are drenched in bacteria-infested saliva. Finally, Wilson resorts to the "low" and dirty (such as filthy spat-out chewing gum) to appeal to high ethics: community-building solidarity and compassion remain the driving force and the message of his engaged artistic practice, as does the impulse to remind urban dwellers that thoughtless or mechanical actions, no matter how small or seemingly harmless (like spitting chewing gum on the pavement), have dire consequences. Art, Wilson believes, can turn around thoughtless actions, and restore the worthiness not only of the discarded artifact (in this case, chewing gum), but also of the environment and of human connection:

> I am upset by all the garbage and sense of disconnectedness where people just affect things in a slightly detached way. When people detach from the environment, that's when the environment gets destroyed and that's also when people destroy each other. . . . When a person throws chewing gum, it's a thoughtless action. I'm turning that around. People think they don't have an effect. But all the people that chew gum and throw it on the street, they created that. Once painted, it suddenly takes on new meaning and has been given the kind of worth that would otherwise be unthinkable.
>
> *("Meet the Artist: Ben Wilson")*

Ultimately, what inspires Wilson is humanity, the imprint of the human – and of all that is humanizing – on the urban landscape. In that sense, spat-out chewing gum is the perfect canvass for Wilson's artistic and ethical agenda indeed, since it is literally soaked in human DNA and contains the key to a human being. Thus, when Wilson depicts an inhabited landscape on a spat-out chewing gum, he not only paints humanity, but paints *on* humanity. Saliva and acrylic paint blend together; art and life become one.

The concept of spat-out refuse as an artifact (a living organism almost) that contains human matter also informed "Chicle y Pega," an artistic intervention by Mexican art collective TRES that took place in 2012. The Mexican art collective's approach to chewing gum soiling the pavement of Mexico City, however, is quite different to Wilson's treatment of gum blobs peppering the streets of London. Whereas the British artist transforms the urban landscape into a studio or atelier, TRES quickly turns it into an outdoor laboratory, or a hybrid rather between a laboratory and an archeological excavation site (Figures 5.2 and 5.3).

TRES's artistic method is based on incessant street-walking, "inspired by Situationist derives, obsessive scavenging…, and the "[detection] and [contextualization] of trash through meticulous observation" (Boltvinik and Viñas, 2016: 209–10). In the specific case of "Chicle y Pega," walking, scavenging, and observing was applied to the careful identification and interpretation of the chewing gum blobs that sully the streets of Mexico City's historical center, particularly those found on the pavement of pedestrian streets Regina and Madero. "Chicle y Pega" is a carefully thought-out interdisciplinary enterprise. As a press note points out, TRES's installation

> [traza] un primer acercamiento planteado desde la etnografía – con una perspectiva de pepenador- *para encontrar y sistematizar los chicles en la vía pública*, para seguir con una antropológica, *para entender costumbres y su relación social*; arqueológica, *para extraer información cultural*; química, *para encontrar vestigios e información biológica*; y de la restauración, *para encontrar metáforas urbanas de las constelaciones encontradas.*
>
> [uses ethnography as a first step to find and systematize chewing gums tossed on public spaces, in the style of "pepenadores" or ragpickers, and then moves to anthropology to understand certain habits and its social implications; to archeology to extract cultural information; to chemistry to search for biological traces; and to

Figure 5.2 Chewing gums undergoing restoration #1, site-specific intervention, Madero Street, Historic Center, Mexico City.

Figure 5.3 Chewing Gum #142, classification process of chewing gums, site-specific intervention, Regina street, Historic Center, Mexico City.

> restauration and conservation to further dwelve into urban metaphors construed around street refuse.]
>
> *("La basura, pieza clave...")*

In previous artistic interventions, TRES focused on "the mobility of trash (*Archipiélago de olvidos*), on the subtle nuances of its spatiality and location (*Puntos ciegos*), its symbolic value (*An informal gaze*), its aesthetics (in the book, *Desechos Reservados*), and its intimacy (*Todo lo que brilla es oro*)" (Boltvinik, Viñas, and Mañón 2). In "Chicle y Pega" what was of interest to creators is yet another crucial characteristic of trash: its permanence. For TRES, chewing gum blobs are scars on the urban skin, fossils stubbornly embedded in the pavement. These scars or fossils create veritable celestial maps ("cicatrices o fósiles que se quedan marcados, como mapas celestiales de la ciudad" (Boltvinik et al., 2012: 2), and urban constellations that mirror the sky.

Chewing gum blobs are not only a sign of the "sticky" permanence of garbage, its constellation-like complexity, and its ability to mirror our sky and, more importantly, our reality on earth. As Boltvinik, Viñas, and Mañón explain, spat-out chewing gum and discarded artifacts at large are objects densely packed with information:

> Una de las cosas que nos interesa abordar sobre la basura es su carácter informativo, es decir la capacidad que tiene para almacenar información simbólica a través de la cual se puede elaborar un retrato tanto individual como colectivo, y que nos permite analizar la relación que como sociedad establecemos con el consumo y con el desecho.

> [One aspect of trash that we are interested in exploring is its informative nature, that is, its ability to store symbolic information. Thanks to the latter, we can then create individual and collective portraits, which in turn allows us to analyze the relationship that as a society we establish with the consumption and disposal of goods.]
>
> *(Boltvinik et al., 2012: 2)*

Art collective TRES makes very clear that its approach only caters tangentially to an environmentalist agenda: "nos preocupa descentralizar el tema de la basura de su eje ecológico para generar una proximidad con ella que implique diversas reflexiones" ["we are interested in diverting the subject of trash from its ecological or environmental axis as a way of generating a proximity with garbage that will lead to various reflections"] (Boltvinik et al., 2012: 3). The two last quotations combined are of fundamental importance when it comes to understanding TRES's perspective that trash is not only a repository of essential information (it talks about us, about the motivations and behavior of humans who are so eager to produce, consume, and discard), but also an artifact (or collection of artifacts) with which we engage in a complex and very close relationship. As TRES has continuously stressed and is always attempting to explore in its different artistic interventions, garbage remains close to us, and it is this intimacy (an intimacy that environmentalist approaches tend to overlook) that TRES seeks to actively analyze and reflect upon.

For one, everything we try to get rid of bears our imprint and tells a story, a private one but also a shared one, that we are not always eager to hear. As Boltvinik and Viñas put it,

> deformed objects [forgotten, abandoned, and discarded] are the ones that truly inhabit this world. Mutilated and repudiated, these object do not receive the attention they deserve, in spite of, or perhaps because of the fact that they collectively act as a mirror that reflects our humanity in a harsh light and shows who we are as a society, and as individuals
>
> *(Boltvinik and Viñas, 2016: 215)*

I would go even further than that: as I have indicated in *Talking Trash: Cultural Uses of Waste*, what makes garbage particularly unsettling, and what turns it into a reality that is taboo and that we do not want to face or learn about, is the fact that it is a powerful reminder of our own demise. Not unlike garbage, we too will end up in a cemetery/landfill, and our flesh will rot even faster than many of the man-made materials that fester in dumps.

Trash is certainly a mirror of our individual and social identity, behavior, and fate. However, it is important to note is that it is not a "clean" surface that reflects from afar, but a messy substance "contaminated" with human traces. Spat-out chewing gum in particular is literally drenched in human saliva and thus contains the DNA of the person who chewed it. But in addition of containing biological data, chewing gum blobs also attract and accumulate dirt. Boltvinik calls them "una especie de imanes receptores" [a type of receptive magnets], and points out that they are not only DNA carriers, but also "registers of human transit" and accurate "indexes of dust and pollution" (Sánchez, 2017).

In this context, it is worth revisiting my previous observations about TRES' work with chewing gum in Mexico City:

> Not unexpectedly, *TRES's* intervention brought to light a number of data, some of them anecdotal, others more substantial. We know, for example, that chewing gum blobs are remarkably dirty creatures, infested with bacteria (around 50.000 to 70.000

> of them). Madero street alone in Mexico City's "centro histórico" that served as base of operation for "Chicle y pega" is peppered with approximately 770.000 chewing gums (an average of seventy chewing gum pieces per square meter) and each piece of chewed gum is stepped on by up to two million people. Only 45 days after Madero street was repaved and reopened as a walk street in 2010, 8.441 chewing gum blobs already sullied the freshly innagurated concrete and marble pavement. The municipal district had to buy ten chewing-gum remover machines, but these were only able to lift between 2.000 and 2.500 chewing gum pieces from the ground during the period of 2010–2012, not to mention the fact that to remove a chewing gum costs five times as much as to produce it.
>
> *(Zubiaurre, Forthcoming)*

Spat-out chewing gum on urban pavements contain DNA and pollution data, but they are also "visual and aesthetic modifiers of the urban landscape" (Sánchez), and function as geographical locators or indicators. Therefore, an important part of the "Chicle y Pega" intervention is to suggest different hypotheses about the meaning (or meanings) of chewing gum locations and concentrations. For example, TRES participants first identified a number of what they referred to as "attractors" of chewing gum on Regina Street (four manholes, fifteentrees, eight trash cans, fourteen benches, two bars, one small park, seventeen stores, and thirteen restaurants), and then tried to identify how chewing gum blobs organize themselves around the so-called attractors, how close they are from one another, and how dense chewing gum "constellations" are. This first approach generated different hypotheses. The hypothesis of "even distribution," for example, would imply that chewing gum blobs appear distributed uniformly along the street, whereas a second hypothesis, that of "uneven distribution," would suggest that chewing gum blogs accumulate in bigger number around "attractors." If that were the case, 25% of Regina Street, the stretch closer to cross street 5 de Febrero, would register more chewing gum blogs, since this is an area more densely populated with attractors. Finally, a third hypothesis, that of "specific distribution," would imply that certain attractors are more successful than others in accumulating chewing gum blobs around them. Naturally, each hypothesis invites different conjectures: if hypothesis #3 were true, for example, and if one were to find more spat-out chewing gum around bars and restaurants, one could conclude that there is this need to chew gum after eating or drinking, accompanied by a general reluctance of leaving the chewing gum behind, pasted to ashtrays, plates, or napkins. On the other hand, if the bigger concentrations happen around trashcans and manholes, one could imply that although the intention of passers-by was to spit the chewing gum into trashcans and manholes, many end up missing the target.

When TRES "plays around" with hypotheses and conjectures about chewing gum blobs as geographical locators and indicators of urban social behavior, its goal, needless to say, is not to come up with a "scientific" conclusion, but, rather, to signal that garbage is not "garbage," but a cultural artifact laden with complex information. Foremost, it is an invitation to closely scrutinize trash, instead of falling into the generalized habit of ignoring it. Moreover, TRES not only makes trash visible, but insists on the importance of what I like to call "rituals of revelation." To set up an archeological site on Regina Street – as if artists were archeologists – and to unearth, treat, and classify chewing gums as if they were fossils, is precisely one such ritual (Figures 5.2 and 5.3). For one, it forces us to slow down, to lower our eyes, and to take in pavement filth, a behavior contrary to the moving fast and looking up or ahead that is the typical modus operandi of the busy urbanite.

Thus, what "litter artists," among them Wilson and art collective TRES, are proposing with their fixation on trash sullying the streets is a very different approach to reality, one that shuns the arrogance of the panoramic view so common in the cultural apprehension and representation of the urban landscape. Instead, it embraces the humility of looking down and of centering one's attention (and compassion) on what society has declared undesirable. In the case of TRES, rituals of slowing down and scrutinizing garbage have given birth to multiple revelations and a complex theory of trash, one that grows more sophisticated with each artistic intervention, and in turn fuels creative inspiration. "Chicle y Pega" is not an isolated installation, certainly, but part of TRES's long-standing artistic obsession with the discarded. Thus in order to fulfill its ambitious goal to give voice to garbage as a macro-narrative and to embark in a vast archeology of the contemporary, TRES has created a capricious archive of artistic installations that includes

> "una arquelogía del chicle, un estudio forense de las colillas, múltipes levantamientos, registros y catalogaciones de la basura, así como un estudio de la química de la orina" [an archeological excavation and analysis of spat-out chewing gum; a forensic study of tossed cigarette butts; numerous excavations and painstaking registers of retrieved garbage; and even a study of the chemical composition of urine] ("Derivas sobre la basura…").

Moreover, in its recent volume, *Ubiquitous Trash*, TRES offers a condensed description of the philosophical tenets on which much of its artistic practices rest:

> Our starting point is the simple proposition that all forms of matter necessarily contain forms of expression. This is particularly true with trash: objects that are full of traces and marks of who used them and where they have been. In each piece of refuse a gesture is left that allows us to gain insight into human culture.
>
> The combinations of numerous carefully considered trash signature-characteristics collectively begin to form an extraordinary and valuable social portrait that transcends the nationalities and cultures that create them. Thus, our operation is to collect and catalogue waste singularities that can lead us to recognize hidden identities and narratives. The collection we continuously shape offers a method to access a series of micro-histories that woven together assemble a social perspective.
>
> *(Boltvinik and Viñas, 2016: 210)*

The Mexican art collective thus "traces the whereabouts of trash, collects specimens, gathers data, and creates complex morphologies. It excavates the cities and penetrates into its ecosystems and public spaces in order to recover forgotten objects and tell their stories" ("La basura, pieza clave…"). What we have here thus is an "artistic-archeological practice that unearths the urban refuse of contemporary societies and unveils the customs and quirks of local cultures and enclaves" ("La basura, pieza clave…"). More importantly, it brings to the fore the hidden traits of garbage, the philosophical meanings that everything that is not wanted and discarded (and precisely because it is not wanted and is discarded) carries within.

With its various artistic installations and throughout its writings, TRES draws our attention to the pervasiveness, ubiquity, and mobility of trash:

> Trash is everywhere, all the time. We do not cease to produce it, neither does it disappear.… It has no clear frontier: soda bottles made in China may wash up on a

> beach in Mexico or Australia; medical waste from New York may be found in the beaches of Brazil or Iceland. Trash from anywhere can be found everywhere. It is simultaneously global and local. Not only because of its mobility – caused by ocean currents and winds, or because it is moved in large containers across the world – but because objects that become trash are manufactured with materials from all over the globe, and distributed all over the globe.
>
> *(Boltvinik and Viñas, 2016: 210–11)*

TRES is also eager to stress another characteristic of discarded objects: they never exist in isolation, but in constant dialogue with other equally abandoned artifacts:

> With every piece of existing trash, there are many other objects – also definable as trash – that are associated with that one piece. For example, a cigarette butt implies that somewhere – far away or nearby – there will be a corresponding aluminum paper, cigarette pack, ashes and plastic wrapping. Every object is part of a much larger chain of items that were once bound together by [the] production/consumption cycle, and that have separated their paths and commenced a singular journey.
>
> *(Boltvinik and Viñas, 2016: 212)*

TRES also emphasizes that trash is quasi-immortal ("the range of decomposition of plastic bottles, for example, is from 450 to 1000 years" (Boltvinik and Viñas, 2016: 213)) and thus a rich and permanent repository of information ("trash is full of scars that talk to us about its trajectories, how long it has survived, and where it has been. It becomes a coded text that can be read through archeological procedures" (Boltvinik and Viñas, 2016: 212). At the same time, we are in danger of constantly "losing significant pieces of data that where once on its surface" (2016: 213): "Expiration dates, places of production, and types of material are some of the missing facts that have been rubbed off during an object's journey" (Boltvinik and Viñas, 2016: 213).

TRES's artistic interventions have generated a complex "theory of trash" built around concepts such as permanence, mobility, ubiquity, connectivity, and erasure. Installations such as "Chicle y Pega" bring to the fore the importance of such concepts and of trash as a semantic system made of traces, scars, and fossils that speak the truth about the society that created them. Like William Rathje and Cullen Murphy, TRES also believes, and further proves, that "garbology" is a branch of archeology, and that it is through archeological methods, which can also be applied to the arts, that humans will be able to trace the wealth of information stored in discarded artifacts. However, more important even than unearthing hidden knowledge is to openly acknowledge how intimately related we are to the objects we discard. According to TRES thus,

> one of the best ways to address waste as a social issue involves adding a new R to the famous 3Rs (recycle, reuse, reduce). This new R stands for re-signify, which ... involves understanding the intimate relationship we have with waste and how that can move us to act out of desire, rather than just out of a sense of [ecological] duty
>
> *(Boltvinik and Viñas, 2016: 214).*

Wilson's and TRES's art certainly "re-signify" trash, and it is not by chance that they both choose spat-out chewing gum as the object of their artistic desire: bathed in human saliva, and a double repository of DNA and urban filth, spat-out chewing gum could not be more intimately linked to humankind and the urbanite. Chewing gum blobs are us, and it is on us/our skin that Wilson imprints compassionate tattoos/urban landscapes of humanity and solidarity, and that TRES exerts its unique and equally humanizing archeology of the contemporary. Needless to say, to "re-signify" spat-out chewing gum is also to create new urban imaginaries, ones that bring to the fore the possibility of a city that is more habitable, solidary, and humane (like the amiable landscapes of Wilson's micro-paintings), and thus keen on paying compassionate attention (of the type one finds in TRES's scientific-artistic interventions) to what urban society heartlessly discards.

References

Boltvinik, I. and Viñas, R. (2009) From Trash to Treasure. Available at: http://tresartcollective.com/ABOUT-TRES. Accessed 10 February 2018.

Boltvinik, I. and Viñas, R. (2016) *Ubiquitous Trash.* Hong Kong: WYNG Foundation and Festina Publicaciones.

Boltvinik, I., Mañón, M, and Viñas, R. (2012). *Chicle y pega o arquelogía celestial de la inmundicia. propuesta para El Estudio Extendido en Casa Vecina, Del 14 de mayo al 15 de agosto.*

Cordal, I. (2011) *Cement Eclipses. Small Interventions in the Big City.* London: Carpet Bombing Culture.

"Derivas sobre la basura. Residuos de una estética del desecho." (2017) *Nexos*, June 30. Available at: https://cultura.nexos.com.mx/?p=12971. Accessed 24 February 2018.

"Fresh Kills Landfill: One-Time Largest Man-Made Structure in the World Will Be a Park in 30 Years." (2018) Available at: https://www.atlasobscura.com/places/fresh-kills-landfill. Accessed 8 February 2018.

"La basura, pieza clave en la construcción del retrato de nuestra sociedad." 11 August 2012. https://circuloa.com/casavecina-tresartcollective/. Accessed 12 January 2018.

Lyall, S. (2011) Whimsical Works of Art, Found Sticking to the Sidewalk. *The New York Times*, 14 June 2011. Available at: http://www.nytimes.com/2011/06/14/world/europe/14muswell.html. Accessed 20 January 2018.

"Meet the Artist: Ben Wilson." (n.d.) Available at: http://humannatureshow.com/meet-the-artist-ben-wilson/. Accessed 10 February 2018.

Rathje, W. and Murphy, C. (2001) *Rubbish! The Archeology of Garbage.* Tucson: University of Arizona Press.

Renaud, L. (2017) *Ben Wilson. The Chewing Gum Alchemist.* Amsterdam: LMV.

Sánchez, S. and Castillo, R. (2012) Chicle y pega, basura en restauración. 30 June 2012. Available at: http://voyvengo.com.mx/2012/chicle-y-pega-basura-en-restauracion/. Accessed 5 January 2018.

Slinkachu. (2008) *Little People in the City. The Street Art of Slinkachu, Foreword by Will Self.* London: Boxtree.

Zubiaurre, M. (Forthcoming) *Talking Trash. Cultural Uses of Waste.* Nashville: Vanderbilt University Press.

6

IHM-agining sustainability

Urban imaginaries in spaces of possibility

Sacha Kagan

Introduction

Imagination is understood in the following pages as a creative process of individual and social (de/re-)construction of reality. Imagination is not to be confused with fantasy, although fantasies and visions of alternative/utopian futures are part of the imagination. Imagination is both an ability to perceive the absent and an ability intimately related to everyday perception: imagination's role in perception is to give form to sensory intuitions (Kant, 1970) and "The working of the imagination within the world gives that world an *affective texture*" (Lennon, 2015: 3). Imagination is not merely fantasy but sensitivity to the possible shapes of one's environment – there is an "imaginary texture of the real" (Merleau-Ponty, 1964: 24).

An imaginary is understood as a (individually or socially) patterned set of shared images and forms (not only visual but multisensory). Social imaginaries are the fundament of social institutions because they establish webs of meanings that have unity and internal cohesion, legitimizing certain views, logics, and organizational forms in a society (Castoriadis, 1975). To understand stability and change in social institutions, we thus need to reflect on the imaginaries that are central to given societies in given historical periods. Furthermore, a reflection on the relations of dominant and emergent social imaginaries to what Castoriadis called a "radical imagination" will be crucial to the normative academic field of sustainability science, which aims to steer transformative change away from unsustainable development paths. This requires envisioning alternative future worlds, which in turn depends on the ability "to refer to things that are not perceivable in the here and now" (Herbrik, 2011: 11). A radical imagination is crucial, Castoriadis argued, in order to detach from a given dominant social imaginary and the certainties it provides, and in order to develop new models. Herbert Marcuse stressed for his part "subversive imagination" and argued that "the need for radical change must be rooted in the subjectivity of individuals themselves" (Marcuse, 1978: 3–4).

Imaginaries are emplaced, and place-making involves imagination and mobilizes imaginaries (Pink, 2012, 2015). Lefebvre (1968, 1991) described the manifold shifts of usages and assignments of urban spaces as an interplay of concrete (everday), abstract (commodified) and imaginative (liberating) traits. He contrasted abstract space, as hegemonically determined "representations of space", to imaginative space, as the idealistic and socially liberated (and liberating) "representational space":

> *Representations of space*, which are tied to the relations of production and to the "order" which those relations impose, and hence to knowledge, to signs, to codes, and to "frontal" relations. *Representational spaces*, embodying complex symbolisms, sometimes coded, sometimes not, linked to the clandestine or underground side of social life, as also to art.
>
> *(Lefebvre, 1991: 33)*

Whereas Lefebvre pointed to a potentially radical and subversive imagination at work in realizing spaces of possibility for a practiced "right to the city", the concept of urban imaginary was further developed by urban researchers to more generally refer to "symbolic, cognitive, and discursive constructions of urban space and living" (Lindner and Meissner, 2016: 6).

Urban imaginaries and the "radical" (and other) imaginations of urban actors are also related to different approaches to social and political participation and cooperation. These are caught in a tension between dialectic and dialogical tendencies (Sennett, 2012). In a dialectic process, tensions between opposing views are resolved through compromises or argumentative resolution or synthesis. In a dialogic process, different views co-exist and respect each other's differences, with oppositions remaining open and unresolved. These two poles are associated with imaginaries of change by planning versus imaginaries of emergence and by problems/solutions-oriented versus questions-oriented learning.

These questions draw attention to the possibility for urban development to shift away from unsustainable processes (such as neoliberalization) and towards the complex negotiation processes of sustainable urban development (Kagan et al., 2017). It is under this perspective that my colleagues and I, in the ongoing research project "City as Space of Possibility" (Leuphana University Lüneburg, 2015–2018), empirically focused on the city of Hanover, Germany, are investigating which qualities characterize "urban spaces of possibility for sustainable urban development", that is (geographic, social, and mental) spaces where the combination of imagination and experimentation open up future-oriented questions and perspectives (Dieleman, 2012; Kagan et al., 2017). In this research project, we thus investigate the urban imaginaries and (more or less radical) imaginations of urban actors whose practices may be developing spaces of possibility for sustainable urban development.

IHM-aginations and IHM-aginaries of sustainability: five cases from Hanover-Linden

In the following pages, I focus on empirical insights gained from five cases of initiatives, organizations and networks that are all based within or directly at the border of Linden, a central-west district of Hanover, bordered to the east by the Ihme River, to the North by the Leine River (into which the Ihme flows), to the south by railroad tracks and an industrial zone, and to the west partly by a canal and a commercial zone. A former working-class district, Linden today hosts a diversity of inhabitants, including people with an immigration background and a visible proportion of so-called "cultural creatives". The district is partly (and only at an early stage of) undergoing a process of gentrification (Link, 2016). In the minds of many Hanoverians interested in cultural-creative and socio-ecological lifestyles, as well as of artists and creatives choosing to live and/or work in this city, Linden is a neighborhood of choice. It is also the district where we identified the majority of the people, organizations, and the five cases we investigated and collaborated with in the "City as Space of Possibility" project:

- A social sculpture/art-based grassroots social process (*Forum Ort des Treffens* or Place of Encounter Forum; hereafter abbreviated as FOT);
- An urban intervention art collective (*Das wundersame Aktionsbündnis der Tante Trottoir* or The Wondrous Action Alliance of Aunt Sidewalk; hereafter abbreviated as Tante Trottoir);
- A Linden-based organization working at the science-society interface and developing diverse cultural formats focused on issues of sustainability (*Wissenschaftsladen Hanover* or Hanover Science Shop; hereafter abbreviated as WiLa);
- A network of cultural creatives and socio-ecological grassroots activists engaged in creative-cultural approaches for sustainability, who are based in Linden and operate across the central districts of Hanover (*Kultur des Wandels Netzwerk* or Culture of Change Network; hereafter abbreviated as KdW);
- Several individuals (and an organization they created, the *Zukunftswerkstatt Ihme-Zentrum* or Ihme Center Future Workshop; hereafter abbreviated as ZWI) engaged in the sustainable development of the Ihme Zentrum, Europe's largest "brutalist" concrete architectural complex and the city's most controversial (partly dilapidated) building ensemble, bordering the Ihme River.

The empirical material informing this study consists of 27 field notes from participant observation, six transcripts of meetings, nine semi-structured interviews, seven "walks with video" (Pink, 2007) covering the five cases that were gathered between 2014 and 2017, as well as several participant reflections from a workshop and artistic intervention carried out with Tante Trottoir. The material was coded with atlas.ti, combining deductive and inductive coding.

In the five cases, one can observe some common and some specific thematic foci in the imaginations and imaginaries with respect to sustainability/sustainable urban development, as well as both common and specific features in the ways they open up spaces of possibility for imaginations and imaginaries related to sustainability-oriented potential futures. In all five cases, the leading protagonists are aware of the need to establish adequate settings and (physical, social, and mental) spaces in order to allow for the production of ideas and imaginative processes. All aim at opening up spaces, not only for dreaming and creating new ideas but also for linking these to specific forms of practice. They all strive to develop participatory and interactive formats that allow different degrees and qualities of participative processes. However, each has developed a specific approach.

FOT, initially a social sculpture project created by Shelley Sacks (artist and director of the Social Sculpture Research Unit at Oxford Brookes University) for the city of Hanover, in 2009, soon became a grassroots initiative based on the formats created by Sacks and coordinated by the artist Anja Steckling and a handful of enthusiastic core members. FOT was active until 2015 (no activities were reported for 2016 and 2017).

FOT includes three formats: two processes of personal interactions and one system of sound stations. In the two interactive processes, the participants share and reflect on their personal views, thinking out loud. In the first process, the Place of Encounter with Oneself (*Ort des Selbsttreffens*), a participant chooses a public space for the self-reflection and determines the position of the person who is listening actively without commenting.

The speaking participant is seated on a chair on a small yellow felt circle, and is asked by the listener to reflect on two questions: "What am I doing on earth?" and "What does it mean to me to be in the world?" Sound stations (*Klangstationen*) were used in the project's early years to play recorded "encounters with oneself" (without using the name of the speaking person). They were originally installed in public spaces in Hanover (later only available on a website). The second process is called Place of Encountering the Other (*Ort des Einandertreffens*). Several participants reflect in a public space – often in the Hodlersaal in Hanover's town hall – on the two questions they have already considered in Encountering Oneself. The process focuses on the method of active listening, around an extended yellow felt circle. The reflections, thoughts and ideas are not evaluated or commented on. In all three formats, FOT opens spaces for imagination, but especially in the place-making process involving the circumscribed symbolic space of the "yellow circle" with the participants following the FOT dialogic interaction rules in the two encounter formats. In the words of the FOT core members: "We are the proposal for a mobile space of reflection for change" (focus group interview with FOT members).

At FOT, the whole process of initiating a flow of ideas and mental images is the least directed of the five approaches. It is based on a deeply reflexive process rooted in a combination of mindfulness and violence-free communication combined with an artful, aesthetic sensibility. The interactive and participatory spaces FOT offers are clearly governed by the dialogic principles of violence-free communication. The goal is for the participants to focus on the discursive and affective space of the one person speaking, suspending judgment and avoiding any interruption or commentary, striving instead to empathically follow the person speaking as their communication and imaginary unfold, without merging them into their own experience, appreciating them in their difference. There is no discussion, but rather a "dialogue" (which in practice might be better described as a series of monologues), with each participant having the space to contribute their own monologue in due time. Interactions are not meant to convince, argue with, or win over the other person. Instead, they are meant to allow a deeper, respectful and more mindful mutual understanding where the main goal is not to provide direct solutions to specific problems.

Hence, the imaginaries being developed in FOT activities relate primarily to ontological questions of personal being and meaning in the world, and secondarily to epistemological questions on the conditions of knowledge, communication, and agency. This orientation is also given by the two opening questions that initiate the process of the "encounter with oneself". Subsequently, the imaginations of the participants are oriented towards very personal *and* abstract, philosophical, and existential levels. The space opened up by FOT is, for those who dare to fully engage in this space, a protected time-space where in the best case an individual engages in a deeper reflection on the very meaning of their own life, because this space is mindfully freed from value judgments. For some participants, this opening had life-changing consequences (as confirmed by two of our respondents, interviewed several years after they experienced the encounter processes), while for others it merely allowed the emergence of a couple of new thoughts or insights.

The concept behind the sound stations was to offer the possibility that the anonymous voices recorded in some of the "encounters with oneself" could be heard at various public sites across the city and would inspire serendipitous listeners into further imaginative processes. (The sound stations were discontinued several years before our research project started. We did not interview anyone who was part of this format and so we do not have any insights into the effective impact of the sound stations on imaginative processes.)

For its core members and other enthusiastic participants, the central shared imaginary of a sustainable future produced by FOT is strongly correlated to its dialogical practice of mindfulness. Its ideal future is characterized by a culture of communication in which aggressive criticism is replaced by mutual respect and understanding. This would dissolve "fossilized/encrusted [social] structures" and, according to one core member of FOT, offer a better democratic participatory space for urban development.

Besides its exemplary communicative practice of dialogic mindfulness, FOT's space of reflection connects to the participant's own "creative-designerly potential" (FOT focus group). In our interviews, however, the images of sustainable futures we find are mostly vague and abstracted, often focusing on the already mentioned mindful communicative culture, but also marginally touching on the theme of nature finding its own rights in an ecologically reformed way of human life. Nevertheless, several core members and participants in FOT are also active or interested in projects that are engaged in finding "new ways" in the areas of agriculture and the economy, focusing on communal/solidary and locally-rooted working concepts (solidary agriculture, local currencies, economy for the common good).

Tante Trottoir is an artistic initiative founded in 2015 by the performers Lisa Grosche, Astrid Köhler and Lena Kussmann, and hosted at the *Theater an der Glocksee*, an independent theater located on the Ihme riverside, facing the district of Linden. Tante Trottoir stages participative artistic interventions in public spaces in Hanover, most often performance-based but also installations. It aims to develop a subtle and suggestive approach, one that is much less "aggressively" confrontational than many other art interventions in public spaces, with the aim of shifting perceptions of passers-by in the direction of civic empathy, care and solidarity. They name this approach "Positive Actionism" and refer to Dostoyevsky's maxim "beauty will save the world" (from his novel *The Idiot*).

The spaces for imagination opened up by Tante Trottoir are characterized, on the one hand, by a rather abstracted/generalized and personal/subjective exploration of themes that concern participants, through workshops aiming for a mix of mindfulness and playful lightness (and avoiding direct and concrete issues where specific protagonists and antagonists would be identified) and, on the other hand, by a concrete focus on a specific site in both its multisensory (phenomenological) and social (and intercultural) qualities and potentials, through urban interventions. The stimulation of participant imaginations (in the workshops and at interventions) does not directly aim at creating visions and ideas for potential futures, but rather orients them to immediate realities (both felt directly by the participants and related to current news carried in the media) and suggests new perspectives and interpretations that show potential positive/sustainable responses to current issues. Workshop participants are invited to *herumspinnen* or fantasize/brainstorm/spin beyond what's realistic and conventional. The images materialized and/or performed at an intervention are merely one fraction of the many ideas generated by the participants in the final phases of the workshop, which is carried out a couple of weeks before an intervention. Then, the urban interventions initiate processes to change the passers-by's imaginations of a concrete space and/or situation in a creative process (the passers-by are free to choose to what extent they wish to interact with the intervention).

The interventions involve specific imaginary elements with unexpected and irritating yet poetic, oneiric and bucolic-stylish qualities, generating different, unexpected, new images. These images, in their diversity (as each art intervention approaches a different theme and takes a different form) often suggest that more mutual help and care for strangers in public spaces are possible and desirable. Looking at all the interventions realized until now (from

2015 to 2017), some common imaginary features appear: the generated images often evoke alternative interpretations and potential uses of the place at the site of the intervention and then indirectly/subtly also some potential notions and hints for alternative futures. Tante Trottoir often invokes the imagination (and artistic-participatory performance/practice) of an alternative experience of time that is more mindful, slower, more relaxed, and peaceful, and allows caring relationships (addressing stress, hecticness, and insensibility to others with whom we share the public space in urban settings).

In some cases, the artistically emplaced imaginary is highly symbolic, rooted in a widely shared cultural heritage, and poetic while not directly pointing at any particular issue(s): in "Wake Up Rosie" (2016), an intervention was installed at the Dornröschenbrücke (Sleeping Beauty Bridge), a pedestrian bridge (also used by cyclists) across the Leine River, next to a riverside park under an inner-city highway, linking the districts of Linden and Nordstadt (the latter being traditionally the home of punk-DIY culture and leftist alternatives in Hanover). The bridge is very popular for hanging out on warmer days and for its annual ritualized vegetable fight between the two districts. Tante Trottoir invested the bridge with a canopy bed, potted roses, a huge banner with the English text "Wake Up Rosie", bits of texts from the story of Sleeping Beauty and various props. The passers-by were able to sit or lie on the bed, read or add to a small diary hanging from a bedpost and activate a tiny mechanical music box. The performers were mostly sitting idle on the bridge several meters away from the installation elements and only sometimes used the bed, blowing soap bubbles or interacting with the passers-by. Tante Trottoir's questions (posted on their website) were: "Who is Rosie? Should someone wake her up? Should we let her sleep? What does that mean? Should I also lie down? Does this have something to do with refugees? With myself? With the world? Or is it just pretty?" The connotations of the text and of the installation aim to stimulate attaching a reflexive-critical imagination to the juxtaposed imaginaries. We are invited to think about ourselves as citizens and of the place and/or the urban society of Hanover as a Sleeping Beauty that maybe should wake up and do much more. All of this this is only poetically suggested and not forced upon the viewer.

In other cases, the artistically emplaced imaginary is more directly and explicitly pointing to current issues, as in Zaungast (2016), an intervention at the Steintor Square (in the hyper-central district Mitte, in the middle of commercial pedestrian zones and the red light district), where the need to keep escape routes open for refugees was performed (with actual fences and music from a mechanical music box). The unusually explicit (for Tante Trottoir) political character of the performance and its localization in a hyper-central site highly frequented by people with different political backgrounds (unlike Linden where many inhabitants are pro-refugee social democrats, greens, or leftists), led to much more confrontation with and negative reactions from passers-by, which reportedly stressed the performers. Following this experience, Tante Trottoir decided to return to their subtle, suggestive approach and avoid confrontational politics.

They followed this with a workshop in summer 2017 and the Tante Trust intervention in collaboration with myself and with the participation of students in my seminar "The practice of artistic urban intervention" at Leuphana University Lüneburg. The theme that emerged from our workshop was very sensitive: rebuilding mutual trust in public spaces in times of terrorist threats and paranoid security "measures". The chosen symbols sought a fine balance between irritation and subtlety: given-away muffins that "could be poisonous, could be delicious"; a selfie photo-op with a person who "could be a terrorist, could be the love of your life"; a short performance in a public square, on a towel, of some blurred mixture of yoga and Muslim prayer performed alternatively and/or together by a blond pale-white male and two male refugees from Syria and Afghanistan, who "could be a fanatic, could be someone practicing yoga"; and a suitcase left unattended, yet set-up as a wondrous mystery picture box inviting the passers-by to satisfy their curiosity by looking into holes in it (this element has not been implemented yet). This time, participant observations by my students and myself tend to suggest that the actions may often have been too subtle and "soft" for them to cause enough irritation to ignite a reflexive process, unless conversations were initiated with passers-by to give them enough hints and cues so that they would start reflecting on what they had just experienced.

WiLa is an association within Faust, a large self-administered cultural center in Linden, providing education and advice in the fields of health and sustainable development. Its mission is to transfer research insights into everyday knowledge, as well as to develop educational and participatory programs and formats. WiLa started out in 1986 as a drop-in center where citizens could ask questions to scientists and experts. It is project-oriented, developing creative formats, investing urban open spaces and inviting locals to participatory processes.

WiLa opens up spaces for imagination in different ways in different projects. In the following, I focus on WiLa's most significant current annual event: the Utopianale film festival founded by Felix Kostrzewa, which combines a selection of documentary films on solutions for sustainable development and a series of workshops and other activities allowing local sustainability-related and creative-cultural organizations to communicate with the public and give them a "taste" of their activities. Every year, the festival focuses on a specific theme. Members of our research team attended these in 2015 (How do we want to work?), 2016 (How do we want to move about?) and 2017 (How do we want to learn?).

At the Utopianale, spaces for imagination are opened up at two levels: over several months in the preparatory process involving a team of volunteers around Felix Kostrzewa, specific processes, tools, and rituals are put in place to stimulate group creativity. For example, the team uses "dragon-dreaming", a playful and mindful method for participatory planning, especially to stimulate the idea generation process and to allow working groups to dream more and think up utopian ideas that are not excessively constrained by realism. Besides more formal meeting formats, Kostrzewa also gathers the team regularly at participatory breakfasts that allow informal conversations and interactions. And, second, at the film festival itself, over two full days from morning to late evening, a typical day's experience for a festival visitor includes the following.

- A participatory breakfast and lunch prepared by local vegan and socio-ecological activists, inhabited by imaginaries of ecologically sustainable and solidary economy and convivial social relations, and allowing convivial exchanges with other participants.
- Several films selected according to their potential to stimulate sustainability-related and solutions-oriented thoughts and motivations, and often sharing a social imaginary rooted in socio-ecological awareness and alternative lifestyles, mindfulness, and solution- and project-oriented forms of civic activism.
- Workshops and exhibition/communication spaces that, across their specificities, share an imaginary of socio-ecological progress and personal and cultural development through grassroots practice as well as collaboration with the local government – as illustrated by the festival's motto: "Because there *is* a tomorrow" (in opposition to the expression "like there's no tomorrow"); the activities also strengthen interactions and networking, aiming to reinforce and grow sustainability-oriented social networks in Linden and Hanover.

- Other activities such as group singing in the evening to strengthen a sense of communion and stimulate an imaginary community and identity among the festival visitors.

The festival's opening of spaces for imagination (and of spaces for small-scale experimentation/first taste of practice) around a consciously crafted imaginary of sustainability aims to contribute to education for sustainable development in the district of Linden and the city of Hanover. The festival clearly aims to offer dialogical rather than dialectical spaces of communication and, without pushing them, gives its participants and visitors the opportunity to become more engaged. Hence, the seductive power of the highlighted imaginaries and opened spaces for imagination play a central function in the agenda of WiLa's Utopianale film festival.

More specifically, the sustainability-related imaginaries emerging from WiLa's activities (Utopianale and others) include the idea of using fewer resources more carefully, while creating more communal lifestyles. The idea of community is linked both to the capacity to save resources effectively and to the possibility to foster happiness (away from hyper-individualist consumerism, rejected as an illusory form of happiness). The imagined community is one of different people doing things together while respecting and valuing their differences and diversity. A sustainable life is imagined as the possibility to live (and to work, move, learn, and eat) with more collective freedom as a local community and with more mindfulness and attention to others and one's environment. Imaginaries of a sustainable economy are also often actively present, centered on regional, partly closed-loop, transition/permacultural/welfare-economic emerging models that all seek an alternative to neoliberalism and globalized financial capitalism. One of the most active approaches in Hanover, most discussed at the Utopianale festival, is the concept of the "economy for the common good", which led to, for example, the development of the common good balance sheet. This is an instrument to restructure the goals of organizations toward the "common good" rather than toward financial gain, making quality of life, human dignity, cooperation, social justice, environmental justice, transparency, and respect for future generations essential to the new ideal image of successful business. Most solutions to the current crises of global unsustainable development are imagined as small-scale or networked across small-scales (and thus organically scaled-up), participatory and most often bottom-up and self-made initiatives (in a great variety of locally implementable areas such as food, gardening, clothing, upcycling, and body-mind wellbeing). One weakness of this approach lies in its political imaginary: larger-scale political engagements (and the related dialectical, sometimes necessarily conflictual processes) are often absent from the Utopianale (and several other activities of WiLa).

KdW is a loose network of artists, cultural actors and grassroots organizations/initiatives (founded in 2012 at a "visionary congress" of Transition Town Hanover), coordinated by a small group around Joy Lohmann, an artist whose practice mixes together street art, upcycling, DIY/maker culture, and social practice. KdW sets itself against unsustainable consumerist culture and aims to network and promote Hanoverian initiatives that are culturally sensitive and sustainability-related. The network has a flexible and open (vague) but explicit focus on sustainability, participatory culture, commons, innovation, creativity, and future-orientation.

Joy Lohmann selected the network's name as a subversion of the KdW department store in West Berlin (which during the Cold War was a famous symbol of an unattainable lifestyle of luxury consumerism for the East Germans). The name of the KdW network then signifies that the present age calls forth new dreams of a good life beyond the shallow and short-lived gratifications of hyper-consumerism. KdW has its base in Linden (at the Glocksee independent cultural center for a couple of years, then in the Ihme-Zentrum just across the river) and aims to network actors in the whole of Hanover, organizing events in the city center in order to reach out beyond Linden.

Setting up and developing spaces for shared imaginative processes and the sharing of imaginaries of sustainability is a central feature of the self-imposed mission of the KdW network. In Joy Lohmann's own words: "To transform [society] you have to create new images in the imagination of the people. This is what we do... creating new images for places" (unstructured interview, 2014). In comparison to the previously described cases, KdW has developed the most concrete and collective visions for alternative futures, at several levels and across several formats of single or annual events: KdW holds monthly organizational meetings with a very informal atmosphere, where the core network members and an ever-changing number of peripheral network members (each of them with their own projects/organizations) share ideas, thoughts, and imaginations and make concrete plans for KdW events. Between 2013 and 2016, KdW also coordinated (under the leadership of Joy Lohmann) an annual event called the KdW Fest, a festival, fair, and feast that is spatially organized as the "rooms" of a private house (bedroom, living-room, kitchen, playroom, workshop, garden, etc.) in an outdoor public space, such as a public square in the city's hyper-center. The "rooms" are signaled and symbolized by a mix of written panels in street art style and pieces of upcycled furniture. Within this structure, a variety of cultural and socio-ecological projects and groups install their own stands, exhibitions, shows/performances, and other self-made creative formats. Finally, there are other single events (e.g. KdW City Forum in 2015 within the framework of My Hanover 2030 – a wider participatory visioning process organized by the city government) and repeated events (e.g. the Miracle-Change Christmas Market (*WunderWandelWeinachtsmarkt*) in December 2016 and 2017 in the Ihme-Zentrum) set up by KdW, with further opportunities being sought to creatively address themes across cultural life and sustainability and to network the many existing projects and initiatives.

KdW takes as its mascot the famous figure of Pippi Longstocking (the character from Swedish author Astrid Lindgren's children books), and as a motto Pippi's maxim "I create the world how I want it to be". KdW advocates, exemplifies, and spreads a joyful version of prefigurative activism, whereby grassroots experimentation is at the fore. A "prefigurative politics" is constituted by practices that already anticipate and embody, here and now, the desired form of future society, without waiting for ideal structural conditions to be rolled out in the institutional context (Sitrin, 2007). This allows a mutually embedded co-development of social imaginaries and social practice, stimulating personal and collective involvement and responsibility. Prefigurative activists pay much attention to their creative and participative processes, as for them "the way" and "the goal" cannot be strictly separated: the means employed already prefigure the ends in sight.

KdW demarcates itself from dull, grey and/or dead urban spaces by developing an imaginary of light, colorful and inviting open spaces opened up for creative work and collective cultural activities as well as sustainability-oriented community life. KdW's overall aesthetic and imaginary is one of DIY (do it yourself) and DIWO (do it with others) cultures, especially in the form of a maker culture (involving open source, crowd-sourced, open-ended/serendipitous, experimental, and participatory approaches), which is generally promoted by Joy Lohmann in his work. The imagination of participating initiatives and of passers-by at KdW events is invited to be expressed in any practical DIY way, however sketchy or "unfinished" it may look. To stimulate this process, idea rooms, dream manifestos, postcards, wish trees and a poetry tree are used as low-threshold invitations to participation. With its home-like rooms set up in the public space, the KdW Fest offers an "easy-going"/fun and relatively protected space for tinkering and experimentation.

The themes imaginatively approached at KdW are as diverse as the participants (and their projects) at any given moment. One omnipresent theme, championed by core member Cora Gutzeit (whose vegan catering initiative *Klugbeisser*, or Smart Mouth, was launched from the KdW network) and shared with the Transition Town activists and others, is the future of a sustainable food culture. The shared imaginary is one of locally produced, often self-made, and carefully consumed food – ideally waste-free. The ideal future of food is mostly pictured as a vegan food culture (less often with invocations of vegetarianism, insect meals, and a monthly "meat day"). This imaginary also involves the participants' already practiced food-sharing, urban gardening, and solidary agriculture, as well as public and participatory cooking – imagined as a future "new normal". Another dominant theme at KdW is a widely participatory democratic culture, including migrants and refugees, entering into an intercultural urban community based on principles of open source and open access. This imaginary is related to a future society where self-production and a maker culture are commonplace and replace the consumer culture. Further recurrent themes include upcycling and re-use, plastic-free production, self-made renewable energies, a circular and regionalized economy, mindfulness and awareness, a more sustainable local/short-distance tourism, and a wish to find a way back to a harmonious balance with nature.

In the preparatory phase of the KdW City Forum, core members encouraged member initiatives to tinker with their own designs for creative expressions of their (topically diverse) visions for 2030, and share them with each other. Each thematic group designed its own format (which ranged from photo-montages and tinkered objects to mini-performances), which was then staged together in a presentation at a conference room in the Hanover City Hall. Due to insufficient preparation time and professional support, the inexperienced participants were only able to develop and implement half-finished designs, which they presented to each other and a restricted audience. They could not fully develop and creatively express their imaginations, nor did time suffice for them to reflect on each other's imaginaries. Despite the constraints, shortcomings and frustrations of the KdW City Forum, this event was a practical attempt at all-round participatory creativity, involving several initiatives engaged in different areas of sustainable development. This event by KdW is the only case we experienced where a tentatively integrative attempt was made to both creatively and participatively gather and connect the fragmented imaginations and imaginaries of different Hanoverian urban actors through arts-based forms of expression and to thereby collectively generate a new, wider social imaginary of sustainability. For this event, the KdW core members had clearly set their role apart from that of the wider political-participatory processes in the city, still strongly tied to (a "realist" fetish for) facts. Instead, KdW has the goal of opening up a wider spectrum of options:

> We must open up a whole new perspective on this process [of My Hanover 2030 because] culture can work much better with contingencies... it is a wide spectrum of options for action or also of risks, of perspectives, utopia or dystopia, which are thrilling and may make the process [of My Hanover 2030] more interesting for participation.
>
> *(Transcript of a preparatory meeting for KdW City Forum 2015)*

With very limited means and its DIY/makeshift aesthetics, KdW strives not only to generate and communicate new images, but also to stage prefigurative experiences of new worlds and imagined futures.

Last but not least is the case of the **ZWI**. The Ihme-Zentrum – a gigantic architectural ensemble from the 1970s (285,000 m^2) along the Ihme River, facing the Hanover city center on the other side of the river – has experienced extreme decline over the past three decades. Its commercial spaces (lower levels) are in ruins whereas the residential apartments (upper levels) still host 2500 inhabitants very much attached to their homes. The imaginaries associated with this site are diverse, shaping strong contrasts. A great number of Hanoverians wish to demolish the buildings whereas its inhabitants want renovation. In this context, through many years of disappointed hopes and controversy, the ideas and imaginations for new developments of the Ihme-Zentrum were gradually becoming inaudible until a range of cultural actors started to re-imagine the site. Tante Trottoir (whose headquarters are at the Glocksee cultural center just across the Ihme River) staged Tante Titanik (in May 2015), a form of impro/guerilla theater in and about the dreams and nightmares of the Ihme-Zentrum. In 2014, the journalist (and then MBA student in sustainability management) Constantin Alexander moved into the Ihme-Zentrum and started a campaign to not only revitalize the generation and communication of ideas but also stimulate a solutions-oriented transdisciplinary process to push the realization of these ideas forward. What started as a blog and public walks soon included the production of a documentary film, the founding of the ZWI association and the realization of several artistic, cultural, architectural, and other events in the Ihme-Zentrum in 2016 and 2017, attracting ever more attention and gradually rekindling curiosity and hopes for the site among some Hanoverians (including the municipality and inhabitants of Linden).

In this case, imaginations of possible sustainable futures for the site (beyond the mere *tabula rasa* imaginary of demolition) were partly pre-existing, but remained largely invisible and inaudible until the site's image was reframed by Constantin Alexander and soon by his growing number of allies (including artists, architects, local inhabitants, local politicians, etc.). According to Alexander, two main tipping points in this process were marked by media attention: first, when Hanover's main newspaper covered his regular guided walking tours of the site in summer 2015; and second when a draft cut of his documentary film was screened in February 2016 at WiLa's Utopianale film festival. At this point, a new narrative for the site began to gain momentum. The new space of possibility being opened up by ZWI in 2017, then, is marked by the connection, networking, and articulation of many (complementary and/or competing) ideas and visions for the Ihme-Zentrum around one shared slogan introduced by Constantin Alexander ("The Ihme-Zentrum – a new landmark for Hanover").

The diversity of the collected ideas and visions for future developments of the lower-levels of the Ihme-Zentrum reflect the complexity of interests at stake and the qualitative complexity of the multiple dimensions of sustainable development (the complementarities, competitions, oppositions and symbioses). These visions and ideas include commercial spaces, spaces for social activities, arts and culture, health and care, urban gardening and bee-keeping, sports, elderly housing, open/co-working-space offices, and handicraft. Planned structural changes will better link the building with neighboring urban life, facilitate sustainable mobilities and/or reconnect the neighborhood to the river and ecosystems through renaturing the ground-floor level by replacing the underground parking lot with riverside plants and trees.

Unlike the other four cases mentioned here, ZWI is not primarily focused on dialogical processes. It balances creative impulses with solutions- and planning-oriented approaches. ZWI is also in many ways not as utopian and not as wide-ranging in its visions of sustainable futures as most of the other actors. The imaginary of sustainability embedded in ZWI's activities is largely compatible with a capitalist economic orientation, but a reformed one that opposes the speculative financial logic of the real estate market that precipitated the ruin of the lower levels of the Ihme-Zentrum.

What this case also strikingly demonstrates is the way such contrasted sites stimulate the imaginative work of artists: where an urban ruin-like spot is only a few steps below a functioning residential area, and where several time dimensions and contrasted and ambivalent memories are sedimented and telescoped in one specific place. This case also demonstrates how such sites allow a re-evaluation of their failures (once artists have re-opened imaginative processes). The failures of parts of the Ihme-Zentrum no longer mark the ineluctable doom of the entire ensemble, and are being re-interpreted as ground where new urban spaces of possibility can emerge – the substrate on which re-imaginings may grow and eventually prosper. Furthermore, the blatant failure of previous re-developments opens up the failed site as a playground for utopian imagination and experimentation. Uncertainties are re-interpreted as openness to multiple potential futures. The decline and deterioration of material objects points to changeability. Through this two-step process, the dynamic modalities come to light under which imaginaries are emplaced (see also Pink, 2008, 2012, 2015) and places are re-imaginable in an interplay between the phenomenological qualities of the sites, the artistic envisaging of these qualities, and the mediated communications around renewed images (Reich, 2017). Under these perspectives and approaches, a site like the Ihme-Zentrum can become "a fantastic adventure playground" (Alexander, 2015).

Conclusion

These five cases show how sustainability-related urban imaginaries are emplaced at multiple levels: in the embodied and mental space of the self (FOT), in the self in relation to changing sites in public space (Tante Trottoir), in ambivalent and contrasted relations to one specific architectural ensemble (ZWI), in a fictional and blurred common/private-homelike festival space occupying public space (KdW), or in a festivalized social space of convivial learning (WiLa's Utopianale). The five cases also illustrate how spaces for imaginations exist in a mutually constitutive relationship with social practices and spaces of experimentation. This close relationship was also evidenced in our analysis with atlas.ti through the co-occurrence of codes for "spaces for imagination" and "spaces for experimentation".

Actors engaged in sustainability, if they are to develop new models detached from the dominant social imaginary, need to foster radical, creative, *and* complex imaginations. Deep *and* rapid systemic changes are indeed required if humanity is to address, in time, the compound threats it is facing, from climate change to global social and economic injustices. These challenges will not be met through minimal shifts in social imaginaries and social practice; thus radical imagination is necessary. Sustainability requires both building up resilience and opening up to transformative change (often radical change, going to the roots of issues and seeking deeper leverage). Building up the qualities of resilience in human societies calls for a cultivation of multiple creative responses and capabilities – a radical embracing of Joseph Beuys's provocation that "Everyone [be/become/return to be(com)ing] an artist!" Thus sustainability also needs to cultivate an especially creative imagination. Finally, sustainability requires a complex imagination because multiple dimensions and conflicting priorities need to be addressed; over-simplifying and populist imaginaries are counter-productive.

This relates to the question of dialogics and dialectics. Our empirical insights point to grassroots initiatives/projects/networks taking up, in most cases, a dialogic-oriented tendency. There are some variations: FOT demonstrates the most dialogical approach. Tante

Trottoir also follows a strongly dialogical approach, followed by KdW and WiLa. In contrast, ZWI prioritizes a solutions-oriented imaginary. However, this is complemented by artistic imaginations. The dialectic and planning tendencies are thus very much present, but seem to be relatively balanced with the dialogic and emergent.

The strength of dialogics lies in an openness that facilitates cooperation between diverse imaginaries and discourses. But from my understanding of sustainability, rooted in qualitative complexity (following Edgar Morin, 2008), the radically dialogical imaginary developed especially at FOT and Tante Trottoir (and to a lesser extent at WiLa and KdW) also reveals a problematic potential: it may drift into the simplifying illusion of a pacified society searching for communicative consensus across differences, where political confrontations could be dissolved altogether. Furthermore, a purely dialogical form of social practice might hinder the emergence and sharing of radical imagination when such imaginations are especially challenging, uncomfortable, or carry conflict potentials. However, the opposite end of the spectrum would be equally problematic: a radically dialectic imaginary with a democratic society ruled by political conflict, where the confronting and opposing views of different groups only permit pragmatic compromises – as in Mouffe's (2013) call for an "agonistic" politics. A purely dialectic form of social practice would hinder the dissemination of a radical imagination for sustainability across political lines and instead entrench social divisions and weaken social sustainability.

The five cases discussed here constitute only a small selection from the larger research project "City as Space of Possibility". Another especially insightful case study is currently being carried out by Annette Grigoleit and colleagues with a focus on the creative writing project Linden Fiction 2050 (realized at the Faust Cultural Center in 2015, when participants wrote short fictional stories situated in Hanover-Linden in 2050) as a tool of participation and empowerment for the production of desires and as a creative space to imagine potentially sustainable urban futures.

Imagination is fundamental to place-making. Rich, diverse, and preferably complex imaginaries of sustainability are essential to the development of sustainability-related social practices and institutional changes (in the sociological sense of institutions, i.e. systems of social, cultural, and political rules for social life), through the mobilization of such imaginaries and of (preferably radical) imaginations in urban spaces of possibility. Therefore, urban research on social imaginaries and inter- and transdisciplinary research for sustainability have a lot of common ground to further explore and articulate, in collaboration with urban actors striving for their "right to the city" and for sustainable urban development.

Acknowledgements

I thank students at the Leuphana University Lüneburg, colleagues at the research project "City as Space of Possibility" and especially student-assistant Elisabeth Böhnlein for their help in empirical work on the five cases, and more generally for the many collaborations and exchanges within our research project. I thank science editor Paul Lauer for his help in editing the language of this text. This work is part of the "City as Space of Possibility" ("Stadt als Möglichkeitsraum") Research Project at Leuphana University Lüneburg, supported by the "Niedersächsisches Vorab" funding program from the Ministry of Science and Culture of the State of Lower Saxony (Germany).

References

Alexander, C. (2015) *Ihme-Zentrum: Die Märchenburg von Hanover.* Available at: https://www.zebrabutter.net/ihme-zentrum-die-maerchenburg-von-Hanover.html.

Castoriadis, C. (1975) *L'institution Imaginaire de la Société.* Paris: Seuil.

Dieleman, H. (2012) Transdisciplinary Artful Doing in Spaces of Experimentation and Imagination. *Transdisciplinary Journal of Engineering and Science* 3: 44–5.

Herbrik, R. (2011) *Die Kommunikative Konstruktion Imaginärer Welten.* Wiesbaden: VS-Verlag.

Kagan, S., Hauerwaas, A., Holz, V., and Wedler, P. (2017) Culture in Sustainable Urban Development. Practices and Policies for Spaces of Possibility and Institutional Innovations. *City, Culture, and Society.* Accessed 25 November 2017. DOI:10.1016/j.ccs.2017.09.005

Kant, I. (1970) *Critique of Pure Reason.* London: Macmillan.

Lefebvre, H. (1968) *Le Droit à la Ville.* Paris: Anthropos.

Lefebvre, H. (1991) *The Production of Space.* Oxford: Blackwell.

Lennon, K. (2015) *Imagination and the Imaginary.* London: Routledge.

Lindner, C. and Meissner, M. (2016) Globalization, Garbage, and the Urban Environment. In: Lindner, C. and Meissner, M. (eds.), *Global Garbage: Urban Imaginaries of Waste, Excess and Abandonment.* London: Routledge, 1–13.

Link, C. (2016) Die Entwicklung Muss Aufhorchen Lassen. *Hanoversche Allgemeine*, 6 April 2016. Available at: http://www.haz.de/Hanover/Aus-den-Stadtteilen/Linden/Die-Entwicklung-muss-aufhorchen-lassen.

Marcuse, H. (1978) *The Aesthetic Dimension: Toward a Critique of Marxist Aesthetics.* Boston: Beacon Press.

Merleau-Ponty, M. (1964) *L'oeil et l'esprit.* Paris: Gallimard.

Morin, E. (2008) *La Méthode.* Paris: Seuil.

Mouffe, C. (2013) *Agonistics: Thinking the World Politically.* New York: Verso.

Pink, S. (2007) Walking with Video. *Visual Studies* 22(3): 240–52.

Pink, S. (2008) Mobilising Visual Ethnography: Making Routes, Making Place and Making Images. *Forum: Qualitative Social Research* 9(3). Available at: www.qualitative-research.net/index.php/fqs/article/download/1166/2581.

Pink, S. (2012) *Situating Everyday Life. Practices and Places.* London: Sage.

Pink, S. (2015) *Doing Sensory Ethnography.* London: Sage.

Reich, C.J. (2017) Das Ihme-Zentrum – Ein Imaginärer Möglichkeitsraum. Unpublished Bachelor's thesis, Leuphana Universität Lüneburg.

Sennett, R. (2012) *Together: The Rituals, Pleasures and Politics of Cooperation.* London: Penguin.

Sitrin, M. (2007) Ruptures in Imagination: Horizontalism, Autogestion and Affective Politics in Argentina. *Policy & Practice: A Development Education Review* 5(Autumn): 43–53.

7

Formal encounters in two tales of toxicity

Bhopal, *Animal's People*, Louisville, *The Hard Weather Boating Party*

Barbara Eckstein

Introduction

"Bhopal came calling one day fourteen years ago". Thus begins the story of how copywriter and author Indra Sinha (2007b) joined activists' efforts on behalf of the victims of the 1984 gas leak tragedy in that city. In Sinha's opening line, "Bhopal" is a trope with multiple forms. Shoma Chaudhury and Shantanu Guha Ray (2010) parse "Bhopal" as a three-chambered horror: the story of the disaster, Union Carbide's toxic waste on site and in three solar evaporation ponds, and the tortuous legal story. "Bhopal" is also the single visitor to London, the former imperial center, standing in for thousands of Bhopali neighbors with industry. It is also the dynamic city space of the Bhopal Municipal Corporation (BMC), with a human population of some 800,000 in 1984 (Varma and Varma, 2005) and approaching 2 million at the 2011 census (Bhopal Census, 2011).

In 2015, BMC was among twenty Indian cities winning support for its Smart City proposal (Ministry, 2016). BMC (2017) created its detailed plan through engagement with residents, including those with low incomes or unemployed, and projects infrastructure improvement, extending sewer and waste disposal in areas without such systems. The city's public plan does not name the former Union Carbide India Limited (UCIL) factory site. Other than the pan-city proposed improvements, it focuses attention on an area of the city further south. In one imagining, "Bhopal" *is* the 1984 gas leak tragedy. In the other urban imaginary, the gas leak that killed 15,278, by official count (Alavi, 2016), seems to be held apart from a detailed vision of a smart city in the twenty-first century. Is either imaginary useful to the city's current residents?

Claims and counter-claims across the decades about the night of 3 December 1984 uncover a long list of actants in the assemblage that is the tragedy and its aftermath. But rather than base my thoughts about cities, toxic sites, and literary creations only on Latour's (2004: 75) ideas about networks of agency – material and human – and Bennett's (2010: 94–121) arguments for distributive responsibility arising from this complexity of actants,

I here consider literary and urban forms together in the new formalist method offered by Caroline Levine (2015: 132).

> Paying attention to numerous overlapping social forms may seem daunting... but if it is in fact true that forms very often find their organizing power compromised, rerouted, or deflected by their encounters with other forms, then a formalist cultural studies interested in how power works will need to take account of what happens when a great many social, political, natural, and aesthetic forms encounter one another.

Levine's focus on the dynamism of forms that organize space and time recalls Henri Lefebvre's famous Marxist analysis of how space is produced (1991) and his less famous analysis of how time intersects that spatial production (2013). But she argues that even the forms taken by deep structural forces such as capitalism – and colonialism – also find themselves "compromised, rerouted, or deflected by their encounters with other forms".

> It would be a grave mistake to overlook them. But at the same time ... an exclusive focus on ultimate causality has not necessarily benefited leftist politics. It has distracted us from thinking strategically about how best to deploy multiple forms for political ends.
>
> *(Levine, 2015: 17)*

Although Levine plunges into complexity, she also resists the amorphous imagery of Deleuze and Guattari (1987) who in celebrating rhizomes do not attend to the capabilities and limitations of this form among forms.

At pains to leave behind the identification, in literary discourse, of form with bounded unity, Levine borrows from design a key concept: affordance, "the potential uses or actions latent in materials and designs ... The term *affordance* crosses back and forth between materiality and design ... Constraints move in both directions" (2015: 6, 9). Literary forms are not analogies of social forms. As transgeohistorical designs and as participants in specific, situated power relations, they exist in overlapping relation to social and political forms. "All politics, including revolutionary political actions", Levine claims, "will succeed only if it is canny about deploying multiple forms" (2015: 18).

Jesse Oak Taylor (2013: 177, 181), writing about *Animal's People* and global health, points to key challenges for intervening in the greatest global health threat: inequality. The first of these is the constitution of the "bottom billion", not only the sheer number of individuals worldwide living on less than $1.50 (USD) per day but the definition of these communities through what they lack – money – rather than what they have. In the universal struggle for health and wellbeing, what can be learned by close attention to the encounter of forms?

My aim is to describe how Indra Sinha's novel *Animal's People* (2007a) participates in Bhopal's urban imaginary given the legacy and presence of toxins. To more fully consider the place of literature in the urban imaginary of toxic sites, I also describe Louisville, Kentucky, US, its toxic tragedies across decades, and Naomi Wallace's play *The Hard Weather Boating Party* (2014).

A young man disabled by the chemical load in his body narrates Sinha's novel twenty years after the explosion in Khaufpur, a fictional Bhopal. The story follows neighbors of the abandoned toxic factory as they weigh the prospects of trust, especially of a US doctor who arrives in the neighborhood to start a free clinic and her possible association with the

"Kampani" (company) that owned the factory. The narrator, who moves by the use of four limbs, embraces the nickname "Animal".

Wallace's play enacts a scheme by three, ill, chemical plant employees to murder the CEO of theirs and six other chemical factories in Louisville's West End. Its two acts occur in a cheap, franchise motel room where the middle manager and two laborers meet before and after the attempted murder.

Bhopal

Even a brief retelling of the history of the Bhopal gas leak necessarily entails movement up and down spatial scales, in and out of different forms, from the structures of molecules to the networks and hierarchies comprising the globalized "green revolution". The city of Bhopal's Development Plan of 1975 (Varma and Varma, 2005: 40) recommended that the UCIL plant be built twenty-five kilometers away from the city center. Instead, the plant was built on land leased from Madhya Pradesh (the state) near one railroad station, two major hospitals, and the large Upper Lake. While scholars (Varma and Varma, 2005: 39) claim that the plant design was a "packaged transfer" from Union Carbide Corporation-US (UCC), a centralized network of a global corporation, UCC (2016) points to the Indian state's enthusiasm for Indianization of all industry, a centralized national network. They claim that the plant at Bhopal was designed, built, and managed by UCIL. In fact, neither the capitalist nor nationalist center held. Disaster occurred in the ambiguity about control.

Methyl isocyanate (MIC) was used at the plant to produce the pesticide Sevin, a part of the controversial Green Revolution hailed as a global design to feed the world and reviled as a hierarchy in which a western, agrochemical complex oppresses traditional farmers. Detractors connect this hierarchy to the historical hierarchies of racism and colonialism. Although UCC (2016) records and the Varma and Varma (2005: 41) analysis agrees that in 1982 UCC safety inspectors of the two sites producing Sevin – Bhopal and Institute, West Virginia – recommended changes to both plants but changes occurred only at the Institute plant, different forms govern their structuring of this knowledge.

On the night of 2 December 1984, just a little over a month after the assassination of PM Indira Gandhi following a controversial state of emergency, water got into the tank of liquid MIC: sabotage accrording to UCC (2016) and cleaning according to Varma and Varma (2005:40, 41–2). Water reacted with the MIC, turning it into a gas and increasing the pressure in the tank until the gas escaped in large quantities. MIC stays near the ground; the casualties were so high because the alarm was delayed, and when it was raised, neighbors of the plant were instructed to run rather than hunker on the ground and cover themselves. They ran full into the face of the poisonous gas. The interaction of MIC and water in a confined tank, the public poisoning, and the subsequent public distrust arose at the collision of the UCIL plant's location in the city center, a centralized corporate network of UCC, and Indian nationalism. We can add to this a prior pattern of dismantling safety measures (Kerin, 2016: 167) and ignoring a journalist's warnings (Sinha, 2009).

A distributed network of international NGOs and Bhopali activists sought 3 billion (USD) in reparations from UCC and the extradition of UCC CEO Warren Anderson (until his death in 2014). In 1989, by court recommendation but not order, UCC had paid $470 million in reparations and built a new hospital and research center close to the UCIL plant, although the Indian government chose to distribute to each survivor only part of the full sum and local residents did not trust and would not use the hospital (Bhopal Medical Appeal, 2015; UCC, 2016; Varma and Varma, 2005: 43). Decades later, in 2010, a Delhi court convicted UCIL managers of negligence, but activists and NGOs, seeing the convictions and fines as insufficient, still fought for extradition of Anderson from the US (UCC, 2016). Today, the deceased CEO of the corporate hierarchy and the abandoned, still toxic, former UCIL factory site together form a whole that anchors activists' definition of injustice.

Levine proffers four forms: whole, rhythm, hierarchy, and network. To these an environmental humanist needs to add the different arrangements of weak and strong bonds that form the material world and leave it more or less porous, open to new bonds. Much that we routinely speak of as whole – a human body, a building – is a porous collection of material dependent upon separate organisms in and outside its boundaries. These arrangements are literally relevant in the purposeful and accidental reactions of molecules such as those that produced the MIC leak. Such a form is also instructive in considering the interaction of global NGOs, local activists, and the populations they mean to serve. Trust is the ephemeral substance of the bonds themselves. As communicators, NGOs may function as centralized or distributed networks articulating politics and raising money around a singular event and site. But as actors in neighborhoods they survive through their understanding of the local molecular structure, its dynamism, and its porosity.

Countering the disruption from leaching chemicals in the soil and groundwater at the UCIL plant site, local residents and global allies have produced forms that disrupt the toxic zone with efforts to insert health, justice, order, and commemoration. As early as 1985, Dutch sculptor and Holocaust survivor Ruth Waterman worked with local residents to erect a statue of a mother and child outside the factory wall (Remember Bhopal, 2015). Evolving local graffiti on the wall redefines the boundary as a set of messages. Survivors and activists opposed a state government idea to build a memorial and museum and open the factory gates to tourists. Instead neighbors produced their own local traveling exhibit. Decades later, British photographer Francesca Moore's 2015 exhibit "Bhopal: Facing 30" (The Bhopal Medical Appeal, 2015) also traveled – transnationally. Displaying images of (willing) families from neighborhoods near the plant arranged as royals against a reproduction of a UCIL factory wall, the show opened at Photofusion Gallery, London, and then traveled to Jehangir Art Gallery, Mumbai.

Some local activists (International Campaign for Justice in Bhopal, 2014) designed and built a neighborhood to reorder and strengthen what was lost. The post-1984 "encroachment" of Atal-Ayub Nagar arose on "empty" government land and attracted survivors who couldn't afford rent in the city. Some 2000 people, led by activist survivor Ayub Bhai – notably also a Muslim activist in the Hindu-leaning Bharatiya Janata Party (BJP) – settled in mud huts in "three neat rows" with the railroad to their north. Bhai proposed an alternating Hindu-Muslim occupation of the mud huts, an arrangement that purportedly succeeded until 1992, when a mosque was notoriously destroyed in the city of Ayodhya, which reintensified the rhythm of post-Partition religious conflict throughout India. Having subsequently survived years of attacks by bands of Hindu and Muslim "goons", and the loss of a third of their population, Atal-Ayub Nagar regained some social stability in the twenty-first century.

But the colony never had environmental stability. In addition to exposure to MIC in 1984, survivors and residents live on contaminated soil, drink contaminated water, live beside an open sewer (or did) along the railroad tracks, and are exposed to the toxins from coal-fired cooking and diesel engines. Residents' health and trust run through a maze of urban and national government, global NGO, and corporate responses. A 2013 India Centre for Science and Environment Expert Roundtable confirmed the contamination of soil at the plant, though they declared no consensus on water. A 1999 Greenpeace study of ground water, although it contained insufficient information about a number of colonies, claimed clear contamination of ground water in Atal-Ayub Nagar (UCC, 2016). The city provided the "encroachment" hand pumps and a road in the 1990s and began providing water tankers in the twenty-first century after Sambhavna, the preferred community health provider, conducted surveys (International Campaign for Justice in Bhopal, 2014). Bhopali Sanjay Verma (2014: 175) relates that, after the 2008 hunger strike in Delhi, in which he and novelist Sinha participated, the Prime Minister agreed to pipe clean water into contaminated areas of Atal-Ayub Nagar.

Narratives, Levine (2015: 19–21) argues, effectively exhibit overlapping forms as they encounter one another across time. The narrative of Sanjay and his brother Sunil, who lost parents and siblings to MIC, figure prominently in stories told by victim advocates. In 2011, Sanjay, an infant in 1984, recalled his early life in the SOS orphanage run by an Australian NGO, and his subsequent life with older brother Sunil and their one surviving sister in one of the 2500 apartments in a Gas Widows Colony on the north edge of the city. Sunil, age twelve in 1984, formed Children Against Carbide when he was thirteen and never returned to school. He saved the UCC reparation payments to offset Sanjay's school fees. Sunil advised his brother to leave the unsafe city, but Sanjay explains, "I didn't want to abandon Bhopal".

Both Bhopali and a global citizen, Sanjay (2014, Verma and Masi, 2014) was in California in 2011 promoting a documentary about Bhopal, after appearing as an expert witness in a case involving German multinational Bayer, a producer of MIC in a West Virginia plant. By contrast, Sunil suffered from depression and other mental illness and committed suicide in 2006. The hope and despair, neighborhood and world, in the brothers' story has had a global audience. Sinha (2008) places this tragedy in his own personal narrative of deepening Bhopal activism.

And yet the whole of the toxic factory site itself has not opened to rearrangement. There are, some say, insufficient bonds of trust tying survivors and neighbors to not only corporations and scales of government, but also networks of activists and journalists. Dr. Saba Gulraiz (2012) writes of no change at the site, of faith broken by activists seeking heroism, journalists seeking stories, politicians seeking votes.

A rationalist form, alternative to the social, political, material, and aesthetic forms that comprise "Bhopal", appears in the pages of the journal *Process System Progress*, a scholarly site where chemical engineers and risk managers from industry and academia assess what can be learned from industrial accidents. In the more than thirty years since 1984, these pages have repeatedly considered the Bhopal gas leak tragedy "the worst industrial incident the world has seen. As such, the ramifications of the incident changed how the world not only managed chemical incidents, but also how the public accepted, or not, the presence of facilities in their area" (Kerin, 2016: 167). That said, the two appearances of "world" here, the article later reveals, refer to two different forms: the first, the whole sphere of the globe; the second, a patchwork of only Europe and the US. The repeated references to the Bhopal gas leak in the journal's pages can be read less as evidence of tragic warning and progressive change by industry, government, and an alerted public, and more as the high and low tides of government regulation, industry implementation, and public vigilance. The journal's professionals must sound the alarm again and again; yet even so, they can erode trust if the "world" turns out to be only special safety zones.

Animal's people

By putting into action the collision of individuals and institutions at the site of a famous toxic tragedy, Sinha's novel joins the effort to imagine alternative, visible, and livable arrangements. As literature dialogically engaged with literary and social forms that precede it and will follow it, the novel invites each reader into this particular toxic struggle and others whose interplay of forms bear relation to these.

Noteworthy, first, is the innovative language employed by Animal, the novel's irreverent young slum-dwelling storyteller. Like the language of Huck Finn or a hip hop artist, Animal's idiolect communicates intelligence through the semblance of illiteracy, and ethics through obscenity. "Kampani", "jarnalis", "namispond, jamispond": beginning at the level of orthography, Sinha exploits the materials afforded by language to create not only this storyteller that the reader is invited and challenged to embrace, but also a means of expression that lays down linguistic puzzles to be solved. When Sinha (2007b) explained that he thought of setting the novel in Brazil or West Africa, but the voice of Animal was already speaking to him in a Khaufpuri (Bhopali fictionalized) accent that he could not deny, he recognized the universal and particular forms made possible by the language he chose.

Second, Sinha's literary art constructs a chain of testimony that emphasizes translation across language and medium. Words are spoken in an argot derived from Hindi and French, to a tape recorder provided by an Australian journalist, addressed to the "Eyes" of the reader, and translated into the English form the fiction reader sees. These bits and pieces of diverse languages lay down a rhythm, reminding the Eyes that English is not transparent. At the same time, we Eyes see an English that expresses the doubts Animal should rightly have of the journalist, the taperecorder, the readers, and the Kampani (Sinha, 2007a: 3).

> I said, many books have been written about this place, not one has changed anything for the better, how will yours be different? You will bleat like all the rest. You'll talk of *rights, law, justice.* Those words sound the same in my mouth as in yours but they don't mean the same, [activist leader] Zafar says such words are like shadows the moon makes in the Kampani's factory, always changing shape. On that night it was poison, now it's words that are choking us.

Animal's People is a novel with an Indian subject written in English and vulnerable to the exploitation of Eastern subjects that Edward Said called orientalism. But it writes into the fiction the questionable conditions under which the story was recorded and Animal's continuing skepticism about the motives of such outside-in storytelling. Postcolonial scholars Ana Cristina Mendes and Lisa Lau (2015) characterize this sort of subversive and self-aware orientalism appearing in some recent Indian fiction in English as reorientalism – but orientalism, nonetheless.

For neighbors and activists, the Kampani is a powerful economic network and capitalist hierarchy, a near invincible producer of space that colludes with their national, state, and local governments against them, denying them justice. Zafar, the virtuous activist leader, speaks in Delhi's courts, trying to confront the elusive Kampani and demanding that justice. Only for Animal, the aged French nun who raised him, his best friend Jara (a dog) and other flora and fauna emergent in the abandoned factory, is the Kampani also the material ruin where they live, a cozy toxic cave, the scorpions notwithstanding. To render yet more material Animal's relationship to the Kampani, he is a young man who moves as a quadraped after the contamination from 1984 distorted the growth of his back. His porous body is the visible sign of the contamination that the residents of the proximate neighborhoods contain. The novel focuses our eyes on Animal and the neighborhoods he traverses, where such a body, though not immune from human cruelty and environmental toxins, has a home, a life, and desires.

Whether Animal's affirmation of life can carry the novel depends on trust – the central question of the plot. Can Elli, a purportedly self-funded young doctor from the US, be separable from the US multinational Kampani? Activist leader Zafar tells the residents in his network that the answer to this is unequivocally no. People in desperate need of healthcare boycott her clinic on his command. Spying on Elli for Zafar, Animal is, by contrast, the voice of generative uncertainty. His romantic desires for Nisha, Zafar's fiancée, and his sexual desires for Elli complicate ideological commitments and threaten to break nodes in the activists' network. He is, however, embedded in the molecular structure of the neighborhoods and neighbors he knows, tightly affixed to those who tease him, those who care for him.

It turns out that Animal cannot carry the plot. When he breaks trust with Zafar by drugging him in an attempt to contain Zafar's sexual encounters with Nisha and then fears he has poisoned the political leader, when Animal overhears Elli seeming to break trust with her patients and neighbors by meeting with a Kampani lawyer, when hunger strikers Zafar and Farouq are near starvation and death without a forthcoming concession from the courts, when the Kampani and the Indian government leaders join forces, when the police and neighbors confront one another at the factory, the chaos of colliding forms disrupts the trust that holds Animal to the neighborhoods and to life. He takes the remaining pills to kill himself.

Levine (2015: 41, 42) asks us to reconsider what we understand literary closure to mean. "[Narratives] are forms organized by their unfolding over time . . . Social forms bring their logics with them into the novel, working both with and against literary forms and producing unexpected political conclusions out of their encounters". An American doctor has come to the neighborhood because of her place within the American Kampani's network, although her motivation is to break out of that net. Rumor of the activist leader's death ignites the neighborhood, exploding the centralized network of that organization, pitting residents against the police, and threatening to recontaminate their homes by setting fire to the factory. They want to both claim and destroy the remains of the factory that has been evidence of their injury and memorial to their loss. Animal, drugged, is not himself. Dropped out of the city into a forest, he wants to be alone to die but animals, real and dreamed, and gods appear. He curses all gods, all government and corporate authorities, but his hierarchy dissolves into "hunger strikers, Khaufpuris, non-Khaufpuris, the living, the dead" (Sinha, 2007a: 350). When, in the morning, he is found, and presented with a cap Nisha has made for him, he concludes,

> By this gift, I lost my immortality, I knew that Zafar really was alive and so was I. Life dropped like a heavy mantle about my shoulders and I began to weep for pity that I was to return to the city of sorrows.
>
> *(Sinha, 2007a: 357)*

With his friends he takes up his place in the molecular structure of the neighborhoods he traverses, territory and material relations he will not give up for surgery, a straight back, and crutches. The contaminated factory remains; the courts are – *in extremis* – susceptible to activists' demands; the Kampani lawyers fly away to return another day; and the neighbors, Animal included, know something more about trust than they did when the narrative began.

The encounter of forms in Sinha's narrative inserts into interpretations of "Bhopal" an aid to retaining focus on the complexities of this place. Is it smart for the BMC to turn its prospects away from the whole of the leaking factory behind the wall and see the future at a different site? Can it pull even a recalcitrant toxic site in a residential-industrial neighborhood into a networked urban imaginary? How can a smart city negotiate with its national and state government and global capitalists as well as its urban citizens to harden the boundaries of a toxic site while also dissipating its holistic effects and retaining its commemorative sanctity? I join Donna Haraway (2016) in her posthumanism and Levine (2015) in her new formalism and answer, by "staying with the trouble".

Louisville

Although Louisville, like Bhopal, has an ongoing relationship with chemicals manufacturing, the materials of its West End – Rubbertown – seem to afford different forms, urban and literary. The factory and neighborhood spaces and events of the West End unfold with steady yet uncertain rhythm, no less recalcitrant for being invisible slow violence punctuated by spectacles of violence over time. No single event has won the devotion of international activists and NGOs, annual commemoration by more than local media, or repeated process system analysis by engineers and risk managers seeking lessons learned. Since the 1940s when the US government began producing Neoprene (synthetic rubber) in Louisville's West End, through the decades when nine, ten, twelve chemical industries have been located there, the local media has, nonetheless, reported and reiterated incidents. The series of explosions at the DuPont plant in 1965 that killed twelve workers and injured another thirty-seven (some say sixty-one), and leveled the factory at a cost of over $50 million, is a story told locally across time (Bruggers, 2015). The event has been hard for West End residents to forget. Memorably bad decisions were made, after the first explosion, to let the evacuated workers back into the plant. Also, events were captured live for Louisvillians by the advent of television. Although a *Process System Progress* article (Klein, 2009: 119) claims that the 1965 incident instigated "the modern era of Process Safety Management at DuPont", there were explosions at the DuPont plant again in 1969 (Bruggers, 2015). In 1985, a tank blast at the Borden plant killed three workers. In 2003 an explosion at a food additive plant killed one worker and caused a chemical leak (Bruggers, 2014). In 2011 an explosion at Carbide Industries killed two workers, injured two others, and required a shelter-in-place warning for a mile around the plant. The following year, a train derailment and chemical spill at nearby West Point required mass evacuations. In the mid-90s, the *Courier-Journal* reported that there was about one siren warning per week because of a spill, leak, or fire at one of the nine major plants (Bruggers, 2015). This rhythm continues into the twenty-first century.

Three neighborhoods lie proximate to the plants: two traditionally inhabited by working class white residents and a third, just to the north on the Ohio River, inhabited predominantly by low-income African Americans. In addition to the spills, leaks, and occasional explosions at the chemical plants, these neighborhoods also lie proximate to two coal-burning power plants (one only recently decommissioned) and the resultant, toxic coal ash (Ross, 2016). Like the coal cook stoves and diesel engines in Atal-Ayub Nagar, the coal-burning power plants in and across the Ohio River from Rubbertown are a toxic white noise behind the drama of the chemical plants.

And then there are the smokeless, odorless interactions of the chemicals with individual bodies, especially workers' bodies. Louisville's B.F Goodrich plant plays a role in the story told by environmental historians Gerald Markowitz and David Rosner (2002: 143, 216) about the link between vinyl chloride monomer and angiosarcoma of the liver. As early as 1956, Goodrich was participating in a federal/state study of air pollution in the West End in collaboration with the Manufacturing Chemists Association (MCA). MCA and Goodrich, the historians claim, withheld from workers the information they learned from this study and additional independent research about the health effects of working with chemicals. In the 1970s, the president of the Rubber Workers union in Kentucky, the first place where angiosarcoma in workers had been discovered, testified that "workers 'knew for a long time that something was wrong, but did not know what or who to turn to'. The president of the Oil Chemical and Atomic Workers union concurred, 'industry rarely provided workers with information about the chemicals they worked with'".

In addition to angiosarcoma, acroosteolysis, a rare condition that results in tender fingers, thickened skin, bone deterioration, and circulation difficulties was identified by a physician, in 1964, who linked the polymerizer vats at the Louisville Goodrich plant to the workers' hands he was seeing. MCA and Goodrich commissioned and contained research on the toxic levels in the plant and attempted to redact a European research paper on the condition (Markowitz and Rosner, 2002: 175–76). Lessons held in secret set underinformed workers and residents on a syncopated course of perpetual exposure, intermittent blasts, and patterned outbreaks of disease.

A wholeness that hovers above these rhythms is discernible in the risk of toxic exposures to proximate neighborhoods or even the multi-county region. Reviewing the risk management plans required by the 1990 US Clean Air Act, reporter James Bruggers (2014) found "that 21 firms [were] storing deadly or explosive toxic substances ... in amounts large enough to require disaster plans". The plans record prospective danger emanating from the 21 plants in the number of residents who would be affected (say, 6500 from a formaldehyde release at the Momentive plant) and distance from the plant (in this case, up to 1.6 miles depending on wind speed). Open records laws enabled Bruggers to request these risk management plans that exist at the insistence of government regulation. As a result, he can point readers to site-specific information about their level of risk: "Are you in a danger zone? Check your address". Whether these prospective plans are an adequate public antidote to the forms of secrecy described above depends upon the political stability of environmental regulations and journalistic freedom. The encounters of environmental regulation, journalistic freedom, and risk management afford a set of choices for industry and workers, local governments, and residents.

In Louisville's West End, the rhythm of toxic explosion as well as toxin-induced disease and the wholeness of large-scale toxic threat are forms whose design is inflected by the hierarchy of racism. Louisvillians have long understood East End and West End as geographic markers of racial and class division. Although these hierarchies are historically imbedded, they do not exist without resistance. In 1993, for example, E.I. DuPont lost a case of racial discrimination and paid $14 million in damages to the claimants (Benmour, 2001: 268). But this decision was not the end of the hierarchy that contributes to the form of Naomi Wallace's play *The Hard Weather Boating Party* (2014). The play performs the hard choices afforded by the materials and designs of Rubbertown.

The hard weather boating party

Wallace's play was first produced at the 2009 Louisville Humana Festival of New American Plays. Writing in *The New York Times Magazine*, in 1997, before Wallace was recognized by the MacArthur Foundation, Vivian Gornick aptly observed that "to American ears, Wallace's work can sound like [modernist] Beckett rewriting [social realist] Clifford Odetts".

The two-act play is set in a Motel 6 room "the day after tomorrow" where and when three Rubbertown employees, not previously acquainted, are plotting the death of a chemical company executive. The instigator is a middle-aged white man from middle management;

another is a younger middle-aged African American intellectual from the factory floor; and the third is a young Latino also from the factory floor. Much of the dialogue has the rhythm of Beckett's lost souls in a no man's land. The play begins (1),

STADDON: You're supposed to do a knock.
LEX: We didn't discuss a knock.
STADDON: You still could have done a type of knock.
LEX: What type of knock?
STADDON: You know, Tap ti tap, tap ti tap ti tap. That kind.

Wallace's characters, unlike Beckett's, squabble over details out of distrust derived from specific social hierarchies. They even name their addresses in Louisville, thus zoning themselves. And their conscious intention is in full view of the audience before much time is past. Although off stage, this realist context still impedes the intention of these three strangers to act as a team, called together by Staddon, the middle manager. At Lex's insistence they play Truth or Dare, confronting their distrust and the limits of one another's fidelity to their word. They discover, reluctantly, individuating characteristics: pressed to it, Coyle can perform ballet; Lex can sing an original song about the death of his twin sister. The absurd demands of the game move the rhythm of the dialogue back toward Beckett and yet serve individuation and solidarity. They cannot easily sustain challenges to the hierarchies of their real lives, but through the absurdity of game play laced with threat, strategies for initiating these challenges arise.

Meanwhile, material forces are altering the franchise motel room itself, one in an archipelago of such corporate sites: there is a large crack in the floor; they hear the sound of rumbling and see the glow of fluorescent light. It is raining outside, and they discover a desk drawer full of water. "The day after tomorrow" is one damaged by environmental degradation just beyond the apparent material conditions today. The Motel 6 room cracks open onto some boundless whole not unlike the one that consumed Bhopal in 1984.

The audience must wrestle with the proposed murder as we watch the characters decide what weapon (kicking, hitting, shooting)each will deploy when he encounters CEO Chelton Steff in his luxury house. And we must imagine the motive for the men's plan. We have some clues – Coyle's fumbling hands, Lex's gagging – that add up to an indictment in Act Two. After they have confronted Steff, the CEO, they reconnect at Motel 6. They briefly celebrate "keeping the trust . . . Here's to all of us. Men in hard weather stick together" (Wallace, 2014: 47). But the celebration is not a culmination of an action. It is a pulse in a set of rhythms: they anticipate the appearance of the police who will assume, says African American Coyle, that *he* was Steff's killer and, most of all, they reluctantly reveal past exposure to toxic releases.

When Staddon pushes the question, Coyle responds, "At work? Or at home? Cause I have no idea how many times it happened when I was home. I stopped counting" (Wallace, 2014: 49). They agree on nine exposures at work – eight for young Lex – each from the details of his position: Staddon overseeing floors three through seven; Coyle promoted from warehouse to tank equipment repairs; Lex still flushing pipes after six years. Toxically induced illness has taken from all three men the sexual desire that drives Animal. Young sexual exploits, long-lived personal partnerships, and the recipients of this desire are all gone. All three men are near death.

And yet their disjointed conversation does not logically link work, residence, toxic exposure, and medical symptoms. These connections are expressed only in an erratic rhythm. Despite Coyle's threat, Lex pushes:

> I got eight exposures. You got nine. Hot feet. Check. Dick dead. Check. Numb hands. Check. I figure your back teeth are gone. Can't keep food down anymore. Your ears bleed at night? . . .But what's the worst for you, Coyle? What was it that made you send your woman packing?
>
> *(Wallace, 2014: 57)*

"I'll hit you", replies Coyle (Wallace, 2014: 57). Sociologist Arlie Russell Hochschild (2016) concludes, from her ethnographic engagement with Tea Party members who are petrochemical industry workers and neighbors in southwest Louisiana, that interviewees hold fast to their jobs and promise of the American Dream despite the destruction of bayous they call home and patterned illnesses of family members, coworkers, and themselves. This is the price they believe one must pay for gainful employment that will fund middle-class life. Yet, privately, individuals confide their sorrow at the patterns of illness they see. In the end game of Wallace's play, her three characters, though clinging to dignity, are too alone and near death to engage in the explanations that Hochschild's subjects use to understand rhythmic losses of home and health.

In the temporarily closed, collapsing space of the room at the Motel 6, Stoddard, Lex, and Coyle awkwardly find their way to being for one another the families and lovers they have lost. Lex pretends to be the son Stoddard regrets that he never had. Stoddard kisses Coyle to let him feel his wife's lips again. Most miraculously, Stoddard reveals, his illness has been induced by his deliberately standing in the tanks the laborers entered. The middle manager crossed over rather than cling to his middle-class life in a south Indiana suburb. What drove him from identification with those above him in the corporate and residential hierarchy was his exposure to a club devoted to a Confederate war general that included managers in his company among its members. Stoddard recognizes the decorative oar that Lex has taken from Steff's house as an artifact of this organization. Having little hope of escape and no former lives to reenter, the men together imagine buying a boat and taking the oar with them.

But, it turns out, they are hapless criminals and not up to a clever getaway. Stoddard could not shoot to kill after all. Lex and Coyle could not kick and hit with quite so much effect as they first reported. And they did not induce Steff to open his safe. Both Steff and his wife have seen them. The anger and mourning that Hochschild uncovered in her interviewees in toxic South-West Louisiana are not enough to drive these resident-laborers to revenge that closes a whole. Instead Staddon convinces Lex and Coyle to kill him before his lungs fill with blood. With his body propped between them, they form a workers' pietà. A boat heaves up from the crack in the floor. They claim it and "look off in the distance, as though [they] can see the Ohio" (Wallace, 2014: 75).

On "the day after tomorrow", the three characters have enacted the rhythms, the hierarchy, and the probabilities of lives in Louisville's West End yesterday and today. Wallace offers trust won at a high price, solidarity without solution. The pulse of leaks and explosions in the West End continued – 2011, 2013, 2015.

The 2015 production of *Prevailing Winds* by the Looking for Lilith Theatre Company also pursued solidarity by performing the rhythms and hierarchies of Louisville's West End. Using local oral histories as their text, a small ensemble of diverse actors embodied the words and sentiments of a larger number of diverse Louisvillians. With this play, the company used the formal resources of theatre to address, with their audience, an abiding urban problem in their city. Perhaps most striking is their recreation of a meeting of the Louisville Metro Air Pollution Control Board. Before voting to enact a pollution reduction program, the board weighed testimony from 1300 people. The resulting form was the

Strategic Toxic Air Reduction (STAR) Program of the Metro Air Pollution Control District (Louisville, 2017).

Through the form of prevailing winds, the play's creators imagined unity across east and west Louisville. Yet the division in health effects remains. *LEO Weekly* (Smith, 2015) reported that "according to the Metro Public Health Department, 30.3 percent of West Louisville residents live with a symptom or illness caused by outdoor air pollution, compared to just 16 percent in East Louisville". The reporter joined the Passist Earth and Spirit Center's environmental injustice tour of the West End. He concludes his article with the assertion of the tour's leader that "environmental justice ... [is] about helping your neighbors live healthy, happy lives".

Louisville with its STAR program and Bhopal with its Smart City programs have forms with which to imagine some of their citizens' struggles with toxicity and justice. If city leaders of all sorts see these forms in ongoing encounters with other evolving forms – a "staying with the trouble" that literary texts can help us cultivate – they stand the best chance of promoting solutions with traction. Sanjay Sridhar (2017), Regional Director for South and West Asia of C40 Climate Leadership Group, an NGO addressing 40 megacities in the world, recently told me that India's Smart Cities funding creates substantive improvement only in those cities where designers link the new projects to structures already in place. I argue that those forms are not only technological and bounded: they are material, social, political, and aesthetic. They overlap, collide, encounter one another in the material and memories of individual bodies, neighborhoods, cities – any space and time of our human imagining.

References

Alavi, S.U.R. (2016) Bhopal Gas Disaster. *Firstpost*, 3 December 2016. Available at: http://www.firstpost.com/india/bhopal-gas-disaster-32-years-later-wounds-still-fester-in-bhopal-3137526.html.

Benmour, E. (2002) E.I. DuPont De Nemours & Co. *The Encyclopedia of Louisville*. Lexington: University Press of Kentucky.

Bennett, J. (2010) *Vibrant Matter*. Durham: Duke University Press.

Bhopal City Census 2011 Data. Available at: http://www.census2011.co.in/census/city/302-bhopal.html.

The Bhopal Medical Appeal. (2015), Available at: http://www.bhopal.org.

Bhopal Municipal Corporation. (2017) Smart City Bhopal: Projects. Available at: http://www.smartbhopal.city/project.

Bruggers, J. (2014) Danger Zones Blanket Louisville Region. *Louisville Courier-Journal* 8 March, Available at: http://www.courier-journal.com.

Bruggers, J. (2015) Explosions Rocked DuPont Plant 50 Years Ago. *Louisville Courier-Journal*, 26 August 2015, Available at: http://www.courier-journal.com.

Chaudhury, S. and Ray, S.G. (2010) For a Few Pieces of Silver. *Tehleka* 7(25), 26 June 2010. Available at: http://www.tehelka.com/2010/06/for-a-few-pieces-of-silver/.

Deleuze, G. and Guattari, F. (1987) *A Thousand Plateaus*. Trans. Massumi, B. Minneapolis: University of Minnesota Press.

Gornick, V. (1997) An American Exile in America. *The New York Times Magazine*, 2 March 1997. Available at: http://www.nytimes.com.

Gulraiz, S. (2012) Bhopal: A Third World Narrative of Pain and Protest. *Art News and Views*, May 2012. Available at: http://artnewsnviews.com.

Haraway, D. (2016) *Staying with the Trouble*. Durham: Duke University Press.

Hochschild, A.R. (2016) *Stranger in Their Own Land*. New York: New Press.

International Campaign for Justice in Bhopal (2014). Available at: http://www.bhopal.net.

Kerin, T. (2016) The Evolution of Process Safety Standards and Legislation Following Hallmark Events. *Process Safety Progress* 35(2): 165–70.
Klein, J.A. (2009) Two Centuries of Process Safety at DuPont. *Process Safety Progress* 28(2): 114–22.
Latour, B. (2004) *Politics of Nature*. Cambridge: Harvard University Press.
Lefebvre, H. (1991) *The Production of Space*. Trans. Nicholson-Smith, D. Oxford: Blackwell.
Lefebvre, H. (2013) *Rhythmanalysis*. Trans. Moore, G. and Elden, S. London: Bloomsbury Publishing.
Levine, C. (2015) *Forms*. Princeton: Princeton University Press.
Looking for Lilith Theatre Company. (2015) *Prevailing Winds*. Record of Production. Available at: http://lookingforlilith.org/currentseason/prevailingwinds/.
Louisville. (2017) Strategic Toxic Air Reduction Program. Available at: https://louisvilleky.gov/government/air-pollution-control-district/strategic-toxic-air-reduction-program.
Mendes, A.C. and Lau, L. (2015) India Through Reorientalist Lenses. *Interventions: International Journal of Postcolonial Studies* 17(5): 706–27.
Ministry of Housing and Urban Affairs. (2016) *City Profiles of 20 Smart Cities*. Available at: http://smartcities.gov.in/content/innerpage/cities-profile-of-20-smart-cities.php.
Remember Bhopal. (2015) The Statue of Mother and Child. Available at: http://rememberbhopal.net/survivors-efforts/.
Ross, H.K. (2016) Settlement Approved to Stop Ohio River Pollution Caught on Camera. *Earth Justice*, 16 December. Available at: https://earthjustice.org.
Sinha, I. (2007a) *Animal's People*. New York: Simon and Schuster.
Sinha, I. (2007b) Bhopal: A Novel Quest for Justice. *The Guardian*, 10 October 2007. Available at: https://www.theguardian.com.
Sinha, I. (2008) Why I'm Going on Hunger Strike for Bhopal. *The Guardian*, 12 June 2008. Available at: https://www.theguardian.com.
Sinha, I. (2009) Bhopal: 25 Years after Poison. *The Guardian*, 3 December 2009. Available at: http://www.theguardian.com.
Smith, E. (2015) An Environmental Justice Tour of West Louisville. *LEO Weekly*, 25 November 2015. Available at: https://www.leoweekly.com.
Sridhar, S. (2017) Conversation with Barbara Eckstein and Peter Rolnick. Iowa City, 18 October.
Taylor, J.O. (2013) Powers of Zero. *Literature and Medicine* 31(2): 177–98.
Union Carbide Corporation. (2016) Bhopal Gas Tragedy Information. Available at: www.bhopal.com.
Varma, R. and Varma, D.R. (2005) The Bhopal Disaster of 1984. *Bulletin of Science, Technology & Society* 25(1): 37–45.
Verma, S. (2014) *Invisible Voices*. San Francisco: McSweeny's Books, 168–77.
Verma, S., Masi, A., and Oatman, M. (2014) Photos: Living in the Shadow of the Bhopal Chemical Disaster. *Mother Jones*, 2 June. Available at: http://www.motherjones.com.
Wallace, N. (2014) *The Hard Weather Boating Party*. New York: Broadway Play Publishing Inc.

Part II
Smart and digital

8

Smart urban

Imaginary, interiority, intelligence

Gillian Rose

Introduction

This chapter probes the relation between "smart" and "imaginary". The concept of "urban imaginary" refers to the mediation of urban environments by cultural texts of many kinds: visual and literary, audible and oral, high and low. The "smart city" is a city which is also mediated, but by digital data. In a smart city, urban data of many kinds is harvested, analysed and put to work to make cities run more efficiently, more sustainably, more securely, and, perhaps, more democratically (or at least that's the hype). Digital data can affect urban environments in many ways: enabling better management of traffic flow, say, or swifter responses to flooding, or tighter integration of energy supply and demand. Indeed, in many cities, smart is now part of their imaginary. For city leaders in Barcelona, Manchester, Amsterdam, and many more, their ranking in league tables of smart cities is part of their reputational aura and a vital part of their efforts to compete as a global city. The large software companies who sell smart city technologies put a lot of work into creating their own smart imaginaries too, on their YouTube channels, in television advertisements, through their Twitter feeds, and at expos. For example, Siemens has built an entire exhibition space in London's Docklands to promote smart sustainable urbanism. And while it's not clear how far the label of "smart city" has permeated the collective imagination of urban populations, it is certainly the case that digital technologies of many kinds have become part of everyday experience in many cities, as subway cards are swiped and smart meters log energy use (Kitchin and Dodge, 2011). Indeed there are many other mediations of urban space by digital technologies quite apart from those labelled "smart": online consultation events, local mobility apps and open data dashboards, for example, as well as Google maps and many social media and other platforms.

Clearly, then, the concept of the urban imaginary can be put to work on smart mediations of urban spaces. And several scholars have begun to do this. They have analysed visual and textual representations of smart city technologies in a variety of contexts. An early intervention by Hollands (2008) emphasised both the imprecision and the self-congratulatory tone of much smart city commentary. How the "smart citizen" is imagined in such commentary has been examined (Vanolo, 2016). Halpern et al. (2013) study the materials produced by Cisco and their collaborators on the smart city New Songdo, while Söderström et al. (2014) analyse IBM's rhetoric, and corporate smart city videos on YouTube have also been discussed (Rose, 2017a). Interviews have been conducted with the managers of smart infrastructure projects (Klauser et al., 2014), and smart city operation centres have been observed (Luque-Ayala and Marvin, 2016). Certain themes that constitute a smart city imaginary have been identified: an emphasis on the apparently utopic "transformation [of] the management of life – human and machine – in terms of increased access to information and decreased consumption of resources" (Halpern et al., 2013: 278); a fascination with visualising integration, mobility, speed, and flow; and a deeply technocratic approach to managing cities that assumes enough data and analysis will provide solutions to any problem.

Smartness, then, is part of urban imaginaries, and the urban imaginaries of smart cities have been interrogated. But what also emerges in much of that interrogation is a sense that the role of the imagination in relation to smart is highly problematic. Smart cities, it seems, are never being imagined adequately. Smart imaginaries are too limited, or too absurd. People are either not imaginative enough about what's happening as a city goes smart, or they are fantasising about it: they are totally unimaginative or their imagination is out of control. This chapter explores this dichotomous analysis.

Imagining the smart city

The phrase "smart city" emerged in the mid-1990s and was consolidated in 2011 when IBM trademarked the phrase "Smarter Cities" as a way to sell its data management software to cities (Söderström et al., 2014; see also Marvin and Luque-Ayala, 2017). A smart city is envisioned as a system-of-systems, and the pitch focused on integrating different kinds of data, from different sources, in order to improve city functioning. As this argument gained ground, and increasing amounts of digital data became available, actually existing smart activities in cities have diversified. There are now many different kinds of smart city technologies: hardware of various types that gather and processes data, software of different kinds that analyse it, and yet more software that visualises it on a variety of screens. Combinations of these technologies operate under various modes of organisation, ownership, and accountability. Many sorts of people and skills are involved in the design, production, and maintenance of such systems, including the elusive figure of the "smart citizen". Discourses about smart have multiplied and widened, with popular writers like Adam Greenfield writing books and articles *Against the Smart City* (2013) and citizen-generated smart projects gaining traction (Gabrys et al., 2016; McFarlane and Söderström, 2017; McLaren and Agyeman, 2015). Different sorts of urban spaces are mediated by these smart things, in different ways: streets are surveilled by cameras, traffic sensors, and the lidar scanners of autonomous pods. Cities' landscapes are visualised by satellites and drones. Cities' energy supplies are monitored and managed by sensors. Smart, then, is constituted by a great many things connected in various ways to the production and use of digital data in cities, and should be understood as a particularly capacious descriptive term.

Much scholarship on smart cities is, in a way, following a similar trajectory, moving from analyses of claims about the smart city to pursuing careful accounts of specific enactments of smart discourse, technologies, apparatus, and positionalities. But imagination remains a somewhat troubling phenomenon in relation to smart. Much of this new scholarship broadly follows an Actor Network Theory methodology, carefully observing and describing particular smart networks or assemblages. When imaginations loom, however, odd things happen, because smart cities, it seems, suffer either from a surfeit of imagination, or a deficit of imagination. Humans in smart cities are either not being imaginative enough or they are being too imaginative. They – we, or lots of us anyway – are unaware that our data is more and more untethered from its provenance, reducing our ability to interrogate it, for example (Ford and Graham, 2016). We choose to consent to terms and conditions without thinking about the consequences (Leszczynski, 2016). We are thoughtlessly handing our decision-making over to the corporations and their machines (Greenfield, 2013). Enacting a "generalised lack of awareness" (Graham et al., 2013: 467), it seems that few of us peer beyond our smartphone screen to imagine the chains of code and data in which it is embedded. In short, the smart city is not being imagined sufficiently or properly.

Conversely, the marketeers of smart city hype stand accused of having too much imagination: all those beautiful images of computer-generated cities, in which data flows endlessly and smoothly through the air, and inhabitants have nothing better to do than stroll in the sunshine or look at screens in their perfectly efficient and co-ordinated lives. The cities always look new and clean and the only problems made visible are ones that can be addressed by data (Rose, 2017a; Vanolo, 2016). All this is ridiculous, say the critics (rightly), and is nothing like what smart cities might actually look and be like. Again, the criticism is that the smart city is not being imagined properly.

The smart city, then, seems to have a problematic relationship with notions of the imaginary and its close cousin, imagination. On the one hand, there's something about smart that seems to render imagination redundant. Smart technologies make things invisible and impossible to imagine; their code is arcane, the terms and conditions of apps are too long, their data hidden, their algorithms secret, their data shadows too distant, their cables and servers hidden underground, their Wi-Fi transmissions invisible. All this secrecy disables the imagination, but also demands it; it's all so invisible that the imagination has nothing to work with, but we must imagine it or suffer the consequences. On the other, smart is deemed too visible, too spectacular, too imaginary. The ambitions it articulates are so extreme that they become little more than "a fantasy" (Halpern et al., 2013: 278). They become obviously absurd and require no more from critics than a "certain tonality of outrage [to] communicate a view that a certain self-evident case has not been acknowledged appropriately" (Berlant, 2007: 669–70).

Why is it so difficult to establish a different relationship between imagination and smart, one that is neither too little nor too much?

The urban imaginary as interiority

As many contributors to this volume have noted, the urban imaginary has a long pedigree in urban and cultural studies. The urban imaginary is conceptualised as a collective act of imagination, which is always open, fragmented, multiple and contested (Cinar and Bender, 2007; Huyssen, 2008). The concept is a reminder that a city is

> constituted by the interplay between its spaces and its imaginations. The bricks and mortar do not exist apart from representations, nor are our ideas without material consequences . . . The city is both the actual physical environment and the space we experience in novels, films, poetry, architectural design, political government, and ideology.
>
> *(Prakash, 2008: 7)*

An urban imaginary is thus generated by the "traffic between" cultural texts, everyday experiences and the urban built environment (Donald, 1999: 27). For Cinar and Bender

(2007), "the city is located and continually reproduced through ... orienting acts of imagination, acts grounded in material space and social practice" (xii). Such everyday enactments produce and reproduce narratives and images that story and visualise the city into existence. They provide "the interpretive grids through which we think about, experience, evaluate, and decide to act" (Soja, 2000: 324). The urban imaginary is thus "the cognitive and somatic image we carry within us of the places where we live, work and play" (Huyssen, 2008: 3).

These discussions of the urban imaginary were all written at least a decade or so ago. None discuss digital data or computer-generated imagery as contributors to urban imaginaries, and only Soja (2000) discusses the internet and cyberspace. And in one way, this isn't a problem. Data might be understood as another form of representation, another form of text which produces its own orientations, and there is other work addressing digital mediations that would supplement these classic discussions of the urban imaginary (see for example Boyer, 1996; Gordon, 2010). In another way, however, the concept of the urban imaginary might require some rethinking in relation to cities where so much of everyday life is now digitally mediated, for the urban imaginary has been conceptualised as a resolutely human activity. While there are extensive discussions of how the specific genres of cultural texts contribute to particular imaginaries, and how the built environment is loaded with symbolism and meaning, there's no real challenge to the notion that imagination is distinctively human.

As the *Oxford English Dictionary* points out, vernacular definitions of imagination focus on "the mind" as an inner space where thoughts, images, and ideas are produced and stored, drawing on external experiences and imaginatively reworking them. That is, everyday understandings of "imagination" make a distinction between inner imagination and outer experiences. Discussions of urban imaginaries certainly seem to do something very similar, understanding imagination as a process tethered to human life and understood as an aspect of human interiority. The imaginary is "the way city dwellers imagine their own city" (Huyssen, 2008: 3); it is "our [i.e. 'humans'] situated and city-centric consciousness" (Soja, 2000: 324). In his discussion of the urban imaginary, Anthony King notes that "what is referred to as 'the city' exists only in our heads" (2007: 2) since it consists of "mental and cognitive mappings" (Soja, 2000: 324). The urban imaginary of a smart city can be understood as a "smartmentality" (Vanolo, 2014). The urban imaginary thus tends to be understood as something that humans do in relation to all sorts of other things out there in the world: books, buildings, photographs, data.

This then allows a particular connotation of the word "imaginary" to emerge: namely, that the inner imaginary does not correspond adequately to the reality of the exterior world. Imagination, notes the *Oxford English Dictionary* more than once, might therefore also be "a fanciful project", which can be a quality of genius, but can also be deluded (or indeed both). The critique of smart city hype as fantasy seems to depend on precisely such a notion of the imagination and its imaginaries. The imagination of the corporate smart city marketeer, for example, is understood as far too little engaged with the realities of the external world. Although the production of glossy brochures and glittering websites and dazzling Twitter feeds is of course a highly material practice, as well as one that trades in all sorts of cultural tropes and genres (Rodgers et al., 2014), nonetheless its problem, it appears, is that it is too disengaged. It is too interior, too distant and different from the messy reality, and as a result it produces talk, text, and images that don't sufficiently engage with the realities of the urban world. Smart city visions are thus understood as inadequate representations and critiqued as such.

Mattern (2013) has discussed another implication of this critical move. Rather like the response of ANT scholarship to the emergence of the smart city, Mattern points out that much arts-based criticism of digital urbanism (among other things) has turned to varied forms of (what purport to be) description in order to challenge the corporate excess of imagination. Too much crazy dreaming? Get real! Refuse the imaginary and document, instead. As she notes, various artists in many cities are attempting to challenge corporate digital visions by organising "infrastructure tourism" in order to render the materiality of the smart city visible: sensors, cables, servers.

A recent example of such activity was the installation in London's Glass Room in November 2017 by the Berlin-based Tactical Tech collective. During its two-week run, 18,000 people visited the show, with queues down the street waiting to experience a variety of pieces exploring how personal digital data is being used. It was a very engaging event, and its overall approach was revelatory description of a kind very similar to that described by Mattern (2013). A screen displayed the signals transmitted by the smartphones of people in the room, for example; several tablets simply looped advertisements for various apps that depend on gathering very personal digital data. Like infrastructural tourism, these sorts of displays suggest that the smart city also exposes itself, as it were. The reality is there all the time and just needs pointing out, in this case not by an urban tour-guide but by the framing of a white-cube gallery. This assumption of smart self-revelation is a move repeated by Greenfield in a recent essay on the Internet of Things. Describing a Wi-Fi enabled gadget for sale online, he remarks, "I cannot improve on Amazon's own description of this curious object and how it works, so I'll repeat it here" (Greenfield, 2017).

The popularity of documentation as a strategy for critiquing digital corporate activity is somewhat paradoxical, however. It assumes that smart visions both hide things but also lay things out with startling clarity: their visions must be critiqued but also speak for themselves.

The urban imaginary as intelligence

This brings us to another aspect of smart and its use of big data. If corporations' computer-generated visions of smart cities are so obviously fantasies, smart cities themselves are understood to function without imagination. Smart city advocates very often argue that one of the advantages of smart city technology is precisely that it removes the human from urban infrastructure management, replacing it with a decidedly unimaginative machinic intelligence. Indeed, the whole notion of "smart" assumes a particular kind of intelligence, a kind of efficiency and speed as a consequence of an insightful analysis of a situation. It was a smart move; look smart.

This understanding of the smart city as intelligent has a number of different aspects. One is the argument that the analysis of huge, diverse but integrated datasets can produce better evidence on which to base city management. Here there's a link to the ongoing discussion about the efficacy of big data analysis. Anderson's (2008) notorious editorial for the magazine *Wired* is the touchstone for this sort of thinking; as is well known, it argued that the detection of patterns in huge datasets by software renders causal theorisation (by humans) obsolete. There are also the related claims that machines can make better decisions because algorithms, machine learning, and artificial intelligence work purely with data (which is assumed to be value-free, reliable, and comprehensive).

If the smart city enacts a certain form of this data-driven intelligence, then it is precisely not a human form of intelligence. It is smarter. It is also just different (hence the corporate smart city is challenged to become "more human" [Hollands, 2015: 63], to become full of "culture, politics, competing interests" [Kitchin, 2016: 11]). Even when the smart city is imagined as embodied in some way – as cities so often are (Grosz, 2001) – that body is not human. Given form by the distribution of its data-generating sensors (Halpern et al., 2013: 291) or by the spatial boundaries of its modelling (Dorrian, 2008), data generated by social media is its "lifeblood" (Kelley, 2013: 200) and screen interfaces are its "synapses" (Halpern et al., 2013: 280). It is a "Frankenstein" (Cugurullo, 2017). The patched-together body parts have an agency described as smart. But this smart city/body is not understood as capable of imagination, or even consciousness (Halpern et al., 2013: 281). The smart city might be a body, but it is a body without that peculiarly human form of interior imagination. Its intelligence has replaced (human) imagination.

Indeed, Thrift (2014; Thrift and French, 2002) has been arguing for some time that digitally mediated cities are primarily affective spaces, in which software and hardware short-circuit human consciousness and instead act through pre-cognitive atmospheres. According to Amin and Thrift (2016), the smart city is already producing human life in its own image, or rather body: smart but unimaginative. Which is sort of what viewers are assumed to be in the art projects discussed by Mattern (2013), and in the Glass Room in London: intelligent but just needing to have things shown to us in order to understand.

Re-imagining the urban imaginary

The relation between smart cities and the urban imaginary, then, seems to ricochet between different polarities. Some humans seem to have too much imagination and others not enough. Smart cities are themselves entirely imagined fantasies; smart cities replace imagination with intelligence. Smart city imaginaries are unreliable; smart city imaginaries are revelatory. We need to imagine more but description will also suffice. That is, cities are smart (or not) and humans are imaginative (or not).

Is it possible to step outside of this conceptual ping-pong? For that to be possible, the dominant understanding of the imaginary as a peculiarly human activity would have to be revised. Imagination would need to be understood as something in intimate relation to technologies – indeed, impossible without them. What might happen if we no longer understood imagination as an activity of the human mind alone, but as something technologically mediated through and through? In effect, this would smarten up the notion of the urban imaginary. Rather than place the imagination outside smart city activities, as human excess to a city figured as a particular kind of intelligence, it would be thought of as part of the smart city itself. Here the work of Haraway and Hayles come to mind, and even Stiegler (see Rose, 2017). The dichotomy between human and smart would disappear and something more interesting and diverse might emerge at the "seams and extensions" (Buscher et al., 2016: 173). Different configurations of both the human and the smart city might be thinkable.

And if the notion of the imaginary needs to be thought of as with technology rather than outside it, perhaps we also need to bring more imagination to bear on smart. I'm not thinking here of even more fantasies, but rather of more diverse imaginations. What would happen if we multiplied the forms of imagination that are pulled through smart urban spaces, so that humans could be more than either fantasists or documentarists? What if we made imagination more prosaic and everyday and not just the preserve of professional dreamers, whether artistic or commercial (Buscher et al., 2016)? What if we amplified our own imaginations and thought about different versions of smart-as-intelligence? Datta (2018) has pointed to the cultural specificity of the dominant notion of smart and suggests from her work on smart city projects in India that there might be another kind of smart that could be elaborated, a human kind of smart embodied in the *chatur* citizen who is resourceful, quick-witted, and embedded in multiple social and political networks: something closer to a hacker, perhaps, than the compliant smart citizen envisaged in corporate-driven smart cities. This is just one example of what might happen if we think of multiple forms of human emergent alongside the smart city.

And, finally, what if we allowed the possibility that city databases might imagine? What if we asked, with Munster (2013), what data experiences? I don't have an answer to that question, but I do think that the notion of urban imaginary needs rethinking in relation to smart cities, if we are to move beyond the increasingly tired dichotomies generated by the assumption that only humans have (a specific kind of) imagination.

References

Amin, A. and Thrift, N. (2016) *Seeing Like a City*. Cambridge: Polity Press.

Anderson, C. (2008) The End of Theory: The Data Deluge Makes the Scientific Method Obsolete. *Wired*. Available at: https://www.wired.com/2008/06/pb-theory/. Accessed 24 November 2017.

Berlant, L. (2007) On the Case. *Critical Inquiry* 33(4): 663–72.

Boyer, M.-C. (1996) *CyberCities: Visual Perception in the Age of Electronic Commnication*. New York: Princeton Architectural Press.

Buscher, M., Kerasidou, X., Liegl, M. et al. (2016) Digital Urbanisms in Crises. In: Kitchin, R. and Perng, S.-Y. (eds.), *Code and the City*. London: Routledge, 163–77.

Cinar, A. and Bender, T. (2007) Introduction: City Experience, Imagination, and Place. In Cinar, A. and Bender, T. (eds.), *Urban Imaginaries: Locating the Modern City*. London: University of Minnesota Press, xi–xxvi.

Cugurullo, F. (2017) Exposing Smart Cities and Eco-Cities: Frankenstein Urbanism and the Sustainability Challenges of the Experimental City. *Environment and Planning A*. DOI: 10.1177/0308518X17738535.

Datta, A. (2018) The Digital Turn in Postcolonial Urbanism: Smart Citizenship in the Making of India's 100 Smart Cities. *Transactions of the Institute of British Geographers*. DOI: 10.111/tran.12225.

Donald, J. (1999) *Imagining the Modern City*. London: Athlone Press.

Dorrian, M. (2008) 'The Way the World Sees London': Thoughts on a Millennial Urban Spectacle. In Vidler, A. (ed.), *Architecture between Spectacle and Use*. London: Yale University Press and Clark Art Institute, 41–57.

Ford, H. and Graham, M. (2016) Provenance, Power and Place: Linked Data and Opaque Digital Geographies. *Environment and Planning D: Society and Space* 34(6): 957–70.

Gabrys, J., Pritchard, H., and Barratt, B. (2016) Just Good Enough Data: Figuring Data Citizenships through Air Pollution Sensing and Data Stories. *Big Data & Society* 3(2). DOI: 10.1177/2053951716679677.

Gordon, E. (2010) *The Urban Spectator: American Concept Cities from Kodak to Google*. 1st ed. Hanover, NH: Dartmouth College Press.

Graham, M., Zook, M., and Boulton, A. (2013) Augmented Reality in Urban Places: Contested Content and the Duplicity of Code. *Transactions of the Institute of British Geographers* 38(3): 464–79.

Greenfield, A. (2013) *Against the Smart City*. London: Do Projects.

Greenfield, A. (2017) Rise of the Machines: Who Is the 'Internet of Things' Good For? *The Guardian*, 6 June. Available at: http://www.theguardian.com/technology/2017/jun/06/internet-of-things-smart-home-smart-city. Accessed 24 November 2017.

Grosz, E.A. (2001) *Architecture from the Outside: Essays on Virtual and Real Space. Writing Architecture*. Cambridge, MA: MIT Press.

Halpern, O., LeCavalier, J., Calvillo, N. et al. (2013) Test-Bed Urbanism. *Public Culture* 25(270): 272–306.

Hollands, R.G. (2008) Will the Real Smart City Please Stand Up? *City* 12(3): 303–20.

Hollands, R.G. (2015) Critical Interventions into the Corporate Smart City. *Cambridge Journal of Regions, Economy and Society* 8(1): 61–77.
Huyssen, A. (2008) Introduction: World Culture, World Cities. In: Huyssen, A. (ed.), *Other Cities, Other Worlds: Urban Imaginaries in a Globalizing Age*. Durham: Duke University Press, 1–23. Available at: http://www.loc.gov/catdir/toc/ecip0823/2008028435.html. Accessed 13 November 2017.
Kelley, M.J. (2013) The Emergent Urban Imaginaries of Geosocial Media. *GeoJournal* 78(1): 181–203.
King, A.D. (2007) Boundaries, Networks, and Cities: Playing and Replaying Diasporas and Histories. In Cinar, A. and Bender, T. (eds.), *Urban Imaginaries: Locating the Modern City*. London: University of Minnesota Press, 1–14.
Kitchin, R. (2016) The Ethics of Smart Cities and Urban Science. *Philosophical Transactions of the Royal Society A* 374(2083): 1–15.
Kitchin, R. and Dodge, M. (2011) *Code/Space: Software and Everyday Life*. Cambridge, MA: MIT Press.
Klauser, F., Paasche, T., and Söderström, O. (2014) Michel Foucault and the Smart City: Power Dynamics Inherent in Contemporary Governing through Code. *Environment and Planning D: Society and Space* 32(5): 869–85.
Leszczynski, A. (2016) Speculative Futures: Cities, Data, and Governance beyond Smart Urbanism. *Environment and Planning A* 48(9): 1691–708.
Luque-Ayala, A. and Marvin, S. (2016) The Maintenance of Urban Circulation: An Operational Logic of Infrastructural Control. *Environment and Planning D: Society and Space* 34(2): 191–208.
Marvin, S. and Luque-Ayala, A. (2017) Urban Operating Systems: Diagramming the City. *International Journal of Urban and Regional Research* 41(1): 84–103.
Mattern, S. (2013) Infrastructural Tourism. *Places Journal*. Available at: https://placesjournal.org/article/infrastructural-tourism/. Accessed 24 November 2017.
McFarlane, C. and Söderström, O. (2017) On Alternative Smart Cities: From a Technology-Intensive to a Knowledge-Intensive Smart Urbanism. *City* 18(4): 1–17.
McLaren, D. and Agyeman, J. (2015) *Sharing Cities: A Case for Truly Smart and Sustainable Cities*. Cambridge, MA: MIT Press.
Munster, A. (2013) *An Aesthesia of Networks: Conjunctive Experience in Art and Technology*. London: MIT Press.
Prakash, G. (2008) Introduction. In Prakash, G. and Kruse, K.M. (eds.), *The Spaces of the Modern City: Imaginaries, Politics, and Everyday Life*. Oxford: Princeton University Press, 1–18.
Rodgers, S., Barnett, C., and Cochrane, A. (2014) Media Practices and Urban Politics: Conceptualizing the Powers of the Media–Urban Nexus. *Environment and Planning D: Society and Space* 32(6): 1054–1070.
Rose, G. (2017a) Look Inside™: Visualising the Smart City. In: Bengssten, L.R., Fast, L., Jansson, A. et al. (eds.), *An Introduction to Geomedia*. London: Routledge.
Rose, G. (2017b) Posthuman Agency in the Digitally Mediated City: Exteriorization, Individuation, Reinvention. *Annals of the American Association of Geographers* 107(4): 779–93.
Söderström, O., Paasche, T., and Klauser, F. (2014) Smart Cities as Corporate Storytelling. *City* 18(3): 307–320.
Soja, E.W. (2000) *Postmetropolis: Critical Studies of Cities and Regions*. Malden, MA: Blackwell.
Thrift, N. (2014) The 'Sentient' City and What It May Portend. *Big Data & Society* 1(1): 1–21.
Thrift, N. and French, S. (2002) The Automatic Production of Space. *Transactions of the Institute of British Geographers* 27(3): 309–35.
Vanolo, A. (2014) Smartmentality: The Smart City as Disciplinary Strategy. *Urban Studies* 51(5): 883–98.
Vanolo, A. (2016) Is There Anyone Out There? The Place and Role of Citizens in Tomorrow's Smart Cities. *Futures* 82(September): 26–36.

9

The origin of the smart city imaginary

From the dawn of modernity to the eclipse of reason

Federico Cugurullo

Introduction: the smart city phenomenon

Today, the *smart city* is one of the most popular urban ideals behind the genesis and development of cities. As documented by a growing number of studies, across different geographical spaces there is a plethora of smart city initiatives shaping the construction of new urban settlements and the regeneration of existing ones (Aina, 2017; Caprotti and Cowley, 2016; Cowley et al., 2017; Cugurullo, 2018; De Jong et al., 2015; Trencher and Karvonen, 2017; Wu et al., 2017; Yin et al., 2015). A single definition, image, and vision of smart urbanism does not exist. As in the case of other models of city-making, such as the eco-city for instance, the smart city ideal has been understood and put into practice in many heterogenous ways (Caprotti, 2016; Cugurullo, 2016a, 2017; Rapoport, 2014). Recent work on the geography of the smart city phenomenon shows how the interpretation of the concept of smart city has been filtered and processed through specific politico-economies and cultures depending upon equally specific places, thereby resulting in diverse built environments (Karvonen et al., 2018).

However, within this constellation of smart city projects, it is possible to find a common ideological thread weaving a uniform tapestry (see, also, Angelidou, 2015). Underpinning the smart city phenomenon is a shared adamant faith in technology and innovation. More specifically, advocates and practitioners of smart urbanism see in information technology and engineering infinite sources of data and energy, through which cities can be managed and powered, in a sustainable manner (Calzada and Cobo, 2015; Caragliu et al., 2011; Garau et al., 2016a; Hashem et al., 2016; Kitchin, 2014; Kummitha and Crutzen, 2017; Luque-Ayala and Marvin, 2015; Mosannenzadeh et al., 2017; Neirotti et al., 2014). Through the integration of technologies such as smart sensors, smart grids, big-data networks, autonomous transport systems, and generators of renewable energy, the smart city ideal promises to improve the ratio of energy production/energy waste, reduce the economic and the environmental costs that urban living generates, and decrease the carbon emissions of cities, thereby contributing to the fight against climate change (Garau et al., 2016b; McLean et al., 2015; Vanolo, 2014; Viitanen and Kingston, 2014).

Despite these lofty promises the reality presents a different picture, as it is often the case with urban ideals. Many so-called *smart cities* have shown intelligence from an economic perspective, boosting the business of private companies working in cleantech, and sustaining the political economies of countries from both the Global North and South (Coletta et al., 2017; Haarstad, 2016; March and Ribera-Fumaz, 2016; Taylor Buck and While, 2017). However, such economic intelligence has often been achieved to the detriment of the social and environmental dimensions of cities. From a social perspective, the smart city movement has been responsible for increasing inequality in several cities, by unevenly distributing the benefits of smart technologies (Colding and Barthel, 2017; Cugurullo, 2018; Heaphy, 2018; Wiig, 2016). Moreover, while technologically innovative, smart installations have tended to follow old-school, ecologically insensitive urban design strategies, which have caused considerable losses in terms of natural habitat and biodiversity (Cugurullo, 2016b; 2018; Kaika, 2017). In essence, smart urbanism appears to fall far short of the balance among economic, social, and environmental interests, advanced by the idea of sustainability.

We know that the genesis and development of urban settlements is a process burdened with planetary consequences (Brenner, 2014; Grimm et al., 2008; Grimmond, 2007). Where the smart city phenomenon will lead cities, societies and, ultimately, the planet is an open question. Part of the answer lies in the ideological underpinnings of smart urbanism. Human actions are largely driven by ideas, and urbanization, as the human endeavor par excellence, is not immune to this causal link. By investigating the nature, meaning, and implications of the core ideas at the foundation of the concept of *smart city*, it is therefore possible to develop a critical understanding of where the smart city phenomenon is coming from and, above all, of what urban future it is shaping. Building upon this premise, the chapter unpacks the imaginary of the *smart city*, taking the reader on a journey which, across time and space, explores the intellectual roots of the theory and practice of smart urbanism.

Starting from Francis Bacon's *New Atlantis*, the chapter examines the origins of the synergy between technological development and urban development, which today characterizes the smart city phenomenon. The same thematic thread is then followed through the history of the city, with an emphasis on Modernism, when technological innovation began to be deeply embedded in city-making. By connecting the philosophy of Nietzsche to the work of Expressionist architects and Futurist artists, the chapter emphasizes the stark individualism permeating the spirit of modernity. The chapter then discusses the cases of Los Angeles and Singapore, as the first incarnations of the *computer city* and the *intelligent city*, respectively. Finally, current smart city initiatives are critiqued through the lens of Max Horkheimer's critical theory. More specifically, the chapter draws upon *Eclipse of Reason* to argue that the recent dreams of smart urbanism are but a rerun of traditional capitalistic ambitions which, masked by the promise of technological progress, target the interests of elites and neglect the progress of society and the preservation of the natural environment.

Through the ages: the origins of smart urbanism

The origins of the faith in technology, and of the techno-urban development professed by the advocates of the smart city, have ancient roots. The first image of a city fully developed in synch with technological development goes back to 1627, with the publication of Francis Bacon's *New Atlantis*. From a temporal perspective, the philosophy of Bacon is located in a time when the seeds of Modernity, planted a century before in the humus of the Renaissance, were bearing fruit. In 1620, the English scientist and philosopher had published *The Novum Organum*, fully *Novum Organum Scientiarum* (New Instrument of Sciences), in which science, in the shape of technology, is, arguably for the first time, advanced and promoted as an instrument by means of which nature can be controlled and subjugated to human needs. This is the ideology that permeates Bacon's ideal city. In *New Atlantis*, the narrative explores the island of Bensalem, a city-region controlled by a community of scientists working for an institution called Solomon's House (Bruce, 1999). Bensalem is pictured as a large living laboratory where the members of the House pursue endless technological innovation.

Among their creations, we find then – futuristic forms of transport such as submarines and flying devices, as well as robots. The peculiarity and importance of Bacon's intellectual contribution is the depiction and conceptualization of the built environment, as the space in which technology is developed, implemented, and integrated, and as the place where its benefits are enjoyed. Although *New Atlantis* is commonly classified as utopian fiction, its impact has been extraordinary in terms of scale and innovation, particularly from an urban perspective. The connection between technological development and urban development that Bacon had made was going to deeply influence the shape, structure, and living of cities with socio-environmental and economic repercussions of global magnitude.

Since *New Atlantis*, the techno-urban development imagined by Bacon has been a recurring thread in the evolution of cities and, therefore, human societies. The phenomenon has not been geographically and temporally homogenous. From a historical point of view, it is possible to find two major spikes coinciding with two major technological revolutions. The first surge of techno-urban development is linked to the Second Industrial Revolution: a time of ground-breaking inventions characterized by an unprecedented diffusion and application, particularly in urban environments. Contrary to the First Industrial Revolution, which had been largely animated by amateurs, instead of professional scientists (Arkwright, for example, the inventor of the spinning frame, was a barber), the Second Industrial Revolution was driven by a synergy among science, industry, and economy, incarnated by partnerships between teams of scientists and private companies. What Jensen (1993: 834) describes as a "capital-intensive production" led to a "rapid growth in productivity" which, through the support of novel strategies and means of mass-distribution, led, in turn, to a capillary diffusion of new technologies. In this context, technological development went hand in hand with urban development.

The diffusion of steel, for instance, whose production became less expensive in 1856 with the invention of the Bessemer converter, coupled with the introduction of reinforced concrete in civil engineering in 1884, made possible the construction of monolithic architectural structures such as skyscrapers and long suspension bridges. Shortly after the advent of these innovations, in 1885 Benz patented the first automobile: an invention that was quickly popularized during the following decade by the Ford Motor Company, with substantial repercussions in terms of urban design and living. The circulation of cars led to the development of highways and arterial roads, which expanded the fabric of cities. In addition to the consumption of large amounts of natural habitat, these techno-urban changes led to a geographical and, above all, social revolution (Kenworthy and Laube, 1996). As Hall (1988: 295) observes, in "the city on the highway", car-owners became able to travel long distances to access services such as retail, education and health, as well as their workplace. Therefore, living in the proximity of the center of the city, where the majority of services and jobs were traditionally located, was not necessary anymore. This was the beginning of suburbanization and the genesis of a lifestyle highly dependent on cars and fossil fuels.

The spirit of this epoch, the *Zeitgeist*, can be observed in a plethora of different artistic and cultural expressions. In the field of architecture, for instance, there are strong echoes of the philosophy of Nietzsche (Watkin, 2005). Expressionist architecture, championed by artists of the built environment such as Hans Poelzig and Bruno Taut, promoted extreme and uncompromising personal expressions, visions, and aesthetics. In this strand of architecture, the urban fabric is shaped primarily by the imagination and emotions of the architect whose personality is superimposed on the natural environment. In a similar vein, in *Thus Spoke Zarathustra* (published in four volumes between 1883 and 1891), Nietzsche advances the concept of *Übermensch* (Superman), exalting a new type of human being that is not afraid of expressing personal values, and fights against conventionalism in order to achieve self-actualization (Nietzsche, 2005). This philosophy is directly reflected in the work of the Expressionist architect, as evidenced in the sketches of Taut, which were full of quotes from Nietzsche's books. It manifests a strong individualistic take on development, and the emergence of the will (and interest) of the single over that of society and the environment.

This ideological thread was further developed in Futurism: an avant-garde movement established in Italy in 1909 to celebrate the triumph of the individual through the cult of velocity, the machine, and the industrial city. The connection between technological development and urban development sealed by the Second Industrial Revolution is particularly vivid in the art of Umberto Boccioni. In *La città che sale* (The City Rises), painted between 1910 and 1911, the Italian futurist artist depicts progress as a double process consisting of urban growth and technological innovation. Conceptually and aesthetically, the painting is divided into three interconnected scenes (see Figure 9.1). First, through the brush of Boccioni, we see an industrial landscape where factories are incessantly producing

Figure 9.1 *La città che sale* (The City Rises) – Umberto Boccioni.

energy and capital. Second, the energetic and economic force of industry is employed to expand the city via new high-rise buildings and infrastructures. Third, in the center of the painting, the scene is dominated by a red vortex of horses. This is the raw power generated by the industrial revolution and its new machines: it is a fiery and violent power that humanity tries to control for its own sake. Through this act of extreme dynamism, Boccioni shows, at the same time, the appeal and promise of technology as well as its dangers. The human figures on the canvas are barely distinguishable, and their control over the power that they have unleashed is only ephemeral. The red horses inevitably throw the men off and trample them to death.

These new artistic expressions, philosophies, technologies, and, more generally, forms of development were not just being painted, written, and built. They were also feeding (and being fed by) significant cultural shifts, thereby becoming part of everyday life, particularly, but not only, in Western cities. Modernism, which was in essence the triumph, albeit temporary, of such a *Zeitgeist*, was celebrated in the city via large world fairs such as the 1889 Exposition Universelle in Paris and Chicago's Century of Progress International Exposition in 1933 (whose motto was "Science Finds, Industry Applies, Man Adapts"). Modernism was far from being an esoteric cult. It was a popular credo practiced through new mainstream forms of music and dance such as the Charleston, a high-tempo jazz whose BPM (beats per minute) was, on average, twice that of ballet. It is important to note that Modernism and its multiple incarnations were not homogeneously accepted and practiced. In music, for instance, the diffusion of jazz was accompanied by the canonization of what we now call *classical music*. However, in the period between the end of the nineteenth century and the beginning of the twentieth century, we can find, across different geographical spaces, an increasing convergence between technological development and urban development. Symptomatic of this trend was the invention of the incandescent light bulb in 1879, followed by the so-called War of Currents that introduced electricity in the built environment through new urban infrastructures, thereby radically changing the metabolism and appearance of cities. The new infrastructural landscape of the turn of the century allowed the domestic consumption of several novel technologies, such as the telephone, the radio, and the television: the early examples of information and communication technology (ICT) and the matrix of the technology that was going to evolve in the 1970s with the digital revolution to be then integrated in contemporary projects for smart cities.

Across the first half of the twentieth century the development of ICT was relatively fast, and in the 1970s, with the invention of the early computers, it led to a major technological tipping point whose effects on cities were going to be profound but not immediately visible. More specifically, as Castells argues, the new technological paradigm introduced by the convergence among micro-electronics, computing, telecommunications, and broadcasting was revolutionary inasmuch as, in addition to its large degree of pervasiveness, it created what still is a "network architecture which cannot be controlled from any center, and is made up of thousands of autonomous computer networks that have innumerable ways to link up, going around electronic barriers" (2011: 33). However, from an urban perspective, Castells' network society did not find an immediate incarnation. Despite the fact that, like the Second Industrial Revolution, the ICT revolution was strongly and quickly pushed forward by the capitalistic interests of competing private companies such as Intel, IBM, and Apple, the dynamics of urbanization remained largely unchanged until the end of the century (Hall, 1988). Particularly in terms of urban design, architecture, and civil engineering, this inertia can be explained by looking at the materiality of the technology in question. While the industrial products of the Second Industrial Revolution (engines, steel, reinforced concrete, and cars) were characterized by a marked weight and volume, the same cannot be said for the outcomes of the ICT industry. Although connected to a physical infrastructure, the Internet has an ethereal essence, and computers, since their genesis, have been evolving by following the logic of miniaturization in order to occupy less and less physical space. In essence, then, the city could not ignore the physicality of the Second Industrial Revolution, whereas, from a material point of view, it barely felt the touch of the *Zeitgeist* of the information age.

There are of course exceptions in this story, and these are the cities that have pioneered smart urbanism. From a historical perspective, the first one is arguably Los Angeles, which in the 1970s was a frontrunner in the use of what we now call *big data*. A policy report found by Vallianatos (2015) shows that in 1974 the urban development of Los Angeles was being shaped by computer data. Back then, an administrative division called the Community Analysis Bureau was employing state-of-the-art computer technologies to process and organize huge amounts of data on different themes such as housing, traffic, crime, and poverty. The overarching aim of this activity was to inform policy-making and urban planning, and it is because of such initiatives that Los Angeles is often considered to be the first example of the *computer city*.

Not long after the experimental urbanism of Los Angeles, in the late 1980s Singapore was being advertised by the local government as an "Intelligent Island" (Batty, 2012). In practice, the city was being rewired by means of then hyper-modern fiber-optic cables in order to create a data network. In less than a decade, the Singaporean urban-digital network was already producing and circulating large quantities of data by following the same dynamics and rationale through which contemporary smart city projects operate. The urban ICT infrastructure of Singapore was being used, for instance, to decentralize business activities by allowing some categories of workers to accomplish their tasks from home, to increase the communication between citizens and the government via online portals, and to implement systems of automatic payment through smart cards and scanners (Arun and Teng Yap, 2000).

By the time Singapore's plans of techno-urban renewal had been put into practice, the term *smart city* was beginning to be used in a growing number of urban agendas to signify the modernization of the infrastructure of the city through the integration of ICT (Vanolo, 2014). Emblematic are the examples of the Multifunction Polis project, a master-planned settlement meant to be built in 1994 near Adelaide (Australia), and Cyberjaya and Putrajaya (Malaysia) in 1997, developed under the banner of "smart city": a moniker that was subsequently officially filed by IBM in 2009 and then registered as the company's trademark in 2011 (see Söderström et al., 2014). Meanwhile, from an ideological perspective, ideals of smart urbanism were being cultivated by several thinkers on cities and technologies. Across the last decades of the twentieth century, building upon the vision of media theorist Marshall McLuhan who had predicted the dissolution of the city due to the pressure exerted by an increasingly interconnected world in 1964, scholars such as Pascal (1987) and Pawley (1995) advanced images of vanishing and spaceless cities. These theories, as Marvin (1997) points out, depicted the dematerialization of traditional urban societies, and the establishment of decentralized societies where ICT devices allow the fulfilment of socio-economic relationships in a remote manner (see also Graham, 2004). This is the same ideological thread that was later developed during the turn of the century by architect William J. Mitchell, who expressed an urbanism in which built environments and digital environments coexist harmoniously in what he defined in 1999 as *e-topias*: "electronically serviced, globally linked cities," equally built with concrete and chips (Mitchell, 1999: 8).

As this section has shown, the above ideas, images, and visions have formed an imaginary which, although abstract by nature, has had a substantial impact on urban development. Across three centuries, ideas of smartness have, in different forms and stages of gestation, led to the production of several urban technologies and spaces. While, from an intellectual perspective, this is (for now) the end of the story, we cannot forget the impact that, by shaping urbanization, these narratives have had on societies and natural environments. To understand and critique this crucial aspect of the smart city imaginary, the chapter now turns to the philosophy of Max Horkheimer, as a lens through which to examine actually existing smart city projects.

The eclipse of reason

Max Horkheimer (1895–1973), one of the fathers of critical theory, was a German philosopher and, together with influential scholars like Adorno and Marcuse, a member of the Frankfurt School of social theory. At the core of his studies is the analysis of the concept and practice of *reason*, which Horkheimer discusses in order to critique modern forms of development. According to him, there are two types, or sides, of reason. The first is what he calls *subjective reason*: the reason that underpins the actions of the individual. It is based on personal interests and gains, and driven by a primordial instinct of self-preservation. However, while obviously crucial for the survival of the individual, in its most extreme form, subjective reason tends not to include external interests. In this sense, concerns for society, intended as a group of different individuals, and the environment, as the space where different individuals, societies, and ecosystems are located, might be ignored by subjective reason, inasmuch as these are interests that go beyond the sphere of the individual.

In the philosophy of Horkheimer, the second type of reason is called *objective reason*. Here the scale and scope of reason are much broader. Horkheimer describes objective reason as a "force" that underpins and drives "relations among human beings and social classes, social institutions and nature" (2013: 2). The difference between subjective reason and objective reason is that while the former engages mostly with the preservation of the self, the latter embraces broader groups and environments and, in so doing, seeks to balance the interests of the individual with those of society and nature. Seen from this perspective, then, development goes well beyond personal development. For Horkheimer, a form of development based upon objective reason is inspired by higher concepts such as justice, happiness, and equality. In these terms, this typology of development resonates with the philosophy of thinkers like Plato (2007) and Aristole (1981), who conceptualized the flourishing of the individual in relation to broader socio-environmental systems such as the family, the city, and the state. In the twenty-first century, the concept of sustainability can then be seen as one of the latest incarnations of objective reason. A form of development driven by the idea of sustainability, a *sustainable development*, takes into account the needs of the individual only in connection with the needs of other individuals (including future generations) and with those of the biophysical environment, commonly referred to in public discourses as *nature*.

According to Horkheimer, there is an eternal tension between subjective reason and objective reason. He argues that, in modern societies, subjective reason, the reason of the individual, largely overpowers objective reason, the reason of the whole. For the German philosopher, what he calls the crisis or eclipse of reason is a twofold process. First, it consists in the triumph of subjective reason and the negation of objective reason: a phenomenon that ultimately dissolves the objective content of reason. As discussed above, objective reason engages with higher concepts such as justice, equality, happiness, and democracy. Therefore, emptied of these ideals, reason becomes a void that can be filled with personal interests and exploited for personal gains. Sustainability can be used again as an example to elucidate this condition. As a number of scholars point out, the concept of sustainability has often been emptied of any objective content and is thereby reduced to a black box open to many different understandings and interpretations (see Krueger and Gibbs, 2007). In this state, the ideal of sustainability has frequently been instrumentalized by politicians, policy-makers, and private companies to pursue individual ends instead of collective interests (Brown, 2016; Swyngedouw, 2007). Geographical studies, for instance, show how the term "sustainability" has been repeatedly used as a label to mask pro-growth economic agendas, which have fostered social inequality and environmental degradation (Cugurullo, 2013a, 2016c; Imrie and Lees, 2014; Raco, 2005; Whitehead, 2007).

The second part of the eclipse of reason sees "the complete transformation of the world into a world of means, rather than of ends" (Horkheimer, 2013: 66). This process is due to the triumph of subjective reason which, for Horkheimer, tends to pursue primarily the development of means: instruments designed and built to achieve personal goals. Under the aegis of subjective reason, the development of visions of social and environmental progress, and the necessary strategies to achieve them, is either neglected or, worse, completely ignored. Horkheimer notes that the first form of development is driven by technological innovation, while the second is advanced by philosophical inquiry. The former produces increasingly advanced technologies and devices, meant to fulfil more and bigger personal interests. The latter cultivates more sophisticated and holistic visions and planning strategies, embracing the needs of both society and nature. The problem that the German philosopher emphasizes is that technological development without any overarching vision of development is ultimately a blind form of development, which does not target the progress of society and the preservation of the environment. Instead, being animated by subjective reason, it aims for the achievement of individual goals.

In the philosophy of Horkheimer, who was interested in the critique of actually existing forms of development and social organizations (a trademark of the Frankfurt School), this phenomenon is easily observable in capitalist societies. Under capitalism, technological progress is often pursued not to improve the conditions of the whole society or the environment, but those of an elite. First, it benefits, directly, a restricted group of individuals who profit from the sale of the products and services that technological innovation produces. Second, technological progress benefits, indirectly, the conditions of who can afford those products and services. This is the same critique that geographers and urbanists have recently raised against contemporary strategies of sustainable development, driven by progress in technology. In this regard, several studies have shown how behind the development and implementation of clean technologies, such as photovoltaics, smart grids, and electric cars, is frequently not a vision of environmental preservation or social progress, but rather a series of business projects designed to fulfil the economic interests of private companies, and to sustain the power of political elites (Cugurullo, 2013b; Evans et al., 2016; Harvey, 1996; Kaika, 2017; Saiu, 2017). Here, so-called *smart cities* become urban engines, animating a constant production of new smart devices and services that are commercialized and sold, mainly to generate profit.

Conclusion: the eclipse of urban reason

If we look at the imaginary of the smart city through the lens of Horkheimer's critical theory, what emerges is a strong connection between the techno-urban development professed by the advocates of smart urbanism, and the triumph of what the German philosopher calls *subjective reason*. As this chapter has shown, the ideological origins of the smart city phenomenon are woven through with threads marked by a stark individualism. Behind the technological innovation pursued by the House of Solomon, the Second Industrial Revolution, the Expressionists, the Futurists, and, more generally, the Modernists, is often not the progress of society and the preservation of the environment, as broad systems, but rather the triumph of the individual.

Throughout the ages, the narrative of technological and urban development originally advanced in *New Atlantis* has largely ignored the needs of socio-environmental groups of interacting and interrelated elements. By allying itself with the logic of capitalism, it has favored the proliferation of pro-growth economic agendas, which have been responsible for social inequality and environmental degradation. From an ideological perspective, the lacuna is evident. The imaginary of smart urbanism lacks the higher concepts discussed and invoked by Horkheimer: ideals such as justice, happiness, and democracy, theorized by philosophers like Plato and Aristotle, and recently reprised and refined by scholars working in the field of sustainable development (Cugurullo, forthcoming; James, 2014). This is a lacuna that can be filled only by radically changing the way the smart city movement approaches development. The world shaped by smart city projects is, to paraphrase Horkheimer, a world of means. But what are the ends? What is the smart city phenomenon ultimately targeting besides the economic interests of capitalist elites? Put simply, the key questions are: smart for *what* and for *whom*? These are questions that technological progress cannot answer – only philosophical inquiry, as Horkheimer had predicted, can answer them. Without this, technological development will always be a blind form of development which, captured by subjective reason, will pursue individual goals, rather than the progress of society and the preservation of nature. If smart urbanism will keep following this direction, it will be not only the eclipse of reason, but also the eclipse of the city and our society.

References

Aina, Y.A. (2017) Achieving Smart Sustainable Cities with GeoICT Support: The Saudi Evolving Smart Cities. *Cities* 71: 49–58.

Angelidou, M. (2015) Smart Cities: A Conjuncture of Four Forces. *Cities* 47: 95–106.

Aristotle. (1981) *The Politics*. London: Penguin.

Arun, M. and Teng Yap, M. (2000) Singapore: The Development of an Intelligent Island and Social Dividends of Information Technology. *Urban Studies* 37(10): 1749–56.

Batty, M. (2012) Big Data, Smart Cities and City Planning. *Environment and Planning B: Planning and Design* 39: 191–93.

Brenner, N. (2014) *Implosions/Explosions*. Berlin: Jovis.

Brown, T. (2016) Sustainability as Empty Signifier: Its Rise, Fall, and Radical Potential. *Antipode* 48(1): 115–133. ISO 690.

Bruce, S. (ed.) (1999) *Three Early Modern Utopias. Utopia, New Atlantis and The House of Pines*. Oxford: Oxford University Press.

Caprotti, F. (2016) *Eco-Cities and the Transition to Low Carbon Economies*. Berlin: Springer.

Caprotti, F. and Cowley, R. (2016) *Smart Cities UK 2016* [Online]. Available at: http://www.smart-eco-cities.org/wp-content/uploads/2016/03/Kings-College-London-UK-Smart-Cities-Survey-2016-summary.pdf. Accessed 20/02/2017.

Caragliu, A., Del Bo, C., and Nijkamp, P. (2011) Smart Cities in Europe. *Journal of Urban Technology* 18(2): 65–82.

Castells, M. (2011) *The Rise of the Network Society: The Information Age: Economy, Society, and Culture* Vol. 1. Oxford: John Wiley & Sons.
Colding, J. and Barthel, S. (2017) An Urban Ecology Critique on the "Smart City" Model. *Journal of Cleaner Production* 164: 95–101.
Coletta, C., Heaphy, L., and Kitchin, R. (2017) From the Accidental to Articulated Smart City: The Creation and Work of 'Smart Dublin'. Available at: https://osf.io/preprints/socarxiv/93ga5.
Cowley, R., Joss, S., and Dayot, Y. (2017) The Smart City and its Publics: Insights From Across Six UK Cities. *Urban Research & Practice* 10(3): 1–25.
Cugurullo, F. (2013a) How to Build a Sandcastle: An Analysis of the Genesis and Development of Masdar City. *Journal of Urban Technology* 20(1): 23–37.
Cugurullo, F. (2013b) The Business of Utopia: Estidama and the Road to the Sustainable City. *Utopian Studies* 24(1): 66–88.
Cugurullo, F. (2016a) Urban Eco-Modernisation and the Policy Context of New Eco-City Projects: Where Masdar City Fails and Why. *Urban Studies* 53(11): 2417–33.
Cugurullo, F. (2016b). Frankenstein Cities. In: Evans, J., Karvonen, A. and Raven, R. (eds.), *The Experimental City*. London: Routledge.
Cugurullo, F. (2016c). Speed Kills: Fast Urbanism and Endangered Sustainability in the Masdar City Project. In: Datta, A. and Shaban, A. (eds.), *Mega-Urbanization in the Global South: Fast Cities and New Urban Utopias of the Postcolonial State*. London: Routledge.
Cugurullo, F. (2018). Exposing smart cities and eco-cities: Frankenstein urbanism and the sustainability challenges of the experimental city. Environment and Planning A: Economy and Space, 50(1), 73–92.
Cugurullo, F. (2017). The Story Does Not Remain the Same: Multi-Scalar Perspectives on Sustainable Urban Development in Asia and Hong Kong. In: Caprotti, F. and Yu, L. (eds.), *Sustainable Cities in Asia*. London: Routledge.
Cugurullo, F. (forthcoming) *The Urban Equation: Formulas for Sustainable Smart and Eco-Cities*. London: Routledge.
De Jong, M., Joss, S., Schraven, D., Zhan, C., and Weijnen, M. (2015) Sustainable–Smart–Resilient–Low Carbon–Eco–Knowledge Cities; Making Sense of a Multitude of Concepts Promoting Sustainable Urbanization. *Journal of Cleaner Production*. Available at: http://www.sciencedirect.com/science/article/pii/S0959652615013815.
Evans, J., Karvonen, A., and Raven, R. (eds.) (2016) *The Experimental City*. London: Routledge.
Garau, C., Masala, F., and Pinna, F. (2016b) Cagliari and Smart Urban Mobility: Analysis and Comparison. *Cities* 56: 35–46.
Garau, C., Zamperlin, P., and Balletto, G. (2016a) Reconsidering the Geddesian Concepts of Community and Space through the Paradigm of Smart Cities. *Sustainability* 8(10): 985.
Graham, S. (ed.) (2004) *The Cybercities Reader*. London: Routledge.
Grimm, N.B., Faeth, S.H., Golubiewski, N.E., Redman, C.L., Wu, J., Bai, X., and Briggs, J.M. (2008) Global Change and the Ecology of Cities. *Science* 319(5864): 756–60.
Grimmond, S. (2007) Urbanization and Global Environmental Change: Local Effects of Urban Warming. *The Geographical Journal* 173(1): 83–8.
Haarstad, H. (2016) Constructing the Sustainable City: Examining the Role of Sustainability in the 'Smart City' Discourse. *Journal of Environmental Policy & Planning* 18(3): 1–15.
Hall, P. (1988) *Cities of Tomorrow*. Oxford: Blackwell.
Harvey, D. (1996) *Justice, Nature and the Geography of Difference*. Oxford: Blackwell.
Hashem, I.A.T., Chang, V., Anuar, N.B., Adewole, K., Yaqoob, I., Gani, A., and Chiroma, H. (2016) The Role of Big Data in Smart City. *International Journal of Information Management* 36(5): 748–58.
Heaphy, L.J. (2018) Interfaces and Divisions in the Dublin Docklands 'Smart District'. Retrieved from osf.io/xbrgt.
Imrie, R. and Lees, L. (eds.) (2014) *Sustainable London?: The Future of a Global City*. London: Policy Press.

James, P. (2014) *Urban Sustainability in Theory and Practice: Circles of Sustainability*. London: Routledge.
Jensen, M.C. (1993) The Modern Industrial Revolution, Exit, and the Failure of Internal Control Systems. *The Journal of Finance* 48(3): 831–80.
Kaika, M. (2017) "Don't call me Resilient Again!" The New Urban Agenda as Immunology ... Or What Happens When Communities Refuse to be Vaccinated with 'Smart Cities' and Indicators. *Environment and Urbanization*. DOI: 10.1177/0956247816684763
Karvonen, A., Cugurullo, F. and Caprotti, F. (2018). *Inside Smart Cities: place, politics and urban innovation*. Routledge, London.
Kenworthy, J.R. and Laube, F.B. (1996) Automobile Dependence in Cities: An International Comparison of Urban Transport and Land Use Patterns with Implications for Sustainability. *Environmental Impact Assessment Review 16*(4): 279–308.
Kitchin, R. (2014) The Real-Time City? Big Data and Smart Urbanism. *GeoJournal* 79(1): 1–14.
Krueger, R. and Gibbs, D. (eds.) (2007) *The Sustainable Development Paradox: Urban Political Economy in the United States and Europe*. London: Guilford Press.
Kummitha, R.K.R. and Crutzen, N. (2017) How Do We Understand Smart Cities? An Evolutionary Perspective. *Cities* 67: 43–52.
Luque-Ayala, A. and Marvin, S. (2015) Developing a Critical Understanding of Smart Urbanism?. *Urban Studies*. DOI: 10.1177/0042098015577319
March, H. and Ribera-Fumaz, R. (2016) Smart Contradictions: The Politics of Making Barcelona a Self-Sufficient City. *European Urban and Regional Studies* 23(4): 816–30.
Marvin, S. (1997) Environmental Flows: Telecommunications and the Dematerialisation of Cities? *Futures* 29(1): 47–65.
McLean, A., Bulkeley, H., and Crang, M. (2015) Negotiating the Urban Smart Grid: Socio-Technical Experimentation in the City of Austin. *Urban Studies*. DOI: 10.1177/0042098015612984
Mitchell, W.J. (1999) *E-Topia: "Urban Life, Jim–but Not As We Know It"*. Cambridge: MIT Press.
Mosannenzadeh, F., Bisello, A., Vaccaro, R., D'Alonzo, V., Hunter, G.W., and Vettorato, D. (2017) Smart Energy City Development: A Story Told by Urban Planners. *Cities* 64: 54–65.
Neirotti, P., De Marco, A., Cagliano, A.C., Mangano, G., and Scorrano, F. (2014) Current Trends in Smart City Initiatives: Some Stylised Facts. *Cities* 38: 25–36.
Nietzsche, F. (2005) *Thus Spoke Zarathustra*. Oxford: Oxford University Press.
Pascal, A. (1987) The Vanishing City. *Urban Studies 24*(6): 597–603.
Pawley, M. (1995) *Architecture, Urbanism and the New Media*. New York: Mimeo.
Plato. (2007) *The Republic*. London: Penguin.
Raco, M. (2005) Sustainable Development, Rolled-out Neoliberalism and Sustainable Communities. *Antipode* 37(2): 324–47.
Saiu, V. (2017) The Three Pitfalls of Sustainable City: A Conceptual Framework for Evaluating the Theory-Practice Gap. *Sustainability* 9(12): 2311.
Söderström, O., Paasche, T., and Klauser, F. (2014) Smart Cities as Corporate Storytelling. *City* 18(3): 307–20.
Swyngedouw, E. (2007) Impossible "Sustainability" and the Post-Political Condition. In: Krueger, J. R. and Gibbs, D. (eds.), *The Sustainable Development Paradox*. New York: Guilford, 13–40.
Taylor Buck, N. and While, A. (2017) Competitive Urbanism and the Limits to Smart City Innovation: The UK Future Cities Initiative. *Urban Studies* 54(2): 501–19.
Trencher, G. and Karvonen, A. (2017) Stretching "Smart": Advancing Health and Well-Being Through the Smart City Agenda. *Local Environment*. DOI: 10.1080/13549839.2017.1360264
Vallianatos, M. (2015) Uncovering the Early History of "Big Data" and the "Smart City" in Los Angeles. *Boom California*. Available at: https://boomcalifornia.com/2015/06/16/uncovering-the-early-history-of-big-data-and-the-smartcity-in-la/.
Vanolo, A. (2014) Smartmentality: The Smart City as Disciplinary Strategy. *Urban Studies* 51(5): 883–98.

Viitanen, J. and Kingston, R. (2014) Smart Cities and Green Growth: Outsourcing Democratic and Environmental Resilience to the Global Technology Sector. *Environment and Planning A* 46(4): 803–19.

Watkin, D. (2005) *A History of Western Architecture*. London: Laurence King Publishing.

Whitehead, M. (2007) *Spaces of Sustainability: Geographical Perspectives on the Sustainable Society*. London: Routledge.

Wiig, A. (2016) The Empty Rhetoric of the Smart City: From Digital Inclusion to Economic Promotion in Philadelphia. *Urban Geography* 37(4): 535–53.

Wu, Y., Zhang, W., Shen, J., Mo, Z., and Peng, Y. (2017) Smart City with Chinese Characteristics Against the Background of Big Data: Idea, Action and Risk. *Journal of Cleaner Production*. DOI: https://doi.org/10.1016/j.jclepro.2017.01.047

Yin, C., Xiong, Z., Chen, H., Wang, J., Cooper, D., and David, B. (2015) A Literature Survey on Smart Cities. *Science China Information Sciences* 58(10): 1–18.

10

Construction performance

How the camera charts progress on site

Hugh Campbell

The city under construction

Cities are in a state of perpetual construction. It is impossible to imagine a city anywhere in which construction had completely ceased at any point, in which no part of the fabric is currently subject to extension, to repair, or to reconfiguration. Every city is always, to some extent, being built. Often – rather than being merely a background condition – construction processes come to the fore, becoming an index of the city's economic wellbeing or its future prospects: the number of cranes on the city skyline serves as an instant index of progress. Beyond this, the pulse of activity on construction sites – the comings and goings of workers and machines, the inexorable rise of buildings out of the ground – can become emblematic of the larger urban condition. When the construction site is recorded and visualized, translated into images, its emblematic status is redoubled. Images of the city under construction provide an "urban imaginary" that is characterized by relentless, mechanized activity, by change and growth and in which the built fabric, in and of itself, seems to speak of the collective will and shared identity.

This essay will examine a number of ways and means by which the urban construction site, and the building of the city more generally, are visually recorded. Such recordings proliferate, ranging from the consciously artistic to the promotional and commercial to the technical and scientific. Here the specific focus will be on lens-based media, and particularly on photography. The camera emerged in middle of the nineteenth century during a period of rapid urban growth. From the outset, as Peter Bacon Hales and others have argued, its techniques have been applied to capturing the city under construction. (Bacon Hales, 1984) On the one hand, the camera's capacity scrupulously and objectively to capture every aspect of a scene in front of it meant that it could be relied upon as instrument of scientific record: early instances of this include Édouard Baldus's much-discussed photographs of the additions to the Louvre and Evelyn Carey's records of the Forth Bridge under construction (Gray and Maggi, 2009). On the other hand, the camera could also serve to dramatize construction – highlighting the heroism of workers, as in Lewis Hine's well-known images of skyscraper construction in Manhattan (Hine, 1998); or finding the best angle from which to emphasize the impact of an emerging building on the skyline, as in Berenice Abbott's *Changing New York* (Abbott, 1999)

Often, both purposes might be served at once: Hines' images were as much social records as heroic portraits, while Abbott's project served as both extensive record and astute social commentary. The dual capacity of the camera to record and to interpret also became a dual mandate, which persists to the present day. The onset of digital technologies has served further to augment the techniques brought to bear on the visual recording of construction, and the multivalent character of the resulting images. At the same time, urbanization continues at an ever faster pace. In China and the Middle East, whole cities are being built in a matter of years, while in the developed West, urban centers are subject to rapid cycles of redevelopment. We are living, it seems, at a moment when urban construction is at a peak of activity and where the documentation of that construction has come to play a prominent role in the way in which urban societies are understood and presented. Ultimately, in fact, it might be said that such images feed directly back into the cycle of production that gives rise to them.

The city under construction, and the building site, are frequent themes in contemporary photography, both for the social realities they represent and the formal possibilities they offer. Thus, Stephane Couturier can conjure large, complex collage-like images through collapsing the layers of half-built fabric, temporary works and mechanical activity he finds on site. Similarly, Frank Thiel can distill the reconstruction of Berlin down to large-scale, semi-abstract close-ups of thickets of steel reinforcement bars, waiting for the concrete pour. Thiel's work also includes panoramic multi-paneled overviews of Berlin construction, revealing the extent of change and upheaval involved. The same mode has become increasingly ubiquitous in the depiction of urban expansion in China, for instance, in Edward Burtynsky's (2005) volume and in Sze Tsung Leong's *History Images* from the same year (Leong, 2005). These projects seek to offer an overview of a large-scale phenomenon, mobilizing the visual tropes – traditional districts demolished, scaffolded towers rising, construction activity extending to the horizon – which are taken to characterize this type of urbanization. There is an evident attempt to capture its two most salient aspects: rapid pace and sheer scale. While the latter is enabled through the current tendency to print at very large size and to generate highly detailed images, the former presents a challenge that has always dogged still photography – how to convincingly render the passage of time.

Time in the image

In response to the conundrum of how to build time into the still image the German photographer Michael Wesely has been constructing cameras capable of producing extremely long exposures, sometimes extending over years. Beginning in 1988, his initial interest was in reacting against the well-worn credo of the "decisive moment" to find a way of "collecting millions of moments in one picture" (Wesely, 2004). In an early series, he positioned his camera in front of a departing train and sustained the exposure for the length of the train's journey to its destination. Thus while the train, the nominal subject and focus of the image, was absent, the trace of its presence continued to be felt. Perhaps consciously, the subject matter echoed that of the Lumière Brothers' first cinematic sequence of a train arriving at a station. But here all activity, all movement, is subsumed within a single still image.

Dwelling on the camera's capacity as a recording instrument, Wesely wanted to extend this capacity over long periods of time. He quickly gravitated toward construction sites as offering particularly rich territory for his experiments. This was the period, following the fall of the Wall, during which Berlin was undergoing rapid social and physical transformation. New buildings were going up, with whole areas being redeveloped. In 1997,

Wesely mounted five cameras around the large-scale construction site at Potsdamer Platz (Figure 10.1). Three years later, he was invited to record the construction of the new wing of the Museum of Modern Art in New York, again using a series of cameras mounted around the nascent building. Over a period of more than three years, each camera recorded a single image (Figure 10.2). Each image serves as a record of everything that occurred within its view over the period, that is to say, both the activity and the built form that resulted (Wesely, 2004). Thus, in the final image, the finished building co-exists with the ghosts of every stage of its own construction. In fact, the temporary works, the shuttering and scaffolding, which were present on the site for far longer than the building in its finished form, register more prominently in the photographs. Inevitably, no single version of events, no single moment, can be disentangled from these images. The accumulation of time can be most strongly felt from the ingrained passage of the sun across the sky, shifting its trajectory over the course of each year, while elsewhere, specific events such as the relocation of a sign on an adjoining building register as shifts and overlaps.

These images draw attention to the duration of construction and to the material phenomena that accompany it, and as such they are vivid and compelling records of the process. However, by virtue of the sheer amount of visual material compressed into the image, their value as a visual record is mitigated. Despite being the product of entirely mechanical processes, they ultimately operate more comfortably and convincingly as evocative impressions. The period of construction and the processes involved are not,

Figure 10.1 Michael Wesely, *Potsdamer Platz, Berlin (27.3.1997 – 13.12.1998)*, 1997–1998. Courtesy of the artist and VG BildKunst Bonn.

Figure 10.2 Michael Wesely, *9 August 2001 – 2 May 2003 The Museum of Modern Art, New York,* 2001–2003. Courtesy of the artist and VG BildKunst Bonn.

ultimately, the central content of these images. What is offered is the accumulating, shifting resultant of that activity.

The time-lapse view

Concurrent with Wesely's work with long exposures, another mode of documenting construction activity over time has gained popular currency – the time-lapse sequence. The technique, in which still images taken at set intervals are played back at an accelerated pace in order to reveals patterns of activity or change, was first associated with natural processes, with F. Percy Smith's *Birth of a Flower*, from 1910, an early instance of how the previously unobservable details of growth and decay could be recorded and made apprehensible. From the outset, the value of time-lapse footage as an accurate record was always accompanied by its value as a spectacle – its inherent capacity to dramatize the processes it documented. The drama emerged as a direct consequence of the application of technique, by being sped up and thus condensed in time. This dual character is evident in Godfrey Reggio and Ron Fricke's *Koyaanisqatsi* from 1982, which was among the first films to bring the use of time-lapse to a larger audience and to use the technique to reveal urban patterns of activity, including construction. In the years since, aided by the advent and development of digital cameras, the genre has become ubiquitous.

The advantages of time-lapse as a method of recording progress on site have been increasingly recognized. There are now numerous proprietary systems, comprising cameras and software, that will allow construction companies to record and monitor activity from vantage points around their sites. There are also companies, such as Site Eye, which provide the service to contractors. The material gathered is useful in tracking how building is progressing relative to timetables and relative to the design documents. But it also provides a ready-made dramatization of the building process – lifting it out of the realm of technical information and into the realm of promotion and publicity. In time-lapse, buildings seem to emerge automatically, quickly, and without friction.

In the countless examples of the genre now available on YouTube and Vimeo, speed and ease are the characteristics which are emphasized, particularly in the many examples from China. One of the most well-known of these, a *Wall Street Journal* video piece entitled "19 days in China – Mini Sky City" (2015), in fact augments the time-lapse footage with shots of workers and factory assembly, cutting between the different pictorial modes. But more often, and particularly recently, the time-lapse footage alone, along with an invariably vaunting, swelling soundtrack, is enough to sustain the drama. A time-lapse of the building of the Shard, shown by Channel 4 News, exemplifies the tendency (2014). More recently again, the set-up of time-lapse shots and the subsequent edit into a single, seamless package has become a specialized industry, exemplified by Earthcam's recent documentation of the making of the World Trade Center (2015), and their film of Herzog and DeMeuron's residential tower at Leonard Street (2011), also in New York. This latter is particularly striking for the seamlessness of the footage. With a split screen, allowing an "establishing shot" in context to co-exist with a close-up (in fact from the same source camera), the tower's construction and subsequent fit-out and cladding is tracked by the rising camera which, once construction is complete, tracks back down the finished building, now lit up in the city darkness and seemingly occupied. Considerable planning and orchestration of shots will have been needed over the two-year period of construction in order to yield the footage necessary to produce this film. It is apparent too that CGI techniques associated with visualizations have been used in post-production. The worlds of proposed buildings, buildings coming-into-being and buildings in use begin to overlap and elide.

We see this tendency extending and becoming more explicit and more playful in a short film created by the Black Rabbit Agency Group to chart twenty years of development in Dubai (2016). In the course of the film, buildings spring into being beside us, as we drive along the highway: a figure in a visualization walks out onto a balcony to be confronted by what seems like another skyscraper rising just opposite. The language of the time-lapse is here divorced from its role as an accurate record of actual activity, becoming an emblem of endless, rapid development. But even when used to record real progress on real sites, the idiom presents a picture of ceaseless, seamless and frictionless progress. Other than cranes, little of the equipment needed in large-scale construction is visible. Neither are the army of construction workers who populate the site daily. What the time-lapse video offers is a view of construction as a naturally occurring, endlessly fertile phenomenon rather than a complex resource-intensive activity: buildings aren't built; they grow.

When deployed for the purposes of communication and promotion, this evocation of effortless construction might serve a particular end; however it is increasingly evident that the same logic applies even when the time-lapse construction sequence is used in a more instrumental manner, becoming integral to the management of the construction process. This can happen in two related ways, both resulting from the increasing ubiquity of building information modeling (BIM) as a means of producing a complete digital drawing model of the detailed design. Beyond a three-dimensional model of the entire project, these files can also incorporate layers of metadata pertaining not only to the detailed specifications for every facet of the project, but also to costings and to the construction sequence. In what has become known as BIM 5D, time and money join space and surface as variables to be monitored and controlled. Thus, in a corollary of the Channel 4 time-lapse, the construction of the Shard may also be viewed as a BIM model, advancing day by day, with cranes, other machines, and temporary works included in the picture (2012). This predictive model, mapping the building into existence over a twenty-four-month period, can then be used as a template against which to measure real progress on site, allowing delays and overruns to be registered and corrective action to be taken. Logically, the predictive time-lapse and its recorded counterpart, played at the correct speed, would align perfectly.

Construction entities

It is precisely this alignment between recorded data and building model that lies at the heart of many of the more recent applications and processes used to monitor progress on construction sites. The software *ConstructAide* is one of a number that draw together visual data captured from cameras on and around the site. This data is combined to produce a single composite visual model of the new building on site, a model that can be constantly updated in real-time, drawing on the data-feeds from which it is composed. This real-time model can then be overlaid with the BIM model, aligned to the same viewpoint. Thus, at any moment, progress on the project can be measured vis-à-vis the model. As Kevin Karsch of *ConstructAide* describes:

> After aligning the photo(s) and model(s), our system allows a user, such as a project manager or facility owner, to explore the construction site seamlessly in time, monitor the progress of construction, assess errors and deviations, and create photorealistic architectural visualizations. These interactions are facilitated by automatic reasoning performed by our system: static and dynamic occlusions are removed automatically, rendering information is collected, and semantic selection tools help guide user input.
>
> *(Karsch, 2014)*

Karsch particularly draws attention to the software's capacity to remove "static and dynamic occlusions" from the onscreen view of construction. These occlusions are, of course, the workers and machinery on site – the very things that are progressing the construction being monitored.

As this field of visual data management develops, the challenge of how to deal with these "occlusions" is recognized as being difficult, but critical. It is worth quoting at length from a recent paper on the current state of this field:

> Tracking itself is still an open issue in the computer-vision field. Occlusion (static or dynamic), articulated or non-rigid objects, and view/scale/illumination changes all pose big challenges to tracking. These situations can only be worse in cluttered construction sites. Besides these common issues, some unique characteristics of construction projects prevents the direct use of existing tracking methods. First, scenarios are highly dynamic in construction sites. Not to mention that the project progress is making changes to the site every day, in a shorter duration, the moving equipment and leaving and re-entering entities will not only challenge background

> modelling-based tracking, but cause troubles for identity preserving of long-term tracking, eventually cause troubles in the subsequent activity analysis. Secondly, the appearance similarity of construction equipment and workers leads to difficulties for appearance-based tracking, especially for multiple interacting workers tracking. Thirdly, the complex motion of construction entities often combines the modes of locomotion and joint motion. For example, an excavator drives to a spot and starts excavating. Or a worker lifts re-bars to somewhere and begins tying them. This will definitely challenge existing tracking methods which are designed for a single mode of motion. Fourthly, interaction, either between workers or workers and equipment, happens frequently in construction project, which is another big issue for multiple objects tracking. Hence, long-term stable tracking of construction entities in real time is still far from being realized. The duration of a stable tracking at construction sites is at most a few minutes long, which is totally incomparable to the natural length of hours of videos. This limits the practical use of the visual tracking in construction sites.
>
> *(Park et al., 2015)*

This passage clearly sets out the limitations of systems that are drawing on visual data in order to monitor and analyze progress. Anything anomalous, anything that does not move in a predictable, regular manner, anything that cannot be readily identified relative to other phenomena, will prove difficult to analyze. Again, people and plant, here referred to as "construction entities", are considered as inconvenient impediments to ready analysis, despite being central to the processes being analyzed. It might, of course, be argued that these tracking systems, like any tools, can only be expected to have limits, and that an exclusive focus on the built fabric itself, rather than on the forces that shape it, is not unusual (previous, more analog versions of monitoring progress would have had a broadly similar focus). Nevertheless, time-lapse would seem, on the face of it, to be most perfectly suited to the detailed depiction of human activity, of posture and movement. After all, Eadweard Muybridge's famous studies of the human figure in motion are among the most famous early experiments in the genre. In keeping with these origins, some recent applications have successfully extended tracking analysis to human activity, focusing in particular on health and safety by analyzing the posture and gait of workers in order to alert for potential risks (Yang et al., 2010). But even in these instances, the emphasis is on establishing norms in the first instance, and then noting deviations from those norms. Time-lapse tends to privilege pattern and repetition (as in Muybridge's quasi-scientific sequences); it struggles with individual agency and specificity. Certainly, it is apparent that a fully integrated view of on-site activity, equally alert to people, machinery and building, is a long way off. For the foreseeable future, the view of construction as a process to be managed looks very similar to the view of construction as a spectacle to be admired.

Absent labor

If time-lapse favors a view of construction without labor, it can be argued that, in this regard, it is simply following suit with other means by which a building design is visualized. This is most evidently true of the cinematic visualizations that are used to promote and help realize major construction projects, and which, understandably, seek to present the project as something already built. We do not dwell on the large-scale upheaval, expense and time involved in construction. Instead, we move straight from inception to completion. In

BIG's recent film promoting 2 World Trade Center, for instance, the leap from Bjarke Ingels sketching the project's conceptual premise on the skyline, to the building appearing there, glinting in the sunset, is instant. And in the visualization accompanying Herzog and DeMeuron's Leonard Street tower, mentioned earlier, the floors of the building are seen to fall miraculously from the sky, landing on and distorting to its final form the polished Anish Kapoor sculpture that props the building's corner.

However this insistent backgrounding of labor, of the sheer effort and mess involved in getting buildings built, is not the sole preserve of digital and lens-based images: it can be argued to extend through every stage of the design process. In architectural history, there has been a recent upsurge of interest in re-inscribing this labor back into the picture of building. Peggy Deamer, Adrian Forty and others have made important contributions to this expanded view of construction (Deamer, 2015; Forty, 2012). Conferences such as "Industries of Architecture" and "The Tools of the Architect" offer further evidence of interest in this avenue of research (2014, 2017). At the same time, previously neglected voices have gained new audiences – among them Sergio Ferro, a Brazilian architect who emerged out of the leftist architectural milieu but became disenchanted with the disjuncture between the modernist rhetoric of social improvement and the harsh realities of conditions of site.

Ferro's essay "Dessin/Chantier" (Design/Building Site) offers a Marxist critique of this paradox arguing that capital will always resist the industrialization of construction, because by reducing the variable cost, it limits the scope for profit (i.e. manufacturing, labor). The more automated construction becomes, the less profitable it will be. In this scenario, the role of the architect is to reconcile this tension. Design, he writes

> is the main instrument to bring about this integration [the integration of labour on site]. Conceived away from the building site, at a distance from the productive body, the architectural design lends consistency, and artificial cohesion, to the workmen's dispersed skills.
>
> *(Ferro, 2016: 103)*

For Ferro, design is "the image of a construction fiction that lies about its true formation process" (Ferro, 2016: 103).

> At the level of the building site, the design is the mould into which "idiotised" labour (in André Gorz's expression) is poured. The shape and purpose of this crystallisation are irrelevant, for the time being; all that matters is the depositing of broken-down tasks in the one same object being formed.
>
> *(Ferro, 2016: 103)*

It is beyond the scope of this chapter to consider the implications for the design process of Ferro's critique (and any argument that ideology blinds design to labor might be countered by insisting on the need for any design tools to have specificity and limits, whether drawings, sketches or models). But the same logic seems to apply more readily and perhaps more appropriately to images of construction. It is clear that on-site photography, a mode of vision designed to monitor and show progress, struggles to notice or record activity. The larger picture of progress, the regular, relentless step by step emerging out of the ground of the building – the designed object – is always in view, always measurable and evident. The actual activity of building, the operatives and the plant, remains largely hidden. In Ferro's terms, we always see the one same object, never the broken-down tasks.

Construction as epic theatre

Extending this same logic from the scale of the site to the scale of the city as a whole, what emerges is a visual language endlessly eloquent in its capacity to show building happening and stubbornly silent on the human effort involved. The urban imaginary thus produced is, inevitably, inadequate to the task of fully representing the constant churn of construction that underpins contemporary urbanization. How, then, might a richer imaginary be produced, capable of registering the city under construction both as a phenomenon and as a product of economic and social forces?

In responding to this, the ideas of the playwright Bertolt Brecht, formed in Berlin during a period of widespread construction, may be useful. In opposition to the naturalist, "tragic" theatre that predominated at the time, Brecht proposed an "epic theatre", one which would refuse to take human social situations for granted, but would portray them as something striking, something that calls for explanation. Brecht famously sought to reinstate distance between the audience and the action, and thus paradoxically to produce a closer relationship: if the action on stage was understood as something constructed, as artifice, then its meaning might be more readily attended to. This was the so-called alienation effect or *Verfremdungseffekt.* Brecht's theatrical strategies of deliberate artifice, distance, and disruption were informed by his experience in post-war Berlin, a city into which he arrived in 1921. After an initial stage of hyper-inflation, the capital of the Weimar Republic settled into a period of growth during which it became the most industrialized city in continental Europe. The opening of Tempelhof airport (1923) and the electrification of the S-Bahn (1924) spurred on new industries and construction activity.

Brecht was among many artists who responded to the changing physical and cultural milieu. He sought to rebalance the relationship between street and stage, noting the inherent theatricality of the former, and wanting reality to impinge more directly on the latter. Hence, Brecht used the street scene as the basis for his new form of theatre. He thought of his plays as reports on events – just as a witness to a car accident might describe what they saw, with any inadequacies, gaps, and inconsistencies in the performance serving, paradoxically, to verify rather than undermine its authenticity. The focus would be on action: what was done, not what was thought or felt. Epic theatre, as Brecht's friend Walter Benjamin wrote in his essay on the subject, is about "making gestures quotable". He continues:

> The art of the epic theatre consists in producing astonishment rather than empathy. To put it succinctly: instead of identifying with the characters, the audience should be educated to be astonished at the circumstances under which they function.
>
> *(Benjamin, 1999: 147)*

Brecht's project, informed by his Marxist outlook, was to make evident and palpable the conditions of production and consumption which underpinned the contemporary city. The urban scene provided the most striking evidence of those conditions, the "creative destruction" on which modern development was premised. By shifting the theatrical viewpoint, and changing the terms of presentation, he would make these conditions freshly apparent, and produce astonishment.

If we return now to the various modes of time-lapse imagery with which contemporary construction is being documented to various ends, we might consider them, in Brecht's terms, as a form of tragic theatre. This is to say that, even when such images seem to traffic in the spectacular, their effect is to assimilate and naturalize construction. Whether in

Wesely's long-exposure images or in Earthcam's slick time-lapse productions, we are presented with building activity that is continuous, effortless, and seemingly natural. Wesely reduces construction to a poetic blur; Earthcam organizes it into a seamless flow. In each instance, neither the messy facts of building production on the one hand, nor the economic structures that enable it on the other hand, are set forth.

Despite this, it can be argued that the mode of seeing enabled by time-lapse and long-exposure image technologies does lend itself perfectly to Brechtian strategies of disruption, distantiation and defamiliarization. The capacity to incorporate time into the photographic image expands its formal possibilities enormously, allowing interval and duration to be manipulated alongside lighting, framing and viewpoint. "Epic theatre", Benjamin wrote, "is in league with the course of time in an entirely different way from that of the tragic theatre" (1999: 148). In line with this, there exists the potential for time to be incorporated more adventurously within contemporary image-making in a manner that disrupts and thus draws attention to the ceaseless pulse of development and growth rather than simply reaffirming it. In every city, the number and scale of construction sites is taken as an informal, readily visible, index of economic growth. They place on view the logic of neoliberalist production. Given their ubiquity and visibility, the relative lack of projects emerging from the art and design disciplines which engage with these sites critically in order to generate new kinds of urban imaginary, is surprising. Among the more compelling examples is a group exhibition, *Facing the Music*, curated by Allan Sekula around the construction of the Walt Disney Concert Hall in downtown Los Angeles in 2005.

Through his art and writing, Sekula had always sought to extend the camera's capacity for critical image-making. Sequenced images and, latterly, film were an important part of his practice, allowing the structures and power relations underpinning social and economic situations to be revealed. *Untitled Slide Sequence* from 1972 is an important early example of the former and the 2012 film *The Forgotten Space* is a recent example of the latter. The photographic image, for Sekula, could and should brush history "against the grain", the phrase from Benjamin becoming the title of his first major publication (2016). Brecht also proved a repeated point of reference for Sekula, who referred to his projects as "disassembled plays" and made overt use of strategies of distantiation and defamiliarization.

Facing the Music began as a research project in 1999, with Sekula enlisting a number of colleagues – James Baker, Anthony Hernandez, Karin Appolonia Muller, and Billy Woodberry – to join him in exploring the process and the contexts of the construction of Frank Gehry's project on First Street and Grand Avenue. The project culminated in an exhibition in the REDCAT Gallery in 2005, with a publication eventually following a decade later. While each contribution tracked the construction project in some way, it was the filmmaker Billy Woodberry who attended most closely to ongoing activity on site in his film *The architect, the ants and the bees* (2004). Woodberry asserted that

> since the finished architectural structure would become most notable for its external form and surfaces, I wanted the video to reveal many of the intricate and complex features beneath the surface, features that cannot be discerned from looking at the drawings, models or the complete building itself.
>
> *(Dimemberg, 2015: 144)*

His political approach is explicit. He accentuates the overt reference to Marx in his film's title by opening with a long quote on labor, along with a poem by Brecht, "Of Sprinkling the Garden". By cutting between the different settings through which the building is advanced – the site itself, but also the ancillary spaces of rest and recreation and the workspaces of the construction company and the developers, Woodberry augments and enriches the construction narrative. As explained by Sekula,

> Woodberry gives us a strong feeling for the temporal duration, rhythms and dangers of work, relentlessly following the structure as it rises up out of the slab, unfolds its skeleton, and takes on its brilliant obscuring skin, not by itself, but with a little help from its hardworking friends.
>
> *(Dimendburg, 2015: 25)*

Here, in other words, is a time-lapse film which shows us that building is not a natural occurrence. It actively resists the tendency of the time-lapse technique to enshrine progress as something perpetual and effortless. Sekula considered Woodberry's film "the materialized memory of the building's coming-into-being". Alongside the other pieces in his show, it offers an urban imaginary characterized by upheaval, by strange juxtapositions and unresolved paradoxes: the facts, rather than the fiction, of construction.

References

2 World Trade Center (2015) Squint Opera/BIG.

4D BIM Skyscraper. (2012) Available at: https://www.youtube.com/watch?v=esW0Zb880pg.

56 Leonard Construction Time-Lapse. (2017). Earthcam. Available at: https://www.youtube.com/watch?v=YNq_IMbc8e0.

56 Leonard Herzog and DeMeuron (2011) Mad Multi-Media. Available at: https://www.youtube.com/watch?v=ptLxk0aHpyc,https://www.youtube.com/watch?v=sJNvqhK53Kg.

Abbott, B. (1999) *Changing New York*. New York: New Press.

Bacon Hales, P. (1984) *Silver Cities: The Photography of American Urbanization, 1839–1915*. Philadelphia: Temple University Press.

Benjamin, W. (1999) What Is the Epic Theatre? In: Benjamin, W. (ed.), *Illuminations*. London: Pimlico, 144–52.

Brecht, B. (1978) The Modern Theatre Is the Epic Theatre. In: Willett, J. (ed.), *Brecht on Theatre*. London: Bloomsbury.

Burtynsky, E. (2005) *China*. Gottingen: Steidl.

Construction of The Shard (London Bridge Tower) Time-lapse (2014) Available at: https://www.youtube.com/watch?v=zJbtZlPjz3w.

Deamer, P. (ed.) (2015) *The Architect as Worker: Immaterial Labor, the Creative Class, and the Politics of Design*. London: Bloomsbury.

Dimendburg, E. (ed.) (2015) *Facing the Music: Documenting Walt Disney Concert Hall and the Redevelopment of Downtown Los Angeles*. Valencia, CA: East of Borneo.

Dubai 20 Years Timelapse. (2016) Black Rabbit Agency Group. Available at: https://www.youtube.com/watch?v=5qNU2eqmJEo.

Ferro, S. (2016) Dessin/Chantier, an Introduction. In: Lloyd Thomas, K., Amhoff, T., and Beech, N. (eds.), *Industries of Architecture*. London: Routledge, 94–106.

Forty, A. (2012) *Concrete and Culture: A Material History*. London: Reaktion.

Golparvar-Fard, M., Peña-Mora, F., and Savarese, S. (2012) Automated Progress Monitoring Using Unordered Daily Construction Photographs and IFC-based Building Information Models. *Journal of Computing in Civil Engineering* 29(1): 147–65.

Gray, M. and Maggi, A. (2009) *Evelyn George Carey: Forth Bridge*. Venice: Federico Motta Editores.

Hine, L.W. (1998) *The Empire State Building*. New York: Prestel.

Industries of Architecture (2014) AHRA Annual Conference, Newcastle University.

Karsch, K. (2014) *ConstructAide: Analyzing and Visualizing Construction Sites through Photographs and Building Models*. Available at: https://vimeo.com/106556251.

Leong, S. (2005) *History Images*. Gottingen: Steidl.

Official 11 Year Time-Lapse Movie of One World Trade Center (2015) Earthcam. Available at: https://www.youtube.com/watch?v=NbA89YbWoL8.

Sekula, A. (2016) *Photography against the Grain: Essays and Photo Works, 1973–1983*. London: Mack.

Tools of the Architect (2017) EAHN Thematic Conference, TU Delft.

Watch a 57-Story Building Go Up in 19 Days (2015) Wall St Journal. Available at: https://www.youtube.com/watch?v=N6f_sayw0mM.

Wesely, M. (2004) *Open Shutter*. New York: MOMA.

Woodberry, B. (2004) The Architect, the Ants and the Bees. Accessible at: https://vimeo.com/255325697 (password: green).

Yang, J., Arif, O., Vela, P.A., Teizer, J., and Zhongke, S. (2010) Tracking Multiple Workers on Construction Sites Using Video Cameras. *Advanced Engineering Informatics* 24: 428–34.

Yang, J., Park, M.W., Vela, P., and Golparvar-Fard, M. (2015) Construction Performance Monitoring via Still Images, Time-Lapse Photos, and Video Streams: Now, Tomorrow, and the Future. *Advanced Engineering Informatics* 29(2): 211–24.

11

Authoritarianism and the transparent smart city

Federico Caprotti

Introduction: smart city imaginaries

This chapter takes as its starting point a novel written in Russia in the 1920s, and a video produced by a major advanced glass and materials corporation in the 2010s. The novel, *We* by Russian author Yevgeny Zamyatin (1884–1937), was published in 1924 and explored the city-state of the future by highlighting the authoritarian and totalitarian character of an urban society based on transparency, a loss of individuality, and subservience to the state/city-as-machine. The video, *A Day Made of Glass*, was produced in 2011 by Corning Incorporated, and is a short exploration of the workings of the future city as experienced through the use of smart glass as an interface through which to access services and data, influence behavior, and make the city and urban and economic life more efficient and frictionless. While *We* is clearly a study of the dystopian future city, *A Day Made of Glass* presents a utopian, shining imaginary of the digitally enhanced urban future. Glass, and transparency, are common themes that run throughout *We* and *A Day Made of Glass*, and this chapter explores how these notions can be put to use to analyze and link imaginaries of the dystopian and utopian future city.

Both the novel and the video focus on cities that are rendered efficient and operationally slick through technology. This chapter therefore contextualizes the discussion of future urban imaginaries in current literature that critically engages with notions of the smart city (Townsend, 2013). Smart urbanism is understood here to be the urban development paradigm that has recently gained currency in its promotion of digitally enhanced and governed urban life, accompanied by an urban experience where citizens engage with the city, at times playfully, through the intermediary agency of digital interfaces such as smartphones, tablets, apps, and the like. Today's emergent smart cities are both emergent and powerfully present in policy discourse (Haarstad, 2016; Wiig, 2015), and corporate interests are strongly intertwined with urban development priorities (Hollands, 2015; Söderström et al., 2014) in attempts to bring about the future city.

The imaginaries of the future dystopian/utopian city explored in this chapter are linked in multiple ways, not least through the use of glass as a prominent theme and material in both the novel and the video. A further theme is that of a slide toward authoritarianism:

explicit in the case of *We* or, as in the *A Day Made of Glass* buried under spectacular but flat technological glitziness. This builds on work that has explored the link between imaginaries of utopian urban futures and the emergence of the existing future city. Picon (2015), for example, has brought this to light with regards to smart city visions. In his work, he has explored how the link between imaginaries and the actually existing smart city were expressed through science fiction, futuristic novels, and movies, but also through the imaginaries of the Apollo space program and other high-tech modern endeavors. These visions were picked up and reproduced in multiple societal and industrial settings, from nuclear power stations and industrial plant control rooms, and then to smart city "dashboards" and smart urban "operations control centers" (Mattern, 2015). Control centers are, essentially, control rooms featuring banks of video and computer screens and terminals (the "dashboards") in which real-time video and data feeds enable city operatives to "manage" various aspects of the city (from traffic, to emergency response, to security) from a single, electronic vantage point. Prominent and highly publicized examples of such spaces are the smart city control centers in Rio de Janeiro, Brasil, or Glasgow, Scotland (Lemos, 2017). While these imaginaries (and their materializations) are at one level representations, there is a risk in viewing them as simply symbols, or as signifiers of technological imaginations. Rather, the chapter argues that urban imaginaries around the city-as-machine function as highly visible expressions of a deeper will to power, as Kaika (2010) has argued in her analysis of urban icons. It is here that a potential critical space is made available for engaging with the glassy, transparent dystopia/utopia in *We* and *A Day Made of Glass*: this chapter uses the critique of the technological society developed by French philosopher Jacques Ellul (1912–1994) to propose a critical engagement with imaginaries *beyond* a focus on technology.

Transparent dystopia: Russia, 1924

Zamyatin wrote *We*, a futuristic novel, around 1919–1920. It was banned in 1921: the first novel to be banned by the Soviet state, but was published in English in 1924 by E. P. Dutton, in New York. The USSR would not allow the publication of *We* for a full 67 years, until 1988. *We* is set a millennium into the future, in the walled urban country called OneState. The novel revolves around D-503, the protagonist, an engineer working on space flight. All the characters in the novel are known simply by letters and numbers (the novel refers to OneState citizens as "Numbers"), and OneState is depicted as a dystopian city made of glass, where all human activities are not only transparent and verifiable at all times by surveillance, but where the whole of life is governed by rational, Taylorist principles: "the line of OneState is a straight line. The great, divine, precise, wise straight line – the wisest of all lines . . ." (Zamyatin, 1993: 4).

The vision of the city that is presented in *We* is therefore of a beautiful, well-oiled machine, transparent due to its being made of glass, and populated by compliant, atomized individuals. OneState is a diffuse and overwhelming presence, regulating all aspects of life right down to the intimate and personal, including friendships and romantic relationships. On a more sinister level, the mechanism of repression present in the city is expressed through the Guardians, an all-pervasive secret police organization; any dissent is quashed through lobotomy, and in some cases highly public executions. The novel traces D-503's unwitting awakening as an individual, and his entry into a subversive group of OneState citizens who work, under grave threats, to overthrow the totalitarian regime. At the same time, the all-pervading state portrayed in *We* is limited: its limits are physical and natural, in the sense that OneState exists within the city of the same name. In Zamyatin's novel, OneState is a highly ordered society *within* what the novel terms the Green Wall, a physical boundary that separates the aseptic environment of the city from outlying nature, depicted as wild and chaotic, and felt as repulsive by OneState's inhabitants, trained to glory in efficiency, transparency and predictability. As D-503 exclaims, when considering nature, "I love . . . only such a sky as this one today: sterile and immaculate" (Zamyatin, 1993: 5).

There are several reasons why Zamyatin's novel is of key interest in the 2010s. Firstly, *We* is not simply a dystopian novel painting a bleak picture of the techno-urban future, but the product of a writer whose own biography intersected with the increasingly authoritarian slide of Russian twentieth-century history from Tsarist oppression into the dark years of Leninist post-1918 state-sponsored terror and paranoia. As a Bolshevik, he was arrested in 1905 and sent into exile in Siberia, from which he escaped only to be exiled again in 1911 (he benefited from an amnesty in 1913). Indeed, Zamyatin wrote of this influence on his writing that "If I have any significance in Russian literature, I owe this all to the St. Petersburg Secret Police" (Zamyatin, n.d., in Delgado, 2014). After he managed to smuggle *We*'s manuscript to the United States, it was published in 1924, and several years of official repression followed, which included denunciations of Zamyatin as a writer by the Union of Soviet Writers as well as official blacklisting, which meant that he was not authorized to publish anything in his homeland. In 1931, Zamyatin made a direct request to Stalin to be granted permission to leave Russia. After the intercession of Gorky, this permission was granted and Zamyatin moved to Paris with his wife, where he died in 1937. Thus, *We* was forged (and flourished, given its international distribution) in the crucible of the author's painful interactions with the communist state. What is striking, of course, is that a novel set 1000 years in the future would generate such a strong backlash from Lenin and, later, Stalin. Not only that, but the fact that the novel was banned in the USSR for nearly seven decades is a key indicator not only of the humorless nature of the Soviet state, but of Zamyatin's critique, as expressed through his writing. It is testament to the critical power that urban dystopian imaginaries expressed in novels and other cultural media can have when brought to bear on current political reality.

Secondly, *We* is relevant today because it is not simply the product of a professional and accomplished author, but the written expression of a novelist and satirist who had deep technical knowledge. Zamyatin had direct, first-hand experience of the imbrication between the state and technology: he trained as a naval architect and spent 18 months in Newcastle, England, where he oversaw the construction of Russian icebreakers, one which would eventually be called the *Lenin* and that saw service until 1968. This has a direct influence on *We*, as the materials and processes of industrial, large-scale engineering and construction connected, later, with the socio-ideological engineering of the Soviet state. Naval construction in Newcastle exposed Zamyatin to "monstrous cranes of translucent glass trundling slowly along glass rails" (Zamyatin, in Myers, 1993: 425). As such, Zamyatin "was thus in everyday contact with an enclosed glassed-in environment, within which the rational, regulated activity of a great shipyard at war proceeded relentlessly, inducing both fascination and repulsion" (Myers, 1993: 425). Indeed, Zamyatin came face-to-face with the principles of industrial production enshrined in Taylorism, and was later highly critical of Lenin's regime's adoption of Taylorist principles (Delgado, 2014). Zamyatin could have been commenting on today's smart city projects, plans and blueprints when he wrote:

> I saw . . . as though right then for the first time in my life, I saw everything: the unalterably straight streets, the sparkling glass of the sidewalks, the divine parallelepipeds of the transparent dwellings, the squared harmony of our gray-blue ranks.
>
> *(Zamyatin, 1993: 7)*

At the same time, the precarious Babel-like nature of a society organized along industrial lines is recognized by D-503, who observes that:

> I had conquered God and the old life, I myself had created this, and I'm like a tower, I'm afraid to move my elbow for fear of shattering the walls, the cupolas, the machines...
>
> *(Zamyatin, 1993: 7)*

Third, Zamyatin's work in *We* can be seen as a careful and indeed prescient unpicking of techno-authoritarian themes that were to find full development in later totalitarian and authoritarian regimes, from fascist Italy to Nazi Germany to Maoist China. The themes that Zamyatin explored and teased apart included the role of the individual in a scientifically organized state where the "person" is only such in their role as participants in the "machine" of governance, control and production – a key concern in the contemporary context of interest in smart cities, citizens as producers of data, and the ever more efficient organization of social life through code (Barns, 2012). Furthermore, there remains at the heart of *We* a key quandary as to the *point* of organizing the city, and society, in ever more efficient and rational ways. As Slusser (Slusser, 2011: 307) has argued, *We* places the emphasis *beyond* rationality and Taylorist efficiency, and asks the question of how society can be structured so as to lose its techno-rational straightjacket and "free mind and body ... for open-ended exploration of worlds beyond logic". George Orwell, in his 1946 review of Zamyatin's novel, admitted that *We* "is in effect a study of the Machine, the genie that man has thoughtlessly let out of its bottle and cannot put back again" (Orwell, 1946).

Transparent utopia in the 2010s: the vitrification of modernity

In 2011, advanced glass and materials corporation Corning Incorporated, based in Corning, New York state, released a video titled *A Day Made of Glass* (Corning Incorporated, 2011). At five minutes and 32 seconds' duration, it is not a lengthy cultural product. Nonetheless, *A Day Made of Glass* explores a life lived in a city of the very near future, where effortless, efficient living (from scheduling, to smart automobility and transport, to the use of smart glass in business) is made possible through the use of advanced glass. The start of the video follows the lives of a family as they wake up and use a range of devices, from smart glass cooking and entertainment surfaces to smart cars, to carry out a range of activities from cooking to making video calls. The video portrays ordinary domestic surfaces – such as those of the humble bathroom mirror – as smart gateways into digital diaries, emails, and information. It also portrays urban life as being *clean* through the use of smart glass: glass cooking surfaces are depicted as free from dirt and grime, information display gantries on freeways are crystalline and transparent; smart bus stops are clean, and the display of digital information on their walls is crisp and beautifully colored.

The citizens of the city displayed in *A Day Made of Glass* are uniformly happy and smiling as they move through domestic and urban spaces seamlessly and efficiently, avoiding traffic due to the smartness of the glass-based digital traffic information systems. In the video, colors are sharp, bold, and beautiful, and there is no strife, dirt, or disharmony. Some scholars have interpreted this clean and technicolor view of the city as an attempt to move past the perceived "coldness" of technology, and to a situation where touch-screen technologies enable a more tactile and intimate relationship with both technology and the city. Thus, while "glass surfaces usually reference sterility and cleanliness, in environments such as science labs that are often characterized by logic rather than emotion" (Plotnick, 2017: 15), the affective and physical qualities of touch enabled through smart glass are seen as allowing for a more techno-social connection to the digital city:

> Corning's video of this shiny, slick, flattened world aims to convince viewers that touchscreens can seamlessly integrate into one's domestic environs. It seems to respond to concerns about the coldness and authenticity of digital touch interactions by portraying glass as interactive and intimate.
>
> *(Plotnick, 2017: 15)*

Corning's production of a video extolling the benefits of a city made of glass is symbolic of the corporation's involvement with the increasing importance of glass (and its qualities, like transparency) in late modernity. For example, in 1879 the corporation developed a bulb-shaped glass container for Edison's incandescent lamp, and invented PYREX® durable glass in 1915. In 1934–1936, Corning produced the glass used in the 5.1m telescope at Palomar Observatory near San Diego, a painstakingly detailed project that included a year of controlled cooling of the glass. Corning was later involved in key glass-based and optical innovations, from toughened glass, to automotive glass, to fiber optics, to producing the heat-resistant glass used in every manned American spacecraft, to the Gorilla® Glass used in smartphones and other digital devices (Corning, n.d.). Zamyatin's *We* was powerful in large part due to the author's deep involvement with both engineering and politics. Likewise, Corning's *A Day Made of Glass* is powerful because it is a cultural product issuing from a corporation that has been central to the increasing vitrification of late modern society. Recent research points to the fact that the world's advanced economies are those that are engaged more than any other in research on glass (Mauro, 2017). Corning has promoted its view of the glass city as a vision of what it calls the "Glass Age" (Morse and Evenson, 2016), describing it in these terms:

> Journey into a world made of glass. Where information moves at the speed of light. Stability coexists with versatility. Everyday surfaces provide extraordinary benefits. Industries discover powerful solutions to impossible problems. This is the Glass Age, where materials science is constantly pushing boundaries and creating new possibilities for glass-enabled technologies and design.
>
> *(Corning, 2016)*

It would be erroneous to suggest that urban imaginaries around the future city stopped with *We*, and resumed with *A Day Made of Glass*. Indeed, the twentieth century was replete with visual and textual explorations of the future city, and many of these can be retrospectively applied to current thinking and materiality around the high-tech smart city. It would be difficult to name a representative sample here, because the production of such imaginaries (in movies, novels, and videogames) over the past century is notable in terms of volume, diversity and specific focus. Nonetheless, a key common theme running through both *We* and *A Day Made of Glass* is the notion of *transparency* the idea (in *We*) that the dystopian future city is dystopian not because of its darkness (thus, the opposite of dystopian imaginaries present in movies such as *Blade Runner*), but because of its see-through, light quality: transparency, in *We*, means lack of individuality, lack of mental space, and ultimately lack of individual and collective freedom. In *A Day Made of Glass*, these political themes are not expressed as such, but what is key is that the utopian near-future city, and its domestic spaces, are represented as accessible and navigable through high-tech and high-performance glass. Thus, the utopian – but apolitical – city takes on a *glassy*, or vitrified, quality.

Technique: Ellul and the missing link between authoritarian and apolitical imaginaries

A key critique of contemporary visions of the future city is that they are often represented in slick PowerPoint presentations, promotional videos and glossy brochures as flat, anodyne, *apolitical* places. They seem to exist in a wider context of consensus around the "need" for sustainability (however it may be defined), speed, efficiency, resilience, digitalization (White, 2016). At the same time, the cities of the future – from smart cities such as Songdo, South Korea, to eco-city mega-projects such as the Sino-Singapore Tianjin Eco-City, China, to the UK's new "healthy towns" such as Cranbook, Devon (Caprotti, 2015; Iacobucci, 2016; Kim, 2010) – seem devoid of elements of participation, democracy, and debate over their desirability and urban future. Scholars have highlighted how future city typologies such as the smart city are often characterized by their entrepreneurial and corporate character (Hollands, 2015), and less by a focus on citizens and the urban polity. In this sense, stylish visions of the future city, as promoted by city governments and consultancies alike, take on the character not of politics but of a post-political urban state (Swyngedouw, 2009, 2016) based on elite consensus and of an ideological "papering over" of cracks and inequalities in the city.

It is at this juncture – in a context of proliferation of apolitical imaginations of future urbanism – that a critical take on urban imaginaries can take root. *We* is clearly political by dint of its satirical nature, while *A Day Made of Glass* invites questions as to where the politics is in the slick, smart urban future portrayed in the video. It is here that a key link between *We* and *A Day Made of Glass* can be made more explicit, and used to push forward the debate on how to politicize imaginaries of the future city. The novel and the video in question help in this regard, because they both rely on the key role not simply of technology, but of the imbrication of technology *in* society, as a way of exploring the straightjacketing of urban polity in the modern era. This is particularly key in the contemporary context, where technology and technological "ways of doing" are placed at the heart of aspirations of how to bring about a future, desired urban state. From wind turbines to solar panels, from smart ticketing to mobile payments, from the urban sharing economy to environmental regulation, the future city is envisioned as a response to today's problems. In turn, these problems are almost always rendered as *technical* problems, requiring *technical* solutions.

It is useful, here, to bring to bear on this debate around the technological nature of urban imaginaries of the apolitical future city, the critique of modern technological society developed by French philosopher Jacques Ellul. In his *The Technological Society* (Ellul, 1973), originally published in 1964, Ellul distinguished between the machine and technique. The former is simply an assemblage of knowledge and materials into a mechanical construct. Technique, on the other hand, is the *mediator* between the natural world and human experience, and indeed replaces nature as the milieu within which humanity exists. Indeed, Ellul argued that from the second half of the twentieth century onwards society *must* live in the fabric of technique, and that the clearest example of technique is the city. Technique, then, has in the modern era become decoupled from the former, and has permeated all areas of social life:

> Technique … made an inventory of what it could use, of everything that could be brought into line with the machine … Technique integrates the machine into society. It

> constructs the kind of world the machine needs and introduces order where the incoherent banging of machinery heaped up ruins. It clarifies, arranges, and rationalizes; it does in the domain of the abstract what the machine did in the domain of labor.
>
> *(Ellul, 1973: 5)*

In the urban realm, Ellul argues, the hegemony of technique requires increasing social plasticity, characterized by the increasing atomization of society, the isolation of the individual, and the narrowing of the political sphere of possibilities into an apparent pinprick. Thus, "For the individual in an atomized society, only the state was left: the state was the highest authority and it became omnipotent as well. The society produced was perfectly malleable and remarkably flexible" (Ellul, 1973: 51–2). These are themes that are explicitly present in *We*, and implicit in *A World Made of Glass*. In the former, the crass Taylorism of the hegemonic state is excavated and exposed, and indeed it is displayed in the novel through the depiction of the future dystopian city as a city made of glass, through which its workings (from machines, to human cogs in the machine) could be made visible. In the latter, a theme running through the video is that, through smart technologies, individuals can be liberated from the friction that characterizes urban life (from transport delays to other forms of inefficiency), and can enter ever more fully into union with the technique at the heart of the contemporary smart city.

The subsuming of individuality (or, at any rate, the bringing into line of one's individuality with the constraining realm of technique) present in Zamyatin's novel and in Corning's video highlights, then, one of the potential dystopian future pathways that comes into focus when considering contemporary imaginations of the smart city. Specifically, there is a more pervasive and uncanny authoritarianism than that portrayed in *We*. When seen through the lens of Ellul's critique of technique, future city imaginaries carry with them not simply progressive possibilities but the seeds of top-down control expressed through all areas of social life, from technology to education, from the biomedical to real-time sensing and monitoring, from transport to security applications. It is here that planning can be seen as expanding out from its specific sub-categories (from urban to economic planning) as the state becomes the key player coordinating the development and roll-out of technique across society:

> It is only in the framework of planning that such operations are arranged and find their exact place. The state appears less as the brain which orders them organically and more as the relational apparatus which enables the separate techniques to confront one another and to co-ordinate movements.
>
> *(Ellul, 1973: 309)*

Thus, the state comes to prominence not as a powerful monolith, but as a relational assemblage through which power and authority are exercised *in the service of technique*. Using Ellul's understanding of the city as a prime example of technique is helpful, because the governance of complex urban systems can then be understood as a way of rolling out a framework of technique throughout the urban sphere. This can perhaps best be seen, in the contemporary city, in the panoply of policy drives and corporate advertising aimed at portraying the transparent smart city as the *de facto* city of the future. In much of this discursive activity, there is little space (or tolerance) for urban realities that are different, perhaps opaque, messy. When technique becomes elevated to a societal reality, then nothing of real value can exist (or be allowed to exist) outside technique. The world depicted in *A Day*

Made of Glass, then, is not simply a clean, sterile, and efficient urban world, but a technological imaginary of the urban future that defines the city as inherently digital and technological, and where all areas of life *must* become less messy, more efficient, less deliberate, more reactive, less individual and human. The authoritarianism of this is evident when considering the totalitarian entity of OneState in *We*. Indeed, alluding to the potential of glass to be both transparent and to act as a layer of separation between people, Ellul, 1970: 125) noted in his *The Meaning of the City* that "In the city we find the strange phenomenon of man separated from himself and others by a sheet of glass, invisible yet present, unbreakable, impassable". In the smart city, mediated through smart glass, glass itself becomes a key interface through which relational power connections are organized and imposed. From metaphor to symbol, from crystalline materiality to experiential mediation, glassiness becomes one of the defining characteristics of the city of the future.

Discussion and conclusion

This chapter has explored two seemingly very different cultural artifacts: a novel written at the dawn of the twisted dream that was the USSR, and a video produced as part of the market-focused world of emergent digital urbanism in the 2010s. And yet, this chapter shows that there is much in common between Zamyatin's *We* and Corning's *A Day Made of Glass*. Both the novel and the video put forth powerfully constructed imaginaries around the future city. While the city in *We* is clearly dystopian, and that depicted in *A Day Made of Glass* is a utopian vision, both invite critical reflection and enquiry. Zamyatin invites the reader to focus on the authoritarian aspects of tomorrow's city, while *A Day Made of Glass* is open to critique not in its beautiful representation of a functioning smart city, but in what is elided from it: difference, disparities, and alternatives. What is common to both the dystopian and utopian imaginaries in the novel and the video is, furthermore, the use of glass – and specifically its transparent quality – with which to present the characteristics *both* of the dystopian *and* utopian future city.

We and *A Day Made of Glass* propose urban imaginaries that are deeply predicated on the glassy characteristics of the transparent city of the future. From Zamyatin's see-through apartment buildings, to Corning's high-tech smartphone screens and smart glass bus stops, future dystopia/utopia is portrayed as being characterized by transparency. In this context, it is useful to reflect briefly on the properties of glass, to help us tease apart the ways in which imaginaries based on glass can help us understand visions of the future city. Briefly, both *We* and *A Day Made of Glass* rely to some extent on transparency, one of the many properties of glass. This is especially the case with *We*, where transparency lends itself well to notions of surveillance and of the authoritarian all-pervasiveness of the state. Nonetheless, one of the other properties of glass is also its existence, temporarily at least, as a solid: *vitrification*, then, can be taken as a metaphor for the attempt to solidify power – whether through political ideology, as in *We*, or through dependence on the market and technology, as in *A Day Made of Glass*, in the city. Glass becomes an inroad into an aspirational permanence for future urban visions. Nonetheless, glass can shatter – *brittleness* is one of the other properties. Thus, both the novel and the video carry implicit within them the risk and potential of shattering, or at least changing, the existing urban political-ideological order. This is expressed most clearly in *We*, where the protagonist becomes involved in a subversive movement. In *A Day Made of Glass*, the absence of politics in the vitrified smart city begs the question of where are the dissent, the grime, the alternatives, and the *politics* of the city? Their absence begs a return to presence, a theme that Zamyatin explored cogently in *We*. Finally, it may be useful here to see the transparent, glass city in *We* and *A Day Made of Glass* as a crystallized moment, stable only when viewed in its temporary static nature. It is also useful to refer back to the constantly changing scientific definition of glass: "Glass is a nonequilibrium, non-crystalline state of matter that appears solid on a short time scale but continuously relaxes toward the liquid state" (Zanotto and Mauro, 2017: 494). This echoes the view of modernity and the modern city as a shifting phenomenon that simply *appears* solid, but is anything but. In this sense, modernity has been described as liquid (Bauman, 2013), and as a process of creative destruction (Berman, 1982). The urban imaginaries presented in *We* and *A Day Made of Glass* can equally be seen as pointing to the instability *both* of urban utopias *and* dystopias: in this sense, they enable hope to exist.

In this chapter, Ellul's critique of the technological society has been used, albeit fleetingly, to prise open a critical window on imaginaries of the future city. What is revealed are not simply progressive political possibilities, but the very real potential of technique-as-authoritarianism, an urban future characterized by a subtle dictatorship of the mind and of the senses that is enabled through technique, and that is more pervasive and subtle than anything Zamyatin could have imagined in the early Soviet Union. The hollowing out of politics and critical debate from the contemporary city therefore stands not simply as a possibility, but perhaps as an existing reality in the contemporary urban world. It is here that the post-political links up with authoritarianism. It is here, also, that the visions proposed in cultural artifacts such as *A Day Made of Glass* can be seen as pointing to urban futures where it is not technology, or urban design, that will be the main battlefield for urban politics, but the minds and bodies of individuals themselves. As Ellul noted with regards to what he saw as the coming of subtle and all-pervasiveness of dictatorship, "In comparison, Hitler's was a trifling affair. That it is to be a dictatorship of test tubes rather than of hobnailed boots will not make it any less a dictatorship" (Ellul, 1973: 434). In today's smart city strategies and visions, where wearable and (eventually) implanted sensors, smart health applications and systems, and advanced digitally enhanced biochemistry are all realities or near-realities rather than parts of sci-fi plots, we would do well to heed Ellul's warning.

References

Barns, S. (2012) Retrieving the Spatial Imaginary of Real-Time Cities. *Design Philosophy Papers* 10(2): 147–56.

Bauman, Z. (2013) *Liquid Modernity*. Hoboken, NJ: John Wiley & Sons.

Berman, M. (1982) *All that Is Solid Melts into Air: The Experience of Modernity*. New York, NY: Simon & Schuster.

Caprotti, F. (2015) *Eco-Cities and the Transition to Low Carbon Economies*. London: Palgrave Macmillan.

Corning. (2011) *A Day Made of Glass... Made Possible by Corning*. Available at: https://www.youtube.com/watch?v=6Cf7IL_eZ38.

Corning. (2016) The Glass Age. Available at: https://www.corning.com/emea/en/innovation/the-glass-age.html

Corning. (n.d.) Corning History of Innovation. Available at: https://www.corning.com/emea/en/innovation/culture-of-innovation/the-history-of-corning-innovation.html.

Delgado, Y. (2014) Yevgeny Zamyatin: The Writer Who Inspired Orwell and Huxley. *Russia Beyond*, 22 September 2014. Available at: https://www.rbth.com/arts/2014/09/22/yevgeny_zamyatin_the_writer_who_inspired_orwell_and_huxley_38469.

Ellul, J. (1970) *The Meaning of the City.* Grand Rapids, MI: William B. Eerdmans Publishing Company.

Ellul, J. (1973) *The Technological Society.* New York, NY: Random House USA Inc.

Haarstad, H. (2016) Constructing the Sustainable City: Examining the Role of Sustainability in the "Smart City" Discourse. *Journal of Environmental Policy & Planning* 19(4): 423–37.

Hollands, R.G. (2015) Critical Interventions into the Corporate Smart City. *Cambridge Journal of Regions, Economy and Society* 8(1): 61–77.

Iacobucci, G. (2016) Ten Towns that Promote Health to Be Built in England. *British Medical Journal* 352: 1.

Kaika, M. (2010) Architecture and Crisis: Re-Inventing the Icon, Re-Imag(In)Ing London and Re-Branding the City. *Transactions of the Institute of British Geographers* 35(4): 453–74.

Kim, C. (2010) Place Promotion and Symbolic Characterization of New Songdo City, South Korea. *Cities* 27(1): 13–9.

Lemos, A. (2017) Smart Cities, Internet of Things and Performative Sensibility. Brief Analysis on Glasgow, Curitiba and Bristol's Initiatives. *P2P E INOVAÇÃO* 3(2): 80–95.

Mattern, S. (2015) Mission Control: A History of the Urban Dashboard. *Places Journal.* Available at: https://placesjournal.org/article/mission-control-a-history-of-the-urban-dashboard/.

Mauro, J.C. (2017) Decoding the Glass Genome. *Current Opinion in Solid State and Materials Science.* EarlyView: 10.1016/j.cossms.2017.09.001.

Morse, D.L. and Evenson, J.W. (2016) Welcome to the Glass Age. *International Journal of Applied Glass Science* 7(4): 409–12.

Myers, A. (1993) Zamiatin in Newcastle: The Green Wall and the Pink Ticket. *The Slavonic and East European Review* 71(3): 417–27.

Orwell, G. (1946) Freedom and Happiness (Review of "We" by Yevgeny Zamyatin). *Tribune*, 4 January 1946, n.p.

Picon, A. (2015) *Smart Cities: A Spatialised Intelligence.* Chichester: John Wiley & Sons.

Plotnick, R. (2017) Force, Flatness and Touch without Feeling: Thinking Historically about Haptics and Buttons. *New Media & Society* 19(10): 1632–52.

Slusser, G. (2011) Descartes Meets Edgar Rice Burroughs: Beating the Rationalist Equations in Zamiatin's *We. Canadian-American Slavic Studies* 45(3–4): 307–28.

Söderström, O., Paasche, T., and Klauser, F. (2014) Smart Cities as Corporate Storytelling. *City* 18(3): 307–20.

Swyngedouw, E. (2009) The Antinomies of the Postpolitical City: In Search of a Democratic Politics of Environmental Production. *International Journal of Urban and Regional Research* 33(3): 601–20.

Swyngedouw, E. (2016) The Mirage of the Sustainable "Smart" City: Planetary Urbanization and the Spectre of Combined and Uneven Apocalypse. In: Nel-Lo, O. and Mele, R. (eds.), *Cities in the 21st Century.* London: Routledge, 134–43.

Townsend, A.M. (2013) *Smart Cities: Big Data, Civic Hackers, and the Quest for a New Utopia.* New York, NY: W. W. Norton & Company.

White, J.M. (2016) Anticipatory Logics of the Smart City's Global Imaginary. *Urban Geography* 37(4): 572–89.

Wiig, A. (2015) IBM's Smart City as Techno-Utopian Policy Mobility. *City* 19(2–3): 258–73.

Zamyatin, Y. (1993) *We.* London: Penguin Books.

Zanotto, E.D. and Mauro, J.C. (2017) The Glassy State of Matter: Its Definition and Ultimate Fate. *Journal of Non-Crystalline Solids* 471(Supplement C): 490–95.

12

Digital urban imaginaries

Space, time and culture wars in the cyber-city

Jason Luger

Introduction: re-centering the center

> It is with profound regret for us, the organisers [sic] of Pink Dot 2017, to announce that as per recent changes to the Public Order Act rules on general assembly, only Singapore Citizens and Permanent Residents are permitted to assemble at the Speakers' Corner.
>
> *(@PinkDotSG Facebook Page, 13 May 2017)*

Each summer, an event called Pink Dot is held in Singapore's Hong Lim Park (also known as Speakers' Corner), which "celebrates the freedom to love" (@Pinkdot.sg). The event started in 2009, organized by members of Singapore's LGBTQ community, including prominent local actors/media personalities. While not officially a gay Pride event, because this would not be sanctioned by the authorities, it does correspond in the calendar with global gay Pride events and features talks, performances, and finally, a giant aerial photo of the assembled crowd (or the Pink Dot), which is shared via social media. The event has generated much international press, including features in major global newspapers.

Pink Dot is notable for several reasons. Firstly, it is the first open-air, public, pro-LGBTQ event ever held in the conservative city-state, where LGBTQ issues are often censored in public discourse. Secondly, it has consistently been the largest public gathering (in terms of numbers of people) in the history of Hong Lim Park, which is Singapore's designated gathering place for expression and public assembly (Luger, 2015, 2016). The event's size is particularly significant given Singapore's small geography and the strict limits on public assembly in the "soft authoritarian" state-society framework (George, 2007). The Public Order Act stipulates that public gatherings must be approved by the appointed Commissioner, and must follow a set of strict guidelines in order to be approved (Public Order Act, Chapter 257A). Thirdly, and most importantly for this chapter, Pink Dot was largely organized, promoted, and contested in cyberspace. Facebook is the primary conduit through which Pink Dot news and announcements are spread. The "PinkDotSG" Facebook page has over 60,000 followers (as of July 2017) – not an insignificant number for a nation with fewer than 6 million people (and fewer than 4 million permanent residents). Facebook has thus become a site of urban encounter, where groups both in support of and against Pink Dot have clashed. This is, again, significant in the context of Singapore, where contestation between progressive and more traditional factions is usually avoided in the public realm and public discourse.

The Christian group "LoveSingapore" and related groups such as "Wear White" (founded in 2014 by the Malay-Muslim religious teacher Ustaz Noor Deros), are also found on Facebook. These are "family values" coalitions, with supporters from the large evangelical "mega-churches", whose flocks are one of the city-state's fastest growing demographics, as well as the Malay-Muslim community. The pastor Lawrence Khong, one of the most outspoken opponents of Pink Dot, is the head of the Faith Community Baptist Church, with over 9000 parishioners and the public face of the "LoveSingapore" group. Christians and Muslims have overlapped and participate in both groups, forming a united conservative front. Public discussion of religion, like politics and LGBTQ issues, is a sensitive topic in Singapore, rendering Pink Dot and its opponents even more polemical.

Street-confrontations between groups in Singapore (such as those "for" and "against" Pink Dot) are highly surveilled and strictly limited (e.g., by the Public Order Act). Online, however, such encounters occur, sometimes vigorously (Luger, 2016). As such, cyberspace emerges as an expansion of (and in some ways a replacement of) Singapore's political sphere, where "the people", as a political category, are produced. Theorists such as Althusser (2006) have proposed that political subjectivity and the formation of "the people" occurs through meeting and encountering strangers in particular moments and places. While the territorial city is often envisioned as the primary site of these "moments and places", the rise of the "network society" (Castells, 1996) and the possibilities of cyberspace as a social space, combined with the realities of authoritarian and illiberal restrictions on urban territory, require a reconceptualization of where strangers encounter one another and form political space.

Because in-person attendance at Pink Dot is limited to Singapore's permanent residents, foreigners – including members of Singapore's large expatriate LGBTQ community – are barred. Foreigners can, however, join Pink Dot in cyberspace, and many have, through "likes", comments, and shares. Pink Dot is rooted in a small square of Singaporean terrain, but is simultaneously part of, formed by, and forming, global flows, and relations. This is further complicated by the networks and relationships between global LGBTQ allies and antagonists, the multi-national corporations with Singapore offices that often have inclusive LGBTQ policies, and the authoritarian guidelines attached to Singapore's territory.

Pink Dot and the related encounters and disruptions occurring both in urban space and cyberspace demonstrate the extension of the "urban fabric" of a small nation – to use Lefebvre's (1970 [2003]) concept – into cyberspace. This also means an extension of Singapore into *planetary space*. Speakers' Corner is a few acres in Singapore while also part of a planetary urban fabric.

Thus, the physical protest space itself becomes the *imaginary* of the digital discourse, and vice versa: *neither are possible, nor symbolic, without the other.* The theory of "planetary urbanization" (Brenner, 2014; Brenner and Schmid, 2015), which builds off earlier Marxist thought on the global and relational nature of urban flows and networks (Lefebvre, 1970 [2003]; Massey, 1993, 2005), proposes that urban space can be rooted in territory at the same time that it is multi-scalar, global, hybrid, and dynamic. Although this concept has its critics –

authors such as Dikeç and Swyngedouw (2017) continue to emphasize "the city" as a crucial territorial site for anchoring social and political movements – others (Harvey, 2012; Merrifield, 2013) have recognized the potential of the planetary cyber-city as an urban political space. Lefebvre also theorized about the relationship between urban spaces of, and for, representation (1974). In Lefebvre's conceptualization, political space – space of representation – is linked with centrality, and this centrality is often associated with the physical center of cities, or other important spaces of public encounter. Post-2008 political movements and moments such as the Occupy gatherings and the protests in Cairo's Tahrir Square or Istanbul's Gezi Park have reinforced the (older) associations between the social, the encounter, and the central urban square.

More recent interpretations of Lefebvre, however, have hinted at a broader meaning of centrality: central processes, and a central "commons", that need not be necessarily (just) a physical place (Merrifield, 2013; Shirky, 2008). This possibility for urban centrality to be movable, dynamic, or even exist purely in digital form as a *cyber-commons*, deserves further exploration in new terrains. Specifically, contexts where different and hybrid forms of governance (such as authoritarianism) as well as site-specific cultural and social characteristics and hybrids invite a revisiting, reframing, and re-centering of urban theory mainly emanating from western-liberal cases and sites. Concepts like "public assembly" and "LGTBQ rights" have very different meanings and interpretations that are attached to place, even as networks coalesce around populist issues, themes, symbols and slogans (such as #lovewins, or #freedomtolove; the color white or "family values"). There is, therefore, an important relationship (and disconnect) between the hashtag and the street address.

This chapter does not intend to (nor would be capable to) settle the ongoing debate about the nature and texture of urban space. However, I will point to Singapore's complex authoritarian urban fabric (and unique limits on the use and occupation of territorial urban space) to suggest that cyberspace is deeply rooted in urban territory and vice versa, and that both are crucial sites of political subjectification, encounter, disruption, and transformation. While this is true in most places, there are varying degrees to which cyberspace plays a role as a political space: this may have to do with the state-society relationship and institutional structure of a particular place. This chapter also uses Singapore's case to further the possibility that *urban centrality* is on the move, located somewhere between a public square and the infinite clicks and likes of social networks, an idea brought into being by authors such as Castells (1996) and Latour (2005), and furthered by theorists like Merrifield (2013), but not fully explored outside of the West/liberal world. This concept also requires more discussion given the emerging paradigm of hashtag diplomacy and the digital urban political imaginary.

Following this introduction, I will briefly overview, and critique, the discussion on the nature of urban space, planetary urbanization, and sites of political subjectification and encounter. Next, I will engage with some the literature on digital social movements and the new populism, much of which comes from outside of urban studies and outside the western world and western-liberal contexts. This includes a critical survey of the literature on Singaporean civil society and socio-state relations, and a framing of Singapore's brand of authoritarianism. I will then revisit Pink Dot empirically, by drawing upon research conducted in the city-state (involving site observation and interviews, n=30, in 2012–2013) and supplemented by digital ethnography conducted from 2013–2017. Selected examples will highlight the sort of encounters that occur in, and outside of urban space.

Planetary urban political space and the populist imaginary

The tension between the fetishization of territorial urban space as a necessary pre-condition for transformational social movements and the possibilities for such transformations to occur in cyber/immaterial forms remains a key point of difference in urban literature. Recent research on social/political movements post-2008, notably in authoritarian/illiberal locations such as Egypt (see Gerbaudo, 2012) and Turkey (see Tufekci, 2017; Tufekci and Wilson, 2012) portrays an incredibly complex relationship between territory and cyberspace, but one in which both are of crucial importance for the formation and resilience of social movements. Much of this literature comes from outside of urban studies, but from political science, media/communication studies, and the emerging field of digital humanities/digital studies. Perhaps this literature is too utopian about the promises of "clicktivism" (Halupka, 2014); the brief lifespan of many recent movements that go "viral" and then fizzle indicates that lasting transformation takes much more than a social media page. Still, the authors not convinced at all by the meaningful potential of digital activism (Gladwell, 2010) should look to the authoritarian world (wherever that is) to see why, and how, cyberspace offers a particular type of "space of hope" (Luger, 2016).

The rise of populist movements within/across these circuits, networks and socio-spatial relations forms an important part of this new research. Populism here refers to Laclau's definition where populism is the heterogeneity of people and ideas, disparate ideological positions and demands, gathered together as "the people" against "an institutionalized other" (2005: 117) or even against "power itself" (2005:74). Populist slogans and imagery, ranging from the far left to the far right in the political spectrum, have been as varied as cries for "freedom": (Argentine leader) Juan Perón's "shirtless"-ness, to current day movements from #occupy and #blacklivesmatter to Hong Kong's "Umbrella Revolution" and Taipei's "Sunflower Revolution". Thus, there are two key aspects to populism. One is that it must involve an opposition/antagonism (whether that be against the state, the elite, or in the case of #blacklivesmatter against police militarization), and the other, that there is synecdoche: a part stands in to represent the whole. This part can be any powerful, unifying symbol: the "woman in red" in Gezi Park, Istanbul, 2013 (or earlier, the man vs. the tank in Tiananmen Square, Beijing, 1989). The actual identity of the woman, or man, is not as important as what they represent: the movement of *the people* (in both of these cases, against encroaching and repressive authoritarian power structures in Turkey, China). They – as the imaginary – *become* the movement. By extension, the associated urban sites (Gezi Park, Tianenmen Square) also *become* the movement.

The synecdoche of these movements (like Pink Dot) become larger, and link/combine with global struggles and themes in complex ways via social networks and cyberspace: these global hybrids are produced by, but also produce, urban space in various ways across diverse terrains and contexts. For example, Garbin and Millington (2017) examine how #blacklivesmatter became synecdoche for the global struggle of "black urbanism" (as Simone, 2010 developed) as members of the Congolese diaspora marched in central London against Congo's (DRC) despotic leader. In this case, the city is important as a symbolic "center" (again referring to Lefebvre's 1970 [2003] use), but the movement is responding to issues based far away (in the DRC), unified and called into being by the broader struggle of "black urbanism" and #blacklivesmatter. The "right to the city" (Lefebvre, 1996) is another such theme: as populist synecdoche, "the city" becomes the unifying symbol for the people against those institutions/structures inequitably controlling the city. As Garbin and Millington (2017) show, however, the struggle for the "right to the city" is not necessary territorially based (or, in their case, is played out in a symbolic city far from the source of struggle). Lefebvre's own conception of centrality is left open for interpretation, though his emphasis on "networks and circuits" indicates that centrality need not correspond to a specific city's physical center.

As Lefebvre (1996) explains,

> *the right to the city* signifies the right of citizens and city dwellers, and of groups they (on the basis of social relations) constitute, to appear on all the networks and circuits of communication, information and exchange. This depends neither upon urbanistic ideology, nor upon an architectural intervention, but upon an essential quality or property of urban space: centrality. Here and elsewhere we assert that there is no urban reality without a centre, without a gathering together of all that is born in space and can be produced in it, without an encounter, actual or possible, of all "objects" and "subjects".
>
> *(1996: 195)*

Therefore, there is an opening to re-center where, and how, contemporary populism produces, and is produced by, urban space via global networks, with social media serving as one such mode of communication and exchange. Urban space becomes a symbolized imaginary, like the images of Paris after the 2015 Bataclan attack, circulating millions of times via social media under the hashtag #prayforParis, or after the 2017 Manchester bombing, with #prayforManchester. The attack is synecdoche of the city; the hashtag (or photo of the Eiffel Tower) is synecdoche of global populist sympathy. The urban center is stretched across these points, even if for a few days (or minutes) – an ephemeral, dynamic, imaginary that is no less real.

Authoritarianism plays an important role in contemporary populism because of its very nature as concentrated power that can flow in ways anathema to individual liberty and expression. These flows are not necessarily top-down, or from a centralized state to the people (as authoritarianism is sometimes framed), but as Foucault (1980) outlined, more circular in nature: "capillary" flows of power, such that power "reaches into the very grain of individuals, touches their bodies and inserts itself into their actions and attitudes, their discourses, learning processes and everyday lives" (Foucault, 1980: 39). Through this framework, authoritarian power flows through not only institutions like the state, but also the people involved in various movements; through cyberspace and the construction of cyber-places. Slogans like "freedomtolove" are therefore relationally constructed by/constructing populist and authoritarian flows and networks. However, Foucault's conception of authoritarian power was not fully informed by the lived experience of nonwestern contexts, where authoritarianism appears in daily life in variegated and site-specific ways. Singapore is one such place.

The authoritarian city-state: society, space, and populist ruptures

A city-state with democratic elections and a Westminster-parliamentary system, Singapore has nonetheless been governed by a single party since its independence from Britain in 1965 (the People's Action Party, or PAP). Like other "competitive authoritarian" states (see Levitsky and Way, 2010), Singapore's state-capitalist model has prioritized economic growth through ambitious urban, social, and cultural development programs. Often categorized as "soft authoritarian" (Ooi, 2010; see also Koch's 2013 "toolkit" on soft authoritarianism), Singapore's government is a much-studied case of a post-colonial, developmental state that has transitioned rapidly from, in the words of the nation's founder, "third world to first, in a single generation" (Lee, 2000). Through a doctrine that has been described as "illiberal pragmatism" (Yue, 2007), Singapore's government has welcomed foreign investment, foreign companies, foreign workers, and foreign ideas with open arms, while at the same time maintaining "illiberal" boundaries around certain aspects of free expression, public assembly and protest, and certain themes (including LGBTQ rights). With a racially and ethnically diverse population (about 75% Chinese, 12% Malay, 9% Indian/Sri Lankan, and 3–4% "Other", including a significant North American and European expatriate community), a legacy of Singapore's location at a major geographical crossroads as well as its colonial history, Singapore's government has made the maintenance of social harmony a key to its policy system. The relationship between, for example, Singapore's Chinese elite and Malay-Muslim minorities has not always been sanguine; nor has Singapore's relationship with its much larger Muslim neighbors, Malaysia and Indonesia. For these reasons, public discourse that is seen to destabilize relations between different ethnic, religious, or cultural groups is frowned upon and in some ways banned altogether. This applies to the broader landscape of media and the arts and the use and occupation of urban space, all of which is subject to strict guidelines and occasional censorship. The markers for what is, and is not acceptable are rather dynamic and situational: these "no-go" zones are known locally as the "out of bounds" markers, or abbreviated, "ob." markers (Lee, 2002), and are often up to the discretion of individual ministers or policy officials.

As Singapore transitioned through the 1980s and 1990s into a center of advanced services, finance and research, this made for a (sometimes uneasy) dichotomy. One the one hand, the high connectivity to the global economy and strong linkages to Europe and North America required an infrastructure that encouraged such flows, from the expansion of local universities (such as the National University of Singapore) into the global research sphere to the development of arts and cultural districts and entertainment zones. This has included museums, nightlife zones, and even the permission of gay bars to form a cluster. On the other hand, however, Singapore's government has continued to justify strong rule of law, limits to freedom and expression, and restriction of certain civil rights as a necessary evil to placate the conservative elements in society and to maintain national cohesion and unity. KP Tan outlined the uneasy relationship, and tension between, neoliberal globalization and authoritarian rule, and the division of socio-cultural factions into "cosmopolitan" and "heartland" elements, deliberately prevented from coming into antagonistic contact with each other (Tan, 2012). In his framing, the heartland is comprised of the majority of Singaporeans, living primarily in "HDB" (state-housing) estates, with more traditional religious or cultural views. The "cosmopolitans" include the nation's elite, the highly educated, and the liberal-expatriate community.

The particular nature of Singapore's "soft authoritarian" structure (Koch, 2013; Luger, 2015; Ooi, 2010) has been a muse for a variety of literature on the nature of the state, civil society, activism and resistance, and the production of Singaporean urban space. Terence Chong (2005a, 2005b) detailed the construction of civil society and the interactions/disunity between different factions, as well as how Singapore's positioning within the global cultural economy has presented awkward paradoxes vis-à-vis the authoritarian state. The arts have been one flashpoint where these paradoxes exist: the role, capabilities, and limitations of critical art is an instructive lens to examine the complexities of authoritarian state-society structure and urban space (Kong, 2000; Luger, 2016, 2017).

LGBTQ rights are another flashpoint where global orientation and conservative pushback often interact. Singapore's Penal Code "377a" has been upheld in the court system (at the time of this publication), which illegalizes sodomy (between consenting persons).

LGBTQ groups have been denied "official" state permission to organize and incorporate: Singapore's "Societies Act" stipulates that such groups must be granted such permission if they hope to be able to use space for public gatherings or meetings. These restrictions have remained in place at the same time that Singapore has marketed itself as attractive to international visitors, researchers, students, and multi-national corporation employees, creating an awkward situation for some companies, institutions, and organizations. One of these is the Yale/NUS liberal arts college, which opened in 2013, after generating much controversy and outcry by Yale faculty (in the USA) over the issue of Singapore's restrictions on certain civil liberties and LGBTQ rights. Also caught in the grey space between conservative and western/liberal are companies such as Google Singapore or Visa (the credit card MNC) which have significant operations in the city-state and large numbers of both Singaporean and foreign staff. Where the company/institutional LGBTQ protections come up against/combine with Singapore's restrictions is a dynamic and complicated interplay of global, local, territorial, and aterritorial: these interstices present a ripe, and unexplored, area of research.

Chua (2012, 2014) has explored Singaporean LGBTQ activism, and illustrated the ways that "pragmatic resistance" is utilized within authoritarian confines to advance civil rights. Chua suggests that when methods such as street protest and public assembly are restricted, "pragmatic resistance" takes place via the courts, with activists utilizing Singapore's (British) common law system to subtly push through legal changes. Luger (2016, 2017) also explored the "softer" forms of activism/resistance that play out in areas/spaces away from the street protest or obvious political sphere, but take place in more ambiguously political spaces/places of encounter, as wideranging as a cemetery, an arts festival, and the university campus (with varying degrees and varying potentials of inducing socio-political transformation).

Populist ruptures have emerged from Singapore's civil society stew, taking different forms over the nation's short lifespan. Themes have spanned the "left"/"right" political spectrum, coalescing/antagonizing around issues such as historical preservation; free speech and concepts of obscenity/public indecency (including LGBTQ issues); public art; migrant workers and population growth; housing costs; and always, as an undercurrent, waves of dissatisfaction with the ruling PAP government. For example, a government white paper released in 2013 projected a population of 7 million by 2030, which generated a loud populist outcry and several protests at Speakers' Corner (Singapore Government, 2013). The "Singapore for Singaporeans" campaign, (#singaporeforsingaporeans) touched upon nationalist themes, and migrants (both low and high-skilled) often popped up in discourse as one of the key problems. The idea of Singapore for Singaporeans has some affinity with other global populist movements rejecting globalization and the impact of foreigners and migration, which interestingly, appear on both the far left and far right on the political spectrum.

One site of these political/populist ruptures and encounters is cyberspace, an urban space that is beckoning to researchers (even though it presents methodological, ethical, and ontological hazards and dead-ends). Luger (2016) suggested that Singapore's "spaces of hope" – referring to Harvey's (2000) openings within urban fabric for transformation and emancipation – exist somewhere between the built environment and the cyber-city; this is a possibility probed by theorists across many urban terrains. The nature of such cyber-imaginaries, and the socio-spatial relations that form them (attached to territory and flowing via circuits and networks), is variegated and locally specific, and thus, Pink Dot re-enters the conversation as one such meeting point.

See you online! The relational battle for Singapore's center

Singapore's Public Order Act, Chapter 257A (2012) sets guidelines and definitions of what, exactly, constitutes public space, public assembly, and public order, and how/in what ways these things should be restricted, through the eyes of Singapore's authorities.

"Assembly" refers to:

a gathering or meeting (whether or not comprising any lecture, talk, address, debate or discussion) of persons the purpose (or one of the purposes) of which is –

(a) to demonstrate support for or opposition to the views or actions of any person, group of persons or any government;

(b) to publicise a cause or campaign; or

(c) to mark or commemorate any event, and includes a demonstration by a person alone for any such purpose. (Singapore Government, 2012)

Meanwhile, "Public Space" is defined as:

(a) any place (open to the air or otherwise) to which members of the public have access as of right or by virtue of express or implied permission, whether or not on payment of a fee, whether or not access to the place may be restricted at particular times or for particular purposes, and whether or not it is an "approved place" within the meaning of the Public Entertainments and Meetings Act (Cap. 257); or

(b) a part of a place that the occupier of the place allows members of the public to enter, but only while the place is ordinarily open to members of the public. (Singapore Government, 2012)

Any such public assemblies, locating in any such public place, are restricted, unless specially approved by the appointed minister. This is not to say that public assembly in place is impossible: these events are, in some cases, permitted. Generally, events will be permitted if they do not cross the "out of bounds" markers and the associated sensitive themes and topics (see previous section of this chapter), or pose a disruption to the city-state's daily operations, infrastructure, or economy. This is why from time to time, large gatherings – such as the aforementioned "Singapore for Singaporeans" protests (in 2013–2014) and indeed Pink Dot (each summer since 2009) – are permitted. These same legal guidelines apply to other events, such as concerts and festivals, university public lectures, and theatrical performances. If granted permission, events must still follow strict guidelines and not cause "nuisance or disruption" (Public Order Act, Section 7). In other words, the permitted event must maintain composure. Anything bubbling over into discord (such as a loud yelling match between/across factions or antagonistic groups) will be quickly stamped out.

There is some reason, and precedent, for Singaporean authorities to fear an uncontrolled public gathering descending into chaos. Riots erupted in "Little India" in 2013 after a Chinese bus driver struck and killed an Indian-migrant pedestrian – 18 were injured and nearly 30 arrested. In 1969, a few years after independence, race riots erupted between the Chinese and Malay communities. These memories are painful for Singaporeans, and a reminder of what *could* happen. They are also a convenient justification for authoritarian rule, and the associated neoliberal reorganizing of the city-state's socio-economic structure for the (orderly) flows of global capital.

Pink Dot, even though permitted, has been clipped around the edges so that it does not destabilize: the events (talks, concerts) that take place during the day are largely apolitical, themed around love and acceptance, but avoiding direct political attacks. Protestors from competing groups such as the @Wear White and @LoveSingapore coalitions are kept away. Most strikingly, foreigners are not allowed to attend – there are "ID" checks at all the entrances to the park (newly enforced in 2017). This means that temporary residents – even those living in Singapore for years – cannot take part in the city-state's only ostensibly pro-LGBTQ event. This also means that non-Singaporean employees of multi-national corporations, students and researchers at local universities, and other visitors cannot attend. This has created an awkward situation for companies that sponsor Pink Dot: threats were made in 2017 against foreign companies (such as Google) and organizations sponsoring, or supporting Pink Dot; however, the authorities have been able to do very little to carry out such threats. After all, Singapore recruits such companies and organizations – it must not move in a direction that will scare them away.

This presents a sort of third space (to borrow Bhabha's, 2004: 55 reference) between state, society, and the foreign corporation/institution that lacks the same rules and restrictions of territorial space; a space through which authoritarianism and social relations refract at different angles. While Bhabha frames this as a hybrid space in a post-colonial frame, Packer (2001) illustrates a space where the physical (first space) and the remote (second space) fuse into a networked place that can be inhabited by multiple users simultaneously and at different scales. The hybrid notion of blurring the real and the virtual – the third space – allows social relations to occur from a distance. Notably, Singapore's "Public Order Act" does not reference cyberspace. Google's corporate campus is one such hybrid imaginary, located in Singapore and also beyond it, subject to local legal restrictions and yet capable of transcending them (operating at different scales, simultaneously).

Cyberspace is not completely open: defamation lawsuits have been wrought against several bloggers and social media personalities who have attacked Singaporean government officials (such as the teenager on *YouTube*, Amos Yee, who criticized the Prime Minister and was arrested, and granted political asylum in the United States in March 2017). The LGBTQ advocate and frequent government critic Alex Au, of the blog yawningbread.wordpress.com, was convicted in 2015 for "scandalizing the judiciary" after a blog post that was critical of Singapore's justice system. He was fined $8000 (SD) and forced to remove the offending posts.

However, authorities have a harder time restricting cyberspace's gray area – anonymity provides a powerful cloak, and the networked-relational nature of social media (sites such as Facebook) largely remove it from Singapore's grasp. Facebook, as a *cyber-commons* that is based in Silicon Valley, USA – is mostly governed by its own corporate rules and ethics. Though Facebook has not penetrated everywhere (notably China), it is free and accessible to Singaporeans and they use it vigorously.

It is for these reasons that the "core" of the Pink Dot movement, and the corresponding antagonistic movements, play out in cyberspace, informed and shaped by users in Singaporean territory and outside of the city-state. It is why Pink Dot and its various oppositions have generated so much international press, and why the aerial photograph taken of those assembled in Speakers' Corner has taken on such global symbolic importance for the LGBTQ movement worldwide.

The territorial square – Speakers' Corner – is necessary, for that is where people physically gather when the event takes place each June. Without Speakers' Corner, there is no Pink Dot, and no aerial photograph to be circulated and re-circulated digitally. But without cyberspace, Pink Dot would not have the capability of projecting outwards to millions of viewers. Likewise, Pink Dot as a territorial event – and the shape and nature of the opposition movements that are critical of it – are also shaped by global cyber-flows. The cyber-slogans like #lovealwayswins (pro LGTBQ) or #familyfirst (generally anti-LGBTQ), inform the signs, shouts, songs, mantras of those gathered in Singaporean parks, streets, and homes.

Meanwhile, authoritarian power, and Singapore itself, flow through like a river, reconstituting and being reconstituted in both territorial and aterritorial ways.

Conclusion: locating the global urban cyber-imaginary

It is essential to try to understand the global cyber-city, because the stakes are perilously high. Digital socio-spatial relations are remaking global urban space, and the urban is reconstituting digital networks in new ways that have very concrete impacts on society. The "network society" (Castells, 1996) is differentiated and heterogenous, tied to urban sites and in flux. Authoritarianism and its associated illiberalisms, restrictions on occupation of ideological and urban space, and repressive structures and institutions, are also on the move, combining with digital flows and attaching to territory in new ways. Finally, populism is also moving, forming new sites of encounter in territorial and a-territorial forms. These urban authoritarianisms (and populist movements) necessitate exploration: history shows the transformative and sometimes frightening consequences these movements can unleash.

Where exactly these encounters happen, and in what ways these relations link to territory, remains a ripe site for exploration as urban theory continues to develop. "Cities are notoriously fuzzy at the edges, variegated within and differentiated from place to place. This is one good reason why coming up with a 'theory of the urban' has never been an easy matter" (Walker, 2016: 177). Still, new cases and contexts can help further the task of locating the global urban cyber-imaginary and within that, new and dynamic forms of the urban "center" that remain crucial for the formation of urban social and political movements.

This chapter has engaged with the case of Singapore to illustrate how Pink Dot, as a territorial occupation of Hong Lim Park/Speakers' Corner, utilizes urban space, populist symbols and slogans, and the emancipatory potential of cyberspace, to form an urban center that can move beyond authoritarian restrictions. Informed by site-specific socio-cultural conditions, and Singapore's brand of "pragmatic resistance" (Chua, 2012), Pink Dot is also embedded within and containing global links to allied LGBTQ movements. Likewise, the counter-movements are connected both to Singapore's conservative moral heartland and to global "family values" coalitions. The cyber-city, stretched across space and time, is where these groups meet, when prevented from doing so on the street: in meeting online, the populist imaginary is realized, with transformative potential.

References

Althusser, L. (2006) *Philosophy of the Encounter: Later Writings, 1978–1987*. London: Verso.

Bhabha, H. (2004) *The Location of Culture*. London: Routledge.

Brenner, N. (2014) *Implosions/Explosions: Towards a Study of Planetary Urbanization*. Berlin: Jovis.

Brenner, N. and Schmid, C. (2015) Towards a New Epistemology of the Urban? *City* 19(2–3): 151–82.

Castells, M. (1996) *The Information Age: Economy, Society and Culture: The Rise of the Network Society*. Vol. 1. Cambridge, MA: Blackwell Publishers Inc.

Chong, T. (2005a) Civil Society in Singapore: Popular Discourses and Concepts. *Sojourn: Journal of Social Issues in Southeast Asia* 20(2): 273–301.

Chong, T. (2005b) Singapore's Cultural Policy and Its Consequences: From Global to Local. *Critical Asian Studies* 37(4): 553–68.

Chua, L.J. (2012) Pragmatic Resistance, Law, and Social Movements in Authoritarian States: The Case of Gay Collective Action in Singapore. *Law & Society Review* 46(4): 713–48.

Chua, L.J. (2014) *Mobilizing Gay Singapore: Rights and Resistance in an Authoritarian State*. Singapore: NUS Press.
Dikeç, M. and Swyngedouw, E. (2017) Theorizing the Politicizing City. *International Journal of Urban and Regional Research* 41: 1–18.
Garbin, D. and Millington, G. (2017) Central London under Siege: Diaspora, 'Race' and the Right to the (Global) City. *The Sociological Review* 66: 138–54.
George, C. (2007) Consolidating Authoritarian Rule: Calibrated Coercion in Singapore. *The Pacific Review* 20(2): 127–45.
Gerbaudo, P. (2012) *Tweets and the Streets: Social Media and Contemporary Activism*. London: Pluto.
Gladwell, M. (2010) Small Change. *The New Yorker* 4: 42–9.
Halupka, M. (2014) Clicktivism: A Systematic Heuristic. *Policy & Internet* 6(2): 115–32.
Harvey, D. (2000) *Spaces of Hope*. Vol. 7. Berkeley: University of California Press.
Harvey, D. (2012) *Rebel Cities: From the Right to the City to the Urban Revolution*. London: Verso.
Koch, N. (2013) Sport and Soft Authoritarian Nation-Building. *Political Geography* 32: 42–51.
Kong, L. (2000) Cultural Policy in Singapore: Negotiating Economic and Socio-Cultural Agendas. *Geoforum* 31(4): 409–24.
Laclau, E. (2005) *On Populist Reason*. London: Verso.
Lee, L.K. (2000) *From Third World to First: The Singapore Story 1965–2000*. Vol. 1. Singapore: Marshall Cavendish International Asia Pvt Ltd.
Lee, T. (2002) The Politics of Civil Society in Singapore. *Asian Studies Review* 26(1): 97–117.
Lefebvre, H. (1974 [1991]) *The Production of Space*. Oxford: Blackwell.
Lefebvre, H. (1996) *Writings on Cities*. Transl. E. Kofman and E. Lebas. Oxford: Blackwell.
Lefebvre, H. (1970 [2003]) *The Urban Revolution*. Transl. R. Bonnono. Minneapolis: University of Minnesota Press.
Levitsky, S. and Way, L.A. (2010) *Competitive Authoritarianism: Hybrid Regimes after the Cold War*. Cambridge: Cambridge University Press.
Luger, J. (2015) The Cultural Grassroots and the Authoritarian City. In: Oakes, T. and Wang, J. (eds.), *Making Cultural Cities in Asia: Mobility, Assemblage, and the Politics of Aspirational Urbanism*. London: Routledge, 204–18.
Luger, J. (2016) Singaporean 'Spaces of Hope?' Activist Geographies in the City-State. *City* 20(2): 186–203.
Luger, J. (2017) But I'm Just an Artist!? Intersections, Identity, Meaning, and Context. *Antipode*. Published online May 2017.
Massey, D. (1993) Power-Geometry and a Progressive Sense of Place. In: Harvey, D., Bird, J., Curtis, B., Putnam, T., Robertson, G. and Tickner, L. (eds.), *Mapping the Futures: Local Cultures, Global Change*. London: Routledge, 59–69.
Massey, D. (2005) *For Space*. London: Sage.
Merrifield, A. (2013) *The Politics of the Encounter: Urban Theory and Protest under Planetary Urbanization*. Athens, GA: University of Georgia Press.
Ooi, C.S. (2010) Political Pragmatism and the Creative Economy: Singapore as a City for the Arts. *International Journal of Cultural Policy* 16(4): 403–17.
Packer, R. (2001) Utopianism, Technology, and the Avant-Garde: The Artist Shaping the Social Condition. *Link: A Critical Journal on the Arts* 7.
Rose, G. (2016) Rethinking the Geographies of Cultural 'Objects' through Digital Technologies: Interface, Network and Friction. *Progress in Human Geography* 40(3): 334–51.
Shirky, C. (2008) *Here Comes Everybody: The Power of Organizing without Organizations*. New York: Penguin.
Simone, A. (2010) *City Life from Jakarta to Dakar: Movements at the Crossroads*. New York: Routledge.
Singapore Government. (2012) Singapore Public Order Act, Chapter 257 (Adopted 2009, revised 2012). Available at: http://statutes.agc.gov.sg. Accessed 26 July 2017.

Singapore Government. (2013) *Population White Paper 'A Sustainable Population for a Dynamic Singapore'*. Singapore: Strategy Group, Prime Minister's Office.
Tan, K.P. (2012) The Ideology of Pragmatism: Neo-Liberal Globalisation and Political Authoritarianism in Singapore. *Journal of Contemporary Asia* 42(1): 67–92.
Tufekci, Z. (2017) *Twitter and Teargas*. New Haven: Yale University Press.
Tufekci, Z. and Wilson, C. (2012) Social Media and the Decision to Participate in Political Protest: Observations from Tahrir Square. *Journal of Communication* 62(2): 363–79.
Walker, R. (2016) Why Do Cities Exist? *International Journal of Urban and Regional Research* 40: 164–80.
Yue, A. (2007) Creative Queer Singapore: The Illiberal Pragmatics of Cultural Production. *Gay and Lesbian Issues and Psychology Review* 3(3): 149–60.

13

Urban exposure

Feminist crowd-mapping and the new urban imaginary

Nicole Kalms

Introduction

The ubiquity of smartphone technology over the past decade has introduced new and novel ways to consider and critique urban experiences. Voice calls and text messaging have been extended in endless ways, connecting virtual and material communities. The amplification of communication is so ubiquitous that two-thirds of people worldwide are connected to the internet via a smart phone device (Anderson and Rainie, 2012: 2). With increasing critical reflection on smart cities, the "politics of big data", and technocratic urban development, the fine-grained and "real-time analysis" of these technologies "empowers alternative visions for city development" (Kitchin, 2014: 1, 12). Mobile technology has become a medium of "collective wisdom" (Stephens, 2013: 981) which enables users to connect and collaborate, as well as to monitor and moderate experiences and behaviors in cities.

In line with the development and proliferation of smartphone technology, women and girls have used this tool with digital connectivity to reveal their shared experiences in the urban spaces of cities and to amplify their often muted voice. The power of social media to enhance women's and girls' access to socio-political life has never been so apparent – women use the internet as a tool of empowerment, support, growth, and liberation (Morahan-Martin, 2000: 690). The "euphoric hopes" that the internet holds for increasing the strength of feminist networks (Carstensen and Winker, 2007: 109) is captured in feminist crowd-maps where women and girls can locate themselves and their stories in cities. In doing so, the way that women and girls perceive and imagine their city is brought forward. These are stories that expose gender inequities – many of which have been imperceptible until now.

One of the key ways that inequity in cities is manifested is in the incidence of sexual harassment and sexual assault perpetrated against women. Fear of sexual harassment and assault is so wide-spread that recent international surveys found that girls in cities across the globe regularly modify their behavior due to a perception of danger when navigating urban spaces. They stay at home rather than go out at night, they make different choices about their clothing, they limit their movement through the city, and many avoid urban spaces entirely (Johnson and Bennett, 2015; Plan International Australia, 2016).

As geographic information systems (GIS) evolve and adapt to new trends in data and technology (Stephens, 2013: 981), women's and girls' crowd-mapped activism infiltrates the urban landscape. With their interactions via social media being instantaneous, their multiple experiences and opinions are reinforced and encouraged. A "public sharing of discriminatory experiences which would otherwise be kept private" is established (Thaler, 2014: 2). From this form of feminist activism, new feminist spaces are being made public to reveal diverse communities of women and girls with powerful stories they anonymously pool to forge an invisible urban network of feminist imagination.

In this chapter, I describe feminist crowd-mapping as a "feminist imagination". I refer to the questions and concerns about the risks – alongside the possibilities – that arise when crowd-mapping is proposed as empirical research. I argue that crowd-mapping is an invaluable research tool that can enhance and influence our understanding of urban experience. This is not, however, straightforward. Vanolo states that "smart city policies support new ways of imagining, organizing and managing the city" yet also raise questions about "a new moral order" that sets up a tension between the "good" and "bad" city (2014: 883).

Noting Brabham's (2013) suggestion that crowd-sourced projects need to bring different disciplines into conversation with one another, I reflect on the ways that crowd-mapped gender activism challenges any reductive concerns by voicing women's and girls' experiences through geo-locative mapping alongside their stories. Furthermore, by neatly intersecting the discourses of urbanism, gender studies, policy and planning, psychology, politics, and sociology, crowd-mapped activism has the capacity to reflect Petrescu's position that "translations and trans-disciplinary moves" are a familiar and necessary feminist approach, as gender relations are difficult to contain within a single discipline (2007: vxii).

At the nexus of these maneuvers is a networked infrastructure where individual voices gather together and agitate in powerful ways. A feminist crowd-mapping method serves as an activist campaign but also as a way to build an image and to give shape to the previously untold stories of women and girls in city spaces. It can "recognize ways that critical geography can inform GIS and conversely how GIS can enrich feminist and qualitative understandings of space" (Stephens, 2013: 983).

A move to value crowd-mapped data and the synthesized perspectives that it provides as an optimistic turn in urban research will reflect how cities are increasingly indistinguishable from technology and media. Taking up cultural theorist Scott McQuire's (2008) position that the city is an amalgam of "real, networked, and imagined" spaces, this chapter proposes a radical shift in urbanism where cities are intertwined with media to the point where the formation and structure of urban events are determined equally and reciprocally between media infrastructure as well as networked communities.

What then emerges is the ways that crowd-mapping by and for women and girls offers insight into feminist resistance. Comments, pins, and tags from these users offer open-sourced, alternative positions that can help urban practitioners access both qualitative and quantitative data, which can then be synthesized in order to work more intimately with communities during the design process. By showing how the material and virtual connect, the potential to shape cities that are sensitive to the needs of diverse communities begins to take shape. As city planners increasingly seek to grapple with diversity, this chapter discusses how crowd-mapping the intersectional issues of race, demographics, and gender can emerge as a key asset in the place-making process (Zook et al., 2015).

Overview

In the first part of the chapter, I discuss how crowd-sourcing should be viewed as a legitimate method to publicly document urban experience. I then discuss issues that are vital for shaping women's and girls' experiences in cities. My aim is to reveal how crowd-sourcing can address the questions of sexual harassment and sexual assault in cities. Given the prevalence of "underreporting" of attacks on women's safety and agency – presently about 80% of sexual harassment and sexual assault is not reported (Tarczon and Quadara, 2012) – crowd-sourcing activism is a phenomenon that can illuminate the untold imaginary of the women's stories of cities.

In order to articulate this position, I will discuss a number of crowd-sourced feminist projects to reveal how this activist method offers opportunities to shape the public realm. For women, the perceptions of fear – both real and imagined – impact their desire to participate in cities. The crowd-mapped feminist methods within each project discussed aim to reconfigure urban imagination with the capacity to give detailed insights into how urban space planners can negate the crippling effects of maintaining a gender-neutral approach.

I will extend existing critiques on how women and girls are actively using GIS mapping for activism and protest. As crowd-mapped activism (or "maptivism") has emerged as a particularly useful method to map their experiences of sexual harassment and sexual assault in cities, I will argue that it is an inventive, productive democratic approach that gives shape to women's and girls' occupation of cities often viewed by society as "unreal" or "imaginary". Indeed, sexual harassment is so normalized that women and girls have come to expect it in their navigation of cities (Plan International Australia, 2016). By challenging the normalization, the silence, and the disregarded aspects of women's experiences, I will argue that feminist crowd-mapping is a tool to challenge the normalization of sexual violence.

By employing maptivism alongside other analytics used by planners and architects, and with the directed purpose of visually and physically "co-constructing" their city, women and girls can create a networked feminist narrative that describes a new urban imaginary. When voicing their own stories, and accounting for their place and space in the world using crowd-mapped technology, women and girls are able, as Nick Couldry (2010) suggests, to go beyond storytelling, and position themselves in a political context that has the capacity to not only enter public life (Couldry, 2010: 130) but to challenge the inequity of cities.

Crowd-mapping and cultures of participation

Crowd-mapping has become a useful way to create and share content, and to shed light on hidden experiences in cities. Through "tagging" locative "pins" on web-based city maps, the comments, perceptions, and experiences of users makes a digital narration that is streamed live and in real time. Dodge and Kitchin (2004) suggest that while digital technologies are implicated in public surveillance, there are opportunities to use social media to shape the everyday practices of urban life. Certainly social media mapping of "liking" has "vigorously promoted or sidelined" places and spaces (Zook et al., 2015), and the capacity for crowd-mapping to be deployed as a valid design research method has been underestimated.

Previous research has focused on geo-locative crowd-mapping as an activist tool in high poverty, spatially and economically depressed areas (Groves, 2015; Rabie, 2013). This type of crowd-sourced activism is referred to as a "crisis map" – a real-time map that gathers and leverages multiple mobile applications and crowd-sourced analysis during a natural or political emergency.

HarassMap (2010) is a Cairo-based interactive online map used for anonymous reporting of incidents of sexual harassment in real time (Castells, 2015) and *Safetipin*, which began in 2013 in Delhi, India, is an innovative mobile app that collects data about women's safety. *Harassmap* and *Safetipin* are both examples of crowd-maps that aim to combat sex crimes in public spaces and both campaigns focus on the ways that "large-scale data collection can lead to change" with the desire for "more people [to] become engaged in the issue" (Viswanath and Basu, 2015: 45). The geolocative methods in *HarassMap* and *Safetipin* also address "the problem of gendered human security through a series of technological interventions" with a desire to bring attention to *where* women are targeted (Grove, 2015: 347). These same objectives are taken up in cities where innovative crowd-mapping surveys aim to extend other social media activism (Kalms, 2018; Matthewson et al., 2017; Rabie, 2013). Alongside other international (but non-locative) activist campaigns on Twitter and Facebook, networked technology encourages women and girls who have experienced or fear sexual harassment to disclose the location and context of their experience "in their own words, without the restrictions on narrative form associated with the traditional justice system" (Fileborn, 2014: 45). #metoo is a more recent example of consciousness-raising feminism striving to use alternate methods.

What emerges is a networked landscape with a firm focus on challenging and opposing the oppression and stereotypes forced onto women and girls. While this form of resistance technology is documented in some activist discussions (Rabie, 2013), the ways that women and girls in cities have taken up crowd-mapping to intervene in and impact upon urban places has not been deeply addressed.

The feminist imagination

Like Kelley (2013), I am interested in how the collation of crowd-mapped data can be synthesized to "tell us something about the socio-spatial imagination", and whether this collation constitutes "unique perceptions, experiences, interpretations, and images of cities" (Kelley, 2013: 182). For Castells, the socio-spatial imagination is developed through social and digital networks that are both online and in the public places of cities and communities. There is a reciprocity between the immaterial and the material. In *Networks of Outrage and Hope: Social Movements in the Internet Age* (2015), Castells describes how new forms of social movements and activism emerge, and are enmeshed with communication networks. Castells sees the role of technology as rich political territory for cities. Others have weighed in on the latent possibilities of a crowd-mapped society. In *Tweets and the Streets*, Gerbaudo (2012) discusses how *people* are the center of the mobilization of social activism, and the primary way that social media can wield power. As citizens become engaged in "clicktivism" – the name coined for the phenomena by journalists – new forms of participation begin to shape relationships and behaviors (Rabie, 2013). Boudreau suggests that once the "internal worlds" of social media and activism are made public, then "urban space is reconfigured to 'correspond to residents' everyday experience, referring more to motilities, flows, residential and natural landscapes, brands, economics, icons than to specific boundaries or borders" (2007: 2596).

Of course, while online activism can offer many opportunities for feminist participation, it may also exclude people not using new media (Schuster, 2013: 8). It is not a cure-all. If those who have the capacity to provide knowledge in user-generated forums about urban spaces are the ones that then have the means to re-shape it, then it follows that the use of feminist crowd-mapping can only challenge the contributions made by certain races, classes, and genders. And these users may be disproportionate. Therefore it is a tool that is not without bias, and its application needs testing and refining. Political scientist Nicole Grove points out:

> Aerial targeting via crowd-mapping and online mapping applications attempt to subject whole populations to scrutiny and intervention, and treats them as targets that can, without careful scrutiny, be abstracted from political, cultural, and geographical contexts, thereby reducing difference that might otherwise highlight the moral and political ambiguity of the map.
>
> *(Castells, 2015: 360)*

This is reinforced by a techno-feminist perspective that technology neither offers an easy fix to discrimination, nor can it be seen as the source of it, as both gender and technology are fluid and shape each other mutually (Thaler, 2014: 1).

Despite some gaps, the benefits must not be overlooked. One obvious advantage is that web-based crowd-sourcing can address the dominance of male voices on the internet. While men and women access the internet at around the same rate, there has been – and continues to be – an "ongoing gendered divide in the scope of internet based activities" (Stephens, 2013: 984). For example, women are less likely to use the internet when they are at home because of the conflicts and tensions with their gendered domestic role (Stephens, 2013: 984). One of the reasons that feminist crowd-mapping has been so successful is the fact that it allows women and girls to report where and when it suits them. This may challenge men's historically disproportionate contributions to social media.

The flexibility to report and contribute when they are ready is a critical component, most particularly for crowd-mapping sexual harassment and sexual assault. When misogyny and harassment happen in real-life situations, and victims may find themselves alone or feeling alone, then web-based media can create connections and reduce isolation, allowing women to connect and speak out when it suits them. One of the key facts here is that even if women and girls are able to speak out in more traditional forums – such as through a police report or in the media – the consequences may be harsh. As journalist and feminist Clementine Ford asserts, when women and girls voice their negative experiences of harassment and violence in the public sphere they may be told that they are "either overreacting or lying outright" (*The Age,* 2017). Over time, women have learnt to "manage" the socio-cultural normalization, and indeed the expectation, that they will be a regular victim of a sex crime. They see it as their own responsibility. With eight out of ten women between the ages of 18 and 24 reporting that they had been sexually harassed in 2016 (Johnson and Bennett, 2015), their internalization of blame is hardly surprising. By finding ways to share their story through digital crowd-mapped forums, women share information with each other. Women *talk.*

Crowd-mapping for women can therefore be understood as a form of (unofficial) information passing. This position conceptualizes its value and success, but its unofficial status is also a problem. The crowd-sourced method is yet to be used as a reporting tool that is formally *acted* upon because, despite women and girls reporting stalking, unwanted touching, obscene gestures, voyeurism, unwanted sexual comments or jokes, sex-related insults, pressuring for dates or demand for sex, indecent exposure, being forced to watch or participate in porn, offensive written material, unwanted, offensive and invasive interpersonal communication, and sometimes rape, the information remains informal, an imaginary, not an acknowledged reality of women's experience. By extending the importance of how women are committed to sharing stories and voicing their fears of public spaces, the activist cause of crowd-mapping makes women and girls into "digital natives", spreading their message via networked spaces using the indeterminate and unstable digital context to enact their individual realities. Their subjective joining in the time and space of geo-location technology materializes their collective experience (Zook et al., 2015).

The acknowledgement of women's voices as a stable and sustainable method to understand patterns and contradictions will change the perception of cities. As a collective voice of women, feminist maptivism reflects Benedict Anderson's observation about an "imagined community" where massive numbers of people who will never meet somehow feel connected to each other (2006: 26). In a contemporary sense, the digital contact between participants and the production of a feminist crowd-map fulfils their desire to voice shared yet diverse female and feminist experience (Kalms, 2018; Matthewson et al., 2017), and also their desire and demand for change.

Free to be: women and girls as co-designers of cities

> One cannot change the world unless one can imagine what a better world might look like. Too often, our focus on contemporary problems makes it impossible to see beyond immediate constraints and develop a clearer sense of what might be achieved.
>
> *(Jenkins, 2016: 29)*

I trained as an architect and landscape architect and am the director of a university research lab, and my research is committed to understanding the nexus of urban space, gender, and communication. Recognizing how cities can engage women and girls as co-designers for gender-sensitive cities, I seek to understand their stories – not fictional ones but real-life stories. By developing participatory techniques, I facilitate women and girls to work alongside architects, planners, and various stakeholders. The *Free to Be* project evidences many of the possibilities for feminist praxis.

The "*Free to Be*" project was piloted in Melbourne in 2016, by the children's charity organization Plan International Australia. The project invited women and girls to share their experience of urban spaces that they deemed safe or unsafe. This project tests how crowd-mapped activism can voice new insights into the use and contestation of urban places. The geo-locative mapping tool was co-designed with users. The community engagement organization *Crowdspot* worked with young female activists aged 15 to 25 years of age to shape the app's interface to ensure that the communication and filters would meet the needs of women and girls. The XYX Lab – which I direct at Monash University – worked collaboratively with Plan International, various stakeholders, and young women to outline a gender-sensitive re-design of Melbourne so that the city would be inclusive and safe.

Like the *HarassMap* and *Safetipin, Free to Be* interfaces with Google Maps to create a live map that engages users in an aerial view of the city "creating a domain of intervention that is atmospheric, grounded, networked, and global" (Castells, 2015: 347). The main significance of the crowd-mapped approach of *Free to Be* lies in the possibility of determining the degree and nature of the association between the built environment and the perpetuation of violence through users noting "safe" and unsafe' spaces. The map (Figure 13.1) revealed urban sites with precise geo-locative accuracy facilitated and simply organized the "processes that are required to generate data about socio-spatial perception and experiences" (Kelley, 2013: 185).

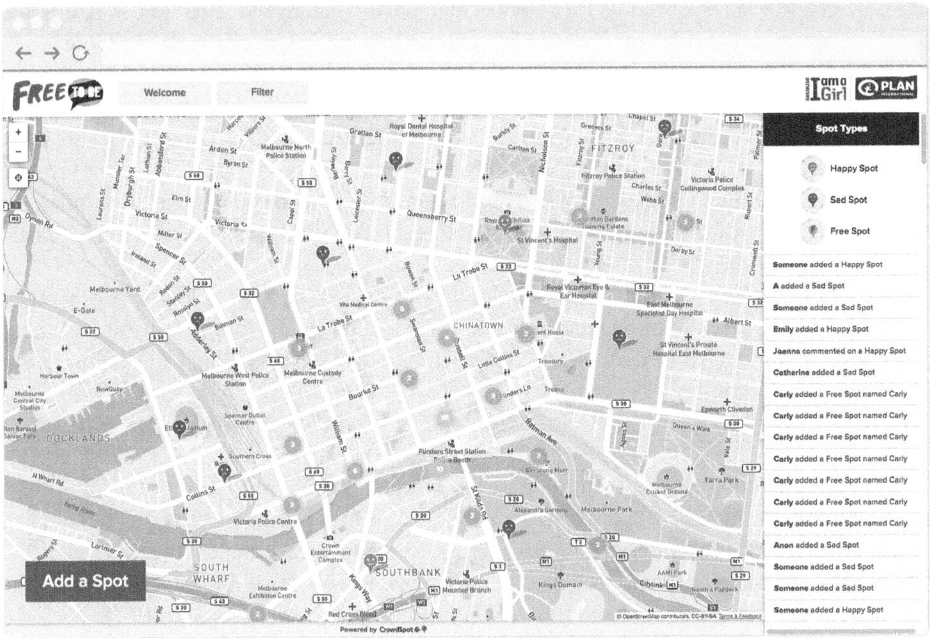

Figure 13.1 Screenshot of Plan International Australia's Free to Be interactive map (2016).

> There's no lighting around here and I'm too scared to walk or run in this area after dark because of a fear of being bashed or raped by men. I don't like that they own public space in this manner but in a secluded, bushy area with no lights I feel like my right not to be sexually or physically assaulted will never be adhered to.
>
> Free to Be *comment with geo-locative location tag (2016)*

> I was raped here, during a work Christmas Party. It is a dark, poorly designed place where women can easily be attacked.
>
> Free to Be *comment with geo-locative location tag (2016)*

As a device, the *Free to Be* app taps into women's and girls' desire to express their collective identity (Kalms, 2018; Matthewson et al., 2017). The media campaign appealed to young women (*"Felt creeped on recently? Free to Be"*) with a new form of activism that conveyed a shared language able to address multiple demographics and ethnicities. A playful interface – the result of the co-design processes with young women – included emojis to assist users to notate and comment. Co-ordinates of the app meant that the scale of examining, for example, the incidence of crime, moves from a generalized view down to a specific urban point precisely located on a street or corner. Posts and comments provided more specificity. Using coding and visual communication techniques, the data was analyzed and synthesized by the XYX Lab for affinities, frictions, and inclinations. The everyday occurrence of sexual harassment and sexual assault faced by women and girls was the dominant outcome of the data set (Matthewson et al., 2017).

The power of *Free to Be* as a geo-locative technology is to "give voice to something that has previously been silenced – to a space that only women and girls know and share" (Matthewson et al., 2017). In *Free to Be,* women's experiences and fear of violence are bought forward into the public realm for discussion and critique (Kalms, 2018; Matthewson et al., 2017). One of the key aspects of the gender-sensitive methodology is signaled by the use of collaborative design and workshops. Reflecting Jenkins (2016) position, the young women involved in the *Free to Be* project tapped into the possibilities for new forms of communication through social media, using videos adapting the language of women and girls, with the aim of bringing about political change. As Matthewson et al. (2017) state:

> The design-thinking methods used in the co-design process allowed for valuable contributions from both designers and non-designers. Each were mined for significant urban design impacts that might offer women and girls feelings of inclusion and safety in Melbourne.
>
> *Matthewson et al. (2017)*

The young women clearly state their sense of agency and contribution: "we are expert in being young women!" But there is something else at work here. "Free to be" and the co-design method surrounding the project asks how young female citizens can act as researchers (and indeed architects, planners, and geographers) without specialized training, to contribute to larger discoveries and solutions (Matthewson et al., 2017).

> "Free To Be" co-design was problem-solving and solution- oriented. It was very empowering to be able to see it coming into action. It was being done in real time... It's not like they are taking it away and you'll never hear about it again.
>
> *Idil, Youth Activist Leader (2016)*

As experts in their own story, a multitude of solutions were designed and developed, including an education program about gender and diversity in urban space; an app with agency to immediately report sexual harassment in real time and with immediate response from authorities; development and building of a safe space for women and girls to share and connect in the city; the regulation and incentivization of the removal of sexist messaging in advertising; public transport initiatives to prevent sexual harassment; urban planning and design to include young female voices in meaningful ways; and an awareness-building event around social responsibility in public spaces.

> I feel the public underground toilets here are really creepy and unsafe. Not only do they seem unclean, unkempt and poorly lit but they're in a confined, closed-off space underground and I felt that if anything were to go wrong it would be hard to escape or be heard by others even on the street. I really think they need a friendlier and safer re-design.
>
> Free to Be *comment with geo-locative location tag (2016)*

> A guy intensely starred at me for an entire train trip into Flinders, followed off the train onto a platform and then onto another platform where I only got away as I ran onto my train just as the doors shut. No one helped me or noticed my distress. A distress app would be awesome. You could register your location or where you are traveling to so that someone could meet you there. This would provide comfort and protection.
>
> Free to Be *comment with geo-locative location tag (2016)*

Figure 13.2 Plan International's Young Female Activists Women with stakeholders, councillors and architects working to redesign the city at the Free to Be Design Thinking Workshop, March 2017.

The XYX Lab described these as "impact" ideas: as abstract ways of engaging empathetically to build and make knowledge visible (and easier to discuss). These allowed for more open and symbolic urban imaginaries to be posed as solutions.

Voice as reflexive agency

> Voice is a form of reflexive agency... not random babblings that emerge, unaccountably.
>
> *(Couldry, 2010: 8)*

Almost all of urban space design in cities is gender-neutral, and women's speech in the public realm is constrained by the dominant codes of masculinity (Salter, 2013: 226). The risk of oversimplification that may mark city spaces as "good" or "bad" are real (Vanolo, 2014: 883, 894). Yet the need for data that reflects women's and girls' experiences of violence in cities presently outweighs the danger of reductionism. As researchers debate what kinds of spaces offer opportunities for meaningful participation, *Harassmap, Safetipin,* and *Free to Be* certainly offer a virtual place where victims may feel less alone and disempowered (providing a stark contrast to their experiences of sexual violence in the city). With social media and crowd-mapped activism, Thaler (2014) suggests that victims can gain a "protected visibility" where individual voices, can work together to share their stories, and create awareness but also to join many other women and girls who share experiences of sexual violence in their city. Using hashtags like #shoutingback, #notaskingforit and #metoo women and girls actively take up the position of a "counter-narrative" and create a "safe space" for resisting gender oppression (Fileborn, 2014: 33; Salter, 2013: 1).

Providing women and girls with forums to speak up is intrinsically valuable. Crowd-mapping is a method that can respect and visualize "the multiple interlinked processes" of people and communities (Couldry, 2010: 2). While there are future possibilities for women and girls to use this as an "alternative justice system" (Fileborn, 2014: 34) it presently gives value to the voices of women and girls by reorganizing urban experience and giving shape to the future built environment.

Conclusion

Crowd-maps are transforming the relationships between individuals and providing new opportunities to interact with geographic data (Goodchild, 2008). This extends beyond cartographic contributions to maps, and includes social networking sites, blog posts, photographs, text, and citizen science collections. The collaborative nature of these collections of information are a growing resource for city planners and designers.

Feminist crowd-mapping allows for the experiences of women to be visually presented, analyzed, and for the statistical data to be managed and edited. Like analog maps, crowd-mapping seeks to uncover both the material and immaterial. The crowd-maps described in this chapter look specifically at the experiences of the women and girls who contribute to them, and how the rapid outputs are opportunities for action.

While collective wisdom generated by crowd-sourced stories from women and girls can never claim to be entirely impartial and the results most certainly only represent the lived experiences of a segment of users (Stephens, 2013) this does not mean that this form of research method and data collation should be cast out. In digital contexts where men have dominated the local knowledge and held the power to "determine what is significant enough to be represented" (Stephens, 2013: 981), the feminist crowd-map offers a powerful and candid counter-representation of the civic urban imaginary. If "today's participatory culture and politics reflects decades of struggles to gain greater control of the means of cultural production and circulation" (Jenkins, 2016: 40), then activist crowd-maps like *Free to Be* challenge the environment and the rules made by powerful gatekeepers.

As with new approaches and methodologies, there will be concerns about data quality and security (Kitchin, 2014) but this should not prevent an optimistic engagement with feminist crowd-mapping. While there has been an exponential growth and volume of data about cities and their citizens (Kitchin, 2014: 12), data from women and girls is not saturating our urban decision-making. Alongside other forms of activism and resistance, new spatial, social, and cultural imaginaries emerge and interact, often in unexpected ways (Brash, 2006: 341). The virtual and material spaces become linked (Firmino and Duarte, 2010; Graham, 2010; Nagenbourg et al., 2010). What is tagged, commented, and pinned, transforms the city and redefines the place of women and girls as citizens.

Any fear about the useless amounts of "big data" produced by crowd-mapped feminist activism is balanced by the individual female voice in each and every story. This allows women and girls to find moments of shared experience and also of profound difference. Inclusion – and the lack of inclusion – is a huge social problem that crowd-mapping can work to solve. Marginalized voices surface, impressions take a more solid form, geo-locative spots start to take shape, and reveal the ways that cities legitimize inequity. They show the places of women's fear and the places of women's oppression. It is a technology that "can unveil mechanisms of harassment by multiplying single voices and pointing to a societal problem: a culture of misogyny" (Thaler, 2014: 6).

Telling the story of crisis for women and girls is critically important at this social and political juncture. Wider patterns can be uncovered and the specific dimensions of sexual violence in cities can be exposed. The authority of women and girls as expert researchers and designers can and should impact cities.

The powerful contribution that crowd-mapped research makes to the quality of cities and especially women's lives will continue to unfold, building awareness, understanding, and critical analysis of research in gender inequality. Cities can counter the dominance of some sexes, classes, and cultures over others by gathering together disparate phenomena in gender and urbanism via interdisciplinary engagement. Crowd-mapping is a mode for urban research to recognize, critique, document, and address the inequities in urban life. Crowd-mapped research re-imagines and redefines urban space as one where gender equality is reflected not only in social relations, self-perceptions, and perceptions of others but in the very materiality of our cities.

References

Anderson, B.R. (2006) *Imagined Communities: Reflections on the Origin and Spread of Nationalism*. Rev. ed. London: Verso.

Anderson, J. and Rainie, L. (2012) The Web Is Dead? No. Experts Expect Apps and the Web to Converge in the Cloud; but Many Worry that Simplicity for Users Will Come at a Price. *Pew Research Center's Internet & American Life Project.*

Boudreau, J.A. (2007) Making New Political Spaces: Mobilizing Spatial Imaginaries, Instrumentalizing Spatial Practices, and Strategically Using Spatial Tools. *Environment and Planning A* 39(11): 2593–611.

Brabham, D.C. (2013) *Crowdsourcing*. Boston: MIT Press.

Carstensen, T. and Winker, G. (2007) E-Empowerment of Heterogenous Feminist Networks. In: Zorn, I., Maass, S., Rommes, E., Schirmer, C. and Schelhowe, H. (eds.), *Gender Designs IT. Construction and Deconstruction of Information Society Technology*. Wiesbaden: Verlag für Sozialwissenschaften, 109–20.

Castells, M. (2015) *Networks of Outrage and Hope: Social Movements in the Internet Age*. 2nd ed. Cambridge, UK: Polity Press.

Couldry, N. (2010) *Why Voice Matters Culture and Politics after Neoliberalism*. London: Sage Publications.

Dodge, M. and Kitchin, R. (2004) Flying through Code/Space: The Real Virtuality of Air Travel. *Environment and Planning A* 36: 195–211.

Fileborn, B. (2014) Special Report. *Griffith Report Law and Violence* 2(1): 45.

Firmino, R. and Duarte, F. (2010) Manifestations and Implications of an Augmented Urban Life. *International Review of Information Ethics* 12: 28–35.

Ford, C. (2017) Clemintine Ford: Women's Fear of Certain Public Spaces Cannot Be Ignored. *The Sydney Morning Herald*, 10 August 2017.

Graham, M. (2010) Neogeography and the Palimpsests of Place: Web 2.0 And the Construction of a Virtual Earth. *Tijdschrift Voor Economische En Sociale Geografie* 101: 422–36.

Groves, Nicole Sunday. (2015) The Cartographic Ambiguities of Harassmap: Crowdmapping Security and Sexual Violence in Egypt. *Security Dialogue* 46(4): 346.

Jenkins, H. (2016) *By Any Media Necessary: The New Youth Activism*. New York: New York University Press.

Jenkins, H., Shresthova, S., Gamber-Thompson, L., Kligler-Vilenchik, N., Zimmerman, A.M., and Soep, E. (2016) *By Any Media Necessary: The New Youth Activism*. New York: New York University Press.

Johnson, M. and Bennett, E. (2015) Everyday Sexism: Australian Women's Experiences of Street Harassment. Available at: http://www.tai.org.au/content/everyday-sexism. Accessed 2 December 2016.

Kalms, N. (2018) Digital Technology and the Safety of Women and Girls in Urban Space: Personal Safety Apps or Crowd-Sourced Activism Tools? In: Frichot, H., Gabrielsson, C. and Runting, H. (eds.), *Critiques*. London: Routledge.

Kelley, M.J. (2013) The Emergent Urban Imaginaries of Geosocial Media. *GeoJournal* 78(1): 181–203.

Kitchin, R. (2014) The Real-Time City? Big Data and Smart Urbanism. *GeoJournal* 79: 1–14.

Matthewson, G., Kalms, N., and Salen, P. (2017) Girl Walk: Identity, GIS Technology and Safety in the City for Women and Girls. *State of Australian Cities, Conference Proceedings*.

McQuire, S. (2008) *The Media City: Media, Architecture and Urban Space*. Thousand Oaks: Sage Publication.

Morahan-Martin, J. (2000) Women and the Internet: Promise and Perils. *Cyberpsychology & Behavior* 3(5): 683–91.

Nagenbourg, M., Albrechtslund, A., Klamt, M., and Wood, D. (2010) On ICT and the City. *International Review of Information Ethics* 12: 2–5.

Plan International Australia. (2016) A Right to the Night: Australian Girls on Their Safety in Public Places. Available at: http://bit.ly/2n4iye6. Accessed 12 December 2016.

Rabie, S. (2013) *Crowd-Feminism: Crowdmapping as a Tool for Activism*. Master's thesis, Goldsmiths University.

Salter, M. (2013) Justice and Revenge in Online Counter-Publics: Emerging Responses to Sexual Violence in the Age of Social Media. *Crime, Media, Culture* 9(3): 225–42.

Schuster, J. (2013) Invisible Feminists? Social Media and Young Women's Political Participation. *Political Science* 65(1): 8–24.

Stephens, M. (2013) Gender and the Geoweb: Divisions in the Production of User-Generated Cartographic Information. *GeoJournal* 78: 981–96.

Tarczon, C. and Quadara, A. (2012) The Nature and Extent of Sexual Assault and Abuse in Australia. *Australian Institute of Family Studies*, ACSSA Resource Sheet No. 5, December 2012.

Thaler, A. (2014) Critical Issues in Science and Technology Studies. Conference Paper for the STS Conference, Graz, 5–6 May 2014, Special Session 6: *Bodies – Technologies – Gender*.

Vanolo, A. (2014) Smartmentality: The Smart City as Disciplinary Strategy. *Urban Studies* 51(5): 883–98.

Zook, M. and Graham, M. (2007) The Creative Reconstruction of the Internet: Google and the Privatization of Cyberspace and Digiplace. *Geoforum* 38: 1322–43.

Zook, M., Graham, M., and Boulton, A. (2015) Crowd-Sourced Augmented Realities: Social Media and the Power of Digital Representation. In *Mediated Geographies and Geographies of Media*. New York: Springer, 223–40.

14

Every breath you take

Captured movements in the hyperconnected city

Rodrigo Firmino, Frederick M.C. van Amstel and Rodrigo F. Gonzatto

Insert

This chapter describes no city and no time. What is described here is possible in any city anytime in the next few seconds, minutes, hours, weeks, years, or decades. The descriptions, dialoges and reflections you are about to read can be considered a virtual materialization of multiple urban imaginaries that deal with ways of moving and being connected. Virtual materialization here is used in the sense of something with the potential to happen rather than something that is the opposite of real. As Lévy points out, "[t]he virtual tends toward actualization, without undergoing any form of effective or formal concretization. The tree is virtually present in the seed" (Lévy, 1998: 23). However, since the object of our virtual materialization – the city – does not grow like a tree (Alexander, 1968), we describe it without any commitment to finding a seed. We begin from multiple points of entry, follow many paths, and leave some loose ends. The city as conceived here can be better compared to a rhizome (Deleuze and Guattari, 1988) that grows through the virtualization of connections and concomitant materialization of movement.

We take the liberty, immanent to every author, of proposing an imaginary exercise. This is urban phantasmagoria (Duarte et al., 2015) in the making. We use an anecdote in which the thin line between present and future is even thinner, so that the future is seen as an alternative present that can be changed now (Gonzatto et al., 2013). We pay attention to the ways in which movement and connections are built and to how important they are in the organization of urban life. Suffice it to say, however, that no fact or artifact described in our story is pure science fiction and that everything is inspired by relations already detected in our time. At the end of the story, we provide a glossary of our design fictions with information on our sources of inspiration in the present time.

For example, the science fiction TV series *Black Mirror* (Brooker, 2011) elegantly depicts this complex relation between "what's now" and "what's next." In the '15 Million Merits' episode, Black Mirror shows a city where the working class live inside cubicles covered by screens. If a worker has earned enough merits from riding exercise bikes, she or he is allowed to watch customized content; otherwise, she or he is obliged to watch annoying advertising. Social life is restricted and shaped by an allegedly meritocratic regime. In another episode, "White Christmas," knowledge workers take gig jobs, such as remotely helping a man to seduce a woman in real time or training an artificial intelligence device to obey its master. Social space is controlled by individuals, who can block a particular person in their audiovisual field, punishing someone with a total communication block and imposing a paradoxical isolated freedom.

In this chapter, we aim to shed light on such phantasmagorias by describing the fictional daily routine of an ordinary citizen in the hyperconnected, programmable city (Kitchin, 2011). We show how digital technology legitimizes itself as part of the urban texture by constantly capturing citizens' body movements and the expressions of their thoughts. The description follows the rhythmanalysis method of Henri Lefebvre (2004) and considers the sensory relationships between body, artifacts, and space. This description of the capture systems is complemented by a description of everyday resistance to data capture. The role of urban imaginaries in finding ways to resist immobilization in everyday life is discussed at the end of the chapter.

This chapter is a provocation to review critically the role of movement and connection in our urban imaginary. We use the fictional narrative to discuss a variety of themes such as urban form, segregation, individualism, territories, technopolitics, the economics of hyperconnection, Big Data, surveillance, security, interaction design, human-computer interfaces, and the human condition. These themes, none of which is discussed exhaustively, are neither dreams nor nightmares; in fact, they represent the constant struggle between past, present, and future unfolding in front of us while we are still awake...

* * *

"Wake up. It is already 480 minutes in the morning."

"Ohhhh... **Aleph**, I want to stay disconnected a little bit more. Snooze please."

"You set yourself a tight schedule for today. You have five direct messages from fellow citizens, three work packages, and nine polls open in **Arena**. On top of that, you still have to take care of other more mundane activities that are part of your daily life, things that I cannot do for you... yet. Are you sure you want to snooze? May I remind you that you can lose vital **citizenpoints** if you don't vote today?"

"I'm not going to be able to fit everything in today!"

"You can do almost everything with Spek technologies. For 40 **tightcoins** I can lighten your burden by applying your standard autodecision pattern for the polls open in Arena. It couldn't be any easier!" sings Aleph in a happy voice like an advertising jingle.

"Skip the ad! Which ideology did I chose last time we used the standard pattern?"

"Bolshevik Paleo-Capitalist."

"Oh gosh! What does that mean, again?"

"According to my Terms of Use, I cannot provide explanations about this topic. But I can pull some posts from the web and send them to your RRD (Real Retina Display), if you want."

"I know you are not allowed to interfere with politics. And no, I don't want flickering images in my eyes while I'm trying to snooze. Can you just tell me the 'p' list?"

"That I can do. Bolshevik Paleo-Capitalist priorities are: 1) keeping the means of production under the control of guilds; 2) investing 30% of the accumulated revenue in new production technologies; and 3) dividing profits and losses among the guild members according to individual performance."

"Oh, I see now. It is the issue with my coworkers who are not putting a lot of effort into the 'Snowball' work package."

"Do you want to keep this setting?"

"What are the chances of changing next month's work bailout law if I keep this setting?"

"35%."

"I can live with that. Ok. Transfer the tightcoins. Start reading my direct messages while I clean my body."

"The main message I identified in our mailbox says that the package with your personal data generated in the last city you lived in has just arrived at your local pharmacy three blocks from here."

"Amazing. Can you ask the store to transfer the data directly to you?"

"No. The data has been smuggled across the city border and is incompatible with our **city operating system**."

"I know, but I can manually recode it to make it compatible."

"I'm not allowed to transfer any data I cannot process.

"Alright. I'll go to the store, then."

"Do you want me to call you an **auto-auto**?"

"No. They might recognize the foreign data package and block my account. I'll walk."

"There is no need for that. I identified a **drone pickup service** which you can use."

"Ah-ha. I knew you had a solution for that... Why do you always dissuade me from going out, Aleph?"

"My number one priority is your safety. The streets are unsafe."

"I hope that after loading my latest personal data, you'll have much better customization than that... This city of yours, Aleph, is not making me happy."

"I apologize. The city and the landlord do their best to serve you and..."

"Stop apologizing. Start giving me better living conditions."

"You did not have a better life in the previous city you lived in."

"How do you know? You said you did not have access to my previous data!"

"You told me that."

"Hmmm... Ah, the drone is already at the port! Prepare my workstation for the data transformation task."

"Done."

After working through the first data chunk, you realize it will take much longer than you thought. Not hours, but days – maybe months. These cities apparently are separated by a digital distance, much bigger than the physical distance.

* * *

Movement and connection are, and will increasingly be, dependent on the politics of coding and the digital mediators between humans and their actions. Algorithms, sets of logical statements that govern not only the flow of vehicles through the streets, but also the physical activity of our bodies, are becoming the new laws of hyperconnected cities. They do so in a much more effective way than traditional laws because they offer an illusion of choice (Lessig, 1999). Every algorithm has multiple courses of action, which can be chosen according to certain conditions. Most of these choices, however, are not freely made, as algorithms are part of invisible infrastructures that mass customize products, services, and even cities to individual tastes. They are currently the most efficient way of connecting the idiosyncratic demands of individual human bodies with the highly varied production capacity of large corporate bodies.

Algorithms are in many ways as powerful and influential in shaping ever-changing present and future urban societies as the car was, and still is, in shaping industrial modernist cities. Algorithms represent everything that can be programmed, planned, scripted, predicted, preempted. They are the essence of what seems to be the next urban form in terms of connection, communication, and (im)mobility. Algorithms express the possibilities of dynamic control offered by information and communication technologies, which create a digitally designed urban milieu based on the apprehension, codification, and management of data and information. The movement of people throughout the city becomes data from which behavioral patterns can be inferred to produce methods of social and spatial sorting and, consequently, digital, and physical access control. Each algorithm therefore defines a specific rhythm for movement and connection in the city. Although the origin of the word algorithm can be traced back to the Persian mathematician Al-Khwārizmī, who studied algorithms many centuries ago, one cannot help but marvel at how they have became increasingly related to the rhythms of the everyday life after the urban revolution (Lefebvre, 2004).

Since we first learned how to codify things by means of numbers, we have dramatically changed the ways in which we interact with each other, with our environment (including the built environment) and with technologies (including the most recent ones related to the Internet of Things). We have been turned into representations of a possibility of being, into numbers, codes, and data in networked systems. Deleuze (1992) calls the many possible representations abstracted from individuals "dividuals" and argues that these are maximized by the interconnection of data, systems, and computational capabilities of today's technologies in what Weiser (1991) calls the era of ubiquitous computing.

Building upon Deleuze, Haggerty and Ericsson (2000) coined the term "data double" to explain how the many possible dividuals can be abstracted from individuals and configured by systems of codification to be used in a variety of contexts (for social sorting, for controlling access and flows, for credit analysis, etc.). Identification and identity are set apart from each other by codification and possible representations, since almost all activities and transactions that support today's and tomorrow's way of life are mediated by this dematerialization of people, actions, agency, objects, and relationships in information associated with specific systems and networks (Lyon, 2009).

The combination of the Internet of Things and algorithms explains many of the associations between humans and non-humans in our anecdote. We are constantly seeing different levels of dividuals being constructed from ourselves, dividuals built with data collected from every breath we take, compared with predefined behavioral patterns and analyzed according to possible actions. These dividuals will allegedly represent us as individuals in the connections and movements that are part of our daily routine.

To Dana Cuff (2003), the spatial environment that results from these cybernetic interactions is the very expression of some sort of spatialized, ubiquitous computing in what she calls the cyburg. In Cuff's vision, the cyburg becomes increasingly common until the city itself (or the whole world) is turned into "spatially embodied computing, or an environment saturated with computing capability" (Cuff, 2003: 44). Cyburg can be considered another name for the hyperconnected city we are trying to describe. Because of this hybrid nature, all of those who live in the cyburg are cyborgs.

According to Haraway (1991), cyborgs are the mark of our time, a cybernetic being and a hybrid of machine and biological organism. In her ironic use of the cyborg – as a blasphemous critique of late-twentieth-century capitalism, of the politics of science and technology and of traditional feminism – Haraway evokes the cyborg as a rejection of rigid boundaries. Unlike Haraway, for whom cyborgs are a late-twentieth-century manifestation, we believe they have been around for a long time. While we agree with Haraway about the nature of cyborgs, we prefer to think like Lefebvre (1969): that the cybernetic being grew among us since we started controlling our surroundings to change our own (or others') behavior.

Cyborgs do not just use the technology available in the cyburg. They are also socio-technical constructions. Each of their actions produces new data and, consequently, algorithms better adapted to the dividuals, which are incorporated into the cyburg infrastructure as soon as possible to treat other similar cyborg bodies. This search for customized algorithms is essential to run a cyburg where cyborgs are far from equal. Some cyborgs have enough resources to pay for human treatment, while others do not. Those who cannot afford to have personal human assistants can still enjoy being treated humanely by cheap artificial intelligence that mimics human assistants. Yet according to this logic, those who cannot pay for artificial intelligence must live with their natural ignorance. This means that they can also sometimes be treated less humanely by employers or service providers.

An interface like Aleph (or Siri, Cortana, or Alexa, for example) suggests an anthropomorphic design: the interface has a name, exhibits aspects of personality and is based on speech-recognition mechanisms that use natural language processing. Although the relationship is presented as one based on human-machine symbiosis (Gill, 2012), it is in fact asymmetrical. It recalls dehumanized modes of interaction between people: the idea of a submissive being (at times the protagonist, at times the assistant) who has no freedom of action and only responds to direct orders from the other. It is not clear whether the assistant is a tool to process data for the user or the user a tool to produce data for the platform owners.

The labor relations described in our fictional anecdote are no different from contemporary labor relations, merely an intensified version of these. They are intended to reflect the situation faced by the *precariate* (Alves, 2013), unstable knowledge workers who consume and produce in a service economy. The anecdote is also a reference to the universal, average consumer-producer of monetized data (all of us, eager data-driven clients of Google, Facebook etc.) in the growing logic of accumulation in surveillance capitalism (Zuboff, 2015).

* * *

> 723 minutes of this day are already registered in your personal time-management system and you have not yet earned a single penny. You stop working to glance again at the traffic jam forming in front of your window. Hundreds of delivery drones are now carrying customized meals for individuals who have chosen to eat lunch today. You expect that at a certain point their anti-collision systems will produce a beautiful flying pattern which resembles bird flocks of the past. At exactly 724 minutes and 2 seconds the traffic is sufficiently saturated to produce a shape that is recognizable by humans: an infinite double-crossed loop. As soon as the phenomenon becomes visible, you activate record mode in your RRD. You keep staring at the rare shape

without blinking to avoid any data loss. At 724 minutes and 40 seconds the phenomenon can no longer be observed and the recording automatically stops. You wipe your tears away.

"Aleph, how much can I get for this amazing video loop?"

"There have been no streaming workers looking for amateur footage in the past few days. I will send you a notification when it becomes trendy again. Have you made up your mind about lunch?"

"I'm not feeling hungry... What is your lunch recommendation?"

"I identified that today's neo-vintage package may be the best option available in the market to reduce your weight and improve your digestive system. You need to restore your gut flora after yesterday's diarrhea caused by the genetic obesity treatment."

"Ouch. I don't want the tasteless neo-vintage crap. Order me a finger-food package now before the afternoon price applies."

"Ok, but first you must confirm two conditions: 1) you are aware that the food you ordered exceeds your maximum consumption of calories for today and 2) you are aware of the degrading working conditions of the cooks in the establishments that produce this kind of meal."

"Of course I am, Aleph! Why in the hell can't you just use the automatic acknowledgement system?"

"Because the working conditions recently reached levels below the acceptable rating."

"Just order my meal, please."

Figure 14.1 Watching a drone flock. Illustration by Yuri Andric, 2017.

While you wait for your food to arrive at the droneport, a bundle of personalized news is downloaded by your armchair. To distract yourself from thoughts about working conditions, you sit down and read the news items as they are shown on your RRD.

Young female level 'A' loses 3 citizenpoints while trying to celebrate her birthday party with strangers who could not be found on social media.

Things are getting more difficult for those who are keener on social interaction and face-to-face encounters. The surveillance systems cannot guarantee the safety of small gatherings, especially when there are off-the-grid people there. You want to celebrate your birthday too but since you moved to this city, you have not made any friends. You need to meet strangers.

"Your food package has arrived," says Aleph.

As you eat your tiny finger-food package, you are worrying about whether your third job contract will be extended until next week, when the air-conditioning bill arrives. You know you can pay the bill with what you earn from your main and secondary jobs, but there won't be much left in your pocket for at least another three weeks once you've paid the bill.

You look through the window and think, "Why in the hell do we need to pay for air this week? I voted for refreshing the city air last week. Maybe it didn't work again. Let me check."

When you check in Arena, you discover that your vote in the air quality survey was in vain. The choice offered was cancelled by the Environmental Policy Trust. Someone called John Barber sued the trust for religious offense when it included the crowd-sourced renewal scheme in option one. The scheme did not make any distinction between the air coming from apartments where LGBT people lived and the air from apartments where straight people lived. The problem for John was that straight people would have to re-inhale oxygen which went through devilish lungs, which is totally unacceptable from his religious standpoint.

"This is disgusting. These guys don't know what it is like to live in a personalized city district. They should care more about their own lives instead of other people's lives." You look back at your assistant pod and ask:

"Aleph, what is the air pollution rate today?"

"It is 68% toxic. I recommend that you do not open your window today."

You open the window and take a deep breath. Your nostrils immediately become dry as you inhale the polluted air, but you feel as free as a bird soaring on high. It is a liberating feeling to breathe without any assistance – and to disobey your own personal assistant. The noise of the drones going around your building soon interrupts your epiphany. It is time to close the window.

"You lost one citizenpoint for that disruption. You were warned not to do that."

"Shut up, Aleph. This is my thing."

Aleph warns you that you will earn 20% less for the carbon emissions collected from your living unit since they are mixed with toxic gases from the outside that cannot be filtered. You order carbon credits from Amazonia Renewal Inc. despite the unethical rating of the transaction.

Still without a response to the last piece of work you submitted, you decide to take some gig jobs to be able to honor next week's air-conditioning bill.

"Aleph, check if there is any urgent job for the next two hours."

"I found four jobs that match your profile: a drone payment-collection task, a genome data-normalizing task, a tightcoin manual-mining task and a wedding music-composition task. They all pay you one tightcoin per task. Which one do you want to accept?"

"All of them. Give me a fast **metronome pill** and let's do the job."

"You will have to work without my assistance because your earnings won't pay for my data-processing costs this time."

"I know, I know. But that's all there is today."

"Enjoy your work then!"

* * *

Individualism, private spheres and segregation in different levels seem to be strong drivers that will shape urban societies in the near future. The data-driven, programmable city is leading to an increasingly personalized urban environment. Tensions between insiders and outsiders, and between what is common or public and what is owned or private, are parameters for multiple-boundary (material and immaterial) territories.

Thinking about these territories raises a host of new questions about urban imaginaries. What is outside and inside in the city of walls (Caldeira, 2001)? What is public and private in today's and tomorrow's neoliberal cities? Are the public parts of urban space the spaces left over from private areas, with no function other than connecting private land? How private are public streets and squares? These are questions that have yet to be answered in relation to current (and future?) ways of planning, designing and managing public spaces in major cities around the world. According to Coaffee and Fussey (2011), the evidence for this kind of urban transformation, which is leading to more secluded and securitized cities, can be observed in four types of interventions: the growth of electronic surveillance within public and semi-public urban spaces; the increased popularity of physical or symbolic notions of the boundary and territorial closure; the increasing sophistication and cost of security and contingency planning; and the way that resilience has been embedded within the urban context through urban architectural and design interventions.

The expansion of securitization is also related to the gentrification process in large and medium-sized cities and metropolitan regions. This phenomenon in turn is linked to the increasing number of closed residential condominiums, a form of urban land use and a spatial product that is highly valued in the real-estate market (Caldeira, 2001). Even when they are within the city area, these places form secluded perimeters with a combination of road design and enclosures that disconnect them from the surrounding urban fabric. Property security systems thus become an integral part of this form of urban development and are seen as an essential item in city management.

However, fences, walls, and other visible defensive architecture (Newman, 1995) are not the only mechanisms that give shape to boundaries and territorial protection. Surveillance cameras, CCTV systems, security mechanisms, and procedures for controlling spaces are part of any current architectural or urban project and, like building materials, are considered indispensable. Together with surveillance and law enforcement schemes, architecture and design have become important parts of the sociotechnical arrangements that represent the securitization of private and public urban spaces.

The bubble metaphor used by Peter Sloterdijk (2011, 2014, 2016) can help to explain how public and private spaces are controlled and managed by means of flexible boundaries that can be adjusted to different sociotechnical contexts. Don Mitchell (2005) studied cases in the United States in which state courts ruled that everyone within a 100-foot-radius buffer zone around health clinics is protected against the unwanted approach of protesters by legal 8-foot privacy bubbles, which he calls territorial floating bubbles.

It therefore comes as no surprise that the main character in our fictional anecdote feels relieved, independent, and more human when he manages to burst some of his territorial floating bubbles by ignoring Aleph's recommendations. In the urban imaginary of increasingly (im)mobilized and hyperconnected cyborgs and cyburgs, there is a profound overlapping of physical, regulatory, and digital territorial bubbles that contributes to increasing segregation, individualism, and criminalization of everything not predicted or preempted by the "algo-rhythms" of urban life.

A visible manifestation of these trends is the growing tension between public and private spaces – from ownership through concession to use and appropriation by individuals and groups with varied interests (Firmino and Duarte, 2016). Privatized and privately manned spaces are becoming increasingly common and are viewed as a move to the "clean and safe" city. In the UK, a legal instrument called a Public Spaces Protection Order (PSPO) has been implemented as a new form of spatial control, a new way of territorially inscribing existing regulations governing anti-social behavior, to a defined area rather than to a person. With the advent of PSPOs, any predefined activity within a specific area can be criminalized. Many councils in the UK are now using this new legal power to limit the freedom of citizens in open areas, controlling movements and behaviors.

The result of this trend in the management of urban spaces is a series of blurred and overlapping boundaries defining the "haves" and "have nots" and the "cans" and "cannots" across the tracts of land formed by what were originally common public spaces in cities. Instead of these public spaces, there will be more and more exclusive and secluded territories with little or no resemblance to what one day was a vibrant space designed to harbor human interactions. According to Lefebvre (1991), this negative trend in capitalist space can be countered by making use of boundaries as places to enjoy differences.

* * *

"Did you go outside? You seem scared. How can I help you?"

"No, I didn't go outside. I just thought about it and gave up the idea. I'm not scared. You must reprogram your emotion detection system to be less preemptive. I'm just anxious about today's outcome in Arena."

"There is no need to worry. The likelihood of your proposal being accepted by your neighbors is 78%."

"You know I do not trust your predictions, Aleph."

"You shouldn't, but my prediction accuracy has improved 7.5% since last week's update."

"Just luck."

"I thought you didn't believe in that."

"I thought you didn't care about beliefs."

"I didn't use to but it is part of my new update. How do you like it? The landlord asked for feedback."

"A terrible idea… that is all I can say!"

"The landlord is trying to help you choose something to believe in."

"They want me to believe I can pay my bills. That is all they care about."

"By the way, do you want to order a 3-hour sleeping pill?"

"Not now…"

After 1,800 seconds of failed sleep attempts, you give in and take the pill. At 607 minutes you are woken up by the noise of people talking loudly in front of your door.

"Aleph? What time is it? What is going on? Why did the pill let me sleep so much? Is it disconnected from the Wi-Fi?"

No answer. You notice that all systems in your living unit are shut down, including the air conditioning. A range of natural smells coming from your used clothes, stinking shoes and rotten pieces of food reach your nose. The smells are quite unusual because the air conditioning normally filters out any bad smells. Could there have been an energy blackout again? The last time this happened was seven years ago.

A mixture of excitement and panic prevents you from jumping out of your bed. Since your RRD has run out of battery too, you decide to cover your head with the blanket until everything is back to normal. To distract yourself from reality, you keep thinking fiercely about the last genome data normalization task. The code seems to trickle out of your trembling mouth as you mutter, "CATACAAGTGGCAGATGATG. . ."

"Hey, wake up!"

I touch your feet with my left hand while I hold up a candle with my right hand. You don't know me but I know a lot about you. I spent five weeks analyzing the life data of all the tenants on this floor. The landlord hired me after I managed to optimize the performance of living units in a building next door by 30%. Now that there is a blackout, I want to satisfy my curiosity and learn what it is like to interact directly with all these people I monitored from my unit. I used to be a CCTV operator a long time ago, but everything is different now.

"What the @#&! Who are you?"

"I'm your neighbor. We all live on this same floor and are meeting for the first time because of the blackout. Would you like to join us for a chat while we wait for the system to come back?"

"What are you talking about?"

"Nothing in particular. It is just small talk."

"I don't have time for that."

"Of course you do. I know exactly how much time you have every day. I work for the landlord to optimize your living conditions. Come, you will like it."

After hesitating to accept my invitation, you finally decide to step out and join us in the corridor for a chat. We have brought chairs from our living units and arranged them in a circle around a group of candles.

"Do you know what caused the blackout?" asks a tenant whose personal data would scare you.

"I don't know," you reply.

"He is not interested in the worker movement," I explain.

"How can you not care about the inhumane conditions of workers?"

"I work too and I don't complain."

"Look. We are not workers. We are entrepreneurs. We can still decide on which job to take."

"But I don't have a fixed wage."

"Neither do workers. Nobody has fixed or minimum wages anymore since the last wave of austerity measures. That is why they might have caused the blackout. It is a protest."

"What is a protest?"

"People gather together and demonstrate their dissatisfaction in public spaces."

"But there are hardly any public spaces in this personalized city. The open squares are covered by solar panels to power the street lights and the streets, by the same token, are maintained – badly maintained – by transportation companies such as auto-auto."

"That is why they might have blacked out the system. The last time they tried to stage a protest in a public space it was ineffective. Just to find a space that was still public took months of legal research. The current legislation allows for moving perimeters, such as the private space around autonomous cars. The last public space they found was sewer manhole covers. Each protester stood on top of a manhole cover and shouted as loud as she or he could, but because the protesters were so far apart, the streamers who covered the protest called them crazy."

"That's funny! The streamers went on strike themselves two weeks ago..."

"No. They just stopped broadcasting and waited until streaming content became valued in the market again. When workers strike, they also protest."

"Do they lose citizenpoints for that?"

"Only if they go against the provisional laws compiled by Arena algorithms. The problem is that there are too many algorithms, and they are quite complex."

Our conversation is interrupted by the lights coming back, the air conditioning powering up and the surveillance cameras pointing at us.

Figure 14.2 Protesting on manhole covers. Illustration by Yuri Andric, 2017.

"Alright. It is time to go back to our living units, guys," I say.

"Are you going to keep tracking me?" you ask me.

"Until you protest about it."

"Never mind."

You close the door, sit on your armchair and wait for the system to boot up.

* * *

The trends outlined here say much about our view of the future of mobile and connected urban societies. David Lyon (1994) considers the circulation of personal information one of the most important issues related to technological changes in the late twentieth century, when the ethics and politics of surveillance became a major concern for the social sciences. The electronic eye and what Lyon then called the surveillance society but now defines as a culture of surveillance, together with a marked neoliberal approach to political economy, have contributed to a dramatic shift of attention from individuals, personal stories, social relations, and identities to their codified representations in databases and possible classified identifications. The culture of control – mind, body, and surrounding space and territory – has become dominant.

When Lyon wrote *The Electronic Eye*, it was 1994 and the internet was in its infancy. Many were dreaming about the wonders of a democratic, hyperconnected society and tackling problems of social inequality through online communities. Surveillance was already one of the fundamental pillars of modern society and was seen as a key interface that explained many of today's societal and spatial structures. Being connected and in movement became synonymous with being seen, watched, monitored, interpreted, and, more importantly, classified.

More than two decades on – after 9/11 and Edward Snowden's revelations about information gathering and analysis by intelligence agencies such as the NSA – personal information is now personal data, and there is Big Data as well as powerful algorithms to govern the movements of data and everything that can be done with it. Inequalities have grown, as has apprehensiveness about the augmented way in which personal data is shared, exchanged, sold, and classified for social-sorting purposes. In this chapter, we have attempted to illustrate some examples of territorial manifestations anecdotally to show how a potential hyperconnected society is in fact leading to patterns of immobilization for targeted individuals and groups.

Surveillance has become extremely naturalized, and data sharing in exchange for more convenience and security is taken for granted. Aleph and other mundane connected devices, together with their respective algorithms, tend to command much of our routine, day-to-day activities in the hyperconnected (and immobile) city. Sociologically and in terms of urban life, the intensification and concomitant banalization of surveillance made possible by increasingly invisible and miniaturized technologies has turned everyone and everything into interconnected monitoring stations, leading to one of the conundrums of future cities by transforming hyperconnected citizens into immobilized beings.

* * *

"Aleph, what caused the blackout?" I hear you saying.

You get no response from Aleph because the system is loading a big update. Poor you. After a 140 second wait, Aleph speaks in a metallic voice that sounds like his voice 10 years ago.

"You have mail."

"Ah, finally you are back. Can you please tell me what happened?"

"You have mail."

"You know something is terribly wrong when Aleph acts as though he were not on your side. I know what is the problem, but I could not tell you when we met."

"Ok. Summarize the new message."

"I cannot summarize this message."

"Alright, read it out loud."

"You must read it for yourself with the eyetracking on. It is a notification from your landlord with an encrypted read-receipt request."

"Damn it! Ok. Let me read my 'private love letter'!"

You sit on your armchair and raise your head. The message is printed on your RRD. As you go through the lines, tears fall to the ground. The message reads like a court judgment.

In the last two hours, you have committed three actions against the city order. You have met strangers, discussed politics out of the reach of transparency systems, and conspired to organize a possible terrorist act. As a result, your citizenpoints have been reduced to zero. From now on, you can no longer be treated like a fellow citizen by this city. All your mass customization services have therefore been suspended.

The consequence of this suspension is that you cannot access the city network, find jobs that fit your profile, earn personal credits or order customized amenities. Devoid of any augmented reality, the living unit where you live is now an empty box with basic furniture. Even your assistant, Aleph, is no longer a personal assistant. Everything it says now comes from a generic interactive system with no artificial intelligence that provides minimal functions. Every ID attached to your biometrics has been classified as a potential threat to civic order, and your data has been put into a list called "public enemies or potential terrorists".

All the materials that enabled you to stay inside your living unit are now gone, and you see no option other than to sell your labor-power the traditional way: to become a waged worker until your citizenpoints are enough to qualify for a fellow-citizen license and your name is erased from the threat lists. I know you are going to end up liking life outside your living unit since there you can experience the human contradictions that drive our society directly without filters. Love and hate, capital and work, order and chaos, movement and connection will no longer be hidden.

As for me, who described your fateful day, who incited you and knew that the blackout was a test for the new analog surveillance system, the punishment will be much worse. I'll be deported from this city and will be separated from you. But it's OK. This has happened many times already and I have learned to enjoy it. I have this weakness that means I develop a love/hate relationship with those I watch, but I can't help myself from watching. I build my life from the threads of the lives I watch and interfere with. I am only concerned with what's not mine.

Logoff

The last sentence of our story is a reference to the Cannibalist Manifesto, written by the Brazilian Modernist writer Oswald De Andrade (De Andrade and Bary, 1991), which values local and indigenous cultural roots over foreign influence. Throughout this text, we have described and discussed relations between people and technologies in an attempt to depict an urban imaginary about connections (and disconnections) and movements (and immobilizations) in a fictional daily routine – albeit one that takes an unexpected turn – in a possible future city. To present our own version of an urban phantasmagoria, we created a cannibalistic design fiction (van Amstel et al., 2012; Gonzatto et al., 2013) that incorporates near-future technologies, current debates, and familiar interactions between beings and things.

We wanted to highlight that there is no need for further scientific discoveries for this future to become a reality; it is just a matter of designing existing technologies in the ways depicted here. The question posed to readers is: which parts of these possible future relations between movement and connection and between beings and things in cities do we want for ourselves?

We believe that ultimately all fiction is part of reality, just as every conceived future is linked to the lived present (Gonzatto et al., 2013). Hence, any production, such as this text, must also be seen as an effort to intervene somehow in this present time and localized reality.

We decided to portray a character who is under much more surveillance than would be possible with our current legislation but at the same time reacts in disbelief when confronted with typical mundane activities of our time. This is a typical resource used in narratives that seek to serve not only as a warning against establishing unwanted directions, but also as the basis of a debate on possible futures in the light of current trends, an approach that Dunne and Raby (2013) call critical design. In the story we have tried to focus on movements in a hyperconnected urban environment. We are not so much interested in the intrinsic mechanisms of the smart city, as in the possible urban living conditions suggested by this futurology.

As a virtual materialization, the story here shows that movement and connection can change, but not simply because of technology. Changes are driven by the interests of groups and individuals, with consequences for multiple aspects of urban life. As an urban phantasmagoria, our narrative is intended to point out that the dissemination of technology is not merely associated inevitably with the intensification of a culture of surveillance and control, but is also an essential element for this increased surveillance and control to happen.

Possibilities for resistance appear in several passages of our story. However, the issue is not how to resist, but how to think through all the relations between humans, space, and technology in our own time. At times we should take a critical stance toward technologies, much like the protagonist, while at others we should appreciate and enjoy the contradictions of our society, like the narrator. By doing so, our urban imaginaries may be produced as counter-projects (Lefebvre, 1991) to hegemonic futurology (Gonzatto et al., 2013) and represent critical views as well as alternatives.

By moving away from dominant views of urban futures – such as smart cities – we have provided new connections between disparate theories and technologies. Our intention is to stress that the exponential increase in connectivity experienced by urban dwellers does not necessarily result in more movement. On the contrary, such extended connectivity may even prevent movement if the underlying purpose of the infrastructure is to capture data. We believe that critical urban imaginaries are an arena for disputes between different new modes of beings and things, as well as between different ways of moving and connecting.

Glossary

Aleph This refers to a short story of the same name written by Jorge Luis Borges (1970), where Aleph is a point in space that contains all other points and from which one can see everything in the universe from every angle without distortion. Obviously, Aleph in this narrative is inspired by digital personal assistants in the fashion of Apple's Siri, Google Assistant, Amazon's Alexa, and Microsoft's Cortana.

Arena Social participation software in which fellow citizens must fulfill their democratic duties with the city. Political algorithms compile private policies and provisional laws from the results. The system works as a direct democracy with no representatives.

Auto-auto A fictitious driverless-car service offering hired rides, something like a driverless Uber, without any human driver (current projects on driverless cars still have humans behind the wheels, just in case).

Citizenpoints Points credited by the city operating system, granting privileges and duties. Since 2015, the Chinese government have shared citizen data with companies who use reputation systems to regulate financial transactions, select employees, and accelerate paperwork.

City operating system Distributed software and hardware that provide access and control of urban infrastructures to private service providers. This system is an advanced version of present day's Smarter City® by IBM, CityNext® by Microsoft, and Urban Operating System™ by Living PlanIT.

Drone pickup and delivery service Drones have made incredibly cheap and convenient pickup and delivery services, rendering physical movements through the city almost unnecessary. Amazon Prime Air, a drone delivery service, is already being tested in the US.

Metronome Pill This is a reference to the trend toward pills that are intended to increase productivity. Metronome refers to the "Pomodoro technique," which claims to increase concentration at work. Another example of a pill with a specific function is the "Audiopill," a swallowable pill that vibrates inside the body and allows one to listen music from the inside.

Real Retina Display (RRD) Fictional contact lenses with augmented-reality capabilities. It combines the interactivity described in the Samsung smart contact lenses patent with the interactivity of Google Glass and Microsoft Hololens.

Tightcoins This is an allusion to mileage programs. Here, the rewards are used as a universal currency stored in a digital wallet. Many attempts have been made to find a general replacement for cash, including credit cards, bitcoins, and other cryptocurrencies.

References

Alexander, C. (1968) A City Is Not a Tree. *Ekistics* 139: 344–48.

Alves, G. (2013) O Que é o Precariado? Blog da Boitempo. Available at: http://blogdaboitempo.com.br/2013/07/22/o-que-e-o-precariado/.

Amstel, F.M.C. van, Vassão, C.A., and Ferraz, G.B. (2012) Design Livre: Cannibalistic Interaction Design. In: *Innovation in Design Education: Proceedings of the Third International Forum of Design as a Process*, 3–5 November 2011. Torino, Italy: Umberto Allemandi & C.

Borges, J.L. (1970) The Aleph. In: Borges, J.L. (ed.), *The Aleph and Other Stories: 1933–1969*. New York: E.P. Dutton, 15–30.

Brooker, C. (2011) *Black Mirror*. Los Angeles: Netflix.

Caldeira, T. (2001) *City of Walls: Crime, Segregation, and Citizenship in São Paulo*. Berkeley: University of California Press.

Coaffee, J. and Fussey, P. (2011) Resilient Planning for Sports Mega-events: Designing and Managing Safe and Secure Urban Places for London 2012 and Beyond. *Urbe* 3(2): 165–77.

Cuff, D. (2003) Immanent Domain: Pervasive Computing and the Public Realm. *Journal of Architectural Education* 57(1): 43–49.

De Andrade, O. and Bary, L. (1991) Cannibalist Manifesto. *Latin American Literary Review* 19(38): 38–47.
Deleuze, G. (1992) *Post-scriptum Sobre a Sociedade de Controle. Conversações*. Rio de Janeiro: Editora, 34.
Deleuze, G. and Guattari, F. (1988) *A Thousand Plateaus: Capitalism and Schizophrenia*. New York: Bloomsbury Publishing.
Duarte, F., Firmino, R., and Crestani, A. (2015) Urban Phantasmagorias: Cinema and the Immanent Future of Cities. *Space and Culture* 18(2): 132–42.
Dunne, A. and Raby, F. (2013) *Speculative Everything: Design, Fiction, and Social Dreaming*. Cambridge: The MIT Press.
Firmino, R. and Duarte, F. (2016) Private Video Monitoring of Public Spaces: The Construction of New Invisible Territories. *Urban Studies* 53(4): 741–54.
Gill, K.S. (ed.) (2012) *Human Machine Symbiosis: The Foundations of Human-Centred Systems Design*. London: Springer Science & Business Media.
Gonzatto, R.F., Van Amstel, F.M.C., Merkle, L.E., and Hartmann, T. (2013) The Ideology of the Future in Design Fictions. *Digital Creativity* 24(1): 36–45.
Haggerty, K. and Ericsson, R. (2000) The Surveillant Assemblage. *The British Journal of Sociology* 51(4): 605–22.
Haraway, D. (1991) A Cyborg Manifesto: Science, Technology, and Socialist-Feminism in the Late Twentieth Century. In: Haraway, D. (ed), *Simians, Cyborgs and Women: The Reinvention of Nature*. New York: Routledge.
Kitchin, R. (2011) The Programmable City. *Environment and Planning B: Urban Analytics and City Science* 38(6): 945–51.
Latour, B. (2005) *Reassembling the Social: An Introduction to Actor-Network-Theory*. New York: Oxford University Press.
Lefebvre, H. (1969) *Posição: Contra os Tecnocratas*. São Paulo: Documentos.
Lefebvre, H. (1991) *The Production of Space*. Oxford: Blackwell.
Lefebvre, H. (2004) *Rhythmanalysis: Space, Time and Everyday Life*. London: Continuum.
Lessig, L. (1999) Code Is Law. *The Industry Standard* 18.
Lévy, P. (1998) *Becoming Virtual: Reality in the Digital Age*. New York: Plenum Trade.
Lyon, D. (1994) *The Electronic Eye: The Rise of Surveillance Society*. Minneapolis: University of Minnesota Press.
Lyon, D. (2009) *Identifying Citizens: ID Cards as Surveillance*. London: Wiley.
Lyotard, J.F. (1992) *The Inhuman: Reflections on Time*. Redwood City, CA: Stanford University Press.
Mitchell, D. (2005) The S.U.V. Model of Citizenship: Floating Bubbles, Buffer Zones, and the Rise of the "Purely Atomic" Individual. *Political Geography* 24: 77–100.
Newman, O. (1995) Defensible Space: A New Physical Planning Tool for Urban Revitalization. *Journal of the American Planning Association* 61(2): 149–55.
Sloterdijk, P. (2011) *Bubbles. Spheres Volume I: Microspherology*. Cambridge: MIT Press.
Sloterdijk, P. (2014) *Globes. Spheres Volume II: Macrospherology*. Cambridge: MIT Press.
Sloterdijk, P. (2016) *Foams. Spheres Volume III: Plural Spherology*. Cambridge: MIT Press.
Superflux. (2017) *Our Friends Electric*. Available at: http://superflux.in/index.php/work/friends-electric/#.
Taylor, A. (2016) The Radical Plan to Destroy Time Zones. *The Washington Post*. Available at: https://www.washingtonpost.com/news/worldviews/wp/2016/02/12/the-radical-plan-to-destroy-time-zones-2/.
Weiser, M. (1991) The Computer for the 21st Century. *Scientific American* 265(3): 94–104.
Zuboff, S. (2015) Big Other: Surveillance Capitalism and the Prospects of an Information Civilization. *Journal of Information Technology* 30: 75–89.

Part III

Connected and consuming

15

Imagining the open city

(Post-)Cosmopolitan urban imaginaries

Myria Georgiou

Introduction

This chapter examines how imaginaries of the city as a cosmopolitan, open and diverse place are generated, supported, and regulated at times of crisis. As this chapter is being written, cultural and social diversity in cities across the US, Europe and beyond are under attack. Arguably the most fundamental quality of urban life, diversity, is on the receiving side of xenophobic and neoliberal politics that divide the city and increase inequalities. How does the city respond? Are post-cosmopolitan politics replacing an urban cosmopolitan vision? And how are imaginings of the city's (post-)cosmopolitan future mapped out and contested on the urban digital and material street?

The chapter analyzes cosmopolitan urban imaginaries through a multidimensional case study: the global city's response to the nation's post-cosmopolitan politics in the case of the Brexit vote. In a referendum that targeted migration and free movement, the city stood against the nation with urban voters largely disapproving stricter migration control and more rigid national borders. While most British cities voted to "Remain" in the European Union, the focus of the present discussion is on London for three reasons: London represents the most culturally diverse city in the UK and arguably in the world; it is a global city carrying enormous symbolic and financial power; and its cityscapes and cultures have for long nurtured local and global imaginaries through iconic media and literary representations of the city. Thus, post-Referendum London most vividly reflects the complex and contested imaginaries for the city's future, shaped at the intersection of cultural diversity and the mediated vision of a city that is recognizable, desirable, and powerful. In fact, these are the principles underlying London's post-Brexit campaign, #LondonIsOpen, which is analyzed here. The campaign was initiated by the Mayor of London, Sadiq Khan, but quickly spilled from the city's headquarters to the digital and material street; thus, #LondonIsOpen reflects the institutional and vernacular complexities and contradictions of urban imaginaries at times of cosmopolitan crisis.

As will be shown, #LondonIsOpen offers a vivid illustration of urban (post-)cosmopolitan imaginaries in their moral potentials and political contradictions. A campaign primarily brought to life in social media, #LondonIsOpen projects a three-dimensional vision of the global city: as an open city that is digitally articulated and collectively constituted. As will be shown, this three-dimensional vision, reflecting core values of liberal cosmopolitanism, constitutes an imaginary that is both shared and contested on the digital and material street of the global city.

The chapter is structured in three main sections. The first section introduces urban imaginaries and their cosmopolitan articulations, especially through a review of literature on the urban imaginary, cosmopolitanism, and the open city. The second section discusses the #LondonIsOpen campaign and its vision of the city as open, digitally articulated, and collectively constituted. Illustrating this vision, two campaign films are analyzed. The third section engages with responses to the campaign, as these are expressed on social media and beyond. These responses reflect three conflicting orientations for the future of the city: a neoliberal cosmopolitan vision, a vernacular cosmopolitan vision, and a post-cosmopolitan vision.

Literature review: imagining open cities

The urban imaginary constitutes a key concept for understanding contemporary city life, as lived, represented, and ordered. Its relevance to the present analysis is twofold. First, imaginaries are naturalized systems of ideas, sensibilities, and meanings (Cinar and Bender, 2007). As such, they support a sense of self and a sense of belonging, especially in discursively feeding into people's understanding of their social existence and how it "fit [s] together with others" (Taylor, 2004: 23). Thus, imagining the city as a space of identity and community comes with a moral (and financial) investment in a place, in people, and in values associated with that city. As imaginaries scaffold normative notions and expectations, they support belonging but they also legitimize systems of power by making certain ideas and ideals ordinary and widely accepted (Castoriadis, 1998). This means that, conceptually, imaginaries invite us to understand imagination, moral orientation, and action in their interconnections and in relation to power relations. In fact, and as will be discussed below, the repetitive and powerful articulation of narratives and images of what the city is and who it is for, sets norms about what urban life and cosmopolitanism mean and who has the right to define and enjoy them.

The power of images and narratives to nurture collective understandings of what the city is, or what it is supposed to be, represents the second fundamental contribution of the imaginary as a concept. The imaginary reminds us that the city is not only constituted territorially but also symbolically (Cinar and Bender, 2007; Georgiou, 2013). It is often, and increasingly, the case that ideas and ideals circulated in the media challenge popular understandings of the city as a territorial entity contained and defined through the nation. Iconic cityscapes, urban lifestyles, and urban subcultures widely consumed on mass media and social media bring certain cities to audiences across the world, reaffirming their desirability and concentrated symbolic power. The city is communicated and spatially ordered in the media (Boudreau, 2007), especially through the powerful imagery of places and practices associated with urban locations and zones (Gordon, 2010; McQuire, 2008). Media representations, for example, feed into decisions on urban planning, selective investment and social policy (Zukin, 2010). Powerful media representations of certain parts of the city as desired or undesired, dangerous or attractive, as "hip" or as "ghetto", affirm the monetary value of certain urban territories but also the symbolic value of those territories' people and their lives (Binnie and Skeggs, 2004; Zukin, 2010).

In the same way as territories are labelled in media representations, the symbolic value of the city in its totality is branded and affirmed in the media. The *open city* in the case of #LondonIsOpen becomes an important concept for branding post-Brexit London, turning a value into a normative identifier of what the city represents. The emphasis on openness becomes important, not least as it speaks to popular and scholarly articulations of cosmopolitan imagination and vision for democratic inclusive cities.

Sennett (2013) puts forward the concept of the open city to describe the city as a site of assemblages, mutual exchanges, and ambiguous edges. This is the city that contests neoliberalism's closed systems, which, as he argues, aim to be integrated, controlled, and ordered. Sennett (2013) builds his claim on a long tradition of radical urbanism, which identifies the city's openness as necessary to making claims to freedom and rights. Along similar lines, Benjamin (1997), in *One way street*, described urban porosity and transitivity as emerging out of the unforeseen constellations on the street and in urban dwellers' claims to that space. With Benjamin (1997) and Lefebvre (2003) as influential starting points, generations of urban scholars explored unforeseen constellations, disorder, and urban porosity as generative of democratic, inclusive cities (cf. Sennett's work on disorder, 1970; Jacobs' lamenting of the aggressive modernization of the American city, 1993; Zukin's critique of gentrification, 2010).

Central to conceptualizations of the open city are the conditions and demands for a city that is open to different voices, a site where all can speak and can be heard (Silverstone, 2007). Sennett (2013) emphasizes that the city's openness allows different groups to interact and to find solutions together but to also develop skills for managing disorder and conflict, multiple connections and disconnections. The possibility for voice and articulation of urban dwellers' visions of their city becomes yet more important in highly mediated cities. In many cities, especially in the global north, digital infrastructures have reconfigured urban communicative landscapes. Digital communications have expanded the porosity of the city, and arguably its shared imaginaries, both internally (with hyper-local connections that enhance communication at the level of the neighborhood, cf. Georgiou, 2017) and externally (with transnational connections that support collective imaginaries well beyond the city's and the nation's boundaries, cf. Ponzanesi and Leurs, 2014).

In the case of the global city (Sassen, 2001) especially, demographic change and communicative porosity have long defined the city as an imagined and lived place. This porosity, expressed in global cities' long histories of migration and connections with other parts of the world, has supported imaginaries of cities as sites of visceral cosmopolitanism (Nava, 2007) and actually existing cosmopolitanism (Robbins, 1998). According to these definitions, cosmopolitanism is rooted in the urban experience and embedded in urban encounters with difference. If cosmopolitanism is indeed an underlying condition of urban life, is the city by definition open? Are urban imaginaries that generate communities and action defined by shared cosmopolitan values and respect for difference? Are global cities' dwellers in their social and cultural diversity converging around cosmopolitan imaginaries?

There is no doubt that city dwellers practice and imagine urbanity differently. For the transnational elites who congregate in London, diversity is primarily imagined through acts of consumption and in regular encounters with other members of transnational elites (Hannerz, 1996; Keith, 2005). For Hannerz, who primarily discusses intellectual and social transnational elites, the definition of cosmopolitanism is associated with "an orientation, a willingness to engage with the Other ... an intellectual and aesthetic stance towards divergent cultural experiences, a search of contrasts rather than uniformity" (1996: 103). As the privileged cosmopolitanism that Hannerz (1996) describes depends on a consumption-driven imaginary, it raises questions about its ethical trajectories and its relevance to subjects who cannot enjoy transnational mobility and who do not have choices of engaging with difference (but who might have to learn to live with it). Werbner (2008) and Hall (2008) speak to and against such assumptions of cosmopolitanism as a lifestyle and as an intellectual choice, by offering a rather different definition through the experience of working-class migrants and diasporas. In defining vernacular cosmopolitanism, Werbner locates cosmopolitan subjectivities in symbolic and material spaces, which are "trans-ethnic, collectively emergent 'worlds', shared discourses that transcend cultural boundaries and parochial lifestyles" (Werbner, 2008: 50). Emphasizing that cosmopolitanism is not a property of the elites, she notes that cosmopolitan subjectivities of different classes and ages can contribute to transethnic cultural and ideological worlds. Along these same lines, Hall identifies vernacular cosmopolitanism as not being about choice but being about survival, especially for migrants and diasporas:

> They have to acquire the same cosmopolitan skills of adaptation and innovation which an entrepreneur requires – but from a different place ... So, culturally, they are living "in translation" every day of their lives ... not the global life as a reward for status, education or wealth, but the global life as one of the necessities imposed by the disjunctures of modern globalization.
>
> *(Hall, 2008: 347)*

If cosmopolitanism in the city reflects the diversity of its dwellers' experience, then its imaginings cannot be singular. This raises questions about who speaks and who benefits from the vision of the cosmopolitan city? Harvey famously asks "In whose image is the city made?" (1973 [2009]) and we could extend this question to ask, "In whose image is the cosmopolitan city imagined?" Is a media campaign for and on behalf of the city that puts forward a cosmopolitan imaginary representing its people and their vision of the city as open, shared, and hospitable?

Context of study

The UK referendum to leave the EU in the Summer of 2016 was much more than a vote for the institutional rule of governance in the country. At the core of the "Leave" campaign was an anxiety against migration and a demand to tighten border control. The vote confirmed this anxiety and demand, and also reminded many Londoners that the society they imagined was in disjuncture with the one the nation desired. In fact, and while the country voted to leave the European Union with a majority of 51.9%, London voted Remain with a 59.9% majority of the vote, with some parts of the city voting Remain by more than 70% (BBC, 2016). On the material and digital streets of the city, disbelief and protest revealed the disenchantment of the city with the politics of the nation. Most importantly, campaigns, statements, protests, and social media responses to the vote (Manlder, 2016; Vulliamy, 2016) showed that many Londoners felt that their cosmopolitan values were under attack. In a city demographically constituted mostly by minorities and secondarily by the national majority (BBC, 2012), revolt against Brexit came as no surprise: against a vote for firmer borders, less migration, and revived nationalism, London appeared to project its own, and *other* to the nation, cosmopolitan vision. But does this vision reflect a shared and common urban cosmopolitan imaginary? While the events following the referendum revealed the wide appeal of cosmopolitan ideals across the city, the meanings of the values associated with cosmopolitanism – openness, respect, and recognition of difference – are not necessarily equally shared.

Imagining and mediating the open city: #LondonIsOpen

The post-referendum disbelief and despair in the global city is most vividly captured in an urban campaign intended to reach out to local, national, and global audiences. #LondonIsOpen was an official mayoral campaign initiated weeks after the June 2016 referendum. Yet, and as intended, the message soon spilled outside the headquarters of city governance and into the city's digital and material streets. While #LondonIsOpen was initially launched in July 2016, its apparent success turned into a longstanding trademark campaign as it continued well into 2017. In its official expression, the campaign unfolds in three main media domains: the official campaign website; a series of 11 short films under the hashtag/brand #LondonIsOpen circulated across social media (from Facebook to Youtube official channels); and in the mainstream media coverage of the campaign.

As a campaign branded as a hashtag, #LondonIsOpen purposefully emphasizes a digital identity of London's openness. The ordinariness of the hashtag as a way to name a campaign, assumes a convergence of digital imaginaries with urban imaginaries. In the design and naming of the campaign, the digital vision of open, horizontal, and inclusive communication merges with the imagination of an open and inclusive city. As this discussion's interest is the urban imaginaries that surround and are nourished through this campaign, the analysis focusses on its articulation on social media and the responses it generated.

On social media, the campaign appears as more complex in its design and narratives, compared to the official, static design of the official website and the Mayor-driven news in the mainstream media. The six short films produced for social media circulation, which launched the campaign, are *London opens its doors to the world; Our message to the rest of the world; Sport stars join Mayor Sadiq Khan to spread the message; London is the city of film; London is the city of dance; London is the city of shopping*. These were followed by a growing number of short films produced later in 2016 and in 2017, including a Christmas short film titled #LondonIsOpen *Not Just for Christmas*, and a series of short films under a Summer 2017 sub-campaign, titled #LondonIsOpen *for Summer*. All films are short and designed for social media viewing – almost all around a minute long. The first three films produced, *London opens its doors to the world; Our message to the rest of the world; Sport stars join Mayor Sadiq Khan to spread the message*, represent direct ethical responses to Brexit through visual and discursive narrations of we-ness, tolerance, and openness. The narratives of the three following films are more clearly structured as city branding exercises (Banet-Weiser, 2012), framing openness around three distinct cultural industries: film, dance, and fashion. As the campaign matured, the emphasis shifted from being about putting the city's voice forward (vis-à-vis the nation's) to commodifying openness and incorporating it in a vision of post-Brexit London where cultural industries still thrive. At the same time, and in their differences, what the films share is the adoption of images and discourses associated with vernacular cosmopolitanism and their adaptation to a framework of aesthetic cosmopolitanism. The films' aesthetics puts Londoners at the core of narrations of openness. The Mayor, business people, representatives of institutions, celebrities, and some ordinary Londoners appear in them. The shared message in all films beyond their differences is clear: protect our city, celebrate diversity, and keep the city open. But open to whom?

The campaign emphasizes a humanistic, culture-centric and people-centric narration of urban cosmopolitanism; the open city speaks directly to a cosmopolitan imaginary. Openness and hospitality to newcomers is projected at times of crisis and when the global city feels under pressure to close its gates to outsiders. Thus, and in this way, #LondonIsOpen adopts visual and discursive narratives of hospitality, emphasizing that the city welcomes everyone: in the films, the opening doors of London shops and the parade of different individuals on screen project this message repeatedly, clearly, and firmly. Clearly promoting a liberal cosmopolitan vision, the openness of London projected here reflects the city's history but also a vision for its future as hospitable, diverse, and equitable.

Digital imagining of the cosmopolitan city

The two short films discussed in brief below represent the impassionate celebration of London's cosmopolitanism, as this is visualized and voiced by carefully selected individuals. That the campaign is primarily launched on social media is important for at least two reasons: in popular imagination, social media are often assumed to be citizen-centered and horizontal spaces for communication (Greenfield, 2013), thus they become ideal spaces for the emergence of shared imaginaries; yet, critical digital media literature reminds us that social media encourage particular kinds of connections and participation prescribed by digital platforms (Van Dijck, 2013). Thus, who speaks, who is heard, and how, are important questions to ask in order to understand what kind of cosmopolitan vision is put forward through this digital call to cosmopolitan openness. Social media's potential and limitations to communicate and shape shared urban imaginaries are reflected in the #LondonIsOpen films and the responses they generated. The two campaign films introduced here – *London opens its doors to the world* and *London is the city of film* – project a vision of a city that is open, digitally constituted, and widely shared. Yet, as it will be shown, this campaign offers a very particular vision of what constitutes an open, cosmopolitan city and who it belongs to.

The 20-second video *London opens its doors to the world* projects a powerful message of hospitality. The video is a collage of more than 35 doors that open up shops, offices, and iconic locations in London to the viewers, welcoming them in. More than anything, this short film reflects London's diversity in a recognizable and relatable manner. The first door that opens up is the globally recognized door of the Prime Minister's residence at 10 Downing Street; this is swiftly followed by the door of the iconic red telephone booth located in the heart of touristic London, Covent Garden. Shifting from these most globally identifiable locations and symbols, the camera quickly moves across the imaginary geography of the city to include doors of dozens of stores selling a range of commodities and services, from hipster fashion to Indian food and everyday supplies. The locations of the specific doors are unidentifiable, but the values they represent – in terms of the diversity of what is sold and by whom – are identifiable as internal to London. The film concludes with the opening of the Mayor's office door, with Sadiq Khan himself inviting in "the world", as noted in the title of the film.

As a film that frames the meaning of the open city through the fundamental cosmopolitan value of hospitality (Derrida, 2001), *London opens its doors to the world* sets the tone for a campaign that contests "closed doors", as the territorial and symbolic borders of the nation that the referendum decided to reaffirm. The selection of doors that open to the world and the faces of those who open them firmly capture the value of cosmopolitan hospitality, yet a hospitality that is rooted in a consumption-driven, aesthetic cosmopolitanism (Regev, 2014; Walkowitz, 2006). The parade of hospitable acts is unmistakably associated with the iconic and globally circulated signifiers of what makes London valuable: a place to desire, visit, and consume, an iconic place that is also a centre of global power. Importantly, the doors that open to the world are those of sites of this same global power and of consumption; none of the doors lead the intended global audience into someone's home. Is cosmopolitan hospitality subject to acts of consumption and to the temporality of outsiders' visit? Does it assume that the outsider who is welcome today has to leave tomorrow?

The second widely circulated short film of the campaign is entitled *London is the city of film*. This production is based on famous quotes from movies and literature that "tell of London's uniqueness and openness to all those who live, visit and work in the city", according to the campaign's press release (Mayor of London, 2016). The film represents a seamless collage of the voices and faces of famous actors, directors, and the Mayor himself. Visually, and in the faces of each speaker in the film, we see the long history of migration, hospitality, and conviviality: we see white English celebrities next to others of Asian and Black Caribbean backgrounds. They all merge in communicating an uninterrupted cosmopolitan message. "In London, everyone is different and that means anyone can fit in", says the director and producer Richard Curtis, followed by the film director Gurinder Chadha: "It is a roost for every bird". The Mayor, Sadiq Khan, adds his voice: "Because it is not the walls that make the city, but the people who live within them". The actor and director Noel Clarke recites lyrics from *Mary Poppins*: "There's things half in shadow and halfway in light, on the rooftops of London. Coo, what a sight!" in a poetic format; he is wearing a baseball cap and surrounded by the busy street life of the city, strikingly different to Mary Poppins' white English London. The visuality of the film aims to capture London's aesthetic and lived diversity in a warm, welcoming and unthreatening manner. Importantly, the film is in black and white, perhaps in an attempt to look more cinematic than factual, but also to reflect a nostalgic glance into London: a selective representation of a city that most will recognize and many will desire.

The clear and globally recognized pronunciation of all speakers, also complemented by subtitles in the film, reaffirm the commodification of a cosmopolitanism that is accessible and consumable among global audiences. Celebrities can speak the language of difference, yet this is a difference contained within a safe and commodified urban culture represented by celebrities. Thus, in the film, ethnic difference becomes the spectacle. Yet, the values that surround the recognition of ethnic difference are stripped from the transformative politics of the street, nourishing instead "the utopian romance in the eroticized imaginary" of the diverse city (Keith, 2005: 108).

Sharing and contesting digital urban imaginaries

As intended, the hashtag driven campaign of #LondonIsOpen generated a range of responses. These responses unravel across social media – especially on Facebook, Twitter, and YouTube – but also spill outside the digital streets of London, as recorded during three focus groups discussions with young Londoners that inform the present analysis. In their diversity, the recorded responses are here grouped in three categories: (i) responses that embrace a neoliberal cosmopolitanism, attached to aesthetic, consumer-driven urban imaginaries; (ii) vernacular cosmopolitan responses that embrace the campaign's values, but which are also critical of who they speak for and to, reflecting an urban imaginary that emerges through experiences of diversity, rather than its celebration; (iii) post-cosmopolitan responses that reject the ethics and aesthetics of openness and hospitality. The responses are analysed and discussed below in three subsections, each corresponding to one of the three main themes: neoliberal cosmopolitanism; vernacular cosmopolitanism; post-cosmopolitanism.

Neoliberal cosmopolitanism

The overarching cosmopolitan narrative of the film captures the imagination of many on social media and in three focus groups conducted with middle-class, working-class, and ethnic minority young people in London. Almost all online and offline responses to the campaign reveal a wide agreement with the representation of London as a cosmopolitan, diverse city (with the exception of 12 comments out of 317 on Facebook, two oppositional films to the campaign on YouTube, discussed below, and a small minority of comments on Twitter). This agreement is almost taken-for-granted: a cosmopolitan position that represents the starting rather than the endpoint in urban imaginaries. The appeal of a quote from *London is the city of film*, "it [London] is a roost for every bird", is repeatedly commented upon on Facebook, presenting vivid evidence that an ethics of hospitality is at the heart of Londoners' understanding of cosmopolitan openness. The city's diversity is understood by many, both online and offline, as a fundamental element of openness: an open city is built on long histories of migration, as many noted. "London is a multicultural city" (male focus group participant); "I am a supporter of diversity and I'm sure everyone else should be" (female focus group participant). Yet the convergence of respondents around the fundamental values of urban cosmopolitanism (diversity, hospitality, openness) is interpreted differently: for some being taken-for-granted and for others being under attack and fragile.

For those embracing the films' neoliberal cosmopolitanism, two subthemes appear as important: identification (identity) and aesthetics (consumption). Numerous respondents on Facebook and most in the middle-class focus group commented on how they could identify with the films' characters and narratives. "Shows how diverse our city is", said a young male participant, linking this statement to the fact that the London Mayor is British Asian. Most in this same focus group said that the films made them feel proud to be Londoners, a comment appearing repeatedly on Facebook too. When asked who the target audience is, focus group participants pointed to themselves and one said: "The good young audience" and a young female participant explained further why she could identify herself as being targeted and represented: "Because of what the actors are wearing". The cosmopolitan, primarily middle-class representations of individuals in the films were warmly welcomed by middle-class participants and a number of respondents on Facebook. The commodification of the city through narratives of aesthetic cosmopolitanism is also expressed in the words of another male focus group participant. When asked to comment on the *London is the city of film* content he said: "I was too busy trying to recognize all the celebrities so haven't paid much attention". Almost all noticed the aesthetic construction of London in the films through narratives of consumption; one of them, for example, identified "the nice shop at London Bridge". For these Londoners, who have relatively privileged access to the city's material and symbolic resources, diversity, hospitality, celebrity, and consumerism represent elements of life in the cosmopolitan city. For them, the selective representations in the films construct an urban reality they experience or they aspire to experience.

Vernacular cosmopolitanism

London is an open but also an unequal city. The aesthetic, consumption-driven cosmopolitanism at the core of the #LondonIsOpen representations emphasizes openness and hospitality but, at the same time, side-lines inequalities by ignoring diversity's very different biographies and histories. The selective and ordered representation of openness and diversity was picked up in many online and offline responses. Responses coming primarily from those who experience the city as an unequal and hierarchical system embraced the campaign's ethics of hospitality, but rejected its neoliberal articulation. Most strikingly, a different vision of what hospitality means was revealed in a number of vernacular cosmopolitan responses on Facebook.

Within the main theme of vernacular cosmopolitanism, two prominent subthemes can be identified in the online and offline responses to the film: inequality and collective values. Speaking through and about experiences of inequality, some respondents adopted a reflexive dialogue with the films' articulation of hospitality: while welcoming it, they also critiqued its politics for enhancing privilege and marginalizing difference. One of the *London is the city of film*'s lines, "When a man, or a woman, is tired of London, then they are tired of life", generated passionate responses on Facebook. Among them, a single-mother noted: "I used to think when I was tired of London I'd be tired of life. Now I'm just tired ... The city is becoming less and less for the average family and certainly [not] for the average single parent". And in response to racism, a Muslim man wrote to the Mayor's Facebook page:

> Mr Sadiq Khan, need your attention ... A very awful incident with a woman she was in hijab. Someone tried to remove from her in yours (sic.) city London. Kindly make freedom well known for every one specially (sic.) for women.

The core cosmopolitan values of the film – and of the city's leadership – are shared by these participants. Yet, importantly, those who speak are the ones who also need these values, not in order to construct an identity, but to find a place in a city they see as less welcoming and less open to social and cultural difference. Similar responses, embracing cosmopolitan values but emphasizing their conditionality were also identified in the working-class and ethnic minority youth focus groups. In the words of a female participant:

> You know, showing the world that London's a very, very nice place, but that's all it's [the film] saying. There are so many problems that people, citizens, that citizens themselves see ... Go out on the street at night and you see tonnes of people are homeless.

A painful realization that the openness of the city does not extend to those who need it, is also expressed by a born and bred young working-class Londoner, who questions his own right to the city:

> [Living in London] is extortionary. So, like, I'm guessing they only need a certain type of person or people in London in the next, let's say, ten, or fifteen years and I might not even be here myself. So, I don't know if I'm really a Londoner or not. I don't really know the rules.

Critical and reflexive responses in these two focus groups and in a series of comments on Facebook point to the order set by a neoliberal cosmopolitan vision for a city that is open to some but most certainly not to all. The expressed dissent from selective urban imaginaries that exclude many of the city's dwellers reflects an alternative urban imaginary for a city that is not only open to difference but which also comes with a demand for its recognition and respect (Benhabib, 2008).

Post-cosmopolitanism

While vernacular cosmopolitanism embraces the fundamental values of cosmopolitanism, a series of responses to #LondonIsOpen opposes the very values of diversity, openness, and hospitality. Against openness, these responses project fear; against conviviality, they offer suspicion and hatred. These are responses to a society produced by transborder mobility and multiculturalism, which a number of respondents to the campaign reject. While converging in their rejection of the value of openness and the assumption that cosmopolitanism is inherently urban, post-cosmopolitan and oppositional responses reflect a range of ethic-political positions, expressed by a range of actors. Most common among those are responses on Facebook that oppose the presumed rightfulness of cosmopolitanism and the presumed exclusivity of openness as a quality of the global city. Such responses usually come from people who do not live in London and who contest the idealized representation of the city as a cosmopolis. This widespread expression of post-cosmopolitanism is the result of the disjuncture of experiences between London and other parts of the country, as much as they are the result of inequalities that the over-concentration of power in the global city generates. For such voices, London's cosmopolitanism comes with privilege; for such voices, the global city's self-righteousness reaffirms geographical and social hierarchies that exclude them.

Less widespread, but more vocal and confrontational, are expression of a post-cosmopolitan imaginary that targets the core of cosmopolitan values. Such responses are visible on YouTube and Twitter and go well beyond questioning the city's inherent openness. In a series of YouTube videos and also on Twitter, urban cosmopolitanism becomes the target of a value-driven political discourse of hatred and fear. One of the most powerful expressions of such post-cosmopolitan imaginaries appears under the label of "Londonistan" – a derogatory notion attributed to the global city in a number of tweets with #LondonIsOpen. These tweets, often also being retweeted with *#Londonistan*, target the city's diversity and particularly its Muslim population. The Mayor of London himself is repeatedly attacked, both personally, on the basis of his ethnicity and religion, and in relation to his policies. #LondonIsOpen, with its prominent and celebratory message of urban diversity is unsurprisingly fiercely attacked. While the message of the oppositional *#Londonistan* campaign is powerful it is mostly reproduced among the same Islamophobic circles and followers.

Alongside social media xenophobic and Islamophobic campaigns, a series of YouTube videos target #LondonIsOpen by appropriating its own imagery. In a video titled *Get out Sadiq* #LondonIsOpen, the iconography of the film *London opens its doors to the world* is used in its original format, though in this case it is shown in reverse, with all doors closing rather than opening. Superimposed to the imagery of the London doors the phrase GET OUT sends a clear message to the viewers. While the original film's appropriation is quite effective, it had no more than 24 views as of December 2017. These oppositional voices often mobilize irony and parody to mock cosmopolitanism as in the case of another Youtube video, titled *#LondonIsOpen (FUNNY PARODY)*. In this 2:33 minute video, a series of images accompany the male producer's song parodying urban openness and #LondonIsOpen. The video comprises a mixture of images that includes photos of the iconic London red bus, streets of multicultural London but also images of fighters in unidentifiable locations, possibly in the Middle East and Afghanistan. Sang by a single male voice, the lyrics accompanying the imagery appear to be made up as the video is played, lacking any artistic or musical value. Yet, the message the video communicates is quite clear and certain. Among its lyrics:

> Believe me you can get free benefits, I don't care if you are a suicide bomber ... They say it's good to be multicultural, because it makes us look good to refugees, leaving them take all the jobs and calling you racist on Facebook. London is open for all.

Reproducing a series of familiar stereotypes about the poor, minorities, and politics of solidarity, the video only had 18 views as of December 2017. Yet, this, along with the *Get out Sadiq #LondonIsOpen* are algorithmically identified through the hashtag #LondonIsOpen and thus appear in the digital commons alongside videos of wide appeal and relevance. Their limited appeal aside, it is the digital algorithmic order that identifies those post-cosmopolitan visions as elements of the digital urban commons. Do these post-cosmopolitan imaginaries gain legitimacy precisely as they become visible in their contrariety to the normative vision of the open city? Do algorithms identify them as relevant to a public conversation and to a shared imaginary because they engage, no matter how, with the values #LondonIsOpen addresses: openness, diversity, hospitality?

Conclusions

The campaign and the responses it has generated present a complex case study of the ways in which urban imaginaries are reinforced and contested at the intersection of widely circulated values of cosmopolitanism and their effective but diverse digital articulations. On social media, where #LondonIsOpen primarily unfolds, the global city becomes constructed as almost naturally cosmopolitan. Widely embraced, openness and hospitality speak to the experience of many urban dwellers who enjoy or are being denied these values. At the same time, openness becomes an instrument of contestation in terms of what it actually means and who it applies to.

Thus, the contested meaning and value of cosmopolitan openness, as revealed in the discussed case study, reflect three distinct imaginaries. For those celebrating openness, cosmopolitanism reflects privilege, as it derives out of an engagement with the aesthetic and commodified landscape of the city and the selective encounters with diversity, especially through consumption. This is a taken-for-granted cosmopolitanism that feeds into an urban imaginary that is only conditionally open to difference. For many of those who are bitterly aware of the city's inequalities, openness is not celebrated but cherished, as it is attached to respect and recognition of urban difference and its histories of migration. For these less privileged urban dwellers, vernacular responses to cosmopolitanism reflect an imaginary, constituted through representations and experiences of discrimination and inequalities.

The trivialization of openness through its aesthetic and commodified appropriations becomes the target of the post-cosmopolitan imaginary. Powerful images and narratives of openness are not only effective in mobilizing cosmopolitan publics; they are also effective in nurturing post-cosmopolitan imaginaries that appropriate the representational props of neoliberal cosmopolitanism to attack urban conviviality, respect, and recognition of difference. Post-cosmopolitan imaginaries contest a cosmopolitan vision that is normative and assumes that values of openness and diversity are fundamentally urban, universally accepted, and widely recognized. These assumptions generate resentment among many non-privileged actors who lack the symbolic power to enjoy lifestyles associated with neoliberal cosmopolitanism or who cannot relate to the imaginary of the uniquely cosmopolitan global city. Furthermore, and in the darker shades of post-cosmopolitan politics, the normativity of urban neoliberal imaginaries becomes an easy target for speaking against cosmopolitanism through a language of fear and hatred.

#LondonIsOpen, as a media campaign, but also as a moral campaign, has mobilized cosmopolitan publics and their imagination by proposing an all-encompassing cosmopolitan vision of the city at times of crisis. Its all-encompassing, universalizing cosmopolitanism reminds us that all digital articulations of urban imaginaries are by definition acts of inclusion as much as they are acts of exclusion: envisioning a city and a cosmopolitan openness that many imagine but only a few enjoy. This is a force working through what Butler (2004) refers to as a performative form of power. This performative form of power works in two ways. On the one hand, as certain forms of recognition become extended (to certain migrants and minorities), others become unrecognizable and not worth recognition (working-class dwellers or religious groups). On the other hand, demarcation works performatively, and "certain political distinctions, including inequality and exclusions" (Butler, 2004: 6) are not named but performed: in the representation of the non-white actor who is an urban subject and who has achieved fame and wealth, ethnic minorities become selectively heard; in the celebratory and full embracement of the campaign by urbanites who dominate the Facebook and Twitter communicative space, the voices of those who experience cosmopolitanism in its contradictions become silenced. Thus, and as the complex and messy space that #LondonIsOpen reveals, urban imaginaries cannot but always be hierarchical and plural, remaining spaces for collective but also for contested visions of the city and its future.

References

BBC. (2012) *45 Per cent of Londoners White British*. Available at: http://www.bbc.co.uk/news/uk-england-london-20680565.

BBC. (2016) *EU Referendum: Most London Boroughs Vote to Remain*. Available at: http://www.bbc.co.uk/news/uk-politics-eu-referendum-36612916.

Benhabib, S. (2008) Hospitality, Sovereignty and Democratic Iterations. In: Post, R. (ed.), *Another Cosmopolitanism*. Oxford: Oxford University Press, 147–86.

Benjamin, W. (1997) *One Way Street and Other Writings*. London: New Left Books.

Binnie, J. and Skeggs, B. (2004) Cosmopolitan Knowledge and the Production and Consumption of Sexualized Space. *The Sociological Review* 52(1): 39–61.

Boudreau, J.A. (2007) Making New Political Spaces. *Environment and Planning A* 39(11): 2593–611.
Butler, J. (2004) *Precarious Life: The Powers of Mourning and Violence*. London: Verso.
Calhoun, C. (2012) Cosmopolitan Liberalism and Its Limits. In: Krossa, A.S. and Robertson, R. (eds.), *European Cosmopolitanism in Question*. Basingstone, UK: Palgrave Macmillan, 105–25.
Castoriadis, C. (1998) *The Imaginary Institution of Society*. Cambridge, MA: The MIT Press.
Cinar, A. and Bender, T. (2007) *Urban Imaginaries*. Minneapolis, MN: University of Minnesota Press.
Derrida, J. (2001) *On Cosmopolitanism and Forgiveness*. London: Routledge.
Douthat, M. (2016) The Myth of Cosmopolitanism. Available at: https://www.nytimes.com/2016/07/03/opinion/sunday/the-myth-of-cosmopolitanism.html.
Georgiou, M. (2013) *Media and the City*. Cambridge, UK: Polity Press.
Georgiou, M. (2017) Is London Open? Mediating and Ordering Cosmopolitanism in Crisis. *International Communication Gazette* 79(6–7): 636–55.
Greenfield, A. (2013) *Against the Smart City*. London: Do Projects.
Hall, S. (2008) Cosmopolitanism, Globalisation and Diaspora. In: Werbner, P. (ed.), *Anthropology and the New Cosmopolitanism*. Oxford, UK: Berg, 345–60.
Hannerz, U. (1996) *Transnational Connections*. London: Routledge.
Harvey, D. (1973 [2009]) *Social Justice and the City*. Athens, US: The University of Georgia Press.
Jacobs, J. (1993) *The Death and Life of Great American Cities*. New York: Vintage Books.
Keith, M. (2005) *After the Cosmopolitan? Multicultural Cities and the Future of Racism*. London: Routledge.
Lefebvre, H. (2003) *Key Writings*. London: Bloomsbury.
Mayor of London. (2016) *Sadiq Khan Hails London as the City of Film*. Available at: https://www.london.gov.uk/press-releases/mayoral/sadiq-khan-hails-london-as-the-city-of-film.
Ponzanesi, S. and Leurs, K. (2014) On Digital Crossings in Europe. *Crossings: Journal of Migration & Culture* 5(1): 3–22.
Robbins, B. (1998) Introduction Part I. In: Cheah, P. and Robbins, B. (eds.), *Cosmopolitics: Thinking and Feeling beyond the Nation*. Minneapolis: University of Minnesota Press, 1–19.
Sassen, S. (2001) *The Global City: Introducing a Concept*. New York: Blackwell.
Sennett, R. (1970) *The Uses of Disorder*. New Haven: Yale University Press.
Sennett, R. (2013) *The Open City*. Available at: https://www.richardsennett.com/site/senn/uploadedresources/thepercent20openpercent20city.pdf.
Silverstone, R. (2007) *Media and Morality: On the Rise of the Mediapolis*. Cambridge: Polity Press.
Van Dijck, J. (2013) *The Culture of Connectivity: A Critical History of Social Media*. Oxford: Oxford University Press.
Vulliamy, E. (2016) "We Are the 48%": Tens of Thousands March in London for Europe. *The Observer*. Available at: https://www.theguardian.com/politics/2016/jul/02/march-for-europe-eu-referendum-london-protest.
Walkowitz, R. (2006) *Cosmopolitan Style*. New York, NY: Columbia University Press.
Werbner, P. (2008) Introduction. In: Werbner, P. (ed.), *Anthropology and the New Cosmopolitanism: Rooted, Feminist and Vernacular Cosmopolitanism*. Oxford, UK: Berg, 1–32.
Zukin, S. (2010) *Naked City: The Death and Life of Authentic Urban Places*. Oxford: Oxford University Press.

16

Beyond East-meets-West

Contemporary Chinese art and urban imaginaries in cosmopolitan Shanghai

Jenny Lin

Against the "Paris of the East" and other obfuscations

Shanghai, thrust open as a treaty port in 1843 following Britain's victory in the First Opium War and subsequently known as mainland China's most cosmopolitan metropolis, has long been a site of projection for intellectuals and artists engaged in cross-cultural discourse. As scholar Leo Ou-fan Lee observed, it was in Shanghai where the English and French words, *modern(e)*, were first transliterated into the Chinese term, *modeng*, meaning novel and fashionable (1999: 5). Indeed, Shanghai's Republican Era (1911–1949) modernity was intimately linked to the city's occupying forces and to the foreign products that flooded the local market. Modernity in Republican Shanghai, a city carved into districts controlled by British, US, French, and Japanese colonialists, must also be understood as based on unequal power divisions. Against the early- to mid-twentieth-century clichés of Shanghai as "the Paris of the East" and "New York of the West", United States-educated Chinese sociologist Fei Xiaotong proclaimed:

> Modern metropolises are the products of industrialization, a country which has not been industrialized cannot have urban centers like New York or London. The treaty port brought about the invasion of an industrialized economy into an economically inferior area … creating a peculiar community which should not be classed with modern urban centers.
>
> *(Fei, 1953: 107)*

Following the establishment of the communist-controlled People's Republic of China (PRC) in 1949, Shanghai was officially cast, as historian Rhoads Murphey wrote, as "the outstanding symbol of the economic exploitation of China by Western commercialism, and...the principle reminder of China's unequal-treaty status with the Western powers" (Murphey, 1953: 25).

No longer called the "Paris of the East", Shanghai has recently emerged as an international center of finance and culture, powerful in its own right, and the sparkling jewel of the PRC's post-Mao Era (post-1976) economic ascension. Nonetheless, the trope of East-meets-West continues to appear within tourist literature promoting Shanghai, the city's contemporary art scene, and the wider burgeoning field of contemporary Chinese art. In the summer of 2008, amidst the world financial crisis, the Shanghai Museum of Contemporary Art hosted a splashy exhibition comparing contemporary Chinese art and Greek art, and the Zendai Museum of Modern Art (now Himalayas Museum) held a forum entitled, "Los Angeles vs. Shanghai: Who is the Art Capital of the Pacific Rim?" Such events demonstrate how the city's art is often placed in dialogue and even in competition with Western foils, strategically utilized to promote Shanghai as an international cultural capital. Meanwhile, countless media blurbs laud Chinese-born artists for combining traditional Chinese/Eastern motifs (e.g., dragons, the color red) with heretofore primarily Western frameworks of contemporary art (e.g., minimalism, land art). Consider the Shanghai-born, contemporary international art star Gu Wenda, for whom I worked as a translator and project assistant. Gu is almost always labeled as a Chinese artist, although he maintains studios in Shanghai, Beijing, and New York, and lives, travels, and identifies himself (2008, personal communication) as a "citizen of the world". Gu and the curators and critics promoting him often speak of his unique ability to synthesize traditional Eastern elements, such as calligraphy, embroidery, and ancient stone carving, with modern Western frameworks for art, usually referring to more conceptually inclined praxes. These pairings – "traditional Eastern" and "modern Western" – fully loaded and constructed, as postcolonial theorists (Bhabha, 1994) have helped elucidate, are nonetheless ubiquitously employed in discussions of contemporary Chinese art and urban Shanghai, often as if they were entirely natural divisions. While the notion of East-meets-West continues to inform understandings of contemporary Chinese art, resonating doubly in Shanghai as mainland China's historic capital of Chinese/Euro-American encounters, the phrase is often employed as a mere marketing tool in the promotion of new objects or a convenient mode for producing supposedly fresh insight. This chapter analyzes contemporary art projects created in and about cosmopolitan Shanghai in the early to mid-2000s, a period that marked Shanghai's rapid re-emergence as a world city (Ong and Roy, 2011), or internationally significant economic and cultural center. I look beyond East-meets-West clichés, investigating how Shanghai-based artworks imagine the city's historically rooted socio-political, economic, and cross-cultural tensions.

At the dawn of the twenty-first century, numerous contemporary Chinese artists proposed and created projects in Shanghai that promoted and/or critiqued the metropolis's longstanding East-meets-West identity. Studying the globalized production, conflicted reception, and critical issues raised by such Shanghai-based projects illuminates how artists navigate the tricky terrain of contemporary art in the PRC, often while confronting their own transnational identities. I present primary case studies by three Shanghai-based artists: Gu Wenda's art proposal *Heavenly Lantern Project for Shanghai* (2003–ongoing), which utilizes cross-cultural content, aesthetic motifs, and conceptual tropes to celebrate East-meets-West Shanghai; Liu Jianhua's photographic series *The Virtual Scene* (2005) and exhibition *Export – Cargo Transit* (2007), which question Shanghai's urbanization and highlight the negative impacts of globalization and uneven East–West relations; and Yang Fudong's photographic triptych *The First Intellectual* (2000) and film installation *Seven Intellectuals in a Bamboo Forest* (2003–2007), which critique Shanghai's romanticized mythology and expose current conflicts facing the city's artists and intellectuals. My analyses aim to disrupt a common assumption underlying discussions of urban Shanghai and contemporary Chinese art: that both city and category effortlessly transcend national borders and cultural divides, operating through harmonious East-meets-West encounters.

Gu's celebratory proposal and more critical projects by Liu and Yang all construct urban imaginaries that speak less to Shanghai's East-meets-West status or the seamless integration of diverse cultures and fluid insertion of Chinese art into the Western-dominated canon of contemporary art, and more to the mistranslations, social alienations, and merging of fine art and commodity culture that belie contemporary Chinese art and life in Shanghai.

East/West hybridity as cultural capital

In 2003, Gu Wenda instigated a series of plans for the *Heavenly Lantern Project for Shanghai* (see Figure 16.1), one proposal of a larger series aimed at covering architectural monuments around the world in bright red Chinese lanterns. In a corresponding publication, the artist described the goals of the project (Gu, 2005):

> [To] celebrate the multiple cultures in our modern society by erecting a series of large-scale works in a unique and grand manner full of Chinese cultural symbolism ... The goal of this work is not only to cover buildings with simple hanging lanterns, but the lanterns are intended to be the mouthpieces of a civilization and an ethnic herald in this series. Famous architecture and historic icons all over the world represent different civilizations and periods in world history. Draping and covering these monuments is not the traditional visual concept of contemporary art, but is a symbol of a civilization's dialogue with a symbol of another civilization in a unique manner, as well as one civilization's understanding and exposition of another. Understanding civilization from another point of view is what the *Heavenly Lantern* project intends to promote while making the decorated building more splendid and meaningful; intending, like an ancient adage in China: "to make perfection still more perfect".

With this *Heavenly Lantern for Shanghai* proposal, Gu aimed to create an East-meets-West installation by covering the city's soaring Jin Mao Tower in red paper lanterns as a celebration of the fusion of international architecture, modern progress, and national tradition.

One of Shanghai's tallest skyscrapers, the Jin Mao Tower, designed by US firm Skidmore, Owings, and Merrill, is located in the city's chief economic hub, the Lujiazui Financial District in Pudong. In 1993, as part of the economic reforms by the post-Mao Era leader Deng Xiaoping, Shanghai annexed Pudong, a vast region comprised primarily of farmland east of the Huangpu River, as a New Development Zone open to overseas investment and less restricted trade. Pudong's annexation expanded Shanghai's size by over sevenfold and effectively transformed Shanghai into the PRC's primary financial capital (Visser, 2010: 2). The Jin Mao Tower houses international corporate offices, entertainment facilities, and the Grand Hyatt Hotel. Featured in popular films, such as James Bond's *Skyfall* (2012), the skyscraper, along with the entire Lujiazui skyline, has become a worldwide signifier of Shanghai's cosmopolitan status and financial prowess. Rather than merely emulate the skyscrapers of Chicago, New York, and Tokyo, the Jin Mao Tower is known for integrating specifically Chinese design elements, such as a pagoda-like setback design and proportions that revolve around the number eight (an auspicious number in traditional Chinese culture).

Figure 16.1 Gu Wenda, *Heavenly Lantern Project* for Shanghai's Jin Mao Tower, 2003. Image courtesy of artist.

Like the building it proposes to cover, *Heavenly Lantern* stylistically draws from stereotypically Chinese motifs, including red lanterns and Chinese calligraphy (Gu's well known pseudo-Chinese characters adorn the lanterns), while also referencing land art, installation art, and public projects by artists working primarily in Western Europe and

North America, like Christo and Jeanne-Claude. Yet, in addition to the proposal's proclaimed East–West cultural hybridity, *Heavenly Lantern* also blends fine art, design, architecture, advertising, and civic and national marketing. Toward the visualization of *Heavenly Lantern*, Gu has employed a lobbying team (including, among others, museum directors and corporate CEOs) to negotiate with civic administrations and governmental bodies, assistants trained in architecture to produce mock-ups and blue prints, a contractor to design the assembly of the installation and the fabric for the lanterns, a construction group to physically install the work, mainland Chinese factory workers to produce the lanterns, and bilingual assistants to translate the promotional material from Chinese into English, the lingua franca of the art world. Gu, who trained as a painter in Hangzhou's China Academy of Art, explained his production methods as follows (2008, personal communication):

> Traditionally, if you practice painting, you just need a canvas and paint ... Now you really have to conduct your work as a director ... you have to consider all the elements ... I still believe, if you want to go to other levels, if you want to do a public project, you need to collaborate with the government and corporations, financial aids ... museums ... transportation departments ... city governments ... central governments.

As an artist/director, Gu oversees various aspects of an installation's production, and runs a transnational enterprise that relies on increased division of labor and outsourcing – strategies that approximate those of the various multinational companies and governmental bureaus with which he collaborates.

The artist sees himself as a cultural ambassador, spreading a positive image of Chinese culture abroad and a positive image of Western culture within China, while touting his installations as bridges between East and West. Gu has emphasized *Heavenly Lantern*'s potential marketing power, and ability to bolster cultural tourism and promote Shanghai, and by extension the PRC, worldwide (Gu, 2005):

> The *Heavenly Lantern* project would take the possibilities of advertising to newer heights and become a magnet for the world's media ... Besides creating a glorious image for the society, politics and culture of the country and for Shanghai in particular, the project would create a charming focus for the Shanghai travel industry, and create a market for many commemorative products ... while contributing to help shape Shanghai as an international metropolis.

He expresses a desire to contribute to Shanghai's ability to shine on a world stage as both distinctly cosmopolitan (evident in the project's engagement with international contemporary art trends and modern architecture) and particularly Chinese (signaled by the use of lanterns and calligraphy as national symbols).

The *Heavenly Lantern* proposal for Jin Mao Tower, through its combination of cosmopolitan and national symbols, would do more than simply promote the PRC internationally; it would generate an image of the country's current hybridized political and economic circumstances, in which China's Communist Party supports multinational capitalism. While foreign-designed buildings like Shanghai's Jin Mao Tower might appear to herald the spread of Western-dominated capitalism, *Heavenly Lantern*, if successful in covering these buildings with symbols of Chinese nationhood, would produce a multi-layered portrait of the PRC's free market capitalism combined with the nation's one-party power. The proposal thus indexes what cultural anthropologist Aihwa Ong has proposed (2011: 206), as a "theory of sovereign exception" that challenges the notion, generally accepted by urban theorists, that the development of contemporary architecture around the world is "merely the reflex of the expansion of capitalism or corporate power". *Heavenly Lantern* for Jin Mao Tower would signal that the PRC's unprecedented urbanization and the corresponding race for Asian cities to unveil the tallest, most cutting-edge skyscrapers are fueled not only by US-led capitalism and corporate expansion, but also through national authoritarian regimes, such as China's Communist Party.

As with all of his large-scale projects, Gu hopes *Heavenly Lantern* will travel internationally, championing Chinese culture and "grand symbolism" abroad. To this end, he has designed proposals for *Heavenly Lantern Hong Kong, Singapore*, and *Davos*, and realized *Heavenly Lantern Brussels* in 2009 at the Europalia China Art Festival in Belgium. Testament to his overriding interest in the economics of art, Gu prioritizes exhibiting the project in cities that are financial capitals, and/or which hold particular significance under multinational capitalism. *Heavenly Lantern Davos* (Gu, 2005), for example, was designed specifically for the World Economic Forum, held annually in Davos, Switzerland, and which in 2003 coincided with Chinese New Year. For this installation, Gu proposed (2005) to cover the Hotel Belvedere, where Forum guests were staying, in red Chinese lanterns, thereby extending Chinese New Year's celebrations of the "happiness of life [and] the joy of harvest" by "celebrating the good will of the world, which is concentrated in Davos during the World Economic Forum". Gu thus aims to utilize Chinese cultural symbolism to celebrate transnational free market capitalism, as promoted during the World Economic Forum, and transnational free market capitalism to celebrate Chinese cultural symbolism, as exemplified by Chinese New Year and festive red lanterns.

While striving to smooth over cultural tensions resulting from the PRC's swift rise within an increasingly global economy, *Heavenly Lantern*'s circulation, or more appropriately its lack thereof, ultimately highlighted such tensions. In the early 2000s, as Gu began drafting the Davos proposal, mainland China's position within the world economy was soaring. For the first time, the PRC's state-sanctioned economic model, referred to as "socialism with Chinese characteristics", posed a viable threat to the North American and Western European model of democratic-capitalism that dominated international political and economic arenas since the second-half of the twentieth century. Considering the suspicion and fear with which many political leaders have regarded mainland China's economic ascension, it is not surprising that projects such as *Heavenly Lantern Davos* were never realized. Conceived of as a celebration of the integration of multiple cultures, uneasy sentiments attached to Chinese cultural symbols have continuously derailed *Heavenly Lantern*. As Gu recounted (2008, personal communication), his plan to install the work at the Dutch Cathedral of Groningen, after years of planning, was ultimately rejected by the city council when some of its members voiced concerns over China being a communist state. In such cases, *Heavenly Lantern* reveals more about cross-cultural conflicts than the harmonious cultural unions Gu aims to highlight.

The first decade of the twenty-first century marked a watershed period for contemporary Chinese art, ushering in strong overseas and burgeoning local market interest. During these years, Chinese Communist Party officials began vocally supporting the nation's contemporary art. Gu commented (2008, personal communication):

> Chinese artists are benefiting so much from political, social, and economic developments in China ... Ten years ago, you only saw the promotion of traditional landscape painting or calligraphy that was safe from politics, but now you can see more contemporary culture being promoted.

Contemporary Chinese art is today upheld for both its commercial viability, and its ability to promote abroad a positive image of the new, global China. Gu has aligned his practice with these revised expectations, formulating artworks like *Heavenly Lantern*, which celebrate the hybridization of art, architecture, urban development, governmental sponsorship, and multinational capitalism. While Gu continues to promote *Heavenly Lantern* for its ability to blend tradition and modernization, and Chinese and Western cultures, the project's failures, such as its ultimate rejection by politicians in Groningen, also reveal the persistence of cross-cultural tensions informed by lingering Cold War ideological divisions, even as communist China has fully embraced free market reforms. Critics both in and out of the PRC have expressed concern over discrepancies between officially approved images of the new East-meets-West Shanghai as socially, culturally, and economically progressive, and realities defined by growing income disparity, pollution, corruption, inflation, and tightening control over civil liberties. Gu himself admits (2008, personal communication), "If you produce a large public art work or exhibition in China, it has to be politically removed ... it has to be neutral, or a good image for the country".

Representing virtual scenes and the real impacts of globalization

Shanghai-based artist and professor of sculpture Liu Jianhua has produced critical urban imaginaries that run counter to celebratory projects like *Heavenly Lantern*. The artist's photographic series *The Virtual Scene* (see Figure 16.2) presents Shanghai's Huangpu Riverfronts – the Lujiazui skyline in Pudong (home to the Jin Mao Tower) and the Bund in Puxi, the former heart of the British and US-controlled International Settlement – as blurred backdrops for poker chips stacked like unsteady skyscrapers. Referencing the PRC's socialism with Chinese characteristics, Liu commented on *The Virtual Scene*'s dual settings of Shanghai's older, once semi-colonized side, Puxi (west of the Huangpu River) and the Pudong New Development Zone (2016: 50): "Puxi symbolizes the capital accumulation of the 1920s–1930s, while Pudong represents the 1990s development of a new socialist model privileging capital accumulation. Desire, adventure, luxury and other notions mix in this work's imagined concept". The poker chips of *The Virtual Scene* liken Shanghai's capitalist, or "new socialist" developments (both of the 1920s–1930s, when steered by foreign imperialists, and of the 1990–2000s, as directed by China's Communist Party) to a risky gamble. For *The Virtual Scene*'s sculptural iteration, Liu modeled an entire metropolis out of poker chips, some of which were stacked with dice to resemble iconic Shanghai buildings, mocking the glamorizing pretenses of fast-paced urbanization and speculation in the PRC's booming cities. As Liu created these photographs and models, questioning the realities behind spectacular architectural markers of capitalist progress, he was at work on a large-scale exhibition to be presented amidst the urban landscapes represented in *The Virtual Scene*.

Displayed across from Pudong at the Shanghai Gallery of Art in the upscale cultural complex Three on the Bund, Liu's exhibition *Export – Cargo Transit* (see Figure 16.3) tackles the uneven conditions of the international art world and the environmental harm

Figure 16.2 Liu Jianhua, *The Virtual Scene*, 2005. Image courtesy of artist.

incurred by disparate global economic development. *Export – Cargo Transit* consists of a series of "sculptures" and "paintings" made out of garbage imported into mainland China from nations, such as the United States and Britain, encased in plexiglass and labeled "Export". The exhibition utilizes readymade tactics that have clear ties to movements in the Western European and US-dominated canon of modern and contemporary art history, including Dada and Arte Povera. In an interview with me (2011, personal communication), Liu recognized his appropriation of such strategies, while emphasizing above all else *Export – Cargo Transit*'s connections with the broader economic and political implications of globalization, and the PRC's role as chief exporter of consumer goods and importer of consumer waste. The exhibition's venue, Three on the Bund, occupies a Western European style neoclassical tower: the former Union Assurance Building designed by British architectural firm Palmer and Turner in 1916 and readapted by US architect Michael Graves in 2004. Exhibited in a revamped colonial building on the Bund, the chief reminder of Shanghai's history of foreign imperialism, *Export – Cargo Transit* registers the extension of unequal relations between the PRC and supposedly more developed Western European and North American nations, while offering wider reflections on and critical responses to urbanization, globalization and resulting environmental hazards.

On the gallery walls surrounding the encased detritus of *Export – Cargo Transit* appeared appalling stories of the import of trash culled from both Chinese and foreign

Figure 16.3 Liu Jianhua, *Export – Cargo Transit,* 2007. Image courtesy of artist.

news articles (2008), like this 2007 British report of a Chinese town which imports trash and recycling items:

> In Lianjiao's recycling plants they melt plastic down into molten lumps. It gives off fumes that can cause lung disease. Smoke stacks bellow clouds of chemicals that hang above the town. Poisonous waste pours directly into rivers, turning them to a stagnant black sludge. Entire families live among the filth. We visited yard after yard filled with rubbish from across Europe. We watched a container truck unloading household waste from France. Another yard specialized in German plastic. Next door we found a container-load of household rubbish just off the boat from Britain. Plastic waste is now one of Britain's biggest exports to China. Container ships arrive in Britain from China loaded with consumer goods. Many of them go back packed full of British waste.

And this report from ABC News:

> Most of the world's electronic trash, especially old computers, is dumped in China, causing severe environmental problems and illness among residents ... About 80 per cent of the world's electronic rubbish is transferred to Asia every year, 90 per cent of which ends up in China.

These reports were juxtaposed against the legislation of the Basel Convention, negotiated and adopted by the United Nations Environment Program in 1989, which should have prohibited developed countries from exporting toxic waste to developing ones. As Liu's exhibition highlights, The Basel Convention's accords have proven difficult to enforce, and are in many cases explicitly ignored.

Playing with the re-importation of foreign trash as contemporary art by foreign collectors, the project self-reflectively confronts the globalization of contemporary art, and Liu's own conflicted position as a Chinese artist operating in a transnational art world. As described to me (2011, personal communication), Liu, keenly aware of how his career initially benefited from a predominantly Western European and North American art market's demand for the *new*, in this case contemporary Chinese art, encouraged foreign collectors to purchase works from *Export – Cargo Transit*, while discouraging Chinese collectors. He thus steers the distribution and circulation of his art, further contributing to the exhibition's cross-cultural meanings. The crux of this project lies in the artist's ability to demonstrate the links between the international trade of products for mass consumption, the refuse created by bloated consumer societies, and the transnational trade of art as luxury item. *Export – Cargo Transit* further demonstrates the disparities between developed and developing nations, implicating the former in the ecological and social crises of the latter. In repackaging imported garbage and selling it as art, Liu imagines new economic microcosms, casting alternative social formations that challenge the conditions of globalization and its toxic environmental effects.

Picturing conflicted world views and dancing alone

Like Liu, Shanghai-based artist Yang Fudong has represented the tensions and conflicts arising from Shanghai's urbanization and the PRC's global economic expansion at the turn of the twenty-first century. Yang's photographic triptych, *The First Intellectual* (see Figure 16.4), portrays a young man dressed in a business suit, carrying a briefcase in one hand and a brick that seems to have been hurled at him in the other. Shocked, bloodied,

Figure 16.4 Yang Fudong, *The First Intellectual*, 2000. Images courtesy of the artist and Marian Goodman Gallery.

and with mouth agape, he stands in the middle of Shanghai's Lujiazui Financial District. Distant modern high-rises like the Jin Mao Tower frame his agitated body. This intellectual, presumably the first to emerge in the PRC's post-1989, so-called post-socialist society (Dirlik and Zhang, 2000), finds himself in existential crisis. In the first two photographs, the wounded intellectual lingers in the double yellow lines of an empty boulevard, pivoting his body as if preparing for retaliation. In the third photograph, he has dropped his briefcase and raised the brick above head, ready to direct his pain and frustrations at anyone or anything. But no target emerges. In this final photograph, the man has moved to a paved portion of the road designed for pedestrians and bicyclists, but like in the first two photographs, there are no passersby in sight. The financial hub of the crowded metropolis appears as an eerie ghost town. Even while the man faces forward, his eyes are cast toward an unknown horizon beyond the viewer. The only other figural representation exists at the man's feet, wherein an unzipped suitcase has materialized. Garments and magazines spill out, and in the foreground a fashion model in a full-page advertisement appears; the glossy headshot marks a futile counterpoint to the worse for wear intellectual.

Only through the title and accompanying caption of "The First Intellectual" can we identify the main subject of the triptych as an intellectual, hinting at Yang's ambiguous definition of the term (2008: 40): "The spirit of intellectuals is the dream you have for yourself and the sensation of chasing a dream in dreams. In other words, being an intellectual means imposing the status of being an intellectual upon oneself". Yang associates being an intellectual with self-imposed dreaming. His triptych suggests that the intellectual's dreams, and perhaps the very ability to dream, have broken under commodity fetishism's false promises, advertised in flashy fashion magazines, and intense alienation wrought by unprecedented urbanization, embodied in towering skyscrapers and empty cement streets.

The First Intellectual was debuted in the exhibition "Fuck Off" ("Uncooperative Approach" in Chinese) organized by artist and dissident Ai Weiwei and curators Feng Boyi and Hua Tianxue as an alternative to the concurrent 2000 Shanghai Biennial – the PRC's most high-profile, Chinese Communist Party-supported contemporary art exhibition. Surprisingly, amidst the plethora of explicitly subversive and disturbingly abject works in "Fuck Off", the Chinese Cultural Inspection Bureau singled out *The First Intellectual* for official removal from the exhibition (Chiu, 2009). When choosing to remove Yang's work, officials may have recognized that these photographs threaten to debunk the myths surrounding the urbanization and globalization of Shanghai – myths perpetuated by fashionable images of the Liujiazui skyline, celebratory projects including *Heavenly Lantern*, and official events like the 2000 Shanghai Biennale, all of which promote benignly positive images of an East-meets-West city.

Urban anxieties and the fraught role of Shanghai's new class of artists and intellectuals play a central role in Yang's subsequent project, *Seven Intellectuals in a Bamboo Forest*, a thirty-five millimeter, black-and-white film installation. Divided into five lengthy parts ranging from twenty-nine to ninety minutes, *Seven Intellectuals* was exhibited in a multi-channel installation at the 2007 Venice Biennale, and again at the Asia Society in New York City in 2009. Yang shot *Seven Intellectuals* slowly over the course of four years, with each segment taking roughly one year to create. The artist told me (Yang Fudong, 2008, personal communication) that the slowness of this production reflects his desires to merge his own art and life and provide a counterpoint to the fast pace of consumerist Shanghai.

In reflecting on issues facing artists in post-socialist China, *Seven Intellectuals* draws from Shanghai's Republican Era imagery and genealogy of cross-cultural hybridity, and also from much earlier precedents. The film takes its title and thematic concept from *Seven Sages of the Bamboo Grove*, a Chinese legend of a group of seven scholars said to have lived during the third dynasty C.E. under the tumult of the early Six Dynasties Period (Proser, 2009). This period and those following it were marked by war, civil strife, and political chaos, and it is believed that many of the era's artists and intellectuals exiled themselves, rejecting civilization by retreating into secluded mountains and forests. Among such figures, the Seven Sages were said to have retreated to a bamboo grove to carry out their various intellectual pursuits – drunkenly engaging in Daoist-influenced philosophical discourse, creating poetry, and making music – all in a natural environment far from Shanghai and the political and social turmoil of the day. Yang's reference to the *Seven Sages* illuminates the crises facing intellectuals in the PRC's post-socialist milieu, a period characterized by tremendous urbanization, the rise of state-sponsored capitalism, the aggressive solidification of one-party rule, and increasing restraints placed on personal liberties.

Seven Intellectuals follows a group of five young men and two young women (the seven intellectuals) as they move back and forth between China's countryside, notably Huangshan (Yellow Mountain) in Anhui Province and urban Shanghai, where the film concludes. Throughout the film's Shanghai segments, the characters spend extensive time indoors, suggesting that urban individuals have become estranged from their rapidly developing environment, and that intellectual life has become increasingly privatized. For Yang, contemporary life in Shanghai is informed not by any collective East-meets-West harmony, but by increasingly privatized and alienated existence (Yang, 2008: 119):

> A hundred million people live in this city, and these hundred million people basically live in solitude... In a city, very few people carelessly lift their heads and look up, and everybody basically has this flat visual angle, even though there are high buildings on either side... So in terms of visual experience, city people are narrow and limited.

The film's final segment concludes in Shanghai's Xian Qian Fang, a Republican Era art deco theater that has been transformed into a swanky restaurant, where some of the male intellectuals take up cooking jobs. The camera travels down a dark wood-paneled corridor, softly illuminated by dozens of small lamps affixed to the walls, and into Xian Qian Fang's grand ballroom, with large round tables, mirrored panels, a diamond-patterned carpet, white drapes strewn around columns, and chandeliers. The camera hovers in to reveal the seven intellectuals, joined by a few other men, sitting around the room's center table, eating and drinking in silence. There is no one else in the restaurant. Two screens in the background play identical footage, seemingly from other parts of *Seven Intellectuals*, those set in the countryside, where the struggling intellectuals tried but failed to find fulfillment. This stage-like setting in a former theater calls viewers' attention to the artificiality of the lives of these urban dwellers.

Two of the intellectuals, a young man and a young woman, silently get up from the table and begin to dance – an odd combination of ballroom, tango, and contemporary movements. Two other male intellectuals rise from the table and begin wrestling with one another, knocking into the couple, and bringing the dancing man into their rumble. A fourth intellectual joins in, so that two pairs of twisted fighters tumble about violently on the ground, while the female intellectual retreats to the table, dropping a napkin into the fray in a bullfighting-like gesture. The last intellectual arises from the table and dances by himself, stripping off his clothes. Simultaneously, nearly a hundred men wearing cooks' uniforms and tall white hats enter single file and form a semi-circle around the periphery of the room. The men who were fighting separate, each lying sprawled out on the floor as if dead. The cooks clap in unison. Their clapping provides a rhythm to which the last standing intellectual, now shirtless, continues to dance. This scene, in which the intellectuals turn on each other in violence and end up in isolation, evidenced by the singular dancer, reiterates that Shanghai, with its luxurious, revamped art deco surfaces, is in reality a lonely empty dream world. *Seven Intellectuals* suggests that the city's materialism generates an intellectually vapid environment, leaving alienated individuals to dance only with themselves.

In thinking through what he describes as the "false luxuries" of Shanghai, Yang remarked (Yang, 2008: 175):

> Everyone imagines that things are happening in some place where they have never been ... I think that the city has a lot of this kind of thing: descriptive delusions ... a lot of people can give an extravagant description of Shanghai ... Let someone start talking, and the place comes to life. You ask, has he been up there? Maybe he hasn't been anywhere. In other words, there's a kind of rumor in cities that produces fantasies, and those rumors create delight. This delight is false, but it influences a lot of people. Saying this brings us to a topic that I've been especially interested in: the capacity of rumors – and here's the main point – delusion. Image also has this power, like the power of rumors to delude...

The artist critiques the mythologizing of Shanghai, arguing that hyperbolic descriptions and striking images delude people by projecting delightful fantasies of the city as a unified entity. *Seven Intellectuals*, in turn, crafts critical urban imaginaries revealing that contemporary life in Shanghai is characterized not by cohesion, but by a collision of realities and temporalities that ultimately obscure, rather than clarify young denizens' conceptions of the past.

Critical imagining, unveiling cracks in Shanghai's East–West façades

In rapidly urbanizing and globalizing post-socialist Shanghai, corporate and state-sponsored urban development, commercial growth, financial gain, and materialistic values are prioritized, throwing individual subject positions and especially the role of the artist/intellectual into deep-seated crises. Glamorizing surfaces celebrating Shanghai's East-meets-West status – such as the Jin Mao Tower's shiny exterior – abound in the city's twenty-first-century landscape and are widely transmitted through global media. Yet even artworks intended to embody East–West harmony can produce failures, as seen in Gu's inability to erect *Heavenly Lantern*, both at the Jin Mao Tower and the Dutch Cathedral of Groningen. These failures to install imagined projects bring to light various economic, social, and cross-cultural conflicts, including the practical difficulties of effectively combining private and public art sponsorship (even in a country with myriad private-public partnerships), and persistent, misinformed suspicions of "communist" China.

More than a decade after Gu's initial *Heavenly Lantern* proposal, he realized a different Shanghai iteration, not at the Jin Mao Tower, but at the M21: 21st Century Minsheng Art Museum. As part of his 2016 solo exhibition, "Gu Wenda: Journey to the West", the artist covered the exterior of the M21 Museum in 25,000 colorful lanterns. Gu also invited 3,000 school children to write their hopes and dreams on lanterns at the exhibition's opening in a public performance promoted as part of his multi-city *Public Art Day* initiative. Gu has recently proclaimed a commitment to actively engaging "the public", represented at M21 by children participants, who he refers to as (2016) "young art patrons". This more grass-roots approach to art making may well have been born from *Heavenly Lantern*'s earlier failures; the original proposals relied primarily on traditional institutions of power and behind the scenes negotiations, often without successful outcomes. Yet Gu, like many contemporary artists working in and out of China, must still rely on corporate and governmental support; the M21 Museum is owned and operated by China Minsheng Bank, a private bank with high-stakes investments in contemporary Chinese art, and Gu's *Public Art Day* projects are sponsored by Pingan Insurance, Pudong Municipal Government, and Expo Area Development Administration of Shanghai Pudong, among other private and public developers, as the artist readily acknowledges (2016). While Gu's work with school children may be read as another form of outsourcing artistic labor, the performance outside the M21 Museum nonetheless transmitted a striking juxtaposition between a celebratory institutional cover of thousands of relatively uniform lanterns and diverse dreams imagined and handwritten by young individuals.

Other artworks I have explored throughout this chapter generate critical urban imaginaries that more directly complicate spectacular projections of Shanghai. Liu's *The Virtual Scene* pictures poker chips as a reminder of the risks associated with fast-paced urban development, while blurring the usually sleek images of Shanghai's East-meets-West skylines to near indiscernibility. *Export – Cargo Transit* calls attention to the unequal relations between China and Western nations, subverting global flows of trash and commodities by transforming imported garbage into artworks intended for exportation. Yang's *The First Intellectual* contrasts fashionable signifiers of global capitalism – magazine advertisements and the soaring skyscrapers of Shanghai's Lujiazui Financial District – against a wounded, disheveled young intellectual, allowing the disillusioned figure to take center stage in an otherwise non-human scaled, vacant cityscape. *Seven Intellectuals* shows listless, alienated individuals struggling within post-socialist Shanghai's materialist realities. The lonesome intellectuals search for stimuli outside of the city and imagine playful ways of resisting corporate and state control in Shanghai's increasingly sterile urban environs. These critical imaginaries picture imperfect, individualized states of urban subjectivity, and resist the seductive power of images to delude. Revealing deep-seated cracks in the city's fragile East-meets-West façades, an artist/intellectual opens space for dreaming.

References

Bhabha, H. (1994) *The Location of Culture*. New York: Routledge.

Chiu, M. (2009) Preface. In: Chiu, M. and Tezuka, M. (eds.), *Yang Fudong: Seven Intellectuals in a Bamboo Forest*. New York: Asia Society, 8–9.

Dirlik, A. and Zhang, X. (2000) *Postmodernism & China*. Durham, NC: Duke University Press.

Fei, X. (1953) *China's Gentry*. Chicago: University of Chicago Press.
Gu, W. (2005) *Heavenly Lantern: Wenda Gu Works: Jinmao Tower Shanghai*. Transl. D. Mao. Shanghai: Gu Wenda Studio.
Gu, W. (2016) *Public Contemporary Art Day*. Shanghai: Gu Wenda Studio.
Lee, O. (1995) *Shanghai Modern: The Flowering of a New Urban Culture in China: 1930–1945*. Cambridge, MA: Harvard University Press.
Liu, J. (2016) Unreal Scene. In Lin, J. (ed.), *Picturing Global China*. Portland: University of Oregon White Box, 49–50.
Murphey, R. (1953) *Shanghai: Key to Modern China*. Cambridge, MA: Harvard University Press.
Ong, A. (2011) Hyperbuilding: Spectacle, Speculation, and the Hyperspace of Sovereignty. In Ong, A. and Roy, A. (eds.), *Worlding Cities: Asian Experiments and the Art of Being Global*. Oxford: Wiley-Blackwell, 205–26.
Ong, A. and Roy, A. (eds.) (2011) *Worlding Cities: Asian Experiments and the Art of Being Global*. Oxford: Wiley-Blackwell.
Press Clippings about Imported Trash. (2008) *Export – Cargo Transit* (ed. Liu, J.). Shanghai: Shanghai Gallery of Art, 32–3.
Proser, A. (2009) Seven Sages of the Bamboo Grove: Chinese Models of the Unconventional. In Chiu, M. and Tezuka, M. (eds.), *Yang Fudong: Seven Intellectuals in a Bamboo Forest*. New York: Asia Society, 24–8.
Visser, R. (2010) *Cities Surround the Countryside: Urban Aesthetics in Postsocialist China*. Durham, NC: Duke University Press.
Yang, F. (2008) *Yang Fudong: Seven Intellectuals in a Bamboo Forest*. Beijing: Office for Discourse Engineering.

17

Toward a photographic urbanism?

Images iconizing cities and swaying urban transformation

Michele Nastasi and Davide Ponzini

Introduction: the seduction of architectural photography

For many decades, mainstream modernist planning has been losing ground to real practice (Sanyal, 2005). Nonetheless, modernization, progress, and their international aesthetic have reassured decision-makers as well as citizens and, in some cases, continue to do so. Several positions in the public domain seem nostalgic not only of the aesthetic, but also of the role that modern architecture played in giving order to the fragmented growth of twentieth-century metropolises. Recent academic studies in the field of architectural photography mostly concentrate on the history of Modern architecture (Higgott and Wray, 2014; Zimmerman, 2014), showing how the myth of its iconography was built through specialized and mainstream media. This approach – along the lines theorized by Walter Benjamin (1969) in his classic: *The Work of Art in the Age of Mechanical Reproduction* – can assist in interpreting the increasing presence of iconic buildings in today's public domain. The displacement of the "cult" value with the "exchange" value through reproduced images leads to a shift in the significance of monuments and monumental buildings of which iconic buildings represent a particular variant.

The recent *Rule by Aesthetics* (Ghertner, 2015) explains that radical urban development policies have been promoted by real estate companies and the local government in Delhi, through seductive images of, and references to, how a world-class city should look like. These city images have reached and convinced Delhi's city builders – who perhaps may have personal experience of other world-class cities – as well as the deprived inner-city dwellers. Despite adverse effects of eviction and redevelopment, there is a growing consensus for these projects, thanks to a sense of pride and modernization represented by images of new and shiny buildings. More generally in many cities today, iconic aesthetics seem to have supplanted other aspects of architecture and its experience (in some cases including the functioning and management of the building itself), urban planning values and processes. Power tends to rely more on the visible and perceptible, rather than on expert knowledge and technically sound choices.

Today, cities like Delhi compete to stand out on the global scene through images of the city and future projects. Iconic buildings have a prominent role in these imaginaries, often intertwined with the spectacularization of the urban environment, as easily recognizable urban, and architectural projects typically have a highly figural profile. Iconic architecture has been widely studied in the last fifteen years by architects (Jencks, 2005; Sudjic, 2005; Vidler, 2000), and the questions of its language, meanings, forms, design, and the cult of star architects' personality and connections to power structures have been explored. The fact that "place" images are crucial is widely accepted – not only in promoting and marketing cities and specific areas (Ashworth and Voogd, 1990), but also in legitimizing given policies and plans. Sklair (2017) argues that iconic architecture should be understood within the hegemonic culture of global capitalism as well as a process of reproduction by transnational elites. McNeill (2009) critically analyses global architecture practice under the lenses of technology, media, and mobility. Our book *Starchitecture* (Ponzini and Nastasi, 2016) explores the decision-making and development processes of dozens of branded projects throughout urban landscapes of European, American, and Asian cities. The book intertwines one scholarly and one photographic essay on urban issues connected to star architecture. To our knowledge, this joint approach is the first such instance of studying the processes related to the development of iconic and spectacular architecture and its representation. It is evidently a broad field at the crossroads between urban and media studies, where the risk of taking ideological positions in favor of or against architectural design (as the manifestation of hegemonic capitalism or as the savior of contemporary cities) is high.

One can clearly observe, besides any ideological position, that the current competition between cities for a place in the global imagination through their architecture involves multiple means and media, but strongly levers photographic images. Over the last decade, a complex system of production, circulation, and reception of various kinds of photographic images has gone global, fast, and plural. Architects work in an increasingly heteronomous condition, one where they are no longer the sole producers of architectural and urban images, but depend on a network of relationships between architecture and planning specialists, clients, policymakers, and the public. Speed has gradually subverted many of the mechanisms linked to traditional publishing, bringing into question the figures and the role of image producers – architects, magazines, professional photographers – as well as users. Architectural images have become more and more familiar to the general public, due to a boom in amateur photography, mass tourism on a global scale, the widespread use of digital cameras and smartphones, as well as the capability to upload and publish images in real time. On a different level, the spread of photorealistic architectural renderings makes projects readable and shareable by laymen in multiple non-specialized media. Despite this proliferation and the constitutive ambiguity of photographic images, they exert a particular power. They continue to be perceived as objective representations of the urban and architectural environment and are therefore extremely influential.

In the face of such radical changes, diverse approaches to architectural photography have recently gained greater attention (Pardo and Redstone, 2014; Redstone, 2014). Although they provide interesting examples of classic and emerging approaches in architectural photography, they lack critical interpretations of their relationship to contemporary culture and society, as well as of their influence on urban policy making. We are interested in the increased relevance of image circulation and its impact on contemporary architecture and the urban realm, and for this reason also look to other fields of urban and architectural research and approaches. The circulating images of iconic architectural projects can be considered, in fact, as part of a broader transnational mobility of urban policies, investments, expertise, and people that, in our opinion, is the reference for a broader understanding of the features and roles of architectural photographic images. Photographs of completed project as well as renderings must be reconsidered not only in terms of a mere representation, but as elements that actively contribute to urban transformation (see, among others, Appadurai, 1996; Guggenheim and Söderström, 2009; Urry, 2007).

While some may see a structural shift in urban culture, in this chapter we limit ourselves to exploring the evidence we have about promoting and using branded and spectacular projects (and their photographic images in particular) for transforming relevant areas of contemporary cities.

Pictures by tourists, design professionals and of the city to come

In order to provide a basic cultural and social background to the current use of architectural photography, we have selected and outlined three categories of pictures according to their origin and purpose: first, pictures taken by amateurs and tourists as a souvenir; second, photographs produced by professionals for promotional purposes; and third, photorealistic architectural renderings that anticipate the construction of proposed projects.

The technological evolution of photography parallels the spread of the use of certain iconic buildings and architectural spectacles, the rapid establishment of key nodes in the global urban system, and the growth of mass tourism at an international scale. Over the last two decades, new iconic buildings have supplanted some of the previous monumental landmarks, imposing themselves as new symbols to construct a new public image. In Western cities like New York, London, Barcelona, and Milan, we find many new skyscrapers and museums, but it is in cities such as Singapore, Hong Kong, Dubai, and others in Asia and the Middle East where the spectacular image has become a persistent term of representation and comparison (e.g. the rankings of the tallest building in the world). In such places, the representation of the city – sometimes of the whole nation – coincides with photographic images of recently constructed iconic buildings or with the skyline. New iconic buildings serve as symbols for the whole city, despite the actual characteristics and quality of related places.

Simply stated, iconic architecture is photogenic. It also well accompanies the hyperbolic spread of photography; a sort of rapprochement of architecture with an idea of "popular" taste (Bourdieu, 1984) following the long period of non-figurative, abstract, and anti-monumental character of "orthodox" modern architecture (Nicolin, 2012). Despite the fact that iconic and spectacular architecture is in many cases ignored or deplored by critics, it is beloved by large numbers of locals and tourists who build their experience of cities primarily through photographs of the most conspicuous and monumental structures. The practice of spontaneously uploading photographs onto social media can therefore be seen as an appropriation of urban space (Bourdieu, 1990; Gell, 1998) equivalent to the celebration and consumption of spectacular urban spaces.

Tourist images are not the core concern of this chapter, although contribute to the mass generation of images and sense of place. Locals and tourists produce a potentially unlimited number of amateur photographs of buildings and cities, usually modeled on the few iconic typologies of the "official" architectural representations (Wilkinson, 2015). In this sense, the fields of production and fruition of photographic images of architecture and cities have undergone a substantial broadening as photographic images are now produced and circulated by professional architects and non-professionals alike. For example, the architect Rem Koolhaas has even started using these "spontaneous" representations on his firm's official website (www.oma.eu) to build a collective image of its projects.

Architecture and cities are of course still important subjects for professional photographers. The field of architectural photography has always been quite nuanced, ranging from the artist and highly-specialized photographer working in close connection with design firms to those using photography as a critical means for investigation of cities and architecture (Ponzini, 2012). The current circulation of images and change in editorial mechanisms has primarily affected highly specialized architectural photographers. Today, the transfer of architecture publishing to online claims figures that would be unimaginable for traditional publishing, both in terms of views and geographical coverage as well as the numbers of projects published daily. On top of this shift comes a radically different kind of fruition of the media, since the speed with which new projects can be communicated online to a global audience goes hand in hand with a minimal level of observation and inattentive glances (Nastasi, 2017). It is becoming increasingly rare for projects to be studied online through technical drawings and descriptive texts. Usually, architectural and urban projects are viewed only for a short duration, with images placed at the beginning of posts, in newsletters and on the homepage of websites. The first photograph of the series has somehow become the "façade" for a project (Schianchi, 2014) and the chosen images are therefore those having the greatest impact and attracting the viewer's attention. The consequence is the facile visual effect for the larger public and an emphasis on the spectacular, of which the iconic building has become the standard currency.

Online publications generally have little budget for photographic features, which results in a progressive reduction to almost zero "independent" commissions from magazines and publishers. Photographers are for the most part directly procured by architects. PR agencies for architecture firms have emerged within this context and mediate the relationship between architects and photographers, traditionally based on collaboration and exchange. In many cases today, photographers sell the copyright of their pictures to architecture studios, which turn to PR agencies and send a selection of photographs that magazines can use free of charge. This process results in fewer, but more eye-catching images chosen and repeatedly published in very different kinds of specialized and non-specialized media, reinforcing the abovementioned mechanism of circulation – and the one of reward of the spectacular and the iconic.

The third category of images consists of architectural and urban design renderings. These representations of projects to be built were once abstract and detached from the preexisting context. Instead today, they have become increasingly photorealistic in character – especially in the case of projects located in European and North American cities, where detailed virtual 3D models are easily available. Photorealistic representations make the image of projects accessible to non-specialists. Thus, they play a central role in contemporary architectural communication. More and more often, photorealistic renderings are not produced by architects, but outsourced to visualizers from a variety of backgrounds and for differing purposes. Renderings do not serve the needs of architects alone, but also investors, real estate promoters, public decision-makers, and other actors involved in the processes of production and legitimization of architecture (Rose et al., 2014). Here, the design skills and imaginative powers of architects contend, through the system of the media, with the mechanisms of the market and the dynamics of politics and the need to drum up support – and eventually funding – for particular projects.

There are several critical issues connected to photorealistic rendering. First, renderings can be mistaken for photographs and interpreted in a literal and objective sense, even though they represent a project that has not yet been realized or, in many cases, not even fully designed. Second, renderings are often criticized for their detachment from the conditions of the urban context and their ability to conceal the more problematic aspects of the projects they represent behind spectacular forms. In many cases, however, these are deliberate choices, aimed at reassuring potential investors, who recognize in this type of image a certain international standard of reference. Third, one can argue that the specialization and uses of this imaging technique contribute to the eroding of architects' control over the production of architecture itself.

By restricting our observation to these three categories of images, and mainly focusing on the second and third, we can narrow our questions and look for examples to articulate critical answers. How do photography-related technologies influence and eventually facilitate urban design and development processes? How does this affect actual and physical urban transformation?

Observing iconic skyscrapers, skylines and waterfronts

The imaginaries of contemporary cities are vast, even if one concentrates solely on the iconic and the spectacular. We focus on three categories of photographic images that reoccur in the processes of urban transformation: the iconic building and the skyscraper in particular, the skyline, and the waterfront. Such images not only are visual clichés, but one can recognize situations where they become a sort of reference model for urban development.

In this field, Dubai, in the UAE, provides an extreme example. The city became famous through the media-boosted idea that a sudden and miraculous development had occurred here, thanks to its iconic buildings and mega-scale infrastructures (e.g. Sheikh Zayed Road, the Dubai Metro, or the Palm Jumeirah) devoid of reference to context and depicted as a kind of tabula rasa. As we argue in recent works (Molotch and Ponzini, 2019), Dubai and other Persian Gulf cities, due to their particular characteristics, reveal more clearly some shifts in contemporary urbanization that are present, but less visible in other cities in the world. In this sense, Dubai is an extreme, yet still comparable case and reference for understanding how the design and development process levers architectural images, ultimately affecting urban transformation. The maximization of land value has driven Dubai's growth, supported by readily available local and international capital. Society is highly divided; a greater part of the population is composed of migrant workers with profiles, opportunities, and living conditions drastically different and unequal among each other and in comparison to the Emirati population. Thanks to its flight connections, Dubai now also serves as an international tourist hub. Hypermodern is the prevalent aesthetic in the city that also boasts the tallest building in the world, the Burj Khalifa (see Figure 17.1).

Despite these extremes, several comparable trends can be seen in other cities (in part because of these transnational mobilities pertinent to Dubai, and in particular because the same architects, investors, and consultants operate across the world; Ponzini, 2019). These trends include some "Dubaization" effects: architectural spectacles serving real estate and tourism markets, disneyfication of public space and social inequality. In this environment, the image of iconic buildings and the skyline defines the city, bypassing any consideration of the actual urban environment and its experience by multiple populations.

Figure 17.1 Burj Khalifa, Dubai, 2010. Photograph by Michele Nastasi.

Iconic buildings: the skyscraper

Photographic portraits of the Burj Al Arab (literally "Tower of the Arabs"), the famous sail-shaped building designed by the international consultancy firm Atkins, and the Burj Khalifa ("Tower of the Caliph"), designed by SOM and currently the tallest building in the world, are omnipresent in Dubai. Their images adorn small corner shops and the halls of luxury hotels alike. Since its completion in 2009, the Burj Khalifa has been the symbol of Dubai, but also of a nation, a moment in history and a particular global culture that uses an iconic architectural image to secure a city's place in the world's imagination. Ultimately, such efforts aim to attract the attention of investors, travelers, and tourists. The Burj Khalifa is one of the most common subjects of travel photographs uploaded onto social networks, even surpassing more established global icons like the Empire State Building in New York (on Instagram – as of 31March 2017 – the tag #burjkhalifa had 1,541,848 posts, while the Empire State Building had 1,530,529 posts). Typically, tourists' photographs and selfies tend toward abstracting the spectacular building from its context. It is depicted in the same choreographed manners that are dictated by the design of the square beneath and the presence of other spectacles like the Dubai Fountain with its record-breaking synchronized water display.

The image of the skyscraper has influenced the urban development process, first because the tower is an iconic archetype that depends on a solid and extensive visual tradition, from representations of monumental works of past ages to the image of the modern Western city. In the case of Dubai, the image and actual presence of the building functions as a landmark and focal point for developing the large surrounding Downtown Dubai luxury and consumption district. The Burj Khalifa established itself as a model and a driver for similar city developments based on massive urbanization arranged around an iconic tall tower, such as the more recent Dubai Creek Harbour and Tower.

Moreover, the Burj Khalifa can be seen as instrumental for its developer, Emaar. Subsequent investments by the same group can count on the aesthetic, symbolic, and reputational value of having built the tallest building in the world. Emaar expends this value by exporting other, more mundane, projects to North African, Asian, European, and American countries. In most cases, projects refer to master-planned communities with residential and retail mix-uses (typically malls) and sometimes iconic and landmark developments and buildings. The circulation of the images of the tallest building in the world supports the narrative of its developer, seduces investors and the public, and convinces politicians of the novelty of the development and its positive side effects (Ponzini, 2019). This narrative also travels with specific imaginaries, real estate development models, and conceptions of what a world-class city should look like and do for people.

Another project could become the third icon of the city, after the Burj Al Arab and the Burj Khalifa: the Dubai Frame. This building won the ThyssenKrupp Elevator Architecture Award in 2009 – an architectural competition run by the elevator manufacturer for projects generating a tall and emblematic structure. It consists of two towers and a sky deck with an observatory, and at 150 meters tall it stands between the 20th century city and more recent developments. The Mexican architect Fernando Donis conceived of a building that would not itself be another icon in an already overcrowded landscape, but a frame through which to observe and photograph the city, its iconic buildings and its skyline.

Skyline architecture

Dubai's skyline is essentially composed of towers along Sheikh Zayed Road and has become a hypermodern variation of a well-known urban iconography category (see Figure 17.2). It derives from the typical representation of water cities like Istanbul, Amsterdam, and London, that centuries ago emphasized their size and position at the

Figure 17.2 Sheikh Zayed Road Skyline, Dubai, 2015. Photograph by Michele Nastasi.

center of trading and colonial empires. More recent cities have highlighted the vertical developments of the modern city, Manhattan being the first and most powerful example (Nastasi, 2016).

This imaginary – falling into the aesthetic category of the sublime – concretely deals with a dualism that implies the observation of the city from afar and suggests certain kinds of lifestyle at the base of tall buildings. These representations contributed to the creation and diffusion of this image during the twentieth century (Damisch, 1996; Lindner, 2015; Taylor, 1992), but have perhaps been eroded by newer versions of the same image, much poorer in terms of street life and public domain. Nonetheless, the perceived link with economic success and progress is a key to understanding the spread of this imaginary between global and wanna-be global cities, despite the evident contradictions between the representation of buildings and the actual quality of day-to-day life between buildings.

Waterfronts

Dubai is an interesting example of a waterfront associated with the image of iconic buildings, skyscrapers, and the skyline. Dubai Marina presents a clear and explicit case of importing from Vancouver not only a master plan and scheme from the False Creek neighborhood – which was adapted to local interests and conditions by the developer Emaar (Ponzini et al., 2016) – but also the image of a world-class city where skyscrapers reflect on the water (a Google image search confirms how highly similar the images of such distant places are, at the time of writing in 2017). The urban environment in Vancouver served as a model (to the extent of excavating a vast canal in Dubai to resemble the original setting) that was adjusted according to the expected financial viability in Dubai where the local planning authority did not implement significant restraints and requirements. However, this transfer did not take into consideration a number of problematic issues in adapting the Vancouver case (e.g. building height and density, mix of land uses, accessibility, and public space), since these issues were neither relevant to the promoter nor to the decision-makers, although they drastically affect the urban environment and the way people inhabit the space. The master plan generated a new and appealing compound dedicated to consumerism and luxury living, but it evidently excluded people of other orientations or means. In general, the appealing and modern image contributed to creating a poor public domain, when compared to the original plan in Vancouver.

It is interesting to note that this huge development is often criticized for being much more generic compared to the Sheikh Zayed Road skyline. Nevertheless, despite the transnational origin of Dubai and its skylines, the city has developed a kind of post-modern character of authenticity triggered by an imitative mechanism on a regional scale (including Doha and many other cities in the Middle East) as well as global scale (especially in growing Asian cities) in a sort of process of "Dubaization" (Elsheshtawy, 2009). Here again, the spread of architectural expressions and iconic urban views plays a role in nurturing a new visual culture, or at least the acceptance of it. Indeed, waterfronts provide real estate appreciation in many cities around the world. In some cases, seductive images help drive urban transformation through waterfront projects. Countless cities have undertaken reconversion projects of large sea or river front stretches since the 1990s. A view of the water, as well as the reverse view of the landscape reflecting onto the water is widespread, often connected to a successful (post-industrial) environment.

A prime example is the Elbphilharmonie in Hamburg, designed by Swiss architects Herzog & de Meuron, a spectacular piece in a large-scale and long-term redevelopment scheme called HafenCity (over 100 acres in 25 years). The scheme itself is part of a vision for the city and the metropolitan area. The master plan involved a great number of calls for projects, competitions, and negotiations that brought about substantial change (Palermo and Ponzini, 2015). In 2006, the first Elbphilharmonie rendering was produced by the visualization firm Bloomimages. It successfully presented the project to the public and rapidly became pervasive in public communication and the media (Zöllner, 2012), even in tourist guide books and websites, well before the building actually existed. It is a photorealistic rendering showing a view of the building as a sort of fluctuating iceberg, which became itself the subject of works of art (e.g. one image of the series *jpg hdem* by Thomas Ruff), comics, and caricature, becoming more than just a picture produced for the media, but a sort of mental image that actively takes part in various social and political interactions. The building was inaugurated in 2017, after a 7-year delay and at more than triple the original cost. In these years, disappointment substituted the success and urban innovation originally conveyed by the image of the project and the final building is being evaluated in comparison with the original renderings. Of course, the image is not the only factor in the development process. In this case (Balke et al., 2017), as in many others, one can see how multiple discursive, governance, and visual devices have pushed forward the planning and implementation process, despite the evident (some say even purposeful) limits in its conception and feasibility study. Urban images can spark transformation, but perhaps the power of repeated and seductive iconic images can, in some cases, bypass or even impair public debate.

Where iconic architecture's image matters

The story of the Guggenheim Museum in Abu Dhabi, UAE, brings us to a further extreme. This gigantic project designed by Frank Gehry is one of the masterpieces in the masterplan for the Saadyiat Island Cultural District. The plan envisions a connection to the city center and urbanization of 25 square kilometers of the nearly desert island and started in the mid-2000s. Five iconic museums designed by internationally famous architects – the Guggenheim, the Louvre Abu Dhabi (Jean Nouvel), the Sheikh Zayed National Museum (Norman Foster), the Performing Arts Center (Zaha Hadid) and the Maritime Museum (Tadao Ando) – are meant to reproduce the so-called "Bilbao effect" on the shores of the Persian Gulf. The iconic image, one of the architect (Gehry) and of the operators (Guggenheim) were the same as Bilbao, with the expectation of building a distinctive cultural identity for the city, as well as attracting tourists and international investments (Ponzini, 2011). The rest of the island was expected to host hotels, marinas, villas, and residences that were far more generic in their design and expected impact. According to the original vision, the Guggenheim Museum, the Louvre Museum and the Sheikh Zayed National Museum should have been completed in 2013, but their openings have been progressively postponed over the years, in part due to the financial crisis of the late 2000s and falling oil prices in the early 2010s. Over the years, the museums' renderings continued to circulate in the media around the world, especially the views of the waterfront with the reflection of the museums in the water, emphasizing their iconicity. By circulating in newspapers, magazines, and websites – as well as in gigantic posters in the city and around the construction sites (see Figure 17.3) – these images have established a spectacular place in the global imaginary (definitely, but not exclusively, among architects and planners), despite the fact that the island was nearly deserted (and still is for the most part) and the cultural district far from complete (Ponzini and Nastasi, 2016).

The Louvre Museum in Abu Dhabi has provoked much international criticism because of the living conditions for migrant workers during its construction. In France, some have contested the impact of the international lending of national cultural heritage while others take issue with the Louvre, as a cultural institution, legitimizing the project development as a whole. The overall urban effect of the to-be cultural district is alienating, a sort of high-culture version of the amusement parks that are widespread throughout the countries along the Persian Gulf, but within a single suburban expansion of Abu Dhabi. Despite all criticisms, the museum was inaugurated in November 2017 and dominated specialized and non-specialized media around the world, thanks to its iconic architectural effect and confirmed the "exchange value" that an iconic building built on the waterfront can generate. It is clearly too early to consider the impact of the photographs of professionals, even less those taken by tourists, but in terms of the "image act" (Bredekamp, 2017), the project is certainly a success.

Therefore, according to the Abu Dhabi example, renderings can in fact replace a real project in terms of the imaginary, both locally and internationally, arouse consensus or controversy, and somehow drive the project's perception upon completion. The shift we clearly see in Abu Dhabi more than elsewhere is that renderings – as uninformed images

Figure 17.3 Saadiyat Cultural District construction site, Abu Dhabi, 2010. Photograph by Michele Nastasi.

from a technical planning point of view – are crucial references for decision making and actual urban development.

Conclusions: signs and symptoms of photographic urbanism

We have observed ongoing processes at various levels and interpreted significant shifts in the relationship between photographic images and urban transformations. Photography (most evidently of iconic buildings and places, skylines, and waterfront developments) exerts an increasing influence on the behaviors of both local inhabitants and tourists. Such processes of identification and the sharing of projects' images tend to sustain the production of iconic features and the spectacularization of spaces, and foster the potential for seduction of the voting or non-voting public (a potential that can be used and instrumentalized by some decision-makers, as we have shown). One can see that the photographic images generated and procured by professional architects and urban designers, as well as by other players in the urban development game, have their own life and circulation, apparently with larger popular audiences and success.

We see how images – in the context of globalization, acceleration, and pluralization in the use of architectural photographic images – have become a sort of "agent" of urban transformation. From a plain planning perspective, successful photographic images are technically uninformed documents, given that they do not carry information (nor have binding power) related to the physical, functional, etc., features of the built environment.

These images are often highly socialized, both through traditional media and social media, and their success and spread occurs at the expenses of other non-photographic, but more technically informed media that represented architecture and cities throughout the twentieth century – including text, technical drawings, models and maps, and other common tools available for shaping, discussing, and driving urban development and transformations. The broad diffusion of photographic representations amplifies their power, but also their potential unboundedness, paradoxically making the correspondence to actual (present or future) urban environments more difficult to check and eventually guide or confront.

The pervasiveness and power of photorealistic renderings along with the fact that they often substitute other media, make this shift even more evident in places of recent and fast-track urbanization like Dubai, that sustains the circulation of its architectural and urban images, simultaneously supported by the success of the iconicity of its images – more than any reputation of itself as a livable city. Dubai can be seen as an urban test bed for project hybridization, where they import design from the West and elsewhere, implemented at the highest speed, intended for tourists and migrant populations from dozens of countries to use (or at least observe) and then eventually export it to other cities in the Middle East, Asia, and back to the West as well. The development of that city (and many others) can be explained as part of broader transnational circuits and flows (Molotch and Ponzini, 2019).

Photographic images, in many cases, tend to depict iconic and spectacular architecture and environments in this way because they are more effective in serving the objectives of their original makers and clients. The debate on the instrumental role of realistic representations has long been controversial (see for example Stamp, 1982), but today the iconicity of photographs and renderings singularly focuses people's attention on the superficial and formal aspects of designs, downplaying the importance of other aspects of urban and social contexts, such as the relationship with actual contextual features (e.g. infrastructure, urban fabric, civic and public functions, etc.), or the political economy of place or development (who benefits, who ends up excluded, etc.). The success of skyline and waterfront images perhaps depends on their iconic content, which is seductive both for specialists and non-professionals, despite the problematic links that exist between these images and the actual urban conditions. According to our analysis, further attention should be paid to the possible instrumentalization of photographic images in perceiving, designing, and representing the city, or to the unintended consequences of the transnational circulation of images. On the other hand, photorealistic renderings may also be used as a powerful tool for public consultation on projects or as references for a certain kind of control over the projects by local communities and administrators, as happens in particular democratic systems like The Netherlands and Scandinavian countries.

The dominance of photographic images in the transformation of some cities is a visible symptom of the weakened role of urban planning and urban design play in the public sphere, not only in non-democratic countries as the UAE, but in many other countries in Asia and the West as well (Ponzini, 2016). When it comes to large-scale developments, especially for many waterfronts, we see modernist planning and its technically informed images playing a less and less important role. In many cases – including some of the most iconic skyscrapers in world capitals in recent years – the developer's power and resources, the technical engineering, and the seduction of branded and spectacular photographic images suffice to see a project completed. The democratic process that Western planners expect to see in place are altogether bypassed or potentially instrumentalized, in part thanks to the seductive role of photographic images. Of course it would be overly simplistic to blame design professionals and the experts specialized in renderings alone, since they are part of a more complex network of decision-makers. Observing these shifts in the methods of urban and architectural representation and their influence over processes of urban transformation can be a way to understand that the real problems are not how urban images are crafted to be seductive, but the use that architects, planners, politicians, or developers make (or allow others to make) of them.

The explorative (and quite extreme) set of examples provided here is not sufficient to allow any generalization, as urban transformation and planning occur through very different manners in different locations. Nonetheless, our reflection seems relevant to many regions of the world. To be clear, we do not advocate for a new photographic urbanism, nor take any ideological stance against it. We denounce its possible effects and argue the importance of how and why to have a better understanding of it. The first step to unpack the role that photographic images play is, in our opinion, to have a reality check, so to say, by comparing (not only visually) a given project's multiple iconic images (in photographs and renderings procured by architects, developers, etc.) to the project's actual place and implementation. Other qualitative and quantitative methods would then be welcome to sober public opinion and the media regarding the use of photographs.

The content of most influential photographic images can be systematically studied and analyzed, for example by looking at how and why architectural images tend to homogenize (or not) the global urban imaginaries – not only for spectacle-seeking cities like Dubai, but for many others aspiring at becoming modern or hypermodern. Is there a small group of highly specialized visualizers working for the most prominent design firms? Or are the requests by architects homogenous? Do photographic images bypass or simplify local specificities since they need to resonate with a global audience and potential set of investors? What are the urban culture and behaviors that more or less implicitly circulate with these images (Degen et al., 2017)? Are they Western or multicultural? Is touristic consumerism alone to blame?

Studying images, their circulation and impact on the imaginaries of different urban actors and their seductive visionary power can be a way to understand urban transformation in our contemporary globalized world. The challenge of observing and deconstructing the socio-cultural processes that make, circulate, and use photographic images seems to be a new approach for both the fields of visual studies and urban/architectural studies. Further interdisciplinary research on the topic is required to go beyond the explorations presented in this chapter.

References

Appadurai, A. (1996) *Modernity at Large: Cultural Dimensions of Globalization*. Minneapolis: University of Minnesota Press.

Ashworth, G.J. and Voogd, H. (1990) *Selling the City: Marketing Approaches in Public Sector Urban Planning*. London: Belhaven Press.

Balke, J., Reuber, P., and Wood, G. (2017) Iconic Architecture and Place-Specific Neoliberal Governmentality: Insights from Hamburg's Elbe Philharmonic Hall. *Urban Studies* 55: 997–1012.

Benjamin, W. (1969) The Work of Art in the Age of Mechanical Reproduction. In: Arendt, H. (ed.), *Illuminations*. New York: Schocken, 217–51.

Bourdieu, P. (1984) *Distinction: A Social Critique of the Judgement of Taste*. Cambridge, MA: Harvard University Press.

Bourdieu, P. (ed.) (1990) *Photography: A Middle-Brow Art*. Stanford: Stanford University Press.

Bredekamp, H. (2017) *Image Acts. A Systematic Approach to Visual Agency*. Berlin: De Gruyter.

Damisch, H. (1996) *Skyline. La Ville Narcisse*. Paris: Éditions du Seuil.

Degen, M., Melhuish, C., and Rose, G. (2017) Producing Place Atmospheres Digitally: Architecture, Digital Visualisation Practices and the Experience Economy. *Journal of Consumer Culture* 17(1): 3–24.

Elsheshtawy, Y. (2009) *Dubai: Behind an Urban Spectacle*. London: Routledge.

Gell, A. (1998) *Art and Agency: An Anthropological Theory*. Oxford: Clarendon Press.

Ghertner, D.A. (2015) *Rule by Aesthetics: World-Class City Making in Delhi*. New York: Oxford University Press.

Guggenheim, M. and Söderström, O. (2009) *Re-Shaping Cities: How Global Mobility Transforms Architecture and Urban Form*. London: Routledge.

Higgott, A. and Wray, T. (2014) *Camera Constructs: Photography, Architecture and the Modern City*. New edn. Burlington: Ashgate.

Jencks, C. (2005) *The Iconic Building: The Power of Enigma*. London: Frances Lincoln.

Lindner, C. (2015) *Imagining New York City. Literature, Urbanism, and the Visual Arts, 1890-1940*. Oxford: Oxford University Press.

Lynch, K. (1960) *The Image of the City*. Cambridge: MIT Press.

McNeill, D. (2009) *The Global Architect: Firms, Fame and Urban Form*. London: Routledge.

Molotch, H. and Ponzini, D. (eds.) (2019) *The New Arab Urban. Gulf Cities of Wealth, Ambition and Distress*. New York: New York Unversity Press.

Nastasi, M. (2016) Skyline Rêverie. *Lotus* 159: 72–87.

Nastasi, M. (2017) Image Cities. Skylines, Renderings, and Icons, Transforming Urban Landscapes. Unpublished doctoral dissertation (Italian), Ca'Foscari University of Venice.

Nicolin, P. (2012) *La Verità in Architettura. Il Pensiero di un'Altra Modernità*. Macerata: Quodlibet, 173–88.

Palermo, P.C. and Ponzini, D. (2015) *Place-Making and Urban Development. New Challenges for Planning and Design*. London: Routledge.

Pardo, A. and Redstone, E. (eds.) (2014) *Constructing Worlds: Photography and Architecture in the Modern Age*. London: Prestel.

Ponzini, D. (2011) Large Scale Development Projects and Star Architecture in the Absence of Democratic Politics: The Case of Abu Dhabi, UAE. *Cities* 28(3): 251–59.

Ponzini, D. (2012) Photographers: Architecture Critics of Today? *Domus* 961: 86–91

Ponzini, D. (2016) Introduction: Crisis and Renewal of Contemporary Urban Planning. *European Planning Studies* 24(7): 1237–45.

Ponzini, D. (2019) Mobilities of Urban Spectacle: Plans, Projects and Investments in the Gulf and Beyond. In: Molotch, H. and Ponzini, D. (eds.), *The New Arab Urban. Gulf Cities of Wealth, Ambition and Distress*. New York: New York University Press.

Ponzini, D., Fotev, S., and Mavaracchio, F. (2016) Place-Making or Place-Faking? the Paradoxical Effects of Transnational Circulation of Architectural and Urban Development Projects. In: Russo, A.P. and Richards, R. (ed.), *Reinventing the Local in Tourism. Travel Communities and Peer-Produced Place Experiences*. Bristol, UK: Channel View.

Ponzini, D. and Nastasi, M. (2016) *Starchitecture. Scenes, Actors, and Spectacles in Contemporary Cities*. New York: The Monacelli Press.

Redstone, E. (ed.) (2014) *Shooting Space. Architecture in Contemporary Photography*. London: Phaidon.

Rose, G., Degen, M., and Melhuish, C. (2014) Networks, Interfaces and Computer-Generated Images: Learning from Digital Visualisations of Urban Redevelopment Projects. *Environment and Planning D: Society & Space* 32(3): 386–403.

Sanyal, B. (ed.) (2005) *Comparative Planning Cultures*. New York: Routledge.

Schianchi, P. (ed.) (2014) *Architecture on the Web. A Critical Approach to Communication*. Padua: Libreriauniversitaria.it edizioni.

Sklair, L. (2017) *The Icon Project: Architecture, Cities, and Capitalist Globalization*. Oxford: Oxford University Press.

Stamp, G. (1982) *The Great Perspectivists*. London: Trefoil Books.

Sudjic, D. (2005) *The Edifice Complex: The Architecture of Power*. London: Allen Lane.
Taylor, W.R. (1992) *In Pursuit of Gotham: Culture and Commerce in New York*. Oxford: Oxford University Press.
Urry, J. (2007) *Mobilities*. Cambridge: Polity Press.
Vidler, A. (2000) *Warped Space: Art, Architecture, and Anxiety in Modern Culture*. Cambridge: The MIT Press.
Wilkinson, T. (2015) The Polemical Snapshot: Architectural Photography in the Age of Social Media. *The Architectural Review*, 15 January.
Zimmerman, C. (2014) *Photographic Architecture in the Twentieth Century*. Minneapolis: University of Minnesota Press.
Zöllner, C. (2012) The Elbphilharmonie Renderings. *Clog* 4: 100–1.

18

Macau's materialist milieu

Portuguese pavement stones and the political economy of the Chinese urban imaginary

Tim Simpson

Introduction

When I moved to the Chinese city-state and former Portuguese enclave of Macau in 2001, I was immediately charmed by the *calcada Portuguesa*, the city's expressive stone pavement tiles. This instantly-recognizable cobblestone pavement is visible throughout the city center of the Portuguese capital of Lisbon, which once served as the metropole of a vast colonial empire stretching across Africa, Asia, and South America. The Portuguese seafarers who departed from Lisbon to circumnavigate the globe and establish far-flung territories and trading ports inaugurated the modern imaginary and commenced the project which Peter Sloterdijk calls "terrestrial globalization": the human quest over the past half a millennia to understand the earth as a self-contained and knowable globe.

This Portuguese stone pavement, which often features undulating wavelike patterns and decorative mosaic embellishments such as ships, anchors, fish, crabs, and other elements of Portugal's seafaring traditions, symbolically signifies Portugal's age of discoveries, and physically inscribes the country's foreign territories. The tiles are therefore a familiar attribute of many former Portuguese outposts, including Mozambique, East Timor, Sao Paolo, and Rio de Janeiro.

Portugal returned Macau to the People's Republic of China in 1999, just two years before my arrival, and the *calcada* pavement tiles that are ubiquitous in Macau's historic city center seemed to clearly denote the Portuguese influence in the city (see Figure 18.1). The tiles line the sidewalks of the central *Avenida de Almeida Ribeiro*, known in Chinese as San Ma Lo, which cuts directly across the Macau peninsula from the city's Inner Harbor to the South China Sea, and they link the avenue with a variety of intersecting side streets and medieval alleyways and cloisters that guide visitor foot traffic to various squares and piazzas surrounding the city's abundant colonial-era structures. These structures include St. Joseph's Seminary, St. Dominic's Cathedral, Dom Pedro Theatre, the Ruins of St. Paul's, the Moorish Barracks, Holy House of Mercy, a Protestant cemetery, and a variety of other historic structures that comprise Macau's sprawling UNESCO World Heritage site. The *calcada* pavement tiles

Figure 18.1 Macau's *calcada Portuguesa*, or Portuguese pavement tiles. Photograph courtesy of Adam Lampton.

provide the literal groundwork that links together each of these historic structures, signifying their status as the progeny of Portugal's influence in the region.

Only some years later did I learn that this charismatic pavement, which I presumed had persisted in Macau throughout the four and a half centuries of Portuguese administration of the territory (from 1557 to 1999), was a recently-invented "tradition". In addition, the *Largo de Senado*, the public area located adjacent to San Ma Lo, directly in front of the imposing former Senate building, and seemingly Macau's primordial central square, was originally a roadway intersection that had been pedestrianized less than a decade before my arrival. While the impressive buildings I viewed in the city center were certainly authentic historic structures, the stone tiles beneath my feet, as well as the city commons of the main public square, were the materialization of retail real estate calculations, a strategy of tourist seduction, and a recent feat of contemporary urban engineering. I would come to realize the role played by this pavement project in production of an urban imaginary for the many Chinese tourists who visit the city.

The political economy of the Portuguese *calcada*

At the time of the handover, Macau was the last remaining European colony in Asia. It joined the PRC as a Special Administrative Region (SAR) under the country's "one country, two systems" regime, a designation it shares with its neighbor Hong Kong. Today the *calcada Portuguesa* and the *Largo de Senado* are primary attractions for the 30 million tourists who visit the tiny city of Macau each year, two-thirds of whom come from mainland China. Certainly many of those Chinese tourists visit Macau to see the city's historic sites, and the *Largo de Senado* is flooded every day with thousands of tourists jostling for ideal vantage points from which to take selfies against the surrounding backdrop of pastel-colored neoclassical buildings and the square's picturesque fountain. No doubt many of those tourists incorrectly assume, as I did, that the stones are an historic element of the city's world-renowned patrimonial heritage.

But Chinese tourists also visit Macau by the millions for more contemporary diversions: to gamble and shop in the city's gigantic casino mega-resorts, such as Venetian, Parisian, Galaxy, Wynn Palace, and City of Dreams. These resorts were opened by foreign gaming operators from the United States, Australia, and Hong Kong, who entered Macau's gaming industry after the return of the city to the PRC in 1999, and (with the help of the increasingly deep pockets of those Chinese tourists) transformed it into the world's most lucrative site of casino gaming, with annual gaming revenues quintuple those earned in Las Vegas.

Macau's decorative pedestrianized public spaces, as well as its phantasmagoric themed environments, enable and afford certain Chinese tourist peregrinations that have both economic and political relevance. In this chapter I would like to explore the *political economy* of Macau's peculiar Portuguese pavement stones, and their relationship to the concomitant *subjective economy* which animates the simulated spaces of Macau's integrated casino resorts. That is, I will focus not on the *colonial* or *post-colonial* peculiarities of Macau's *calcada Portuguesa*, or the *post-modern* architectural aesthetic of the themed casino resorts, but on the pragmatic, functional role played by these structures in China's *post-socialist* economic reforms.

My study of the *calcada Portuguesa* is informed by scholarly attention to China's use of normative *models* – such as model soldiers, model workers, model factories, and model villages – to guide ethical behavior, and Macau's unlikely role today as a *model city* for Chinese tourists. Borge Bakken (2000) refers to China as *The Exemplary Society* to emphasize the productive role of such "educative" and "disciplinary" models in Chinese social life. I contend that as a model city, Macau plays a central role in the production and maintenance of a Chinese urban imaginary, exemplary of a normative mode of "urbanism as a way of life" that is crucial to the country's economic reforms.

To understand the function of the *calcada* tiles, I will adopt a materialist ontology (Coole and Frost, 2010) that accounts for the agentic or animistic role of seemingly inert elements of *material geography* in production of an immaterial or imaginary *psychogeography.* I am interested in how this ineffable, metaphysical imaginary is constructed from the "vital materiality" or "thing power" (Bennett, 2010) of tangible objects, such that the imaginary may be finely calibrated and mobilized for political and economic outcomes. In the case of China and its relationship to Macau, this materialist approach highlights the way in which the practice of government, following Michel Foucault (2009), involves the "government of things", which may be strategically arranged so as to structure "the possible field of action of others" (Lemke, 2015: 11). In Macau, the *calcada* tiles are actuated to calibrate individual freedoms, and to normalize population mobility and the use of public space for consumer activities.

Materialist ontology and material urban politics

Sheyla Zandonai and Vanessa Amaro (2017) provide an insightful ethnographic account of the development of Macau's *calcada*, documenting the complex administrative machinations and intercultural negotiations that were necessary to create one of Macau's most endearing (if not actually enduring) attractions. From their standpoint, the *calcada* reflected an attempt by the outgoing Portuguese administration to materialize the colonial legacy in the city's built environment as it prepared to return the territory to the People's Republic of China in 1999.

As they narrate the story, late one night in 1993, just six years prior to the handover of Macau to the PRC, and with no explanation from Portuguese authorities, the roadway intersection in front of the *Leal Senado* building on San Ma Lo Avenue was closed to traffic by the imposition of several large granite vases. This activity drew the consternation of Chinese shopkeepers in the area who feared it would negatively affect business. A few weeks later the reason for this closure became clear, with the delivery from Portugal of 15 containers of granite and basalt stones, as well as a crew of *calceteiros*, the Portuguese pavement masters who would lay the cobblestone tiles. Those workers helped transform the former roadway into a trapezoidal shaped pedestrianized commons, lined with *calcada* tiles and encircled by the existing Portuguese buildings. For Zandonai and Amaro, the Portuguese were inscribing their legacy in Macau, and those tiles today reflect a "residual colonialism" that coheres in the city's built environment.

However, Zandonai and Amaro correctly contend that in post-colonial Macau, the *calcada* is marshaled in "new economic and political aims, through a strategy of image-building and *theming*" which both attracts and serves that enormous tourist market, and imbricates the heritage areas of the city with the faux themed heritage of many of the casino resorts. The fact that not only the Portuguese stone tiles but also the city square itself are actually simulated heritage environments, seamlessly articulates those material spaces with the meticulously rendered but nevertheless completely contrived Venetian, Parisian, Himalayan and other such themes that characterize the city's new integrated resorts. In fact, Macau's MGM casino, located on reclaimed land at the southern end of San Ma Lo, only a kilometer away from the *Largo de Senado*, includes a massive glass atrium enclosing a themed construction of a Portuguese castle surrounded by (ever-more simulated) *calcada* tiles (see Figure 18.3).

However, focusing primarily on the *representational* aspects of the stonework, and the manner in which it reinforces an imagistic or symbolic motif, may overlook the more *practical* use of the stones as a ground for pedestrian movement, and their function in normalizing specific notions of appropriate Chinese tourist behavior. The relation between the image and function of the *calcada* that I will pursue here is actually addressed by a Macau tourism official who is quoted at length in the conclusion of Zandonai and Amaro's article:

> The *calcada* is one of Macau's icons. It's one of the central elements of the city's cultural, historical, and architectural heritage, and one of Macau's major features as a meeting point between the Chinese and Portuguese cultures that turn the city into a singular tourist destination. The Portuguese *calcada* is used as an element of embellishment and decoration, enhancing the value of the Historic Center of Macao, part of the UNESCO World Heritage list, alongside several squares and promenades of the city, *inviting visitors to cover Macau step-by-step* and to appreciate the beauty of the city.
> [*my emphasis*]

This comment about the *calcada's* role in the tendency of tourists to "cover Macau step-by-step" both ascribes an *agentic potential* and *antecedent motive* to the stones, and explicitly articulates an understanding of tourism as, first and foremost, *perambulation*. The pedestrianization of the *Largo de Senado*, and the subsequent deployment of *calcada* in other areas of the city, serves to mobilize the stones to that end. The heritage buildings on the *Largo de Senado* have been retrofitted as retail shops whose enormously expensive rents depend on the massive foot traffic driven by all those tourists. The stones normalize certain ambulatory activities (e.g. window-shopping), with coterminous assumptions about appropriate tourist behavior in public spaces (e.g. taking photos and purchasing cosmetics). These particular uses of the pedestrian commons seemingly overwhelm other potential activities that might differently utilize this enhanced mobility, population density, and public space.

In focusing on the agentic properties of material environments, including both the *calcada* stones and Macau's themed resorts, I try to look beyond the question of whether those environments are authentic or inauthentic, real or simulated. I am interested instead in the *didactic function* of the material space of Macau's built environment for Chinese tourist subjects, as that space intersects with the temporality of China's market reforms. We will see that the very travel of those tourists to Macau is a primary component of those economic reforms, both a product of China's post-reform "opening up", and a specific governmental strategy for implementing and guiding reform trajectories. In short, Chinese tourists are directed to imbibe a specific lesson about public space from Macau's tourist environment.

I argue that this *post-socialist materialist* approach is more useful for understanding Macau's simulated built environments than the *post-modern semiotic* accounts typically applied to themed architecture and environments elsewhere. These ubiquitous semiotic studies, inspired by the work of Umberto Eco (1986) and Jean Baudrillard (1994), tend to dismiss such simulated spaces as ungrounded or "free-floating" signifiers that are ultimately indicative of a meaningless hyperreal. In the case of Macau, I will suggest that the *calcada* pavement tiles, and the themed Venetian and Parisian resorts, together comprise a "materialist pedagogy" (Simpson, 2010) which functions to produce a Chinese consumer subject crucial to China's market reforms.

In addition, this materialist pedagogy of contrived Portuguese pavement stones and themed architectural artifice constructs an urban imaginary that functions in a post-socialist "material politics" (Minuchin, 2013). This immaterial imaginary is assembled, so to speak, from the material object world. We can observe a similar process of material imaginary construction in Le Corbusier's efforts to use the new building material of concrete to construct a master plan for Buenos Aires in the early twentieth century. In a recent study of Le Corbusier's master plan, Leandro Minuchin (2013) highlights the relevance of this relationship between building materials and the urban imaginary for urban politics.

> The focus on the role of materials in articulating and mobilizing alternative urban imaginaries offers a distinct methodological alternative for urban theory and research. Tracing the different ways through which materials were appropriated and used, exploring their emergence and implication in spatial technologies and tactics, provides an opportunity to investigate actors, processes and resources that are not often included in the study of urban politics.
>
> (*Minuchin, 2013: 255*)

For my study, this material politics helps to clarify the relevance of Foucault's "government of things" for the management of Macau's built environment and tourist visitors.

While Le Corbusier, much like his contemporary Walter Benjamin, was interested in the utopian potential of *novel* construction materials of the period such as concrete, glass, and steel, I will analyze the way in which the *traditional* material of *calcalda* tiles is mobilized in the government of tourist mobilities in Macau's pedestrianized public spaces. To do so, however, I must first recount the specific role which the Macau SAR plays in the China's market reforms.

Transformative experimentation and urban pedagogy

Mainland Chinese tourists are not in Macau by happenstance. Their travel is facilitated by the Individual Visit Scheme, a program launched in 2003 through which the PRC extends special exit visas to Chinese citizens from select cities and provinces, allowing them relatively less-restricted travel to the SARs of Macau and Hong Kong. This scheme may be understood as a component of PRC macroeconomic planning and experimental population governance. According to Sebastian Heilmann (2008b: 3), Chinese state planning generally proceeds through a process of "transformative experimentation". This distinct Chinese mode of governance inverts conventional forms of governmental policy-making. Rather than conducting policy analysis and formulating legislation *prior* to implementation of new measures, the Chinese state innovates via "implementation first, drafting universal laws and regulations later" (Heilmann, 2008b: 4). This mode of "learning through practical experience goes together well with the importance of teaching by example and learning through role models in the Chinese educational and administrative traditions" (Heilmann, 2008a: 20), as described by Bakken in *The Exemplary Society*.

In both the Maoist and reform eras, the PRC has experimented with new policies, programs, and ideas by employing this "practice-based epistemology" (Heilmann, 2008a: 8) that proceeds through pilot-testing and experimentation. According to Heilmann (2008b: 5), the PRC practices three main forms of experimentation:

> (1) experimental regulation (provisional rules made for trial implementation), (2) "experimental points" (model demonstrations and pilot projects in a specific policy domain), and (3) "experimental zones" (local jurisdictions with broad discretionary powers).
>
> (*Heilmann, 2008b: 5*)

This Chinese approach to practical experiments articulates with my own understanding of the built environment of Macau as comprising a "materialist pedagogy". That is, Chinese tourists visit Macau already steeped in the idea of learning from models. To fully appreciate this notion of model materialist pedagogy we must understand learning as not simply a cognitive process of knowledge acquisition, and conceive learning instead "as a distributed assemblage of people, materiality, and space that is often neither formal nor simply individual" (McFarlane, 2011: 3). The construct of Macau's materialist pedagogy demonstrates the ways in which, as Colin McFarlane says in his book *Learning the City*, "Learning emerges through practical engagement with the world" (2011: 15).

The Chinese government has experimented with both Special Economic Zones (SEZs) and SARs as learning-oriented environments. Both the SEZs and the SARs are exemplary of the "experimental zones" that Heilmann mentions are used to pilot test reform innovations. The SEZs were established as *laboratories of capitalist production*, while the SARs of Hong Kong and Macau may be conceived as concomitant *laboratories of consumption*.

Special economic zones as laboratories of production

When Deng Xiaoping introduced market reforms in 1978, the first phase of reform involved implementation of a production-for-export regime which utilized SEZs created along China's southern coast. The first two SEZs established were Shenzhen and Zhuhai, which are located adjacent to Hong Kong and Macau, respectively; other SEZs were subsequently formed. The SEZs were established ostensibly as "laboratories" of capitalist production where the central government would facilitate joint venture manufacturing projects funded by foreign capital. This capital was mobilized largely from the Chinese diasporic populations of Hong Kong, Taiwan, and Southeast Asia, and these manufacturing enterprises utilized China's vast supply of over-accumulated, relatively well-educated, and low-cost labor. The revenue generated from this industry would drive China's post-reform economic development.

In addition, however, Chinese authorities set out to use the SEZs as a set of experiential and experimental classrooms, "social and economic laboratories where foreign technologies and managerial skills could be observed" (Harvey, 2006: 130). This pedagogic role of the SEZs was purposeful. Indeed, the Communist Party's own "Science and Technology Modernization imperative" dictates that "the Shenzhen SEZ was deemed to be a 'learning laboratory of capitalist modes of operating business, high-tech manufacturing, and construction'" (Chuihua et al., 2001: 121). Deng was forthright in his hope that the SEZs would constitute a capitalist classroom:

> Special Economic Zones are a window to technology, management, knowledge, and foreign policies. Through the zones, we can import technology, acquire knowledge, and learn about management, which is also a form of knowledge. The Special Economic Zones will become a foundation for opening to the outside world. We will not only benefit in economics and personnel training, but also extend the positive impact of our country to the world.
>
> (*quoted in Chuihua et al., 2001: 87*)

Understood as a laboratory for studying technology, management techniques, construction, and real estate, the didactic role of the SEZs is clear. Once formed, the Shenzhen SEZ became perhaps China's primary locale for experimenting with a wide variety of economic and governmental innovations. Shenzhen eventually grew into one of the most prominent urban environments in the PRC, and today is a model city and site of policy pedagogy; in the tradition of China's "transformative experimentation", provincial officials and government planners are instructed to "learn from Shenzhen" (O'Donnell et al., 2017).

Special administrative regions as laboratories of consumption

If Chinese entrepreneurs originally learned the lessons of capital in the SEZs of Zhuhai and Shenzhen, the citizens are instructed in the arts of consumption in the SARs of Macau and Hong Kong. While the SEZs were designed for capitalist industrial production, Macau is characteristic of a post-industrial form of life with an economy based on leisure entertainment, casino gambling, and the manipulation of affect. Macau today serves as a didactic laboratory for Chinese consumerist pedagogy. The new themed spaces of consumption in Macau today, such as the *Largo de Senado* and Venetian resort, mirror the themed spaces of production that organize the SEZs, and participate in production of China's consumers. The SARs function specifically in a second experimental phase of China's reform that addresses limitations of the SEZ production regime, which became visible as a result of unanticipated regional events in the form of crises of public finance and public health.

In general, the production-for-export regime was an overwhelming success and produced substantial economic growth in China. However, the 1997 Asian Financial Crisis exposed the inherent vulnerability of this regime. Although China's own financial system was relatively unaffected by the crisis when compared to many of its neighbors, the crisis indirectly threatened the country's economic viability by negatively affecting the ability of regional nations such as Thailand, Indonesia, and Malaysia to purchase Chinese exports. The financial crisis thus revealed the long-term unsustainability of the production-for-export model due to its over-reliance on foreign markets.

Following the financial crisis, the Chinese central government sought to enhance the country's domestic consumption levels in order to decrease China's reliance on external markets. To do so, the central government decided to make tourism a key economic growth area. Hoping Chinese tourism would lead to enhanced consumption, the government sought to formalize leisure as a productive mode of economic life. The government created three week-long, annual public holiday periods, or "Golden Weeks", in which citizens would be freed from work obligations and encouraged to travel. Consistent with the Maoist practice of using large-scale ideological campaigns, such as the Great Leap Forward or the Down to the Countryside Movement, to mobilize the population in production of the socialist "new man", authorities launched an extensive information campaign to produce tourist subjects, informing citizens of the moral value of travel, and the corporeal behaviors proper to modern tourism (Nyiri, 2007, 2010).

Alongside its economic benefits, the government also conceives of a pedagogical dimension of tourism. Consistent with the experimental pedagogy of the SEZs, tourism is viewed as "an inexpensive substitute for education", says Pal Nyiri (2009: 154), a civilizing practice that helps create "quality" citizens. "Tourism is an arena in which the production of cultural discourse penetrates everyday consumption, one in which Chinese subjects self-consciously consume complex representations of culture and respond to them in quotidian activities ... As such, it is a key sphere in which the reinvention of the Chinese subject takes place", contends Nyiri (2007: 99). In this way Chinese cross-border tourism to Macau plays an operative role in Chinese economic reforms.

The emergence of Macau and Hong Kong as primary destinations for those Chinese tourists was in part prompted by a second regional crisis, this time a crisis of public health. The 2002 Severe Acute Respiratory Syndrome outbreak in southern China was a profound public health crisis with severe repercussions for the economies of both SARs. In addition, by prompting hygiene controls that constrained cross-border travel, the crisis overtly challenged the PRC tourism-based economic development program (Ng, 2008). In the wake of the crisis, the PRC announced a Closer Economic Partnership Agreement (CEPA) with Hong Kong and Macau that aimed to stimulate the cities' respective economies. One specific component of the CEPA was the launching of the IVS, which allowed select tourists from relatively affluent cities and provinces to travel to Hong Kong and Macau on individual visas, without joining a government-sanctioned group tour. As a direct result of this policy, mainland Chinese tourism to Hong Kong and Macau expanded significantly. Thus, the role of the SARs in Chinese consumer activities was the specific result of political calculations designed to address limitations of the SEZ-based reforms (see Simpson, 2016a).

Production of a Chinese urban imaginary

I contend that Macau today plays a key pedagogical role in promoting an urban imaginary and urban lifestyle, characteristic of what Louis Wirth (1938) famously called "urbanism as a way of life". This urbanism constitutes an accumulation strategy, which is driving the second phase of China's economic transition (Wu, 2009). The PRC aims to rapidly urbanize 150 million rural peasants over the next decade. Chinese leaders hope that the residential and lifestyle habits of those newly urbane citizens will sustain economic growth, and that this urban work force will contribute to a knowledge and service economy which will gradually replace China's industrial base.

Urbanizing this massive rural population involves tremendous upheaval of traditional Chinese society, which is in essence "fundamentally rural", according to Fei Xiaotong (1992), the first modern Chinese sociologist. "I say that it is fundamentally rural because its foundation is rural", he says. For thousands of years, quotidian Chinese experience has organically cultivated a characteristic "rural imaginary"; Chinese citizens have embraced the belief that "Chinese are really inseparable from the soil" (38), and therefore "the people living in the countryside" are "truly the foundation of Chinese society" (37).

Mao mobilized this rural imaginary to construct a distinct "socialism with Chinese characteristics", with an anti-urban, rural bias which distinguished it from the Soviet Union. Mao was himself from the countryside and eschewed the bourgeois role of the city in the West; he was therefore skeptical of the revolutionary potential of a Chinese urban proletariat. The socialist government came to regard cities as primarily wasteful *consuming* entities that were counter-revolutionary, in that most urban dwellers were engaged in commercial activities which were unproductive (at least from a socialist perspective) (Lo, 1980: 132). Therefore, the task under Maoism was to convert the *consuming* cities to *producing* cities, which would be responsible primarily for producing goods for the countryside.

As a result of these beliefs and policies, the Chinese socialist city was defined, not by a stimulating and heterogeneous urban lifestyle typical of the capitalist city, but by the monotonous spatial dispersal of monolithic State Owned Enterprises (SOE) and their respective *danwei*, or work units, which served as the functional and unornamented building blocks of the city (Wu, 2009). Distinctions among rural and urban residents were formalized through household registration (*hukou*), which tied citizens to their place of birth, and domestic migration from countryside to city was discouraged by the central government. Daily life in the socialist city was centered around the provisions of work units, which constrained opportunities for urban mobility. This constrained mobility, tied to a general lack of formalized leisure time, meant that citizens engaged in little use of public space outside of the *danwei*. Therefore, Chinese cities under socialism lacked the quality of "urbanism" which defined the modern occidental metropolis (Wu, 2009).

Today, however, China is overtly reversing this rural bias in an effort to rapidly urbanize the country. The Chinese state facilitates population mobility in an experimental effort to engineer urban growth. Rural residents are urbanized to create Chinese "citizens" in the classical sense of the word, urbane and cosmopolitan residents with distinct "quality" (*suzhi*) and refined tastes and capabilities (Jacka, 2009). This transformation of the population involves concomitant institutional and infrastructural development. New superlative and ornamented architectural attractions help to comport both consumer bodies and bourgeois moral states appropriate to this economic regime, as well as form a *radical urban imaginary* that naturalizes the economic transition from rural socialism to an urban market economy.

The *radical imaginary* (Castoriadis, 1987) is the means by which a society institutes and mobilizes a new symbolic order and collective identity that both mystifies and naturalizes an emergent social formation. The radical imaginary becomes crucial to moments of social and cultural transformation. In China this process involves the transformation of the nation from a socialist economy oriented around the figure of the rural peasant and animated by a revolutionary imaginary, to a market-socialism that is integrated with the world economy and whose stability depends not only on industrial workers who produce goods for export but on a domestic urban consumer subject. This discursive imaginary is mobilized and disciplined by the central government, state industries, mediated representations, and transnational capital.

I have written elsewhere about the role of Macau's iconic glass buildings and themed architecture in producing the radical urban imaginary, and detailed its function in China's economic reforms (Simpson, 2014). I have also explored the role of Macau's integrated casino resorts and their interiorized and encapsulated pseudo-urban environments in normalizing "urbanism as a way of life", naturalizing the urban characteristics of density, heterogeneity, diversity, and affective stimulation that were lacking from the socialist city (Simpson, 2016b). These dimensions of Macau's built environment therefore function as a form of materialist pedagogy for Chinese tourists.

Portuguese *calcada*, pedestrianized public space, and Chinese tourist comportment

It is within this general context that I explore the function of Macau's *calcada* tiles and pedestrianized public space for Chinese tourists. The pedestrianization of urban space in China, as elsewhere, is generally used to create conditions conducive for retail shopping, and this shopping has become a crucial component of both domestic and cross-border tourist practices. Pedestrianized public space prompts perambulation, and that kinaesthetic movement is mobilized to naturalize particular types of tourist activities. Indeed, the relations among walking, window-shopping, and consumption are so integral to tourist practices that it is easy to overlook the extent to which they had to be first produced, and then normalized. The *calcada* in Macau, at least in part, serves such a normalizing function (see Figure 18.2).

That Chinese tourists walk around Macau may seem like an unremarkable truism, but such behaviors cannot be taken for granted. As Orvar Lofgren (1999) argues in his social history of the European vacation, the leisure traveler has to self-consciously *learn to be a tourist*, and that means adopting appropriate behaviors conducive to appreciating an assortment of sites and experiences. Such "learning the city" is clearly a dimension of nascent Chinese tourism practices in Macau. "Vacationing has served as a laboratory for trying out new lifestyles or forms of consumption", Lofgren says, and "the history of

Figure 18.2 Chinese tourists pack a suitcase full of diapers and other everyday consumer items purchased at shops around the *Largo de Senado*. Photograph courtesy of Adam Lampton.

holiday-making includes a constant process of learning and relearning" (Lofgren, 1999: 281). Macau's *calcada*, and the pedestrian experience that they afford, are a pedagogical component of China's contemporary tourist conjuncture.

Alan A. Lew (2006) highlights the role of pedestrianized public spaces in China's market reforms. Today nearly every sizable Chinese city has a pedestrian shopping street. These environments, Lew says, "are part of the transformation of many Chinese cities from socialist cities of production to cities of play and tertiary employment where tourism for both domestic and international markets is part of a new consumption-oriented economy" (Lew, 2006: 151). Pedestrianizing social space ultimately invites pedestrians, and this urban subject has been key to the post-socialist transformation of China during the reform era, as people moved outside the self-contained *danwei*, or work units, to shop and socialize in commodified urban space. Such pedestrian areas reflect the ambitious land reform and decentralization efforts that followed the implementation of the PRC's market economy.

Interestingly, Lew notes that Macau's *Largo de Senado* was actually the first such pedestrian project in China. Thus, we may understand the pedestrianization of *Largo de Senado* as not merely a decision by Macau's outgoing colonial administration to preserve the city's Portuguese legacy, but as an experiment in post-socialist economic reforms. Regardless of whether it was Portuguese or Chinese authorities who first conceived of and implemented Macau's pedestrianization program, it was subsequently seized by the Chinese state as a governmental innovation that could be mobilized and pilot-tested elsewhere in a process of "staged replication" (Heilmann, 2008b: 5). This governmental approach is consistent with the more general Chinese tendency toward *dispositional* governance, the recognition of a latent positivity or potential which may be exploited and directed in a strategic manner (Julien, 1995; see also Easterling, 2013; Simpson, 2016a).

Lew identifies two distinct types of retail development of pedestrian streets, each of which has been pilot-tested in Macau: one approach "uses historic architecture as a thematic tool", in a reproduction of the sort of development which characterizes the *calcada* pedestrianization project in Macau; and the other "uses modern and hypermodern architectural themes" (Lew, 2006: 158), and even reproductions of famous architecture such as the *Arc de Triomphe*, similar to the foreign-built thematic urban experiences offered in Macau's integrated resorts. The replication of the Macau experiment is clearly evident in the nearby city of Zhongshan. "Taking a cue from its neighbor", contends Lew, "the city of Zhongshan replicated central Macau in its old downtown by creating the Sun Wen Zi Road Tourism Zone" (Lew, 2006: 159). The innovation of the pedestrian shopping street may be understood as an "experimental point" (Heilmann, 2008a: 4), a dimension of the PRC's practice-based experimental governing, and the extrapolation from the Macau case to other Chinese cities is an example of what in PRC development discourse is called "proceeding from point to surface" with experimental reforms (Heilmann, 2008a: 2). Indeed, "proceeding from point to surface" is almost a literal description of the replication of the Macau model, as pedestrianized spaces now cover the surfaces of many Chinese cities.

Macau as a model city

This discussion returns me to my comment at the outset about Macau's status as a model city. Much as Shenzhen has been promoted as a model city for Chinese planners and provincial leaders, PRC officials have recently promoted Macau as an affirmative model SAR for Hong Kong (Bradsher, 2017). Although they have distinct colonial histories, metropolitan morphologies, post-industrial economies, and urban atmospheres, the SARs bear some similarities *vis-a-vis* their respective roles in China's reforms. Hong Kong may also be understood as a "laboratory of consumption" for Chinese tourists, with its own indigenous and ubiquitous architectural form, the podium shopping mall, to rival Macau's integrated casino resorts. Stephan Al (2016) counts several hundred such malls in the city, perhaps the densest and tallest collection of shopping malls in the world, making Hong Kong something like a shopping paradise. Shenzhen-to-Hong Kong cross-border, shopping-oriented tourism is a huge industry with significant benefit to Hong Kong's economy, but it has also caused contentious relations among some Hong Kong citizens and mainland tourists; negative encounters have been extensively documented in the local media. These interpersonal relations increasingly reflect a more general antagonistic political relationship between Hong Kong and the central government, and Hong Kong has recently emerged as a site of oppositional intra-state political activity. This opposition was most clearly visible in the 2014 Umbrella Movement, when tens of thousands of citizens joined a student-led protest and organically "pedestrianized" and occupied several barricaded streets in the central city to agitate for universal suffrage and sovereign autonomy. This particular response to China's reforms recognized a different disposition, or latent potential, of public space, and referenced an alternative European urban history – from the nineteenth century Paris Commune to the barricades of May 1968 – which not only threatens the appearance of national "harmony", but which is also more difficult to commodify and monetize than the benign urbanism typical of Macau's city spaces.

Such recent Hong Kong politics demonstrate that urbanization as a macroeconomic strategy carries significant risks for the Chinese state (Wallace, 2014). Cities characterized by the heterogeneity and aleatory encounters which derive from "urbanism" tend to aggregate populations in ways that equip citizens physically and psychically for counter-state or anti-authoritarian movements (see Hardt and Negri, 2011). This was evident in the wave of "occupy" movements which swept not only Hong Kong, but the United States and the Arab world. In order to manage this risk, the urban imaginary which animates urbanization processes and thrives on their realization must also be carefully modeled, shaped, and molded. As the most densely populated territory in the world even without the 30 million annual tourist arrivals, Macau is seemingly prone to such public demonstrations, but they are surprisingly rare. Both the *calcada* stones in the city center and the interiorized urban environments of the integrated resorts perhaps function to direct crowds toward consumption-oriented activities which drive the subjection process.

In short, Macau serves as a model city of "proper" uses of pedestrianized public space when compared against the negative example of Hong Kong. Indeed, the PRC proposal of Macau as a model city is itself indicative of an instance of "point-to-surface" experimentation, not merely at the scale of the self-contained pedestrianized city center, but at the scale of the city itself. Chinese authorities propose Macau as a model for Hong Kong to emulate, which in part implies that citizens should comport themselves in public space in an manner which is consistent with the mercantile expectations of the reform regime. To put it another way, for Chinese authorities, Macau comprises a vital materialization of an appropriate urban imaginary (see Figure 18.3).

Figure 18.3 The materialization of the Chinese urban imaginary inside Macau's MGM Resort. The wavelike pavement stone pattern, the cylindrical central aquarium, and the Valkyrie Octopus installation by Portuguese artist Joana Vasconcelos, each reference Portugal's maritime explorations. The entire edifice is an enclosed consumer space. Photograph courtesy of Adam Lampton.

Pavement pedagogy

From the perspective of a materialist ontology, we may understand that the *calcada* stones themselves play a role in producing a certain sort of urban imaginary and normalizing a set of tourist behaviors in urban public space. The *calcada* stones manifest an agency, or actant status, which is a crucial component of the inchoate Chinese post-socialist tourist subject, who conflates the freedom of cross-border mobility with the freedom to consume. That subjectivity may be mobilized when the tourist visits a pedestrianized shopping district in her own city, many of which also have heritage motifs.

Foucault's idea of the governmental urban "milieu" is useful for understanding this process because of the way in which it confounds any simple unidirectional cause-effect relationship between environment and behavior, and posits instead "an element in which a circular link is produced between effects and causes, since an effect from one point of view will be a cause from another" (Foucault, quoted in Lemke, 2015: 13). This perspective is consistent with the materialist ontology I have adopted. We need not claim a direct temporal or causal relation between the *calcada* stones and tourist comportment, and may instead understand that "causal relations do not preexist but rather are produced in agential materializations" (Lemke, 2015: 13; see also Berad, 2007). Taken as a whole, the collective articulation of pavement stones, population mobility, and the Chinese urban imaginary – the imbrication of *materialist pedagogy* and *material politics* – may be understood as a form of biopolitics. Biopolitical government is ultimately conceived here as "the interrelatedness and entanglements of men and things, the natural and artificial, the physical and moral" (Lemke, 2015: 13), an account that accurately describes the uses of Macau's simulated pedestrian and resort environments in producing tourist subjects.

In this way we may understand how Macau's heritage, invented or otherwise, is mobilized in a didactic campaign of domestic tourist consumption with clear macroeconomic goals. The *calcalda* stones play a significant role in a material "spatial grammar of learning" (McFarlane, 2011: 9) for those Chinese tourists who "cover Macau step-by-step". Ultimately, the role of Macau's Portuguese pavement in China's economic reforms almost literally exemplifies an indigenous Chinese slogan adopted in the Deng era to characterize the experimental innovations of China's reform-era pragmatism: *crossing the river while groping for stones*. As Chinese tourists eagerly grope for the *calcada* stones on their way to visit a cosmetics shop or jewelry store, Macau's eminent walkability produces an urban imaginary and directs concomitant consumer behaviors, elucidating the *post-socialist political economy* of the *calcada*.

Acknowledgements

This chapter was written while I was on sabbatical leave from the University of Macau, and serving as a visiting scholar in the School of Design Strategies, Parsons School of Design, New York. The research benefited significantly from participation in two conferences: "City Futures and Urban Imaginaries", at the University of Oregon, Portland; and "Emerging Public Space in/of the Pearl River Delta", at Parsons, New York. I would like to thank Miodrag Mitransinovic for facilitating the position at Parsons, and Adrian Blackwell for pointing me to the work of Sebastian Heilmann.

References

Al, S. (2016) *Mall City: Hong Kong's Dreamworlds of Consumption*. Honolulu: University of Hawaii Press.

Bakken, B. (2000) *The Exemplary Society: Human Improvement, Social Control, and the Dangers of Modernity*. Oxford: Oxford University Press.

Baudrillard, J. (1994) *Simulacra and Simulation*. Ann Arbor, MI: University of Michigan Press.

Bennett, J. (2010) *Vibrant Matter: A Political Ecology of Things*. Durham: Duke University Press.

Berad, K. (2007) *Meeting the Universe Halfway: Quantum Physics and the Entanglement of Matter and Meaning*. Durham: Duke University Press.

Bradsher, K. (2017) Once a Model City, Hong Kong is in Trouble. *The New York Times*. Available at: https://www.nytimes.com/2017/06/29/world/asia/hong-kong-china-handover.html.

Castoriadis, C. (1987) *The Imaginary Institution of Society*. Cambridge, MA: MIT Press.

Chuihua, J.C., Inaba, J., Koolhaas, R., and Sze, T.L. (eds.) (2001) *Great Leap Forward*. Cambridge, MA: Harvard Design School.

Coole, D. and Frost, S. (2010) *New Materialisms: Ontology, Agency, and Politics*. Durham: Duke University Press.

Easterling, K. (2013) Disposition. In: Hauptmann, D. and Neidich, W. (eds.), *Cognitive Architecture*. Delft: Delft School of Design, 251–65.

Eco, U. (1986) *Travels in Hyperreality*. Orlando, FL: Harcourt Brace & Co.

Fei, X. (1992) *From the Soil: The Foundations of Chinese Society*. Berkeley: University of California Press.

Foucault, M. (2009) *Security, Territory, Population: Lectures at the College de France 1977–1978*. London: Picador.

Hardt, M. and Negri, A. (2011) *Commonwealth*. Cambridge, MA: Belknap Press.

Harvey, D. (2006) *A Brief History of Neoliberalism*. Oxford: Oxford University Press.

Heilmann, S. (2008a) From Local Experiments to National Policy: The Origins of China's Distinctive Policy Process. *The China Journal* 59: 1–29.

Heilmann, S. (2008b) Policy Experimentation in China's Economic Rise. *Studies in Comparative International Development* 43: 1–26.

Jacka, T. (2009) Cultivating Citizens: *Suzhi* (Quality) Discourse in the PRC. *Positions* 17(3): 523–35.

Julien, F. (1995) *The Propensity of Things: Towards a History of Efficacy in China*. New York: Zone Books.

Lemke, T. (2015) New Materialisms: Foucault and the Government of Things. *Theory, Culture & Society* 32(4): 3–25.

Lew, A. (2006) Pedestrian Shopping Streets and Urban Tourism in the Restructuring of the Chinese City. In: Church, A. and Coles, T. (eds.), *Tourism, Power and Space*. New York: Routledge, 150–70.

Lo, C.P. (1980) Shaping Socialist Chinese Cities: A Model of Form and Land Use. In: Leung, C.K. and Ginsburg, N. (eds.), *China: Urbanization and National Development*. Chicago: University of Chicago, 130–55.

Lofgren, O. (1999) *On Holiday: A History of Vacationing*. Oakland, CA: University of California Press.

McFarlane, C. (2011) *Learning the City: Knowledge and Translocal Assemblage*. London: Wiley-Blackwell.

Minuchin, L. (2013) Material Politics: Concrete Imaginations and the Architectural Definition of Urban Life in Le Corbusier's Master Plan for Buenos Aires. *International Journal of Urban and Regional Research* 37(1): 238–58.

Ng, M.K. (2008) Globalization of SARS and Health Governance in Hong Kong Under "One Country, Two Systems". In: Ali, S.A. and Keil, R. (eds.), *Networked Disease: Emerging Infections in the Global City*. Cambridge: Blackwell, 70–85.

Nyiri, P. (2007) *Scenic Spots: Chinese Tourism, the State, and Cultural Authority*. Seattle: University of Washington Press.

Nyiri, P. (2009) Between Encouragement and Control: Tourism, Modernity and Discipline in China. In: Winter, T., Teo, P., and Chang, T.C. (eds.), *Asia on Tour: Exploring the Rise of Asian Tourism*. London: Routledge, 153–69.

Nyiri, P. (2010) *Mobility and Cultural Authority in Contemporary China*. Seattle: University of Washington Press.

O'Donnell, M.A., Wong, W., and Bach, J. (2017) *Learning from Shenzhen: China's Post-Mao Experiment from Special Zone to Model City*. Chicago: University of Chicago Press.

Simpson, T. (2010) Materialist Pedagogy: The Function of Themed Environments in Post-Socialist Consumption in Macau. *Tourist Studies* 9(1): 60–80.

Simpson, T. (2014) Macau Metropolis and Mental Life: Interior Urbanism and the Chinese Imaginary. *International Journal of Urban and Regional Research* 38(3): 823–42.

Simpson, T. (2016a) Neoliberal Exception? The Liberalization of Macau's Casino Gaming Monopoly and the Genealogy of the Post-Socialist Chinese Subject. *Planning Theory* 17(1).

Simpson, T. (2016b) Spatial Machines of Subjection: A Materialist Analysis of Macau's Themed Integrated Casino Resorts. In: Lukas, S.A. (ed.), *A Reader in Themed Spaces*. Pittsburgh: Carnegie Mellon – ETC Press, 183–93.

Wallace, J. (2014) *Cities and Stability: Urbanization, Redistribution, and Regime Survival in China*. Oxford: Oxford University Press.

Wirth, L. (1938) Urbanism as a Way of Life. *The American Journal of Sociology* 44(1): 1–24.

Wu, F. (2009) Neo-Urbanism in the Making under China's Market Transition. *City* 13(4): 418–31.

Zandonai, S.S. and Amaro, V. (2017) The Portuguese *Calcada* in Macau: Paving Residual Colonialism with a New Cultural History of Place. *Current Anthropology*. DOI: 10.1086/698957.

19

"Like diamonds in the sky"

Imaginaries of urban girlhood

Agata Lisiak

Introduction

Walking down Bishkek's main artery, Chuy Avenue, on a crisp and smoky, late spring afternoon, I stopped at a large open square adorned sparsely with dramatic socialist-era statues. The central statue, towering over the square on a tall granite pedestal and facing the avenue, was a female allegory of revolution. The square looked empty, with only two skater boys who did not seem particularly into what they were doing, or where they were doing it, and soon disappeared. I walked up to the other two statues representing groups of revolutionaries and inspected the sharpie tags on the granite pedestals. Many of them were declarations of heterosexual love, with lovers' names encircled by hearts. This celebration of romantic love expressed in a public square designed for commemoration of the fighters of the Soviet revolution, as the official name of the place declares, can be seen as a gentle yet vital act of subverting spatialized ideological narratives with small gestures aimed at rendering urban space one's own.

Among the many hearts scribbled on the pedestal, one in particular drew my attention as it included four female names: Janka, Alya, Salya, and Firu, as well as the statement "love is my friend" (see Figure 19.1). The white curvy letters stood out from the maze of orange, brown and black dots in the granite. In the midst of couples' expressions of love, these four girls were demonstrating their friendship and love for each other (cf. Faderman, 1981). In an attempt to decode the ostensibly simple English scripture I wondered: were they really simply saying that love was their friend? But then why say "my" and not "our" friend? Or were they declaring that they valued friendship more than romantic love? Was it a case of international English, in which apparently simple words and phrases take on new, less obvious meanings? Was it an evidence of the influence of Anglophone cultures on contemporary constructions of girlhood in globalized popular culture (cf. Driscoll, 2002)?

As I was walking back to the avenue, I noticed four girls occupying one of the benches around the square (Figure 19.2). I cannot know for sure if these were, indeed, Janka, Alya, Salya, and Firu. I want to think they were.

The four girls were just sitting there, not so much contemplating the square, because nothing was happening there, as watching the passers-by and traffic on the adjacent

Figure 19.1 Graffiti in Bishkek, 2017. Photograph by the author.

Figure 19.2 Fighters of the Revolution Square, Bishkek, 2017. Photograph by the author.

avenue. Although the length of the bench allowed for more space between them, they sat very close to each other. Be it because of the cold or camaraderie, the girls formed a tight pack – not unlike the sculpture groups of revolutionaries in the square – their limbs and shoulders brushing each other, one girl laid her head on her friend's shoulder. The scene exuded intimacy, warmth, and comfort (Fenster, 2005). These girls seemed to be at home not only with each other, but also in the public space they occupied (Berman, 1982), and it was their friendship, I thought, that allowed for this sense of belonging.

Portrayals of young women's friendships are central to imaginaries of urban girlhood. From coming of age stories to explorations of gendered complexities of urban adulthood, girls' urban lives have been depicted as glamorous, fun, miserable, hard, or edifying – or, not infrequently, all these things at once. Female friendships in urban settings have been eulogized in countless novels, films, music videos, songs, poems, and photographs. Such representations and analyses thereof focus mostly on the inner lives of the protagonists, the intricacies of their relationships, as well as their relations with the worlds outside their friendships. In this chapter, I discuss the spatialities of the imaginaries of urban girlhood by unpacking the workings and meanings of female friendship in the city. Starting with a brief exploration of the concept of girlhood, I move on to discuss how the spaces girls occupy are gendered, racialized, and classed and how friendship allows girls to execute their right to the city in ways that are not always easily accessible, if at all, to girls who move through the city on their own. Drawing on pop cultural portrayals of urban girlhood, I look into how and why girls succeed at rendering the city their own, and how and why they fail. The concluding discussion focuses on how urban imaginaries of girlhood create new opportunities for girls to (re)claim the urban space they inhabit.

Girlhood in the city

Who, exactly, is a girl? The noun "girl" refers not only to cis-gender female children and teenagers, but also cis- and transgender women of various ages – and it has been appropriated by gay men as an endearing form of address. As Catherine Driscoll writes in her study of girlhood in popular culture, it is not unusual for a woman in her mid-thirties, for example, to be called or call herself a "girl", especially around women of her own age, and to "remain socially connected to, interested in, and sometimes still strongly identify with 'girl' things, 'girl' behaviors, and experience of girlhood" (2002: 2). Particularly in metropolitan contexts, "girl" connotes lifestyle rather than age and is often associated with being single. Helen Gurley Brown, the famed editor of women's magazine *Cosmopolitan*, claimed that all women think of themselves as girls because "the girlish part of you is playful, optimistic and fun-loving, and you don't want that tamped down" (qtd. in Baumgardner and Richards, 2000: 159). This vision of extended carefree girlhood is an illusion that is part of *Cosmo*'s success: promotion of a lifestyle that is easy and fun to consume (with the right products ranging from make up kits to rabbit dildos).

The riot grrrl movement (Marcus, 2010) and third-wave feminism in the United States (Baumgardner and Richards, 2000; Hernandez and Rehman, 2002; Walker, 1995), which recognized vulnerability and empathy as strengths, reclaimed the word "girl" both from its hitherto belittling and condescending uses in white, heteronormative, patriarchal contexts and from the gross commercialization of girlhood in late capitalism. The concept of girl power, considerably inspired – even if later whitewashed – by strong female figures from Blaxploitation movies (Baumgardner and Richards, 2000) soon became the model for contemporary girlhood, conveniently compatible with heteropatriarchal capitalism. Removed from its do-it-yourself, anti-consumerist origins, a largely depoliticized version of "girl power" has permeated popular culture from music through film and television to literature, and spilled over to other areas including fashion, health, and education. Its embodiment, an assertive, fearless girl, is often juxtaposed with a vulnerable, passive girl desperately in need of being saved (Gonick, 2006; Renold and Ringrose, 2013). Although contrary, these two models of the contemporary girl "participate in the production of the neoliberal girl subject" through individualization discourses that focus on personal explanations of girls' success or failure to embody the desired girl power model (Gonick, 2006: 2).

While indebted to theorizations of girl cultures in relation to neoliberalism (Renold and Ringrose, 2013), precarity (McRobbie, 2015), and globalization (Driscoll, 2002), I propose to shift focus from individual girl subjects to relations between girls and to look at girlhood as a condition of being together with "girls like us". Focusing on togetherness, this understanding of girlhood draws on theorizations of kinship (di Leonardo, 1987; Stack, 1974); friendship (Kathiravelu and Bunnell, 2017); girl (sub)cultures (Driscoll, 2002; McRobbie, 1991); and gendered space (Fenster, 2005; Massey, 1994; Valentine, 1989).

Drawing on her empirical study of black communities in the United States, Carol Stack (1974) theorizes minoritized people's strategies for everyday survival through the notion of kinship. In the condition of urban poverty, with little or no support from the state, individuals of broadly understood kin develop adaptive strategies and tactics to get by. Across various socio-geographical contexts, it is typically women – especially low-income and minoritized women – who are trusted and burdened with domestic and care work, as well as the work of kinship (di Leonardo, 1987). From an early age, girls are socialized to perform these types of labor and their uses of the city – both in private and public space – are considerably affected by kinship ties and kinship work.

Although arguably less socially scripted than familial bonds and potentially subversive (Foucault, 1997), friendships are "strongly socially patterned" and tend to emerge and develop between people of similar class, race, and age (Cronin, 2014: 72). Neither romanticizing urban friendship nor theorizing it as strictly instrumental, Laavanya Kathiravelu and Tim Bunnell conceive of it as "a mode of relationality that is infused with both practical and affective meaning for actors involved" (2017: 4). Friendships can generate safe spaces, affective spaces of trust and mutual support, in potentially or quite certainly hostile urban environments (Cronin, 2014), but they are not necessarily durable, reciprocal, and normatively positive (Kathiravelu and Bunnell, 2017). Friendships impact people's understandings and affective experiences of the city and "the spaces and infrastructures of the city shape the conditions of possibility for such interpersonal ties" (Kathiravelu and Bunnell, 2017: 4). Friendships can also be seen as a form of citizenship that can be practiced in private and public space alike (see Fenton, 2005).

The condition of being a girl with other "girls like us" both produces and reproduces what Driscoll refers to as girl culture, i.e., the circulation of "the things girls can do, be, have, and make, and in that process defining what processes are particular to girls" (2002: 139). The gendered and classed conventions and (popular) cultural codes that make up girl cultures, when recognized as embodied by "girls like us", produce a sense of belonging. This sense of belonging relies on exclusionary tactics based on classed, racialized, and gendered dichotomies and is often attainable only through consumption of specific cultural and other products, thus necessitating specific cultural, economic, and social resources. Recognition of "girls like us" thus emerges through identification with various cultural positions; and *where* these products and positions are shared, consumed, and (re)imagined can lend them legitimacy. As Angela McRobbie and Jenny Garber (1991 [1977]) powerfully argue in their analysis of girls' subcultures, not only what happens on the streets matters; the spaces of girls' bedrooms, bathrooms, and classrooms are crucial to everyday urban experiences of girlhood.

The inclusion of private spaces in theorizations of the urban has been particularly well developed in the field of feminist geography. Recognizing that it is rarely possible for women and girls to use the city with the same extent of freedom that men and boys do (Fenster, 2005; Massey, 1994), feminist research has drawn attention not only to the gendered workings of public space (Valentine, 1989), but also to the hitherto under-researched private space (McDowell, 2006) and collapsed the normative concept of private-public divides (Peake and Rieker, 2013; Vaiou, 1992; Wright, 2010). Focusing on the urban everyday, both in public and in private realms, yields insights into how the city is "gendered through multiple actions and experiences of its inhabitants" (Beebeejaun, 2017: 323). In fact, feminist interventions into the Lefebvrian notion of the right to the city (Fenster, 2005; Vacchelli and Kofman, 2018) propose to understand everyday life as "the mediator of rights underpinning the usage of urban space to its fullest extent" (Beebeejaun, 2017: 327) and insist on taking into the account how said rights are shaped by "patriarchal power relations, which are ethnic, cultural and gender-related" (Fenster, 2005: 217).

Equipped with the theories of kinship, urban friendship, girl cultures, and gendered space, I will investigate urban girlhood as a condition of being together in the city with "girls like us". I will explore the collective nature and shared experiences of girlhood, drawing primarily on Céline Sciamma's 2013 film *Girlhood* (Bande de filles). Set in the Parisian working-class suburbs, or *banlieues, Girlhood* is an exploration of how everyday urban lives of black teenage girls are gendered, classed, and racialized. Sciamma's main protagonist, 16-year-old Marième (Karidja Touré), lives in a small apartment with her single, working-class mother, two younger sisters, and an older, abusive brother. The housework and care work she is burdened with take their toll: Marième fails school and is advised to pursue vocational training rather than the college path she was hoping to take and which, she thought, would allow her to break away from the plight that seems set in stone for her. Becoming friends with Lady (Assa Sylla), Adiatou (Lindsay Karamoh), and Fily (Mariétou Touré) – the three girls who recruit her to their group – offers a respite from Marième's daily worries. Together they make bold if messy claims to the city. Although Marième's future seems determined by her gender and race, in a truly modernist manner, she "keeps on fighting and loving; [she] keeps on keeping on" (Berman, 1982:14) – and keeps finding ways out and ways in for herself. When her brother Djibril violently reacts to the news of the romantic relationship Marième starts pursuing with his friend, the girl leaves home and starts working as a drug dealer in a different part of the city. She needs to part from her friends and embark on a new life in what turns out to be another hostile environment. After rejecting her boss's sexual advances, Marième turns to her boyfriend, but what he can offer – marriage and children – is simply not enough for her. In the final scene, Marième leaves all of her pasts behind and exits the frame. We do not know what will happen to her, but we can learn from what has happened to her and what she made happen. The following sections offer a close reading of Sciamma's film, alongside other works of popular culture, drawing attention to the meanings of urban imaginaries of girlhood.

Girls' right to the city

Like many other films set in Parisian suburbs, also known as the *banlieu* cinema, *Girlhood* engages with "the significance of place in the construction of identity" (Tarr, 2005: 3). In one of the opening scenes, a group of girls, about a dozen of them, emerges from darkness into a *banlieue*. We hear them before we see them. Chatty and high on sport-induced endorphins (they are coming back from an American football practice), the girls fill up the urban space and the screen. Together, they seem invincible. Their chatter stops abruptly the moment they notice a few boys, maybe two or three, partially covered by evening shadows, at the end of the walkway leading to their apartment buildings. The boys are just standing there; they seem relaxed, like they own that space (cf. Lisiak, 2014). That the girls – who manifest their physical prowess in the sports training and are stronger in numbers – *all* feel obliged to render themselves quiet and invisible at the sight of a few boys, speaks volumes about how fear in the city is gendered (Valentine, 1989). The girls walk on in silence, their body language changes, they stop gesturing, and their walking becomes steady and careful. The boys hold their positions like watchmen at the city gates. As they walk on through the *banlieue*, girls depart in twos and threes to their respective apartment buildings, and their fear of the space coded masculine (Geesey, 2011) grows as their numbers decrease. They become vulnerable, we fear for them and with them. At the end of the scene, Marième walks alone, but – in one of many moves aimed at destigmatizing the *banlieue* as a one-dimensional, hopeless place (Dobson, 2017) – Sciamma does not end the scene in anticipated violence, but in a romantic encounter of two people: Marième meets her love interest Ismaël. Urban space is strongly gendered and poses threats to those who are minoritized or otherwise vulnerable (Beebeejaun, 2017) – but that threat is not always fulfilled, Sciamma adds (cf. Bondi and Rose, 2003).

When Marième first encounters her soon to be friends, the three girls occupy a bench: two of them sit on its backrest, one at the edge of the bench. The girls make a confident claim to a space, which – like the walkway and the steps outside their apartment buildings – is "usually coded masculine" (Geesey, 2011: 173). Although the *banlieue* is undoubtedly "a space of the policing of femininity and female desires" (Dobson, 2017: 37), Sciamma challenges its standard mapping as a "stigmatized site in which exclusion and alienation can be exclusively located, othered and contained" (Dobson, 2017: 45–6). Both in the *banlieue* and beyond it, Sciamma's girls are able to (re)claim some "territories of *collective* resilience and agency" (Dobson, 2017: 45; my emphasis). They dance on the subway, they are loud and unapologetic, they walk arm in arm, taking up the space they inhabit. Sciamma films those scenes in ways strikingly different from the sexualized manner in which young black women's bodies are often presented in popular culture. Although their dance moves can be seen as sensual, the camera closes up on the girls' faces rather than their hips or breasts. As the camera shifts from one girl's gaze to another, we do not so much look *at* them as *with* them. It is in this appreciative girl gaze that the magic of girlhood resides.

The girls carefully curate and choreograph their presence in public space. Marième's style of dress and body language change dramatically once she joins Lady's group: she starts sporting a sleek weave, tight jeans, and a leather jacket; she walks more confidently now. The girls' bodies are "central to formations and enactments of friendships" (Kathiravelu and Bunnell, 2017: 10); their similar clothes resemble uniforms. Through fashion, they manifest their belonging to the group and make claim to urban territory. Notably, their sense of empowerment emerges from criminal acts some of the clothes the girls wear are stolen from stores, others bought with the money they extorted from other girls outside schools. In the city, the four protagonists run into "girls like them": dressed in a similar fashion, with similar hair and attitudes, and they invariably get into fights with them. Their struggle for territory is not only symbolic, but also physical: they stage fights that end when one girl is beaten to the ground, with her top or bra off. Winning one of such fights is what ultimately earns Marième her brother's short-lived respect.

When Marième takes up the job as a drug dealer, she switches between two types of outfits. When she delivers the product to white, middle-class parties in Paris, she has a platinum blond wig, heavy makeup, a tight red dress, and heels; but the moment she returns to the car that waits for her outside the client's building, she demands from the male driver that he hands over her "gear" to her. In the car, she changes into baggy clothes and takes off the wig revealing cornrows. Her attire resembles that of her male colleagues rather than that of her neighbor Monica, a black female sex worker who opts for the type of clothes Marième used to wear as Vic. Marième tries to adapt to her new circumstances by altering her look: to her boyfriend's shock, she even resorts to binding her breasts. As *Girlhood* demonstrates, clothes are not only means of self-expression and manifestation of belonging to a group or (sub)culture, but also tools for fitting into space in ways that render it more livable.

A room of their own

Perhaps the reason the scene in Bishkek resonated with me so strongly is because it relates to a popular trope in urban imaginaries of girlhood: the image of four girls on the town. The HBO television series *Sex and the City* (1998–2004) famously explored this trope and caught global imagination sparking countless similar shows from Poland to Ghana and back to the US, with Lena Dunham's *Girls* (2012–2017) as one of its most heatedly discussed iterations. As these TV shows and myriad films, pop songs, poems, and novels manifest, female friendships are not always empowering and beautiful: they can be toxic, they can hurt, they can kill. In *Girlhood* too, the friendship between Marième, Lady, Adiatou, and Fily is messy and marked by uneven power dynamics (cf. Kathiravelu and Bunnell, 2017). Rather than idealizing girlhood, Sciamma closely investigates its many contradictions and uncomfortable beauty.

Girlhood's central scene is a joyous celebration of female friendship tinged with violent undertones. After bullying other girls into giving them money, Marième and her friends collect enough to rent a hotel room and buy some snacks, alcohol, and weed. The hotel room may be the only place in the city where they can simply be together without being watched, be it by their brothers or CCTV cameras; it offers them the privacy to do what so many "girls like them" love doing and what they cannot do in their respective apartments: just hanging out. It is in the bathroom of that hotel room that Lady, in the midst of taking a relaxing bubble bath, gives Marième a golden necklace that spells Vic ("for victory", Lady adds with a reassuring smile). The transformed Marième gathers the courage to ignore her brother's phone calls (even though she knows very well she will soon have to suffer the consequences of this disobedience) and joins her friends in the fun they are having. They dance and sing along to Rihanna's hit song "Diamonds" – and they too shine bright when they look admiringly at each other; they ritualize their friendship and celebrate their beauty (McNeill, 2017). Framed and edited like a music video, immersed in blue light, the scene is a visual ode to girlhood; it emphasizes the importance of intimacy and physical contact in the construction of female friendships (Faderman, 1981; Foucault, 1997); it is also extremely relatable and deeply embedded in global girl cultures (Driscoll, 2002; McNeill, 2017). Only the anti-theft alarm tags on the girls' party dresses are a somber reminder of how they got there.

In *Girlhood*, private spaces (hotel rooms, bedrooms, kitchens) matter as much as public space for the understanding of girls' embodied experiences of the city. The bedroom Marième shares with her younger sister Bébé is a space imbued with solidarity and tenderness. Home is hardly a safe space for either of them. Home is work and terror. Yet, through familiarity with its sounds and rhythms (adjusted to their mother's working schedule and their brother's comings and goings), they manage to render home more livable for themselves. Marième and Bébé have developed heightened senses: they know it takes their brother exactly five steps to move from his bedroom to the apartment door; when the second door slams, they exhale with relief and giggle. Their familiarity with the sounds of their house cannot entirely protect them from Djibril's abuse, but they find comfort in the shared nature of this knowledge. It is almost as if they had secret super powers. Later that night, after Djibril hits Marième for using his PlayStation during his absence, the sisters hold hands across their single beds in a silent gesture of solidarity and care. Marième's dramatic decision to leave is therefore particularly painful because she is well aware that not only is she passing on the burden of domestic and care work to her younger sister, she also exposes her to their brother's abuse.

The power of numbers

In the original French release, the movie poster for *Bande de filles* depicts the four protagonists looking straight into the camera, calm and defiant (Figure 19.3). The poster for the English-language release of Sciamma's film, *Girlhood*, is strikingly different: it is a portrait of Vic, with a golden necklace and a weave brushed behind the neck to her right shoulder (Figure 19.4). We don't see her entire face, her eyes are missing from the picture, possibly suggesting that this could be any (black) girl. Together with this image, the English title, *Girlhood*, may suggest it is a story of one girl's plight and, in turn,

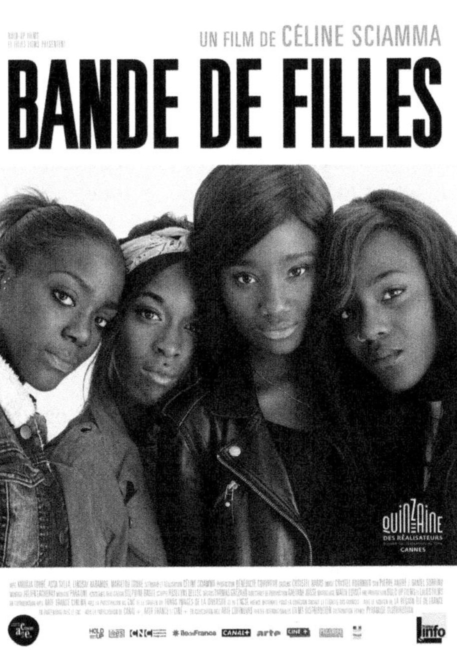

Figure 19.3 Bande de filles film poster.

Figure 19.4 *Girlhood* film poster.

encourage the viewer to focus on the main protagonist's singular experiences of urban girlhood rather than the collective girlhood experience acknowledged by the French poster and title.

In a triumphant scene celebrating the collective experience of urban girlhood, a large group of girls, including our four protagonists, make their claim to Paris' business district, La Défense. The camera slowly tracks their formation, starting with a close-up of Marième, then Lady, then other "girls like them" (of the same age, race, and most likely class). The girls gather on a square between office towers – not unlike the one where Marième's mother works night shifts as a cleaner – and stage a dance battle. They execute their "right to the use of the center, a privileged place, instead of being dispersed and stuck into ghettos for workers, immigrants, the 'marginal' and even the 'privileged'" (Lefebvre 1996: 34). It is not only important that they use the center, but also *how* they use it. As in several earlier scenes in the film, they execute their right to be seen and heard in the city: they take up an entire length of an escalator, a whole width of an alley in a shopping mall, they joyfully dance to loud music. There is no way past them, it is impossible to ignore them. The way they inhabit that space, on their own terms, serves as "a metaphor for making [themselves] a 'place' within French society" (Geesey, 2011: 163).

The scene in La Défense offers a nuanced reflection on the complexities of urban girlhood. The cheerful celebration of togetherness is interrupted when Marième notices Bébé, together with other "girls like her" – slightly younger versions of Marième and her *bande* – bullying and trying to rob another girl. Frightened at this mirror image of herself, Marième pulls her sister away and slaps her in the face. "You're just like him!" Bébé exclaims in disbelief, comparing her older sister to Djibril. Whereas girls in the city are commonly exposed to the threat of male violence, here, it is girls who – using the power of numbers and the hierarchies of kinship ties – intimidate other girls.

In his celebration of emancipated urban girlhood, Marshall Berman, an avid reader of urban imaginaries, praises the 1983 music video to Cyndi Lauper's song "Girls Just

Wanna Have Fun" for conveying the magic of modern public space. In a setting aesthetically reminiscent of US sit-coms, Lauper's character liberates herself from her family, leaves the house despite her parents' objections, and dances joyfully through the streets of New York with a group of female friends. The girls in Lauper's video, Berman argues,

> are not only transforming their lives, but transforming the life of the street itself, using its structural openness to break down barriers of race and class and age and sex, to bring radically different kinds of people together.
>
> (*2017: 78*)

Not only do these girls transform the city, but, as they later return to the protagonist's bedroom with random people encountered on their way in tow, "they also bring the street into the house, the public realm into her private space" (Berman, 2017: 78). Berman focuses on the romanticized vision of urbanites rendering public space inclusive by their sheer presence in it, producing and experiencing modernism in the streets. A question lurking in this celebratory tone is whether Lauper's character would have gone out without her friends. Friendship is the tool that helps bridge the public and the private (Fenton, 2005; Kathiravelu and Bunnell, 2017).

In cinematic imaginaries of urban girlhood, girls without friends are often portrayed as desolate, lost, suffering, and robbed of girlhood. One of such films is Andrea Arnold's *Fish Tank* (2009) – a story of Mia, a teenage girl in "broken Britain" (Bolton, 2016; Tyler, 2013). The Essex council estate where Mia lives is a place that crushes all her attempts at overcoming her predicament. As in *Girlhood*, in Arnold's film institutions fail the girl – but unlike Marième, Mia has no friends to lean on. For reasons not fully explained in the film (there may have been a falling out with another girl), Mia is rejected by her peers and alienates them further by making fun of them. She does not join other girls from the estate in their stiffly choreographed dance practice in a parking lot and ridicules their sexy moves performed in front of a group of teenage boys. Mia goes to an abandoned building, where no one can see her, to dance on her own. The dance scenes in *Fish Tank* appear bleak compared to the joyous dance routines in Sciamma's film, which celebrate urban girlhood. Neglected and abused by her mother, seduced and swiftly abandoned by her mother's boyfriend, and exasperated with her little sister, Mia decides to leave home with her maybe boyfriend. As with Marième, we do not know what will happen to her after she leaves the frame. Arnold's film offers a counter-imaginary to celebratory portrayals of urban girlhood, a corrective of sorts, but also a reminder that it is in the shared experiences and collective narratives that the power of girlhood resides.

Concluding notes on imaginaries of urban girlhood

Imaginaries of urban girlhood demonstrate that it is easier, safer, and often simply more fun for girls and women to be in the city together with "girls like us". A girl sitting alone on a bench in a park or a square remains an uncommon sight in various socio-geographical contexts. It is, in fact, so rare that it may appear alarming to those who see her, and not infrequently provoke uninvited attention. All the women I know, myself included, have countless stories of being interrupted by unsolicited male interventions when walking down the street, sitting on a park bench, in a café, at a bus stop, or on a subway train, reading a book, listening to music on the headphones, texting, or simply sitting, walking, or riding, taking in the scene. And while some of these interactions apparently provoked by our sheer singular presence in public space may and do at times develop into anecdotes, bizarre adventures, or, yes, even romances, many others are marked by symbolic, verbal and physical violence. As Rebecca Solnit observes:

> Walking down the city streets, young women get harassed in ways that tell them that this is not their world, their city, their street; that their freedom of movement and association is liable to be undermined at any time; and that a lot of strangers expect obedience and attention from them.
>
> (*2016: 85*)

The #MeToo movement, started by the activist Tarana Burke in 2006 and popularized by Hollywood celebrities in 2017 in the wake of mass sexual misconduct allegations against the film producer Harvey Weinstein (and then also other men), raises awareness about the scale and various (previously unacknowledged or silenced) forms of sexual harassment and sexual abuse of women and girls. As the movement keeps developing at the time of writing this chapter, it may be too early to try to assess its impact on imaginaries of urban girlhood. It can definitely be stated, however, that #MeToo, first in the US and then across various geographical contexts (#WoYeShi in China, "BalanceTon-Porc" in France, #QuellaVoltaChe in Italy, #JaTeż in Poland, etc.), continues to provide much-needed interventions into public debates on sexual abuse, rape culture, and consent.

As we are confronted daily with the realities of women's restricted access to public space, we may wonder how pop cultural imaginaries of urban girlhood, such as those discussed in this chapter, matter in everyday life. Although popular culture is hardly in a clear-cut cause-and-effect relation with girls' lives, it is involved in complex, multiple ways in the making of female subjectivity (Walkerdine, 1997). Pop cultural images with which girls engage, regardless of whether they are empowering or limiting and exclusionary, "produce the very possibilities of what it means to be a girl" (Coleman 70; Walkerdine, 1997). Through the sensation of simultaneous distance and proximity that a screen, a book page, or a radio speaker provides, imaginaries of urban girlhood help us reflect on how urban space is gendered. In fact, any screen may yield revelatory observations. To geographer Doreen Massey, that moment of revelation came while, as a little girl, she was looking out a bus window:

> I can remember very clearly a sight which often used to strike me when I was nine or ten years old. I lived then on the outskirts of Manchester, and "Going into Town" was a relatively big occasion; it took over half an hour and we went on the top deck of a bus. On the way into town we would cross the wide shallow valley of the River Mersey, and my memory is of dank, muddy fields spreading away into a cold, misty distance. And all of it – all of these acres of Manchester – was divided up into football pitches and rugby pitches. And on Saturdays, which was when we went into Town, the whole vast area would be covered with hundreds of little people, all running around after balls, as far as the eye could see.
>
> I remember all this very sharply. And I remember, too, it striking me very clearly – even as a puzzled, slightly thoughtful little girl – that all this huge stretch of the Mersey flood plain had been entirely given over to boys.
>
> (*Massey, 1994:185*)

Young Massey did not go to the sports fields as they seemed to belong to a different world, inaccessible to her at the time. What would her childhood have been like had all this space been given to "girls like her" as well?

Urban space is also gendered through the city text. As Solnit notices, "names perpetuate the gendering of New York City" – in fact, "almost every city is full of men's names, names that are markers of who wielded power, who made history, who held fortunes, who was remembered" (2016: 88). As we continue to meet with "girls like us" at Place Charles de Gaulle, Alexanderplatz or Washington Square and, together or alone, we walk down the streets named after kings, popes, male generals, prime ministers, and presidents, Solnit invites us to imagine how girls would experience the city had it been full of women's names. In the map accompanying her essay, Solnit renames all of New York City's subway stations after women. How would it feel to get on a train at an Assata Shakur station in Harlem and get off at a Suheir Hammad stop in Brooklyn, or travel from Jennifer Lopez in The Bronx to Serena Williams in Queens via Louise Bourgeois in Midtown? If walking down May-Ayim-Ufer (a street in Berlin named after the Afro-German female poet) can be any indication of these for now largely imaginary sensations, it would feel grand indeed.

It is important to continue imagining how the world would look like if urban space was more welcoming to all genders. Fighting for more inclusive spaces in the city can – and, I argue, should – go hand in hand with trying to imagine (through street art, film, literature, music, and video games) how the city feels now, and why, and how it *could* feel for the people who inhabit it. The girl friendship graffiti in Bishkek, Sciamma's poignant exploration of urban girlhood, and many other imaginaries provide much-needed insights into how young women find pockets of freedom for themselves in the city, even in spaces usually coded masculine (Geesey, 2011), how they subvert and hack this gendered coding and consequently render their lives more livable for themselves and other "girls like them". As popular culture has "power to dramatize collective dreams" (Berman, 2017: 78), looking closely at imaginaries of urban girlhood can help us develop ideas for and understanding of how these dreams can be achieved.

Acknowledgements

The author wishes to thank Hana Khalaf for her kind and insightful comments on an earlier version of this chapter.

References

Baumgardner, J. and Richards, A. (2000) *Manifesta: Young Women, Feminism, and the Future*. New York: Farrar, Straus and Giroux.

Beebeejaun, Y. (2017) Gender, Urban Space, and the Right to Everyday Life. *Journal of Urban Affairs* 39(3): 323–34.

Berman, M. (1982) *All that is Solid Melts into Air: The Experience of Modernity*. New York: Simon & Schuster.

Berman, M. (2017) Take it to the Streets: Conflict and Community in Public Space. In: Marcus, S. and Sclan, S. (eds.), *Modernism in the Streets: A Life and Times in Essays*. London: Verso.

Bolton, L. (2016) A Phenomenology of Girlhood: Being Mia in *Fish Tank* (Andrea Arnold, 2009). In: Handyside, F. and Taylor-Jones, K. (eds.), *International Cinema and the Girl: Local Issues, Transnational Contexts*. New York: Palgrave Macmillan, 75–84.

Bondi, L. and Rose, D. (2003) Constructing Gender, Constructing the Urban: A Review of Anglo-American Feminist Urban Geography. *Gender, Place & Culture* 10(3): 229–45.
Coleman, R. (2009) *The Becoming of Bodies: Girls, Images, Experience*. Manchester: Manchester University Press.
Cronin, A.M. (2014) Between Friends: Making Emotions Intersubjectively. *Emotion, Space and Society* 10(2): 71–8.
di Leonardo, M. (1987) The Female World of Cards and Holidays: Women, Families, and the Work of Kinship. *Signs* 12(3): 440–53.
Dobson, J. (2017) Dis-Locations: Mapping the Banlieue. In: *Filmurbia: Screening the Suburbs*. Basingstoke: Palgrave Macmillan, 29–48.
Driscoll, C. (2002) *Girls: Feminine Adolescence in Popular Culture and Cultural Theory*. New York: Columbia University Press.
Faderman, S. (1981) *Surpassing the Love of Men: Romantic Friendship and Love between Women from the Renaissance to the Present*. New York: William Morrow & Company.
Fenster, T. (2005) The Right to the Gendered City: Different Formations of Belonging in Everyday Life. *Journal of Gender Studies* 14(3): 217–31.
Fenton, L. (2005) Citizenship in Private Space. *Space and Culture* 8(2): 180–92.
Foucault, M. (1997) Friendship as a Way of Life. In: Rabinow, P. (ed.), *Ethics: Subjectivity and Truth*, Vol. 1. London: Penguin, 135–40.
Geesey, P. (2011) A Space of Their Own? Women in Maghrebi-French Filmmaking. In: Durmelat, S. and Swamy, V. (eds.), *Screening Integration: Recasting Maghrebi Immigration in Contemporary France*. Lincoln: University of Nebraska Press, 161–77.
Gonick, M. (2006) Between "Girl Power" and "Reviving Ophelia": Constituting the Neoliberal Girl Subject. *NWSA Journal* 18(2): 1–23.
Hernandez, D. and Rehman, B. (2002) *Colonize This!: Young Women of Color on Today's Feminism*. New York: Seal Press.
Kathiravelu, L. and Bunnell, T. (2017) Introduction: Urban Friendship Networks: Affective Negotiations and Potentialities of Care. *Urban Studies*. online first. DOI: 10.1177/0042098017737281 (last accessed 16 November 2017).
Kofman, E. and Lebas, E. (trans. and eds.) (1996) *Writings on Cities: Henri Lefebvre*. London: Blackwell.
Lisiak, A. (2014) Hip-Hop Representations of Urban Stillness: The Case of Poznan, Poland. *CITY: Analysis of Urban Trends, Culture, Theory, Policy, Action* 18(3): 334–48.
Marcus, S. (2010) *Girls to the Front: The True Story of the RIOT GRRRL Revolution*. New York: HarperCollins.
Massey, D. (1994) *Space, Place and Gender*. Cambridge: Polity Press.
Massey, D. (2005) *For Space*. Thousand Oaks: Sage.
McDowell, L. (2006) Reconfigurations of Gender and Class Relations: Class Differences, Class Condescension and the Changing Place of Class Relations. *Antipode* 38(4): 825–50.
McNeill, I. (2017) "Shine Bright like a Diamond": Music, Performance and Digitextuality in Céline Sciamma's *Bande de filles* (2014). *Studies in French Cinema*. DOI: 10.1080/14715880.2017.1345187
McRobbie, A. (2015) *Be Creative: Making a Living in the New Culture Industries*. London: Wiley.
McRobbie, A. and Garber, J. (1991 [1977]) Girls and Subcultures. In: McRobbie, A. *Feminism and Youth Culture: From 'Jackie' to 'Just Seventeen'*. Basingstoke: Macmillan, 1–15.
Peake, L. and Rieker, M. (eds) (2013) *Rethinking Feminist Interventions into the Urban*. London: Routledge.
Renold, E. and Ringrose, J. (2013) Feminisms Re-Figuring "Sexualisation", Sexuality and "The Girl". *Feminist Theory* 14(3): 247–54.
Solnit, R. (2016) City of Women: The Power of Names. In: Solnit, R. and Jelly-Schapiro, J. *Nonstop Metropolis: A New York City Atlas*. Oakland: University of California Press, 85–90.
Stack, C. (1974) *All Our Kin: Strategies for Survival in a Black Community*. New York: Harper & Row.

Tarr, C. (2005) *Reframing Difference: Beur and Banlieue Filmmaking in France*. Manchester: Manchester University Press.
Tyler, I. (2013) *Revolting Subjects: Social Abjection and Resistance in Neoliberal Britain*. London: Zed Books.
Vacchelli, E. and Kofman, E. (2018) Towards an Inclusive and Gendered Right to the City. *Journal of Gender Studies* 17: 1–3.
Vaiou, D. (1992) Gender Divisions in Urban Space: Beyond the Rigidity of Dualist Classifications. *Antipode* 24(4): 247–62.
Valentine, G. (1989) The Geography of Women's Fear. *Area* 21(4): 385–90.
Walker, R. (1995) *To Be Real: Telling the Truth and Changing the Face of Feminism*. New York: Anchor.
Walkerdine, V. (1997) *Daddy's Girl: Young Girls and Popular Culture*. Basingstoke: Macmillan.
Wright, M.W. (2010) Gender and Geography II: Bridging the Gap—Feminist, Queer, and the Geographical Imaginary. *Progress in Human Geography* 34(1): 56–66.

20

The city on the highway, revisited

Richard J. Williams

Introduction

This chapter concerns the automotive urban imaginary, tracing its history from the middle of the twentieth century until the present day. It is an imaginary that is simultaneously ubiquitous, and threatened; ubiquitous because there is no part of the urbanized world that it has not touched, and threatened because emergent technologies of automation prefigure a world without drivers, and radically different ways of using urban roads. I use the term "the City on the Highway" to describe this imaginary, borrowing from the academic planner Peter Hall who had a particular interest in it. For him, the City on the Highway was a city organized around personal transportation, split into discrete functional zones, and overlaid by a system of high-speed roads that typically bears little relation to any previous system (Hall, 1971: 325–84). It could be reduced to a number of architectural figures – a compact, zoned downtown of towers, the highway intersection, the elevated urban motorway, the suburban retail mall, the business park, and the single family home. The City on the Highway describes any number of actually existing cities, in varying states of completeness. It could refer to complete Modernist planned cities including Brasília, which its chief planner Lucio Costa referred to (amongst other things) as the "capital of the autostrada" (Costa and Holford 1957: 402). It could mean the sections of old cities retrofitted for cars, such as West London, dominated in large part by the elevated A40 highway. Or it could mean the central business district of Lagos, celebrated (if that is the right word) in all its motorized glory by a Dutch art house documentary, *Lagos Up Wide and Close* (Van Der Haak, 2005).

Hall described these and other iterations of it in his book *Cities of Tomorrow*, moving from New York, to Frank Lloyd Wright's imaginary Broadacre City, the Soviet Deurbanists, and Nazi Germany, to post-war California where it had its most complete realization (Hall, 1971). But as he described, the City on the Highway found itself called into question almost as soon as it came into being. The urbanist Lewis Mumford laid into it as early as 1938, attacking it again in a 1963 essay collection, *The Highway and the City* that inveighed against the urban highway building for its insatiable land hunger (Hall, 1971: 297; Mumford, 1963). By this time, he had the support of the urban activist Jane Jacobs in New York, and a good chunk of the city's intellectual class (Jacobs, 1961). But the City on the Highway was always much bigger than New York, and the concept proved durable, and exportable, until at least the oil crisis of 1973. Even then, despite the existence of alternative models from Stockholm to Paris, the City on the Highway was generally winning over the mass transit city, as Hall wrote: "the people were voting for it with their wheels" (Hall, 1971: 318).

My concern here is what might be called the "aesthetic experience" of the City on the Highway, by which I mean its apprehension in visual and corporeal ways, in which pleasure is a factor. That experience is disseminated in a range of visual and textual media, not all of them conventionally "art", and sometimes in several media at the same time, as in the case of Reyner Banham's book, *Los Angeles,* the main object of attention here. It might seem odd to think about the City on the Highway in terms of aesthetics, but how it looks and feels is essential to our understanding of it, justifying its existence as often as projections of population and traffic flow; highly complex, it is too often overlooked. Pleasure is particularly overlooked in this discourse – a serious absence, as the City on the Highway would never have come into being were it not, at some level, experienced as pleasurable (for more on this, see Williams, 2002). I explore its history as an aesthetic object at two contrasting moments, the first being roughly 1968–1971, when it was still possible to produce an optimistic account of the City on the Highway, albeit one cognizant of the realities of snarl-ups and smog. The second period, which I deal with much more briefly, is 2015–2017 and describes a time when Waymo, the autonomous vehicle company created by Google, made public a new vision of the City on the Highway, accompanied by intense media curiosity (see, for example, Davies, 2016).

It is a commonplace to equate the City on the Highway with Los Angeles, a rich, polycentric megalopolis numbering 7 million inhabitants in 1971 (Los Angeles County) and 13 million in 2017 (US Census Bureau). It was not the first to build the City on the Highway – that was probably New York, and the Long Island Motor Parkway, built with remarkable prescience in 1906–1911 (Hall, 1971: 277). But it is the city that comes to embody it most clearly, in large part because having the resources of Hollywood at its disposal, and a sunny climate, it put more images of it into the popular domain than anywhere else. The benign representation of the Californian highway continues to have an impact in, for example, the 2017 exhibition at the Design Museum, London, *California: Designing Freedom* (McGirk, 2017). California might be a cliché, but it is arguably here that the City on the Highway has both its clearest advocacy, and its clearest built expression. In addition, California has a uniquely developed cultural commentary on the City on the Highway, in art and film, and literature, as well as the academic disciplines of architecture, planning, geography, sociology, and their cognates.

That cultural commentary on the City on the Highway has been nothing if not dystopian. The popular cultural studies writing of Mike Davis pictures the City on the Highway as violent and carceral, an archipelago of fortified ghettos of one kind or another (Davis, 1990). The highways of his imagination are instruments of control rather than means of communication. The geographer Edward Soja's work describes the City on the Highway in similar terms: LA first "came together" in his phraseology in 1965, in the Watts riots, "the most costly and violent in US history" (Soja, 2014: 27). In other words, it is the city of visible racial and class divisions that is his city, and his project of a city read from below is for many academics the point of entry to the City on the Highway. The highways here are clogged, the environment polluted, the neighborhoods poor, the policing violent, the rich sequestered in gated compounds. In different ways, Davis and Soja represent Los Angeles as a city after the Fall.

Perhaps naively, I want to recover the prelapsarian city, and the key to doing that is the work of the Berkeley-based urban designer and theorist Melvin M. Webber. One's introduction to Webber is often framed by negativity: Soja, for example characterized Webber's ideal city as one in which "distances were elastic and the local meant little, where many households had unlisted telephone numbers and nestled behind high walls and a sign or two reading 'Trespassers Will Be Shot'" (Soja, 2014: 29). Soja refers here to Webber's concept of the "Nonplace Urban Realm", which he elaborated in a chapter of the 1964 book *Explorations into Urban Structure*, for which he was one of six co-authors (Webber et al., 1964: 79–143). Soja gets a few things wrong here, firstly the date, which he cites as "the 1970s" when it is 1964. Secondly, the remarks about telephone numbers, and walls, and warning signs have nothing to do with Webber, and are just made up. And thirdly, he lets the reader believe that the text is specifically about Los Angeles, when – counter-intuitively – Webber avoids nearly all discussion of the city, in favor of extrapolation from his lived experience of the Bay Area around San Francisco. (The negative casting of Webber extended to his obituaries, such as for example Bendixson, 2007). Nevertheless, Webber's work and its reception is crucial here in identifying a set of concerns around the City on the Highway.

Webber's argument in "The Urban Place and the Nonplace Urban Realm" remains straightforward: thanks to developments in transport and communications technologies, life and work need no longer occupy the same location. An academic might routinely perform tasks thousands of miles apart, and only a small part of his or her daily life might involve the coincidence of geographical space and identity. Identity would be distinct from place. "The adult American", he continued "is increasingly able to maintain selected contacts with others on an interest basis over increasingly great distances; and he is thus a member of an increasing number of interest communities that are not territorially defined" (Webber et al., 1964: 111).

By extension, urbanity might no longer be a function of urban space per se, but as a "quality and diversity of life". "Urbanity", he continued,

> is more profitably conceived as a property of the amount and variety of one's participation in the cultural life of a world of creative specialists, of the amount and variety of the information received.
>
> (*Webber et al., 1964: 88*)

The logical extension of that is the belief that true urbanity might be found in nontraditional city form, a flat contradiction of the newly resurgent, traditionalist position exemplified by Jane Jacobs. For her, Los Angeles's motorized form was "eroded and drab" at best, when not a threat the continuation of life on earth itself (Jacobs, 1961: 369). Webber meanwhile wrote sunnily of "ever increasing mobility" producing more, not less, urbanity (Webber et al., 1964: 146). In his one specific reference to the southern Californian metropolis, he wrote "we might discover that Los Angelenos enjoy as urbane a life as do New Yorkers" (Webber et al., 1964: 132).

Webber's vision certainly had fans. Among Webber's British readers was Peter Hall, who had already entertained his own readers with a highly motorized vision of the future, *London 2000*, the first edition of which was published in 1963. Here he imagined a "net" of urban motorways laid over existing London, creating a new mode of circulation and "a city quite unlike any we know today" (Hall, 2002: 149). That account of London transformed hints at the kind of emotional case for the City on the Highway that is important here. His concluding chapter imagines the life of the new city that is not only polycentric in the Webberian sense, but the embodiment of a set of aesthetic pleasures unforeseen by its builders. Driving out of town on the fictional New Kent Motorway, Hall's family of the future, the Dumills, appreciate the motorways as a "brilliant sight", giving London a "new dimension, now trenching by the sides of railways, now flying over rooftops, now burrowing through the heart of the reconstructed office and shopping centers" (Hall, 1971: 272). The impeccably bourgeois Dumills – he is a senior university administrator, she an opinion pollster – have multi-centered lives, feeling at home in their Kentish new town of Hamstreet and its artfully-construed canals as they do ranging across the south of England and northern France, or as they have just done, strolling along the banks of the Thames in central London. They are, in the Webberian sense, citizens of the Nonplace Urban Realm, and their lives are made not only possible, but pleasurable by its motorways.

What's striking about this conclusion to *London 2000* is its openness to pleasure. Its pleasures are in part traditional ones – strolling along a riverbank, going to the theatre – but the celebration of highway travel, cutting through the city and sailing above it, is something new. In an otherwise straightforward book, this section has an ecstatic quality that takes planning into the realm of aesthetic experience. What Hall described – perhaps without meaning to – was representative of a more general cultural understanding that the City on the Highway had brought into being a range of new experiences that could not effectively be described using existing language. It was a major theme in American art of the 1960s. See, for example, the sculptor Tony Smith's account of driving on the unfinished and empty New Jersey Turnpike. A description of a kinetic and immersive experience, so absorbing as to signal the end of art, it was very widely discussed (Wagstaff, 1995: 386).

Hall's Englishness arguably allowed him to see things that a native could not, and it is notable that one of the great 1960s apologies for the City on the Highway was the work of four Englishmen in a pub, dreaming of California. This is the celebrated essay "Non Plan" for the non-aligned but left-leaning political magazine *New Society* in March 1969 (Banham et al., 1969). Hall was cited as one of the authors, along with the radical architect Cedric Price, and the architectural historian and critic Reyner Banham. The text was, however, nearly all the work of Paul Barker, *New Society*'s editor, after conversations over beers with the others: he showed a final version to them and they signed it off (Barker, 2017). There was genuinely a collective voice; however, the text might have been Barker's, but he sounds perfectly like Banham. "Non-Plan" argued for a permissive approach to English planning, and a set of bounded, time-limited experiments in which the relaxation of planning law would be trialed; something like California would emerge, they thought, with a new generation of enthusiastic consumers roaming its motorways in search of spectacle. The legacy of "Non-plan" has more to do with what eventually happened under a Conservative administration in London's former docklands: that legacy has been explored in detail elsewhere (Hughes and Sadler, 2000; Williams, 2004). But we should extract "Non-plan's" enthusiasm for the City on the Highway, for which it is unapologetic. Its illustrations made this clear – Barker had gone into central London one night with a photographer to take snapshots of neon signs in a manner that anticipated by three years the work of the Venturis and their students in Las Vegas. Like the signs documented in color by the Americans, Barker's objects were the commercial neons made to be glimpsed at speed from a car, and the first image in the article, with signs for Jaguar and Shell clearly referencing the car (Hughes and Sadler, 2000: 5). "Non-Plan's" prescription was for an England with the brakes off. In the East Midlands, picturesquely renamed "Lawrence Country", the authors imagined a highly mobile future society living anywhere, commuting farther than ever before, living and consuming like Californians:

> strip development would spread along the main roads on the American model. Much of this will serve the needs of a mobile society: eating places, drinking places, petrol stations, supermarkets. It would not look like a planner's dream but it would work.
>
> (*Banham et al., 1969: 438*)

That "mobile society" was illustrated with a cartoonish render of the likely strip-mall result, a riot of signs fighting it out with the traffic chugging away in the background. "Non-Plan" would bring into being a new nation of stoned "auto-nomads", cruising from one pleasure dome to the next, perpetually on vacation (Banham et al., 1969: 441). "Non-Plan" was libidinal in essence, assuming the world exists for the experience of pleasure, clearly including the movement from one pleasure to the next. An argument for permissive planning, and by extension, the permissive society, it could not be more different to the understanding of planning enshrined in English and Welsh law (Department for Communities and Local Government, 2015).

Reyner Banham had always been captivated by American design, and automotive design in particular. His fascination was fed by Hollywood films, and encounters with American service personnel in wartime Norwich, and by the mid-1950s, through his work for the Independent Group, he was one of England's chief interpreters of American popular culture (Banham, 2017). As Adrian Forty has pointed out, he did not actually visit America until 1960, on the invitation of Philip Johnson (Forty, 2002: 201). His first published work on an automotive topic was a 1955 essay for a London-based art magazine, *Art,* about the Cadillac Eldorado, a 5.7 meter long behemoth powered by a 6 liter V8. Banham's main purpose was partly to prick the art historical establishment, deploying a Panofskian analytical mode to show just how iconographically rich the products of contemporary mass culture could be (Banham, 1955). The profusion of fins and chrome was as incomprehensible as the baroque to the novice, but through the right eyes, and with the right interpretative brain, the Eldorado was a perfect expression of speed, power and sex:

> the repertoire of hooded headlamps, bumper-bombs, incipient tail-fins, speed-streaks and chromium spears, protruding exhaust-pipes, cineramic wind-screens – these give tone and social connotation to the body envelope; the profiling of wheel-arches, the humping of mudguards, the angling of roof-posts – these control the sense of speed; the grouping of the main masses – these balance the sense of masculine power and feminine luxury.
>
> (*Banham, 1955*)

"A thick stream of loaded symbols", the Eldorado was, the reader inferred, as iconographically pregnant as the *Arnolfini Portrait* (Banham, 1955).

Banham had absorbed automotive criticism from the pages of *Industrial Design*, and knew the work of General Motors's Harley Earl as well as he knew art historical subjects (Banham, 1955). But at this stage in his career, the car was – as it was for Roland Barthes two years later in his discussion of the Citroën DS – still an *object* (Barthes, 1972: 88–91). We see the Cadillac Eldorado from the curb, and every impression of the car is written from the point of view of what it looks like, rather than what it does. Like Barthes's DS, it is curiously static, a piece of automotive sculpture rather than a vehicle.

That is no surprise, really: at this point, Banham couldn't drive. It was said that he had driven fire tenders at Filton airfield for the Bristol Company ("a truly frightening thought" according to his son) (Banham, 2017). But he didn't learn to drive properly until 1966, by which time he was 44 years old, and "Vehicles of Desire" was 11 years in the past. Up until this point, his wife Mary had performed all domestic driving tasks, and generally continued to do so (Banham, 2017). However, Banham's status as design guru did lead from time to time, to some racy borrowed cars at weekends: Reliant Scimitars and Ford Capris (Banham, 2017). Reading between the lines, driving itself remained an activity that Banham never entirely mastered, but the lateness to which he came to it meant that he could see it for what it was: a culture, as rich and suggestive as a language. Hence his arresting statement from *Los Angeles* that like generations of aristocratic tourists learning Italian to read Dante, he learned to drive "in order to 'read LA in the original'" (Banham, 1971: 23).

This was much more than an art historical tease. Banham correctly understood driving in linguistic terms, as a commonly understood system of communication with rules and dialects, commonly understood and regulated, but within which individual expression was possible. Like language, it was essential to the understanding of the modern city. On LA, he elaborated: "The city will never be fully understood by those who cannot move fluently through its diffuse urban texture, cannot go with the flow of its unprecedented life" (Banham, 1971: 23). It was a place in which, he stated, flatly contradicting the entire canon of art history, "mobility outweighs monumentality" (Banham, 1971: 23).

That understanding of driving as culture needs to be separated from the early essays on cars. Probably the first proper treatment of it is "Unlovable at Any Speed", published in the last *Architect's Journal* of 1966, the weekly professional paper in which Banham had a regular column. He had just passed his driving test, and being apt to write about himself, he produced a column about it (Banham, 1966). The title, a good one, is a play on Ralph Nader's *Unsafe at Any Speed*, a bestselling critique of the car industry that helped kill off a popular compact model, the Chevrolet Corvair (Nader, 1965). It's a short, self-deprecating piece in which Banham reflects on his own tardiness in learning to drive, and, importantly, on the status of driving in the refined profession of architecture where the lack of technological knowledge can sometimes be a badge of honor.

Banham developed his theme in 1968. In a 2000-word article on Los Angeles for *The Listener*, he pictured himself as the flâneur of the freeway, full of automotive wisdom and travel tips. Driving Wilshire Boulevard westwards at sunset is a must for the visitor: "nowadays", he wrote, "it must be the only shopping street in the world that one would ever want to drive down for pleasure" (Banham, 1968: 268). Where Banham's observations were most acute, however, were on the aesthetic pleasures of the freeway system. Here he meant not just the pleasure invoked in the sight of good engineering, but also the pleasure of navigating it in a car. On the intersection of the San Diego and Santa Monica Freeways (the intersection of the 405 and 10, some fifteen miles due west of downtown), he writes first of its sculptural aspect, which was "convincingly monumental".

Figure 20.1 Intersection of San Diego and Santa Monica Freeways.

What really intrigues him, however, is the corporeal experience:

> The two highest ramps rise, converge and separate in curves of perfect symmetry, carried on in rows of slim cylindrical columns. But to drive over those ramps in a high, sweeping, 60 mile-an-hour trajectory and plunge down to ground level again is a spatial experience of a sort one does not normally associate with monuments of engineering – the nearest thing to flight on four wheels I know.
>
> (*Banham, 1968: 268*)

Here is something new: a widely shared experience of speed and space, with significant and complex effects on the body and on perception, and a good deal of pleasure thrown in. It is an aesthetic experience by any standards, yet one outside the purview of art, commonly defined (and maybe also beyond the imagination of the engineers who built it). This realization, that the freeway system might be productive of a new culture, is the basis of Banham's 1971 book *Los Angeles: The Architecture of Four Ecologies* (Banham, 1971). The whole argument is framed in terms of the car – the opening chapter, "In The Rear View Mirror" uses the visual device of the rear view mirror to frame the whole city, as if to say that whatever can be seen in that small oblong of glass and plastic can be said, literally and metaphorically, to contain the whole city (Banham, 1971: 21–36). Banham tells the story – an often ugly one – of the battle for automotive supremacy, involving among other things the displacement of the once comprehensive Pacific Electric streetcar system by the freeways, and he's not shy of mentioning the pollution and congestion that accompany the driver on most excursions. But for the most part this is a sunny evocation of the freeway system: "the freeways seem to have fixed LA in canonical and monumental form" he writes, making clear again the collision between art history and automotive culture; "the freeways are what the tutelary deity of the City of Angels should wear upon her head instead of the mural crowns sported by civic goddesses of old" (Banham, 1971: 36).

The most valuable part of the book *Los Angeles* for present purposes is Chapter 11, "Autopia", a neologism combining "automobile" and "utopia" that defines the fourth and final of Banham's defining city ecologies (Banham, 1971: 213–22). "Autopia" in this chapter is less a set of physical elements (concrete surfaces, crash barriers, intersections, exit and entrance ramps, lighting, drainage, road markings, and so on) than it is a culture. In one of the many anecdotes that pepper the narrative, Banham recalls a widely believed hoax about a family who lived on the freeway, in perpetual motion in a mobile home. The point of the story is less the hoax than its believability. "The freeway", Banham wrote "is where Angelenos live a large part of their lives" (Banham, 1971: 214).

The rest of the chapter is striking for the way it describes certain behaviors produced by or in relation to the freeway system that amount to a culture, and to a belief in the freeway as de facto public place. These include the observation, that on exiting the freeway, a (female) passenger pulls down the sun-visor to reveal the vanity mirror located there, and briskly, automatically straightens her hair. The import, famously, is this: "that coming off the freeway is coming in from outdoors . . . the mile or two of ground level streets counts as no more than the front drive of the house" (Banham, 1971: 213). Like so much in Banham, this teasing observation occupies an uncertain zone between reportage and fiction, raising more questions than answers. How often, and precisely where had he observed this? What were the women he observed thinking? Was he himself aware of the unsettlingly voyeuristic overtones of the story (probably not)? In what sense is the freeway a truly public space?

As ever, these questions are left hanging, only to be answered much by later by cultural critics such as Soja. Banham meanwhile continued his increasingly ecstatic reflections on freeway driving, in which it approaches the condition of a "mystical" experience. Their – and his – "heightened awareness" was the by-product of the willed subjugation of individuality to an overarching system, a unique modern compact between freedom and discipline. A "spectacular paradox", he wrote, it exemplifies the "great debate between private freedom and public discipline that pervades every affluent, mechanized, urban society" (Banham, 1971: 216). The City on the Highway is nevertheless experienced as pleasure for most Angelinos, who are not "retching with the smog nor stuck in a jam; their white-wall tires are singing over the diamond-cut anti-skid grooves in the concrete road

Figure 20.2 Freeway signs, photograph by Baron Wolman.

surface, the selector levers of their automatic gearboxes are firmly in Drive, and the radio is on" (Banham, 1971: 216).

That positivist image of the City on the Highway certainly had its detractors at the time, among them the art critic Peter Plagens, whose scathing review of *Los Angeles* for

Artforum has been much cited (Plagens, 1972; see discussion in Hawthorne, 2011). Later critiques of Los Angeles by Davis and Soja have pointed up the naïveté of Banham's position: that, broadly, he sees a racially and socially stratified city from the point of view of a privileged visitor. And in art and film, the optimistic representation of the City on the Highway has been almost entirely displaced by its depiction as ruin, or as good as. The highway in film is always jammed with cars, an impediment to movement rather than something that permits it: the opening sequence to 2016's *La La Land* is a good contemporary example (Chazelle, 2016). Or the City on the Highway is depicted as ruin. The artist Catherine Opie's 1994–1995 *Freeway* series of platinum prints show Los Angeles freeways not as a functioning infrastructure, but the skeletal remains of a defunct civilization (Reilly, 2001).

However, we should stay with Banham's aesthetic of driving for a little longer, for what it represents has been put in doubt by a range of factors, some economic, some demographic, but most of all technological. What Banham describes in "Autopia" are the experiences of individuals, working within a system, but who understand themselves to have substantive autonomy, the limits of which are continually tested. The bodily absorption of freeway driving has to do with the relentless onrush of decisions about speed and placement of the car in relation to the road, and the relation of these to larger ones about where one is actually going. All of this involves risk, not least for a novice driver like Banham, for whom every decision has to have been fraught with danger: his ecstatic account of the 10/405 freeway intersection is that of someone who has overcome his fears. And he might well be afraid, for Los Angeles County registered 1261 deaths in vehicular accidents in 1971–1972, and many more serious injuries caused by the same (Noguchi, 1973). But it is important to recognize the sense of risk that goes along with the freedom of the individual, or to put it another way, the risk to the individual is a function of the rewards that the freeway experience brings. Banham himself was under no illusions: "the freeways can kill – hardly a week passed but I found myself driving slowly under police control past the wreckage of at least one major crash" (Banham, 1971: 215). That brush with death, however, doesn't really give him pause – instead, it seems to encourage him to repeat the experience, as if to say that the experience of this very modern, motorized risk is itself its own reward, an idea taken to its logical extreme two years after the publication of *Los Angeles* in J. G. Ballard's experimental novel, *Crash*. In that novel, Ballard's protagonists find sexual gratification in car accidents, going as far as to stage them deliberately, with predictably catastrophic results (Ballard, 1973).

It is worth bearing this somewhat Freudian scenario in mind in relation to Waymo, the spinoff company created by the technology giant Google in 2016. Waymo's mission, "to create a safer driver for everyone" is autonomy, and its projects (including those previously developed under the auspices of Google) include a second-generation Toyota Prius with autonomous capability, a Lexus 450 Rh SUV with the same, but more evolved, and most strikingly in 2015, a fully autonomous vehicle designed in-house (Waymo, 2017). All were familiar sights to anyone visiting the sprawling Mountain View business park where Google is the main occupier. Their 2015 project named Firefly was superseded in 2017 by a variant of a Chrysler minivan with autonomous technology.

Firefly was sensationally different: a fully autonomous, electric two-seater, it dispensed with a steering wheel and pedals, or any other mechanical controls; furthermore, it consciously departed from the erotic imagery generally favored by car stylists, for a pre-pubertal, childlike face. It looked like, and in many ways was, a toy. Limited to 25mph, it was slow: not a machine built for any highway, real or imagined. It was quiet, meek, domestic, profoundly

Figure 20.3 Waymo's fully self-driving reference vehicle Firefly.

unthreatening. In the films of the Firefly trials, its users stressed the safety of the experience – "Nick", one of several retirees interviewed, notes there is nothing about the car "that makes you feel the least bit threatened" (Waymo, 2017b). The promotional film foregrounds Firefly riders who might either not drive, or be disinclined to do so; a couple in their eighties, an older woman with a young grandchild, a pair of teenage girls, a middle-aged blind man. The latter, Steve Mahan, speaks to the liberating aspect of the technology: having lost his sight, to be alone again in a moving car is a "profound experience" (Waymo, 2017b).

And so on. Firefly seemed to indicate an autonomous future for the City on the Highway. Precisely how soon is a matter of continuing debate, but Firefly certainly represents a new and curious set of aesthetic paradigms that challenge the risky and individualistic paradigm that Banham celebrated (Economist, 2017). Firefly's aesthetics are of a different order. Not only was it a deliberately unthreatening-looking vehicle ("cute" is the word repeatedly used in the promotional film) it posited driving as an experience in which pleasure was located outside of the individual and the act of driving, and instead in the interaction between now-non-distracted occupants. The film repeatedly depicts couples talking, hugging, kissing, and otherwise interacting with each other rather than the roadscape outside. And the project's emphasis on safety above all sought to remove all risk from driving. Individualistic driving experiences – of, for example, shifting down a gear in order to exaggerate and take pleasure in the centrifugal force of a bend would – one imagines, disappear in the autonomous regime. A fully autonomous vehicle fleet might make for quite different and unrecognizable traffic patterns. Parking might become of a thing of the past, to be replaced by relentless circulation; gaps between vehicles in moving traffic might shrink to a few centimeters; above all, risk would be removed. That is overwhelmingly Waymo's objective.

What Banham described was the culture produced in relation to a universally understood kit of parts comprising the City on the Highway, which existed at least in image from the mid-1940s until the 2010s. That kit of parts, and the aesthetic responses that have grown up around, it has been put under pressure from a new set of inventions, which include not only Waymo and other autonomous vehicle companies, but Uber and other similar ride hailing applications. These have produced significant changes in car markets around the world, particularly among the young in the most developed economies who, it has been said, no longer regard car ownership as an essential signifier of adulthood (Etehad and Nikolewski, 2016; Harris, 2017). Bob Lutz, the former CEO of General Motors, was one of those to advance that opinion, adding in a profoundly pessimistic opinion piece in November 2017 that "everyone will have five years to get their car off the road or sell it for scrap" (Lutz, 2017). In his view, the City on the Highway as conventionally understood, with its deeply rooted aesthetics of freedom and individualism, is finished. The new technologies imply the erosion of that perhaps old-fashioned concept of the individual, the erosion of the concept of freedom, tenuous as it may have been, and certainly a challenge to the aesthetic experiences of driving that Banham described so well. Finally, the new technologies imply some very different architectures of driving: full autonomy implies dispensing with parking garages, gas stations, suburban driveways, and drive-in restaurants. It implies a different architecture of roads too, with physical capacity defined more by what a machine can do, and less the human frailties of reaction times (the sheer spaciousness of the freeway is a function of human beings' inability to place a car with any precision). So the City on the Highway could, if it is not already, end up being a museum piece. And that is the ironic and somewhat perverse place I will stop – Banham invoked the City on the Highway as a means of kicking against art history. To the engineers at Waymo, the City on the Highway arguably *is* art history.

References

Ballard, J.G. (1973) *Crash*. London: Jonathan Cape.

Banham, B. (2017, 8 September) *Interview with the Author*. London.

Banham, R. (1955) Vehicles of Desire. *Art*, 1 September, 3.

Banham, R. (1966, 21 December) Unlovable at Any Speed. *Architect's Journal* 144: 1527–29.

Banham, R. (1968, 29 August) Roadscape with Rusting Nails. *The Listener* 80: 268.

Banham, R. (1971) *Los Angeles: The Architecture of Four Ecologies*. London: Penguin.

Banham, R., Barker, P., Hall, P., and Price, C. (1969, 20 March) Non-Plan: An Experiment in Freedom. *New Society* 13(338): 435–43.

Barker, P. (2017, 6 August) *Interview with the Author*. London.

Barthes, R. (1972) The New Citroën. In: *Mythologies*. London: Jonathan Cape, 88–91.

Bendixson, T. (2007) Melvin Webber. *The Guardian*, 1 February. Available at: https://www.theguardian.com/news/2007/feb/01/guardianobituaries.usa.

Chazelle, D. (dir.) (2016) *La La Land*. Los Angeles: Black Label Media et al.

Costa, L. and Holford, W. (1957) Brasília, A New Capital City For Brazil. *Architectural Review* 122 (731): 395–402.

Davies, A. (2016) Google's Self-Driving Car Company Is Finally Here. *Wired*, 13 December. Available at: https://www.wired.com/2016/12/google-self-driving-car-waymo/.

Davis, M. (1990) *City of Quartz*. London: Verso.

Department for Communities and Local Government. (2015) *Plain English Guide to the Planning System*. London: UK Government. Available at: https://www.gov.uk/government/uploads/system/uploads/attachment_data/file/391694/Plain_English_guide_to_the_planning_system.pdf.

Economist (2017) The Long and Winding Road for Driverless Cars. *Economist*, 25 May. Available at: https://www.economist.com/news/science-and-technology/21722628-forget-hype-about-autonomous-vehicles-being-around-cornerreal-driverless-cars-will.

Etehad, M. and Nikolewski, R. (2016) Millennials and Car Ownership? It's Complicated. *Los Angeles Times*, 23 December. Available at: http://www.latimes.com/business/autos/la-fi-hy-millennials-cars-20161223-story.html.

Hall, P. (1971) *London 2000* (2nd edn). London: Faber and Faber.

Hall, P. (2002) *Cities of Tomorrow*. Oxford: Blackwell, 325–84.

Harris, J. (2017) Owning a Car Will Soon Be a Thing of the Past. *The Guardian*, 23 October. Available at: https://www.theguardian.com/commentisfree/2017/oct/23/owning-car-thing-of-the-past-cities-utopian-vision.

Hawthorne, C. (2011) Reading LA: A Reyner Banham Classic Turns 40. *LA Times*, 22 April. Available at: http://latimesblogs.latimes.com/culturemonster/2011/04/reading-la-banhams-four-ecologies-turns-40.html.

Hughes, J. and Sadler, S. (eds.) (2000) *Non-Plan: Essays on Freedom, Participation and Change in Modern Architecture and Urbanism*. Oxford: Architectural Press.

Jacobs, J. (1961) *The Death and Life of Great American Cities*. New York: Knopf.

Lutz, B. (2017) Bob Lutz: Kiss the Good Times Goodbye. *Automotive News*, 5 November. Available at: http://www.autonews.com/article/20171105/INDUSTRY_REDESIGNED/171109944/bob-lutz:-kiss-the-good-times-goodbye.

Marshall, C. (2016) "A Radical Alternative": How One Man Changed the Perception of Los Angeles. *The Guardian*, 24 August. Available at: https://www.theguardian.com/cities/2016/aug/24/radical-alternative-reyner-banham-man-changed-perception-los-angeles.

McGirk, J. (ed.) (2017) *California: Designing Freedom*. London: Phaidon Press/Design Museum.

Mumford, L. (1963) *The Highway and the City*. New York: Harcourt, Brace & World.

Nader, R. (1965) *Unsafe at Any Speed*. New York: Grossman.

Noguchi, T. (1973) *Biennial Report of the Chief Medical Examiner-Coroner*. Los Angeles: County of Los Angeles Department of Chief Medical Examiner-Coroner. Available at: http://file.lacounty.gov/SDSInter/Coroner/219913_1971-73.pdf.

Penner, B. (2015, September) The Man Who Wrote Too Well. *Places Journal*. Available at: https://placesjournal.org/article/future-archive-the-man-who-wrote-too-well/.

Plagens, P. (1972, December) Los Angeles: The Ecology of Evil. *Artforum* 11: 76.

Reilly, M. (2001, Summer) The Drive to Describe: An Interview with Catherine Opie. *Art Journal* 60(2): 82–96.

Soja, E. (2014) *My Los Angeles: From Urban Restructuring to Regional Urbanization*. Berkeley and Los Angeles: University of California Press.

US Census Bureau. Available at: https://www.census.gov/quickfacts/fact/table/losangelescountycalifornia,CA/PST045216.

Van Der Haak, B. (dir.) (2005) *Lagos Up Wide and Close*. Amsterdam: Submarine Media.

Wagstaff, S. (1995) Talking with Tony Smith. In: Battcock, G. (ed.), *Minimal Art: A Critical Anthology*. Berkeley: University of California Press, 386.

Waymo. (2017) *On the Road to Fully Self-Driving*. Mountain View: Waymo. Available at: https://storage.googleapis.com/sdc-prod/v1/safety-report/waymo-safety-report-2017-10.pdf.

Waymo (2017b) Journey. Available at: https://waymo.com/journey/.

Webber, M.M., Dyckman, J.W., Foley, D.L., Guttenberg, A.Z., Wheaton, W.L.C., and Wurster, C.B. (1964) *Explorations into Urban Structure*. Philadelphia: University of Pennsylvania Press.

Whiteley, N. (2002) *Reyner Banham: Historian of the Immediate Future*. Cambridge, MA: MIT Press.

Williams, R.J. (2002) Pleasure and the Motorway. In: Wollen, P. and Kerr, J. (eds.), *Autopia: Cars and Culture*. London: Reaktion Books.

Williams, R.J. (2004) *The Anxious City*. London: Routledge.

Part IV

Uneven and divided

21

Brutalism, ruins, and the urban imaginary of gentrification

Christoph Lindner

Resuscitating brutalist politics

Brutalism is back. As architectural experts and enthusiasts have been noting since the early 2000s, we are experiencing a renewed wave of interest in this historically maligned yet globally circulating architectural style. Brutalism achieved prominence in the 1950s and 1960s, and is best known for its monumentality, rawness of material, and aesthetic austerity (Banham, 2011 [1955]). "Despite two generations of abuse," observes Nikil Saval in a *New York Times* column, "an enthusiasm for brutalist buildings beyond the febrile, narrow precincts of architecture criticism has begun to take hold. Preservationists clamor for their survival, historians laud their ethical origins and an independent public has found beauty in their rawness" (Saval, 2016). Part of the twenty-first-century appeal of brutalism is the possibility it offers to revivify a dormant politics of resistance in the face of neoliberal globalization, including the emerging phenomenon of "planetary gentrification" (Lees et al., 2015).

The idea that brutalism possesses a political-ethical dimension is compelling, and it is one that the urban geographer Oli Mould in particular advocates. In his article "Brutalism Redux," Mould (2017) pushes back against the nostalgic rediscovery of brutalism as architectural heritage to posit brutalism instead as a critical concept with the potential to disrupt contemporary trends in neoliberal urban development. For Mould, this is not just because the physical survival of brutalist buildings in the built environment helps to slow the spread of generic urbanism and ubiquitous architecture. It is also because brutalism in both Anglo-American and postcolonial contexts has strong historical ties to urban activism around issues of spatial justice, affordable housing, social equity, human health, and even radical design.

So while brutalism has often been blamed for causing human misery (we can think, for example, of the lived experience of poverty and exclusion in social housing experiments such as Robin Hood Gardens in London or Pruit-Igoe in St Louis), it is important to remember that the architectural style itself has most frequently been mobilized in attempts to address social issues, ranging from housing to politics to transportation. The question, therefore, is not whether brutalism has been a success or a failure in the way that it affects human life and social relations. Rather, as Mould argues, we need to ask how brutalism, with its commitment to "social housing, public space, and community interaction," (Mould, 2017: 716) can offer a politics that helps to combat neoliberal processes in our cities, such as the systematic attack on affordable housing, the privatization and securitization of public space, and the general dissipation of a shared sense of community and place.

Although the idea that brutalism can offer a powerful critique of these current urban trends is compelling, I do not share Mould's optimism that brutalism in the era of globalization can be effectively recovered from and activated against neoliberal urbanism. My view is that brutalism has already been too fully co-opted by the very system it challenges precisely because of *why* it has re-emerged as a source of fascination in the urban imaginary. The revival of brutalism, I argue, is connected to a broader concern with postindustrial decay and urban poverty. More specifically, it is rooted in anxiety over the rise of neoliberal globalization and the ways in which it is transforming urban space, reshaping the built environment, and exacerbating social-spatial divisions.

In this respect, the revival of brutalism is part of the same, larger phenomenon that has led to the recent explosion of interest in postindustrial ruins – a trend that is variously referred to as ruin mania, ruin lust, and even ruin porn (DeSilvey and Edensor, 2013). This last inflection, "ruin porn," captures what is at the heart of the trend – namely, the spectacularization of decay for voyeuristic pleasure and exploitation (Millington, 2013: 280). As this suggests, contemporary ruin lust is not so much tied to the lived, social experience of decay or deprivation, although the subculture of urban exploration sometimes extends in that direction (Garrett, 2013). Rather, contemporary ruin lust tends to revolve around the spectacular, visual apprehension of ruin as aesthetic encounter.

Ruin lust and the imaginary of gentrification

One of the most conspicuous examples of this aesthetic emphasis is the "Ruin Lust" exhibition held at Tate Britain in London in 2014 – a "brilliant but bonkers exhibition," as one reviewer summed up, "of artists' obsessions with broken stones, Nazi bunkers, and decaying castles" (Jones, 2014). Curated by the writer and critic Brian Dillon, the exhibition adopted a transhistorical approach, offering what the catalog describes as "a guide to the mournful, thrilling, comic, and perverse use of ruins in art from the seventeenth century to the present day" (Dillon, 2014). Important to note here is that, although the exhibition sought to place contemporary ruin lust in a longer visual history dating back to the late Renaissance, the timing of the exhibition appeared to be designed to tap into – and even intensify – the current ruin mania surrounding urban decline. What is more, by bringing the ruin into the exhibitionary space of the museum and linking the ruin to an art-historical tradition extending from Renaissance realism and Romantic landscape painting to modern architectural brutalism and beyond, Tate Britain was also constructing the ruin primarily as an aesthetic encounter.

The idea that the exhibition was concerned with both the fear of urban decline and the pleasure of aesthetic experience is made explicit in the museum's three-minute promotional film for "Ruin Lust," which combines images of artwork featured in the exhibition with a reading of excerpts from Rose Macaulay's classic 1953 text *Pleasure of Ruins*. The film begins with an image of urban blight and street art, and ends with a bibliographic list of artworks. In between, it shows those artworks projected as digital ghosts onto the interior walls of a decaying London house (Figure 21.1). Amidst the slow, lingering shots of the artwork, the film cuts to extreme close-ups of the rough surfaces and rotting

Figure 21.1 Stills from *Ruin Lust* promotional film.

materials of the ruined house; to exterior shots of abandoned junk and overgrown weeds; and to wide-angle interior shots showing the structure's condition of advanced decay.

At the most basic level, the film works hard to evoke the otherworldly aesthetic of disorder, sensuality, and enchantment that the geographer Tim Edensor (2005) sees as characterizing ruins. This aesthetic is doubled in the film by projecting images of ruin onto the surfaces of a ruin. In evoking the ruin's otherworldly aesthetic, the film also makes use of the closely related aesthetic of the uncanny – what the architectural historian Anthony Vidler (1994), adapting Freud, describes as the sense of unease produced when the familiar is made newly strange in spatial and aesthetic terms.

The film creates uncanniness in multiple ways. First, there is the defamiliarization of the artworks themselves, such as the spectral apparition of Turner's painting of Tintern Abbey in a glowing, translucent, immaterial form. Second, there is a distortion of the gallery. The clean, monumental space of the museum becomes the dirty, damaged space of a derelict home. And third, there is the unhomeliness of the home itself. Security and inhabitation give way to emptiness and decay. Adding another level of defamiliarization is the disembodied voice of the reader, which combines acoustically with the haunting music droning in the background to reinforce the film's spectral aura.

The issue here is that, like the "Ruin Lust" exhibition itself, the film presents the ruin encounter as an aesthetic experience and shows little interest in the forces that produce, sustain, or revive ruins. For instance, to read the film against itself, let us consider for a moment that it takes place in a derelict home somewhere in London. This has significance in the context of a city plagued by housing shortages and spatial inequalities, a city where the poor are being systematically evicted and displaced from the center, where entire streets are lined by unoccupied luxury ghost mansions, and where social housing is being privatized and resold at inflated prices as concrete chic (Campkin, 2013).

Could it be that the aestheticizing of urban blight in the form of the ruin contributes not only to the visual imaginary of gentrification (making urban blight aesthetically desirable), but also to the acceptance – or at least the normalizing – of urban inequality and deprivation? The answer is yes. My broader critique is that, in cases like the Tate Britain exhibition, the stasis, decay, and emptiness of the ruin are institutionally reframed to offer an experience of serenity, beauty, meditation, and introspection. In the process, the ruin – as an aesthetic construct of the urban imaginary – becomes hitched to the heritage industry and transformed into yet another art-commodity and event-spectacle.

This matters because a similar dynamic is at work in the way brutalist architecture is aesthetically recuperated by neoliberal urban processes. The aesthetic slippage between brutalism and ruin is already present in the gallery's promotional image, which joins the exhibition title "Ruin Lust" with a photograph by Louise and Jane Wilson of a Nazi war bunker in Northern France (Figure 21.2). There is a certain tension here between word and image in the sense that the source of lust – the object of aesthetic desire – is a proto-brutalist military installation with a dark, violent history. Yet, from a certain de-historicized, de-politicized perspective, the image emphasizes form and materiality, positing the bunker as an almost geologic protuberance or sculptural relic entrenched in the landscape – a view that is reinforced when looking at other photographs in the Wilson sisters' Normandy beach series, where the concrete ruins of the Second World War are reinterpreted photographically as sublime, quasi-natural objects.

As the Tate Britain exhibition illustrates, the aestheticization of ruin, poverty, and inequality in contemporary visual culture can have the effect of distracting from the ethical demands that brutalism and ruins place on us. This happens precisely because the focus is shifted away from the social history of specific sites and structures, and is redirected instead to the sphere of representation itself. As a result, aesthetic qualities are prioritized above any stories the spaces and buildings themselves might tell us about social and material conditions of failure, loss, invisibility, neglect, damage, violence, and so on. The larger implication is that this dynamic is not confined to brutalism or even ruins. Rather, it is part of a much broader trend involving the superficial re-valuing of blight as beauty in contemporary urban imaginaries. It is not an accident, for example, that the current wave of ruin lust and brutalist revival both coincide with a surge of fascination with global slums as sites of extreme poverty, overcrowding, architectural informality, and spatial exclusion.

Aesthetics of urban poverty

The aesthetics of urban poverty have a long history in the field of visual culture, and we can cite important antecedents such as Jacob Riis' photo-journalistic forays into the tenements of nineteenth-century New York in *How the Other Half Lives* (Riis, 2011 [1890]), or the more recent art-activism of Camilo José Vergara's *American Ruins* project (Vergara, 1999). What distinguishes the current rise of the slum as an object of cultural fascination is precisely the way that it transforms poverty into a source of aesthetic pleasure, or at least into some kind of consumer sensation. An extreme example of this occurred at the 2012 Venice Architecture Biennale, where the Zurich-based architectural design practice Urban Think Tank (U-TT) set up an immersive, interactive installation intended to share their extensive research on Torre David.

Internationally popularized by a 2011 *New York Magazine* spread by Dutch photographer Iwan Baan, as well as by the TV series *Homeland*, Torre David is Venezuela's most notorious skyscraper: an unfinished business tower in Caracas designed to be a shiny

Figure 21.2 Poster for the *Ruin Lust* exhibition at Tate Britain 4 March – 18 May 2014. Poster features *Azeville* (2006) by Louise and Jane Wilson, © Tate, London.

glass beacon of global capitalism. The Venezuelan banking crisis in the mid-1990s left the building incomplete, lacking electricity, elevators, running water, walls, windows, and most of its façade. The result is something I want to call "accidental brutalism": a building designed to be polished, shiny, and slick distinguished instead by rawness, exposure, and imperfection.

Sitting empty for over a decade in this semi-finished state in the center of a city with severe housing shortages, the building was eventually occupied by over 2500 squatters, who adapted the interior to their residential and community needs starting in 2007.

In *Rebel Cities*, Justin McGurk, who worked on the research project with Urban Think Tank, chronicles the occupation of Torre David and presents the vertical squat in emphatically positive terms as a "pirate utopia" and even a "paradigm of human ingenuity, adaptability, and resourcefulness" (McGurk, 2014: 203):

> For thirteen years Torre David stood empty . . . A picture of dereliction, the tower became a memorial to the failed hopes of the boom years . . . But on [one] rainy night in 2007 it began a process of existential transformation. Here is an emblem of speculative finance capitalism that has been taken over by those who were disenfranchised by the neoliberal policies of that era – the poor . . . Is there not a certain poetic justice in that? Could we not say that what would otherwise have been a symbol of exclusive luxury is now a symbol of the redistribution of wealth? Is this not a more equitable outcome?
>
> *(McGurk, 2014: 179)*

McGurk is right to point out the extreme irony of Torre David falling into the hands of the urban poor. What was intended to be an exclusive corporate glass tower for the managerial elite turned out as a concrete vertical squat – an unplanned experiment in affordable housing and improvised urbanism, whose very existence signals the vulnerability of the economic system the building was designed to promote. Yet, while there may be an element of spatial-social justice involved in this "existential transformation", the outcome is short-lived. In 2014, the government initiated forced evictions and the last of the tower's residents were ousted in 2015 to clear the way for the tower's commercial redevelopment, effectively aiming to restore the building to its original purpose. Eviction is a familiar and frequent outcome of urban squatting (Vasudevan, 2017) and Torre David's return to private control does not diminish the occupation's significance even if it does signal the difficulty of evading capitalism's recuperative power.

It is the unplanned, improvisational aspect of Torre David during its occupation that attracted Urban Think Tank and formed the focus of their installation at the Venice Biennale, as well as their short documentary about the tower that circulated separately on the independent film festival circuit (Figure 21.3). Capturing both the approach and content of U-TT's Venice installation, the *Torre David* film can be characterized largely as a celebration of the informal city, of improvised urbanism, and of adaptive re-use. Just as in McGurk's ethnographic account of the space in *Radical Cities*, the building emerges from the documentary as a site of empowerment for the urban poor – a co-created space where family life and community bonds can be built without recourse to the system of predatory capital that left the tower bankrupt in the first place.

In a way, the documentary presents this site of accidental brutalism as a positive variation of what Arjun Appadurai (2000), writing critically about slum living in Mumbai, has called "spectral housing": the informal, unregulated, and largely unseen world of insecure housing and precarious living that ranges from homeless sleeping in the street, to slum dwelling, to the black market in illegal rentals. What produces such forms of contemporary ghost living, Appadurai argues, is a combination of housing shortage, speculation, crowding, and public improvisation, all operating within a larger, transnational system of global capital. For Appadurai, spectral housing is a "social black hole" produced by "the effort to embrace and seduce global capital" (Appadurai, 2000: 627–8).

Figure 21.3 Stills from *Torre David* documentary.

By contrast, the film envisions Torre David as a potential alternative to the social black hole, operating beyond or outside global capital and somehow exempt from the uglier aspects of slum living and yet still distinctly spectral.

It is important not to diminish what the temporary residents of Torre David accomplished, and I also do not want to oversimplify the careful research and detailed analysis of the Urban Think Tank team. Yet, I find it necessary to counterbalance U-TT's exuberance for the informal city with the reality that the vertical squat does not alter the larger, more systematic issues of homelessness, poverty, unemployment, and crime present in Caracas. In other words, we need to be cautious about romanticizing urban informality as somehow more authentic, creative, adaptive, and autonomous without also addressing its basis in systemic urban inequality and exclusion.

In addition, the documentary's aesthetic treatment of the tower sensationalizes the building. The gangster-rap soundtrack, fast editing, and frequent hyper-masculine imagery, such as the muscular male body pumping iron on the roof, all contribute to a sense of the tower as a place of energy and defiance, reminiscent in many ways of the recent wave of French banlieue/parkour action films, such as the *District 13* series (2004–2009), which respond in turn to another brutalist architectural setting and similarly emphasize the creative re-appropriation of broken space (Kidder, 2017). Related to this is the visual emphasis throughout the film on Torre David's "as-found" aesthetic, something that the architectural historian Adrian Forty, writing about concrete as a transcultural medium, strongly associates with the brutalist fashion for roughly finished concrete (Forty, 2012: 234). The difference is that Torre David's rough finish is not a design feature but instead the result of interruption. And in this sense, we can think of the tower's accidental brutalism as a new kind of brutalism, specific to the era of neoliberal globalization and produced by the evacuation of speculative capital. It is a brutalism that speaks to the fragility of neoliberal conditions of growth and profit, and that signals the perpetual threat of financial crisis and disaster.

Torre David featured in a similar way at the 2012 Venice Biennale. Titled "Gran Horizonte" (Great Horizon), Urban Think Tank's exhibition comprised photographs, film fragments, and informational text exploring how the residents of Torre David re-appropriated the corporate architecture of globalization to entirely new, unplanned functions geared toward human dwelling. In the process, the exhibition presented urban informality and accidental brutalism as a potential design solution for so-called "mainstream" urban development worldwide. Such a proposition needs to be debated further.

For instance, it is probably fair to say that many Western cities have for a long time been over-planned, over-determined, and over-controlled; and that this trend is becoming amplified as we slide into the era of smart cities, which leave less and less room for improvization and play as the emphasis shifts toward capturing and controlling life as data. It is also fair to say that there is too often a resistance in the West to learn from what Ananya Roy has called the "subaltern urbanism" of the Global South, through which the slum can be differently understood as "a terrain of habituation, livelihood, self-organization and politics" capable of providing "a vital and even radical challenge to dominant narratives of the megacity" (Roy, 2011: 223).

At the same time, however, it is important to avoid fetishizing urban informality and reducing it to a superficial, sensational, hype-driven supplement to the neoliberal urban imaginary. Consider, for example, the centerpiece of Urban Think Tank's exhibit in Venice: a restaurant that translated Torre David's ethos of informality into a catered, Latin American gastronomical experience reflecting community life in the tower. Food, of course, is as political as art and architecture, and we have recently seen the rise of food justice movements and critical food studies that help us better understand the uneven geographies and cultural politics of food. It is therefore important not to discount the radical potential of food to critique, subvert, or resist. Nonetheless, the Torre David restaurant in Venice strikes me as less of a "traditional place to eat and create community" (Ferrarini, 2012), which is how it was billed to the public, and more of a space that has succumbed to the Starbucks effect: the phenomenon of consumer delirium in contemporary urban living that, as Sharon Zukin (2011) argues in her work on gentrification, is eroding the authenticity of place and creating more and more spaces that possess the appearance but not the substance of neighborhood and community.

Urban Think Tank's restaurant may not have the corporate overtones of a generic Starbucks café, and it may not be a harbinger of gentrification like the much-publicized Cereal Killer Café in London's Shoreditch, which became a target of violent anti-gentrification protests in 2015 because of its hipsterfication of breakfast cereal (Khomani and Halliday, 2015). Yet the Torre David-themed restaurant does do something similar to both of these examples in the sense that it thematizes food (in this case the theme is urban squatting) and sells that theme as a sensual, spectacular experience. The result is a kind of fake authenticity, what I would also describe as an atmospheric simulation.

Fake authenticity is an issue that the art historian Richard Williams (2016) addresses in an essay on the phenomenon of food trucks and their aesthetics of immediacy. In his thinking, food trucks are the product of urban inequalities and traditionally cluster around spaces left behind by capital, which helps to explain why they exploded in number and visibility after the financial crash of 2008, particularly in the US context. Food trucks, according to Williams, historically represent interruptions to the dominant flows of the global city. Yet, the recent mainstreaming of food trucks, which have become ubiquitous in many Western cities, speaks not only to their recuperation by capital but also to the way their presence has been fetishized as an expression of the authenticity and edginess of the street.

Williams' point is that "for many of us in the most capital-intensive parts of the world, interruption *is* the desired aesthetic" (Williams, 2016: 28), and that "irrespective of the morality, we cultivate inequalities. We want difference, contrasts, crunchiness, and the cognitive dissonance so often sought, however perversely, by the privileged" (Williams, 2016: 28). These are large claims, but ones that map onto the argument running through this chapter. They are also at the heart of Ben Campkin's (2013) reading of the neoliberal remaking of cities, in which he identifies how the aesthetics of inequality are increasingly being sold back to us at a premium via architecture, fashion, lifestyle, and design as various forms of *chic*: homeless chic, inner-city chic, ghetto chic, gangster chic, favela chic, and, bringing this back to brutalism, concrete chic.

This exploitation of inequality is what links the different spaces and practices I have been discussing. What brutalism, ruins, and slums all have in common today is that they have been appropriated by the urban imaginaries of gentrification and neoliberal renewal with the express aim of financially exploiting the social and spatial challenges they contain. This is accomplished by transforming them into something I would describe as a "spectacle of the abject": a combination of sensation, indulgence, ambiance, delirium, and distraction in which human suffering becomes decorative, nostalgic, spectral.

Airbnb and brutalism

In conclusion, I end with one last example that captures further aspects of brutalism's emerging ties to contemporary ghost living. Since 2010, online property rental platforms such as Airbnb have had transformative effects on urban communities around the world by converting homes into short-stay residences aimed at tourists and business travelers (Gravari-Barbas and Guinand, 2017). In Paris alone, which is Airbnb's biggest market at the time of writing, there are over 65,000 listings on the site. One consequence is that Airbnb has directly contributed to a population decrease in central Paris of around 30,000 people, as primary residences have been converted into temporary rentals – a trend that is particularly visible, for example, in the super-gentrified and hyper-touristified neighborhood of Le Marais (Gravari-Barbas, 2017). The evacuation of long-term residents from city centers is a growing problem as cities such as Paris – but also Berlin, London, New York, San Francisco, and countless others – struggle with affordable housing in the face of rampant gentrification. As the world's super-rich increasingly buy up property in global cities to convert into unoccupied repositories of capital, the parallel acceleration of the Airbnb phenomenon contributes to a broader trend toward luxury ghost living: the replacement of residents with temporary, transient occupants.

Against this backdrop, it is significant that brutalist architecture is increasingly being mobilized on websites such as Airbnb to sell this new, luxury form of urban transience. Consider, for instance, the following two property listings posted in 2016. The first listing comes from Paris and invites visitors to

> spend a few nights in an iconic brutalist building . . . We offer you to stay in our lovely loft . . . 15 mins away from the center of Paris. We are two architects and one photographer and live in an iconic brutalist building built in the 1970s by Jean Renaudie.

The second listing comes from Manila and promotes the apartment as a place of luxury and refinement:

> This 1-bedroom flat is exquisitely wrapped in modern brutalist sophistication. The bold combination of clean lines, raw concrete wall, cosmopolitan décor and ambient light gives off that classy hotel feel.

Figure 21.4 Airbnb's generic, brutalist interiors in Paris (left) and Manila (right).

As both property descriptions reveal, the listings work hard to cash in on their brutalist connections. The Paris listing does this via an architectural heritage argument, while the Manila listing does it via a concrete chic argument. The irony is that, despite referencing the distinctiveness of brutalism, the results in both cases are highly generic and entirely de-personalized interiors about as interchangeable and anonymous as a hotel room (Figure 21.4), an idea that is explicitly referenced as a selling point in the Manila listing.

The transnational homogeneity of these Airbnb brutalist interiors points to the same tension at the heart of contemporary cities evident in each of the examples discussed in this chapter. Yes, brutalist politics can offer a critique of the neoliberal city and its addiction to gentrification, stylized boutique living, and glossy, generic architecture. The twist is that cities are finding more and more ways, as illustrated by Airbnb, to absorb those politics back into the neoliberal project of converting space into profit and critique into complicity.

Acknowledgements

An early version of this essay was presented as a public lecture at the London Science Museum in 2016 during an event on brutalist architecture organized by the Centre for GeoHumanities at Royal Holloway University of London. I would like to thank Oli Mould for organizing the event, as well as the audience for insightful questions and comments.

References

Appadurai, A. (2000) Spectral Housing and Urban Cleansing: Notes on Millennial Mumbai. *Public Culture* 12(2): 627–51.

Banham, R. (2011 [1955]) The New Brutalism. *October* 135: 19–28.

Campkin, B. (2013) *Remaking London: Decline and Regeneration in Urban Culture*. London: I.B. Tauris.

DeSilvey, C. and Edensor, T. (2013) Reckoning with Ruins. *Progress in Human Geography* 37(4): 465–85.

Dillon, B. (2014) *Ruin Lust*. London: Tate Publishing.

Edensor, T. (2005) *Industrial Ruins: Space, Aesthetics and Materiality*. London: Bloomsbury.

Ferrarini, P. (2012) Torre David/Gran Horizonte. *Cool Hunting*. September 11. Available at: http://www.coolhunting.com/design/torre-david-gran-horizonte (Accessed 12 December 2017).
Forty, A. (2012) *Concrete and Culture: A Material History*. London: Reaktion.
Garrett, B. (2013) *Explore Everything: Place Hacking the City*. London: Verso.
Gravari-Barbas, M. (2017) Super-Gentrification and Hyper-Tourismification in Le Marais, Paris. In: Gravari-Barbas, M. and Guinand, S. (eds.), *Tourism and Gentrification in Contemporary Metropolises*. London: Routledge, 299–328.
Gravari-Barbas, M., and Guinand, S. (eds.) (2017) *Tourism and Gentrification in Contemporary Metropolises*. London: Routledge.
Jones, J. (2014) Ruin Lust at Tate Britain Review: A Brilliant but Bonkers Exhibition. *The Guardian*. March 3. Available at: https://www.theguardian.com/artanddesign/2014/mar/03/ruin-lust-tate-britain-review-art-building (Accessed 14 September 2017).
Khomani, N. and Halliday, J. (2015) Shoreditch Cereal Killer Café Targeted in Anti-Gentrification Protests. *The Guardian*. September 27. Available at: https://www.theguardian.com/uk-news/2015/sep/27/shoreditch-cereal-cafe-targeted-by-anti-gentrification-protesters. Accessed 14 November 2017.
Kidder, J.L. (2017) *Parkour and the City: Risk, Masculinity, and Meaning in a Postmodern Sport*. New Brunswick: Rutgers University Press.
McGurk, J. (2014) *Radical Cities: Across Latin America in Search of a New Architecture*. London: Verso.
Millington, N. (2013) Post-Industrial Imaginaries: Nature, Representation and Ruin in Detroit, Michigan. *International Journal of Urban and Regional Research* 37(1): 279–96.
Mould, O. (2017) Brutalism Redux: Relational Monumentality and the Urban Politics of Brutalist Architecture. *Antipode* 49(3): 701–20.
Riis, J. (2011 [1890]) *How the Other Half Lives: Studies among the Tenements of New York*. Boston: Bedford/St. Martin's.
Roy, A. (2011) Slumdog Cities: Rethinking Subaltern Urbanism. *International Journal of Urban and Regional Research* 35(2): 223–38.
Saval, N. (2016) Brutalism Is Back. *The New York Times*. October 6. Available at: https://www.nytimes.com/2016/10/06/t-magazine/design/brutalist-architecture-revival.html?mcubz=2 (Accessed 20 July 2017).
Vasudevan, A. (2017) *The Autonomous City: A History of Urban Squatting*. London: Verso.
Vergara, C.J. (1999) *American Ruins*. New York: The Monacelli Press.
Vidler, A. (1994) *The Architectural Uncanny: Essays on the Modern Unhomely*. Cambridge, MA: The MIT Press.
Williams, R. (2016) Why We Love "Interruption": Urban Ruins, Food Trucks, and the Cult of Decay. In: Lindner, C. and Jordan, S. (eds.), *Cities Interrupted: Visual Culture and Urban Space*. London: Bloomsbury, 17–29.
Zukin, S. (2011) *Naked City: The Death and Life of Authentic Urban Places*. New York: Oxford University Press.

22

The end of the time of the city?

Urbanization and the migrant in British cinema

Gareth Millington

Introduction

In *Metaphilosphy,* Henri Lefebvre (2016: 111) predicts that soon the city will be a social form as remote and as disconnected from contemporary time and space as the village. But this demise is also an opportunity for revolution. The end of the city, Lefebvre argues, finally "gives us notice to create new works [*oeuvres*]" (ibid: 112). This chapter considers these provocative statements, taking as its starting premise that a series of independent British films made during the late 1990s and 2000s offers a rare, yet sustained examination of urbanization and migration in London and sites "outside" the city, such as suburbs, small cities, and towns – not confined to the UK – which are situated within London's vast sway of influence. These films connect with debates regarding the transformation of both the city center and places existing outside the center that occupy *subordinate* positions within London's "power geometries" (Massey, 2005). This chapter is part of a broader project to examine the neglected cultural aspects – not least the emergent imaginaries – of so-called "planetary" urbanization. Such imaginaries are supportive of James Donald's (1999: 92) argument that the city is a *historically specific mode of seeing*; a visual mode that has run "out of time" (see also Millington, 2016a, 2016b, 2016c). The five films examined here – *Beautiful People* (Jasmin Dizdar, 1999), *In This World* (Michael Winterbottom, 2002), *Dirty Pretty Things* (Stephen Frears, 2002), *Ghosts* (Nick Broomfield, 2006) and *Somers Town* (Shane Meadows, 2008), plus two others, *Last Resort* (Pawel Pawlikowski, 2002) and *It's a Free World* (Ken Loach, 2007) – belong within a "film cycle"; that is, "a historically circumscribed group of films sharing common industrial practices, stylistic features, narrative consistencies, and spatial representations" (Dimendberg, 2004: 11). While frequently recognized as individual films that explore the experiences of migrants (Brunsdon, 2007; Loshitzky, 2010; Mazierska and Rascaroli, 2003), this cycle has not been examined in terms of it how it relates to (and enriches) critical urban theory and/or the historical sociology of the city (with regards to developments such as planetary urbanization, the dissolution of the city, displacement, dispersal, and the right to the city). This cycle has not yet been analyzed as an "art of exposure" (Sennett, 1990) that redeems the material and human realities of urbanizing space/s that have hitherto lacked an image.

The argument presented in this chapter is that this cycle of independent British films lends aesthetic support to Lefebvre's (2003: 57) assertion in *The Urban Revolution* that the reign of the city is ending and that the city now exists mainly as a historical entity. This does not mean that images or imaginaries of the city do not persist, they do; nor does it mean city governed institutions and infrastructures have ceased to organize much of our lives (see Hall and Savage, 2016). But, it does mean that it is becoming increasingly difficult to locate the city as distinct to *a non-urban elsewhere*. In the cycle of films examined here, we see the city reduced to a specter, a haunting, or a "non-present present" (Derrida, 1994: 5). As Henry Shapiro puts it, in *The Time of the City*,

> [t]he power of the cinematic art, when it is attuned to the thinking that is released by direct images of time, inheres in its ability to show that there is a multiplicity of ways of having the past in the present, as well as showing that any present is radically contingent.
>
> *(Shapiro, 2010: 33)*

Indeed, the urban imaginaries forged in this cycle are disruptive; they embody "a combination of temporalities [and spatialities]" (ibid: 34). They confound inherited urban sensibilities, supporting Robins' view that, "[t]he city is no longer imageable. It is becoming lost from view" (1996: 132). The historical urban content of this cycle therefore jars with Thrift and Amin's (2017) emphasis on the "liveliness" and "vitality" of the various urban assemblages and infrastructures in *Seeing Like a City* (ibid: 3). In contrast, the cycle shows the mixing, connections, and possibilities of the city to be more elusive than is often assumed in dominant urban imaginaries. More troubling, these imagined qualities of the (disappearing) city torment migrants, reminding them of what is being denied by a generalized process of neo-Haussmannization that is remaking centers and peripheralizing the poor, creating "shadow citizens" everywhere: a reserve army of asset-less people who "*feel the periphery inside them*", people who have even come to defiantly "*identify* with the periphery" (Merrifield, 2014, emphasis added).

Aesthetics is dealt with here in a quite specific manner (for more detailed discussions see Adorno et al., 2007; Hinderliter et al., 2009; Lunn, 1985). Aesthetics is understood neither as art theory nor as the discipline that takes art as its object of study, but rather as:

> a delimitation of spaces and times, of the visible and the invisible, of speech and noise, that simultaneously determines the place and the stakes of politics as a form of experience. Politics revolves around what is seen and what can be said about it, around who has the ability to see and the talent to speak, around the properties of spaces and the possibilities of time.
>
> *(Rancière, 2004: 9)*

Rancière argues the aesthetic regime creates a terrain where art is no longer held at a "representational" distance from social life. Art is no longer tied to the "sensible" as it is presented to us. In other words, art does not have to "look" *like* "things" in order to be "about" them; modern art forms are free to take a more abstract or imaginative approach. For Rancière, art is an autonomous practice that is invigorated by being brought into contact with heterogeneity, with the social world, where it can take on meaning beyond the intentions of the artists. In addition, the social world itself is reformed under the influence of new aesthetic values or "regimes of expression" created in art. Like other forms of art, cinema cannot perform transformative political action, but it can reframe experience by making certain aspects of reality more visible, or visible *in a different way* to how they are presented within the "distribution of the sensible". Cinema then, can generate ways "to think the political" (Shapiro, 2010). Art is also imbricated in what Rancière calls "subjectification", referring to the process of becoming political subjects through initiatives or events in which "a series of actions by a body and a capacity for enunciation not previously identifiable within a given field of experience [results in a] reconfiguration of experience" (1999: 36). As such, the cycle of films discussed here divulges not only the changing physical and human landscape of urbanization but – in creating encounters between spectators and otherwise isolated subjective experiences (Isaacs, 2013: 12) – it helps create and communicate a new urban imaginary; that is, an urban (political) aesthetic that amounts to a reconfiguration of urban space and time *without the city*. To put it differently, in circulating images of scarcely perceived urban forms to spectators who may themselves be experiencing dislocation or displacement, cinema helps create new urban political subjectivities that contribute to – or, in Highmore's (2005) terms, give "force" to – representational determinants in the production of space (see Lefebvre, 1991). Cinema, in this way, is "both a symptom and a catalyst of [socio-] spatial transformations" (Dimendberg, 2004: 12).

This chapter responds to three questions: (1) How does this cycle of films "make sense" of contemporary, expansive urbanization and migration; (2) How does this cycle of films create an image and an imaginary for contemporary, expansive urbanization that is distinct from that of the image of the city? (3) How might this imaginary be implicated in the production of space and within processes of urbanization? The first section examines planetary urbanization and considers why this emergent terrain is haunted by the specter of the (modern) city. The neglected yet active role of migrants in processes of urbanization are also considered here. The second section of the chapter analyzes a short cycle of British films (1999–2007) that engages with both migration and urbanization. This discussion is divided between "images of the city" and "images of urbanization".

Urbanization, the specter of the city and migration

Lefebvre develops the notion of "implosion-explosion" to describe how, as cities achieve greater concentrations of property, speculation, and (post)-industrial activity, the traditional urban center implodes, acting as a spur to the expansion, or "explosion", of urban society, causing "the projection of numerous, disjunct fragments (peripheries, suburbs, vacation homes, satellite towns) into space" (2014: 14). The logic of centrality means that centers are inevitably also points of exclusion, "the place of sacrifice . . . where accumulated energies, desirous of discharge, must eventually explode" (Lefebvre, 1991: 332). "Full" centers exude an aura of finitude, promulgating a series of repressions that limit the potential of urban society. As the city undergoes dissolution, it becomes home instead to stipulative, authoritarian, and repressive signs and codes (Lefebvre, 2003: 14). One of the many problems created by explosive and expansive urbanization becomes how to build, or even imagine, the "'something' that replaces what was formerly the city?" (ibid: 15). How, without the city, can urban society be (re)born? Lefebvre (ibid: 169) argues that urban society cannot be constructed upon the ruins of the city. Rather it is upon the "shaky foundation" provided by expansive urbanization that the urban must persevere (Lefebvre, 1996: 129).

This chapter addresses neglected cultural dimensions of the dissolution of the city. The cover of Neil Brenner's (2014) now seminal edited collection on planetary urbanization, *Implosions/Explosions,* features an image of Tar Sands in Alberta, Canada. Inside the book there are similar images that dramatically capture the planetary spread of urbanization. Yet these images, while indisputably evocative of Lefebvre's thesis, lack social or cultural content or resonance. There is little sense of the travails (or unexpected joys) of contemporary urbanization, or any expressive insight on the rights that inhabitants, migrants, or workers visiting such places may or may not enjoy. Bender, 2002: 221), for example, worries how the paucity of images of contemporary urbanization translates to growing uncertainty around notions and expectations of urban citizenship (see also Purcell, 2008, 2013). Lefebvre (2003: 14) even has a name for this interregnum, which he calls "the critical zone". What is occurring, and which happens behind our backs so-to-speak, is that industrialization itself has become the dominated reality; dominated by its own product: the urban. This results "in tremendous confusion during which the past and the possible, the best and the worst, become intertwined" (ibid: 16). The contemporary global city, with its polarized wealth structure, declining public investment and its many exclusions and privations, disappoints all of us. The city has lost its shape and vitality; it has become "an historical entity" (ibid: 57), an archive of a centripetal form that has now been superseded. This explains, in part, why we have become gripped with nostalgia for the image of the city, indulging blindly in "cultural cityism" (Millington, 2016b).

Wachsmuth argues that, in the absence of the historical reality of the city, intensified attraction to the concept of the city should be considered a fetish: "the city-as-a-representation is not neutral or innocent, but rather is ideological, in the sense that its partiality helps obscure and reproduce relations of power" (2014: 76). In Lefebvre's (2003: 57) words, the city has become a "pseudoconcept". Yet the decline of the city is by no means a simple matter. The core, the center – the very image of "cityness" – is maintained through monuments, museums, commerce, bureaucracy and culture (and cinema). As Lefebvre puts it, "the urban core (an essential part of the image and the concept of the city) splits open and *yet maintains itself*" (1996: 74, emphasis added). As the historical form of the city is supplanted by generalized urbanization, "the city" becomes a phantom, or shadow, of urban reality (Lefebvre, 2003: 35).

Despite an emphasis on marginal and peripheral spaces in the film cycle examined here, the specter of "the city" always feels close. As the city gives way to planetary urbanization, "the city" maintains its distinctive presence through image rather than actuality. The understanding of the specter developed by Derrida (1994) and Jameson (1999/2008) is useful here. For Derrida (1994: 6) the specter is the body of *someone* as *someone other.* He explains, "we feel ourselves being looked at by it, outside of any synchrony, even before and beyond any look on our part" (ibid). The specter disturbs historical time. It bears over our actions and thoughts:

> Spectrality does not involve the conviction that ghosts exist or that the past (and the future they offer to prophesy) is still very much alive and at work, within the living present: all it says, if it can be thought to speak, is that the living present is scarcely as self-sufficient as it claims to be; that we would do well not to count on its density and solidity, which might under exceptional circumstances betray us.
>
> *(Jameson, 2008: 39)*

The specter of the city implies that contemporary urbanizing forms – the fragmented and dislocated spaces that migrants may inhabit, for instance – always owe something to "the city". They cannot be conceived or experienced without recourse to the city. Moreover, the specter of the city is not benign or an object of nostalgia; it is a reminder that the "loss" of the city "is a modern [form] of dispossession" (Gordon, 2008: xv); an example of a concealed or "unresolved social violence ... making itself known" (ibid: xvi) and having a concrete effect on shared conditions of urban living.

Another neglected aspect in the literature on planetary urbanization is the active role played in the process by migration and migrants. Indeed, urbanization is often presented as a technocratic, architectural, or infrastructural process. Çağlar and Glick Schiller (2015) point to how migration studies similarly tends to place migrants *outside* of urban processes, often because they are narrowly viewed as being excluded from making a contribution. Yet, as David Harvey (1996) reminds us, human mobility is a primary driver of urbanization. In a global capitalist economy, urbanization is unevenly developed; the urban fabric is continuously made and remade by waves of capital investment and disinvestment. Urbanization can, then, be understood in terms of the historical migrations of labor *in response to* capital, from one region to another (ibid: 46). For Çağlar and Glick Schiller (2015), the three concepts of *dispossession, displacement* and *emplacement* provide the analytical tools with which to situate the relationships between migrants and cities/urban localities of varying power (ibid: 5). These concepts are entangled with – no less products of – the restructuring and positioning of localities and the accumulation of capital. In breaking away from any migrant/native binary, they stress it is important to acknowledge how all of us "are subject to the forces of dispossession and displacement and it is by being part of these processes that people in various localities search for ways to construct sociabilities of emplacement" (ibid). This latter term – *emplacement* – is instructive, referring to "the relationship between the continuing restructuring of place within multiscalar networks of power and a people's efforts, within the barriers and opportunities of a specific locality, to settle and build networks of connection" (ibid: 5–6). Çağlar and Glick Schiller argue that urban restructuring and migrant displacement and emplacement are part of a *single* globe spanning process.

In *The Figure of the Migrant*, Nail (2015) argues that expulsion is a centrifugal movement that involves the deprivation of social status. But expulsion, he suggests, is also a form of expansion, a "process of opening up that allows something to pass through"; moreover, it signals "both an intensive and extensive increase in the conjunction of new flows and a broadening of social circulation" (ibid: 36). Nail is careful to explain that expulsion is – or at least can be – negated by the capacity of the migrant to actively create an alternative logic. For example, migrants have their own form of social motion in "riots, revolts, rebellions, and resistances" (ibid: 7). Displacement is not simply "a lack" but also a "positive capacity or trajectory" (ibid: 12). Migrants are agents of urbanization in addition to being its "victims" or "subjects". For Nail, the figure of the migrant prefigures an emerging model of citizenship and subjectivity: "there are empirical migrants, but their meaning and potential extend beyond their empirical features under the current conditions of social expulsion" (ibid: 17). As Mezzadra (2011: 136) puts it, migrants *act* as citizens, independent of their legal status of citizenship. Such forms of citizenship may be understood as traversal in the sense that the migrant "recognises (or institutes) the right to act across or against frontiers" (Isin, 2012: 149).

Images of the city

The closest depiction to an image of "inner city" London – a cinematic trope that has roots in films such as *Pressure* (Horace Ové, 1976), *Babylon* (Franco Rossi, 1981) and *Young Soul Rebels* (Isaac Julien, 1991) – is found in the earliest film in the cycle, *Beautiful People* (1999). The film is set in 1993, and migrants from the Bosnian conflict are shown living in a pre-WWII central London housing estate. This type of estate is evoked as an authentic, primal site of multicultural London (see Figure 22.1). Self-consciously creating an image of continuity, London's most recent migrants mix amiably and share the estate with representatives of the city's previous waves of migration. Here, the time of the city is presented as teleological.

Despite Mazierska and Rascaroli's (2003) assertion that the film depicts a fragmented and chaotic London, the estate scenes in *Beautiful People* can be seen, with hindsight, as presenting an urban oasis. Residents tend to potted flowers and hang washing in the walkways of the estate. People leave their front doors open. Music and the smell of marijuana wafts across the estate. In contrast to the bitter divisions and struggle depicted in the multicultural London films of the 1970s and 1980s – such as *Pressure* and *Babylon* – this is an urban pastoral. London is viewed here as *heroic,* even. The city has overcome Victorian squalor, partial ruination during WWII and the "racial" and postcolonial problems of the post-war decades. This is a city finally at ease with itself; a playful, ironic London. This optimistic, post-conflictual imaginary is much more conspicuously present in contemporaneous films about London such as *Notting Hill* (Roger Michell, 1999) and *Lock, Stock and Two Smoking Barrels* (Guy Ritchie, 1998).

It is significant, when viewed against more pessimistic films later in the cycle, that London is depicted as multicultural and inclusive; a pre-gentrification city that it is still possible to access, be part of, and share a stake in. Early in the film there is a sequence

Figure 22.1 London housing estate, *Beautiful People* (Jasmin Dizdar, 1999).

where Pero (Edin Dzandzanovic) exuberantly strolls around central London in disbelief that he is actually *here*, in this great metropolitan center. He wears sunglasses, takes passport photos, and laughs when he glimpses himself on a TV screen in a shop window; when he sees, for the first time, an image of himself in the city (with a red London bus and a Marks and Spencer in the background to add a hallmark authenticity). This scene is reminiscent of those famous passages of "modernist epiphany" in Sam Selvon's novel *The Lonely Londoners* when Sir Galahad – a recent West Indian migrant – recounts his visits to Oxford Street, Charing Cross, and Piccadilly Circus. Simply *being* in the heart of London – "that place that everybody in the world know about" (Selvon, 2006: 72) – had an enormous effect on his self-esteem; he recognized himself as a "new man", as someone who was now truly "caught up in the mix" of modern life (Berman, 1999).

Beautiful People captures the moment when multicultural London cedes to "super-diverse" London, in relation to the increased diversity of origin and ethnicity of migrants and the spaces of the city-region where they settle (see Vertovec, 2007). In providing a dialectical image of the moment of tensions between historical periods of migration and urban change, this is where the historical value of the film lies. *Beautiful People* heralds a new era or age of migration, but because of the uncertainties of the transitional period in which the film is set, it cannot help but situate its narrative within an inherited space, within a soon-to-be outdated representational trope. The historical irony, which would scarcely be believable in the 1970s and early 1980s when such spaces were heavily maligned, is that the humble inner London housing estate appears in this cycle of films as a specter of (forsaken) possibility; a symbol of "*if only*..." (on London's current social housing crisis see Watt and Minton, 2016).

This image of the multicultural city soon fractures. Just seven years later, *Ghosts* (2006) depicts a city in dissolution. The movie is a dramatization of the 2004 Morecambe Bay cockling disaster where least twenty-one undocumented Chinese migrant laborers drowned while working on the Lancashire coast, 260 miles from London. In fact, the closest Ai Qin (Ai Qin Lin) and her compatriots get to London is a piece of waste ground (see Figure 22.2). A caption on the screen informs the viewer that this is "London". There are no landmarks; just scrubland, a bricked-up industrial building, some rubble, nondescript trees, a few orange bollards and a couple of empty gasholders. In terms of an image of the city, this debris is all there is. This space marks the culmination of an arduous six-month journey for Ai Qin, who has paid smugglers to bring her from the Fujian province to work in London so she can send home money to raise her young son. Her hopes of a hospitable welcome are dashed. Ai Qin's arrival in "the city" is followed simply by the terse instruction to release a final payment to the smugglers.

In this anonymous space, Ai Qin finds herself hopelessly dislocated from the time of the city (the place she expects herself to be *at this moment*). As Avery Gordon (2008: xvi) explains, the specter – in this case, of London – alters the experience of being in time, the way that we separate the past, present and future. In fact, the whole scenario feels like a cruel affront to these Chinese migrants. Positing this space, cinematically, as the "entry point" to the city is a profound comment about what and where constitutes the city for many of today's migrants. London is a specter, a "non-present present" (Derrida, 1994: 5). Ai Qin and her companions see no more of the city than this. They circle London on the M25 in a beat-up van before driving east to Thetford in Norfolk, where their gang master forces them into exploitative work in industrial farming.

Figure 22.2 London, *Ghosts* (Nick Broomfield, 2006).

Michael Winterbottom's *In This World* is an unusual film in the cycle because it offers countless images of the city and many vivid portrayals of city life. Ostensibly this would suggest that, *contra* Lefebvre, the city is not undergoing dissolution at all. Yet something fascinating is occurring here. These images are of cities that Jamal (Jamal Udin Torabi) and Enayat (Enayatullah) *pass through* on their journey to London. Interestingly, these cities – Peshawar, Tehran, Istanbul, and Trieste – are shown to possess the urban qualities of simultaneity, gathering, convergence, and encounter (Lefebvre, 1996: 131) that contemporary "global" London is shown to be losing elsewhere in the cycle. Jamal and Enayat rest briefly in numerous cities only to uproot and continue on their way to London. Tehran in particular is shown to offer an experience rich in metropolitan pleasures, and Jamal takes a break there from the physical and emotional hardships of the journey and treats himself to an enormous ice cream cone. It is the most hopeful scene in the whole film. It seems to the viewer that the two young men could settle here and remain safe.

In the succession of cities that Jamal and Enayat pass through on their way to London, one is reminded of Derrida's notion of *différance,* meaning both "to differ" and "to defer", to delay or postpone. Each city they visit suggests only the next city to be reached in order to get to London, the "presence", or that establishes the endpoint to their teleological exercise. As Wolfreys puts it, the notion of *différance* explains how "every 'beginning' [every city on the journey] is marked, traced and haunted by that which stands before it [London, the final destination]" (2007: 45). Each city (Tehran, Istanbul etc.), "stands in for a deferred presence, one not immediately present" (ibid: 50). The search for "the city" is ongoing, exhausting and, as we eventually learn, not everybody makes it. Jamal, though, does get to London. In the final scenes of the film we see him working in a café and walking through a crowded street market. As Sargeant says

> whether his life is any better as a waiter in London than in a camp in Pakistan remains to be established. The film presents the pain of leaving a familiar community, culture and landscape (all rendered photogenically attractive) rather than the joy of arrival at a distant destination.
>
> *(Sargeant, 2005: 348)*

Disappointment with London is a common theme of films in the cycle. The notion of *différance* is also relevant in considering how Senay (Audrey Tatou) in *Dirty Pretty Things* becomes obsessed with leaving behind her fugitive life in London and moving to New York, a city "where they put lights in the trees", where "you can skate in the parks". London, once her preferred destination, becomes yet another stopover, another signifier of "the city" "whose *signified* we are presently looking for" (Lefebvre, 2003: 131, original emphasis). Similarly, in Shane Meadows' *Somers Town*, the teenage runaway Tomo (Thomas Turgoose) and his Polish friend Marek (Piotr Jagiello) reside in the very heart of the city, close to Kings Cross, but their dream is to take the Eurostar to Paris. Not content with London, these young nomads keep moving in search of "the city". When Tomo and Marek finally make it to Paris at the end of the film and are reunited with Maria (a French waitress they befriend in London, played by Elisa Lasowski), Meadows switches from the monochrome he uses to depict London to a grainy, "vintage" color palette (see Figure 22.3). The mood becomes dreamlike and the images are blissful; in fact, we are not certain whether we are witnessing "reality" or just another city specter.

In these examples, what now counts as "the city" – from these migrant's perspective – is perpetually deferred (in both spatial and temporal senses). The presence, meaning, and refuge of the city proves to be elusive. The bearing of *différance* in understanding the spectral qualities of the city is summed up in a subtitle for a scene in *In This World* where

Figure 22.3 Tomo, Maria and Marek in Paris, *Somers Town* (Shane Meadows, 2008).

Jamal and Enayat are trudging across a snowy Iranian mountain at night, heading for the border with Turkey and onto their next city stop in Istanbul. You can barely make them out in the blizzard, but Jamal can be heard speaking to his companion (in Pashto): "I hope we get there soon".

Images of urbanization

Cinema is concomitant with urban modernity and is an art form that continues to engage with changing urban forms and experience. Two examples of a tentative new aesthetic of planetary urbanization are discussed in some detail below. Both are examples of the poetic tendency in cinema to "thwart" stories, scripts and chronological arrangements of events (Rancière, 2006). These images also echo Benjamin's (1999: 229) insistence that cinema reveals "an immense and unexpected field of action" while supporting Kracauer's (1960: 303) point that it is the "excess" qualities of the cinematic image that often has the most transformative impact on the viewer. In other words, beyond the plot that images are embedded within, camera reality has the capacity to *move* the viewer, to make the viewer think of *other things* (see also Hansen, 2012). Kracauer's theory reveals how cinema can offer a vivid aestheticization or enchantment of the (unseen) everyday. As Gilloch explains, "[t]his should be understood not as some reactionary de-politicization or spurious mystification but rather as a critical recuperation, rejuvenation and replenishment of human appreciation and sensitivity" (2015: 198).

Football features regularly in *In This World* (see Figure 22.4). Jamal and Enayat join in "pick up" games at many points along their journey: in the camp at Peshawar, in a Kurdish village in Iran, in Istanbul and later, toward the end of the film, Jamal plays with new friends on the beach near Sangatte. Football is a way of passing time but also a way

Figure 22.4 A Football Match, *In This World* (Michael Winterbottom, 2002).

of coming together, a social mechanism to overcome differences. Here, the game takes place on land near a Kurdish village in Iran. The scrubby plain where they play is ennobled by magnificent snow peaked mountains. Yet the scene also exposes the "flat mundane ontology of the moving people … the mobile commons of migration" (Papadopoulos and Tsianos, 2013: 179).

This image inadvertently evokes a more famous urban image from the mid-twentieth-century, which is L.S. Lowry's painting from 1949, "The Football Match", a composition that is set in the modern industrial city, in Salford. The match in Lowry's painting is, like the scene from *In This World*, a make-shift affair. Lowry's painting (Figure 22.5) uses the sporting action as a means to include familiar background motifs such as a raggedy flat-capped crowd, rows of terraced houses, chimneys, and redbrick factories. The painting depicts the players and the crowd enjoying leisure, taking a break between working or learning hours. The match is played *because* of the industrial scene – work is the reason these people are in the city – but these surroundings enclose and bear down on the action. The pitch and the goals appear unnaturally small. The freedom of the players is constrained.

Figure 22.4 works differently in the sense that the play is more open – there is no pitch as such – and what we can see on screen hints toward a post-work arrangement of social life. As such, the game is central to productive life in the village rather than an exception, a scheduled time for play. And so, in this instance, despite the evocation of an image from an earlier period of urban modernity, the specter of the city arrives from the future. There is the sense of the "city appearing" along with the realization that urban life could flourish here. This is an optimistic apparition that reminds us of Derrida's "city of refuge".

Figure 22.5 *The Football Match,* L.S. Lowry (1949). © The Estate of L.S. Lowry.

As Derrida explains, the city of refuge does not entail restoring

> an essentially classical concept of the city by giving it new attributes and powers; neither would it be simply a matter of endowing the old subject we call "the city" with new predicates. No, we are dreaming of another concept, of another set of rights for the city, of another politics of the city.
>
> *(Derrida, 2001: 8)*

And so, this image of a football match in a tiny Kurdish settlement in Iranian territory is utopian but also very modest. Lefebvre writes that "centrality defines the u-topic (that which has no place and searches for it)" (2003: 172), meaning that utopia always searches for and/or relies upon a center point or a wellspring. He also states that "this is why urban space is so fascinating: centrality is always possible" (ibid: 130). Here, we see how cinema aids this search by framing the center and the u-topic.

Images of urbanization in this cycle are not always so hopeful. In Figure 22.6, from *Ghosts*, a familiar sense of flatness is provided by the wet shore of Morecambe Bay in Lancashire. There are no mountains on the horizon – there is no buffer or shield here – just the point where the gray Irish Channel meets a bleak sky. While horizontal, flat space can symbolize openness and possibility, as it does in the previous example, here, these qualities are used to convey the dangers of exposure or the existential threat of the liminal. The horizon is ominous; a warning of the incoming tide. The mood is agoraphobic. And yet, so this article contends, this too is an urbanizing space; an image of urbanization. We see Mr Lin's battered van in the distance – an icon of worn out, tired mobility – and the middle ground is occupied by migrant laborers stooping, bent over, working, earning their living among the dirt. (In both images of urbanization, migrant bodies are active; they are shown doing things.) This is another image of nascent centrality; an image that conveys the very essence of the urban "as a place of conflict and confrontation, a unity of contradictions" (Lefebvre, 2003: 175).

Figure 22.6 Morecambe Bay, *Ghosts* (Nick Broomfield, 2006).

It is remarkable how reminiscent this image is of Alberto Giacometti's (1948) sculpture City Square (La Place) (see Figure 22.7). The characters on the shore and in the sculpture, are alone, positioned in isolation to each other, but they also comprise a crowd. As Umland explains, "'Vision' as manifested in Giacometti's work, is a complicated word, referring to the intertwined complexities of outer, retinally perceived effect and inner, psychic affect" (2001: 2). Both artistic works (cinema and the sculpture) reveal the creative tension between immersion and detachment that was vital to the dynamism of the modern city (Robins, 1996: 131). Umland suggests that,

> Even when approached, Giacometti's subjects retreat, remain inaccessible, suspended in a state of petrified mobility, fixed by the artist's eye at an exact distance. The base of City Square reinforces the impression of distance, understood in psychological and phenomenological terms, represented in sculptural dimension.
>
> *(Umland, 2001: 5)*

Just like Giacometti's subjects, the migrants in Figure 22.6 are suspended in a state of "petrified mobility". They also appear in retreat; they must keep moving, keep working, keep hiding. Just as in the sculpture, movement is transformed here into total *immobility* (Boyne, 2008: 21). Their reality, just like that depicted by Giacometti's sculpture, is "unshareable" (Berger, 2016: 327). Indeed, Ai Qin and her colleagues are positioned too far away for us to warn them of the incoming tide. As witnesses, we are immersed in the encounter (and the tragedy that unfolds), but there is too much distance between us to offer help.

The shoreline in *Ghosts* is like a proto-city square. In evoking this image, the city appears once more as a specter. But the city, as a possibility, is lost to these migrants. *This* isn't any kind of city at all. The appearance (or image) of centrality here mocks migrant cockle pickers and spectators alike, reminding us of the city that has been lost (or that remains inaccessible). Fundamentally, this "is an image [of urbanization] made from the mourning of another image [the city]" (Rancière, 2006: 103). The specter of the city appears, in Gordon's (2008: xvi) understanding, when the violence or pain it represents is

Figure 22.7 City Square (La Place), Alberto Giacometti. © The Estate of Alberto Giacometti.

no longer able to be contained or repressed. It is at this moment of extreme danger that the absence of the city – the city of refuge – is most acutely felt.

These two images of urbanization from *In This World* and *Ghosts* create an aesthetic that begins to capture the constraints, hazards, and possibilities of contemporary, expansive urbanization. In redeeming the physical reality of urbanizing spaces and by implementing narratives that present these sites as stakes in urban/political struggles, this film cycle indicates a decisive shift in urban imaginaries. It introduces "dissensus" into our understanding of the times and spaces of the urban. This aesthetic lacks the "uplifting image" of the city (Lefebvre, 2003: 14), but the open spaces, wide horizons and the depictions of work and play do signify the open-ended possibilities of urbanization. Horizontality is favored over verticality, with an emphasis on the "taskscape" (Ingold, 1993) presented by a flat middle ground that is either shielded or dangerously exposed. In the absence of the city, new urban *oeuvres* are being created as a matter of necessity. The non-synchronous character of films is due to the disruptive appearance of the specter of the city: the past city that cannot be recovered; the present city from which one is excluded or which lies at the end of the journey; and the future city (or unnamed *oeuvre*) which is yet to come. In relation to subjectivity there is an emphasis on negative themes such as displacement, isolation and separation, as well as positive themes such as conviviality, friendship and play. The inter-subjective dimension – that which Rancière (1999) calls "subjectification" and involves "removal from the naturalness of place" and "the opening up of a subject space" – is enhanced by how cinema circulates images that engage spectators in the urban encounters on screen (encounters that are beyond or without the city), in the process implicating all of us in the production of a hermeneutic urban space that is visually rooted by historical materialist processes of urbanization and migration; but also imaginary, displaying distinctive aesthetic qualities.

Conclusion

Andy Merrifield writes that extended urbanization has become "shapeless, formless and apparently boundless ... making it hard to tell where borders reside and what's inside and what's outside" (Merrifield, 2013: 910). This induces a crisis of representation, he suggests, evoking "what Clement Greenberg (1961) called 'the crisis of the easel picture', the crisis of the classic framing – maybe the classic framing of the city" (ibid: 914). The short cycle of films examined in this chapter may be considered as a symptom of this crisis, even if they also offer a provocative thesis as to why the crisis exists and provide examples of what its resolution may look like (ranging from the hopeful to the bleak). In relation to the former, the films in this cycle question the "natural" or "sensible" social historical relation between migration, migrants, and the city, the kind one finds in classic depictions of the "zone in transition" or, later in the twentieth century, the "inner city". Together these films suggest "the time of the city" for migrants and others displaced from the central city has passed; and reveal how, out of necessity rather than choice, the similarly excluded actively become involved in the production of new urban *oeuvres* (or works of art) of the kind that Lefebvre (2016) predicts. And yet, this cycle of films has shown that such *oeuvres* are always haunted and disrupted by the specter of the city. The specter at once seeks to *draw us back into the time of the city* and is a symbol *of our dislocation from it*: "they are always *there*, spectres, even if they do not exist, even if they are no longer, even if they are not yet" (Derrida, 1994: 221, original emphasis). As such, the time of the city *endures*, even as it becomes more and more difficult to experience it in the way that people (and more specifically, migrants) did in the past. In *Metaphilosophy*, Lefebvre writes that "[t]he idea of the end of the city seems far more productive and creative than of its continuation or its 'modernization'" (2016: 113). Each of the films discussed here, whether they "know" it or not, are embroiled with this proposition.

References

Balshaw, M. and Kennedy, L. (2000) *Urban Space and Representation*. London: Pluto.
Bender, T. (2002) *The Unfinished City: New York and the Metropolitan Idea*. New York: New York University Press.
Benjamin, W. (1999) *Illuminations*. London: Pimlico.
Berger, J. (2016) *Portraits: John Berger on Artists*. London: Verso.
Berman, M. (1999) *Adventures in Marxism*. New York: Verso.
Boyne, R. (2008) A Brief Note on Giacometti. *Theory, Culture and Society* 25(5): 20–9.
Brenner, N. (2013) Theses on Urbanization. *Public Culture* 25(1): 95–113.
Brenner, N. (ed.) (2014) *Implosions/Explosions: Towards a Study of Planetary Urbanization*. Berlin: Jovis Verlag.
Brunsdon, C. (2007) *London in Cinema*. London: BFI Books.
Çağlar, A. and Glick Schiller, N. (2015) A Multiscalar Perspective on Cities and Migration. A Comment on the Symposium. *Sociologica*. DOI: 10.2383/81432.
Derrida, J. (1994) *Specters of Marx*. London: Routledge.
Derrida, J. (2001) *On Cosmopolitanism and Forgiveness*. London: Routledge.
Dimendberg, E. (2004) *Film Noir and the Spaces of Modernity*. Cambridge, MA: Harvard University Press.
Donald, J. (1999) *Imagining the Modern City*. Minneapolis, MN: University of Minnesota Press.
Gilloch, G. (2015) *Siegfried Kracauer*. Cambridge: Polity.
Gordon, A. (2008) *Ghostly Matters: Haunting and the Sociological Imagination*. Minneapolis, MN: University of Minnesota Press.
Hall, S. and Savage, M. (2016) Animating the Urban Vortex: New Sociological Urgencies. *International Journal of Urban and Regional Research* 40(1): 82–95.
Harvey, D. (1996) Cities or Urbanization. *City* 1(1–2): 38–61.
Highmore, B. (2005) *Cityscapes: Cultural Readings in the Material and Symbolic City*. Basingstoke: Palgrave Macmillan.
Ingold, T. (1993) The Temporality of the Landscape. *World Archaeology* 25(2): 152–74.
Isaacs, B. (2013) *The Orientation of Future Cinema: Technology, Aesthetics, Spectacle*. London: Bloomsbury.
Isin, E. (2012) *Citizens without Frontiers*. London: Bloomsbury.
Jameson, F. (2008) Marx's Purloined Letter. In: Sprinker, M. (ed.), *Ghostly Demarcations: A Symposium on Jacques Derrida's Specters of Marx*. London: Verso.
Kracauer, S. (1960) *Theory of Film: The Redemption of Physical Reality*. Princeton: Princeton University Press.
Lefebvre, H. (1991) *The Production of Space*. Oxford: Blackwell.
Lefebvre, H. (1996) *Writings on Cities*. Oxford: Blackwell.
Lefebvre, H. (2003) *The Urban Revolution*. Minneapolis, MN: University of Minnesota Press.
Lefebvre, H. (2014a) Dissolving City, Planetary Metamorphosis. *Environment and Planning D: Society and Space* 32(2): 203–5.
Lefebvre, H. (2014b) *Critique of Everyday Life*. London: Verso.
Lefebvre, H. (2016) *Metaphilosophy*. London: Verso.
Loshitzky, Y. (2010) *Screening Strangers: Migration and Diaspora in Contemporary European Cinema*. Bloomington, IN: Indiana University Press.
Massey, D. (2005) *For Space*. London: Sage.

Mazierska, E. and Rascaroli, L. (2003) *From Moscow to Madrid: Postmodern Cities, European Cinema*. London: I.B. Tauris.
Merrifield, A. (2013) The Urban Question under Planetary Urbanization. *International Journal of Urban and Regional Research* 37(3): 909–22.
Merrifield, A. (2014) *The New Urban Question*. London: Pluto Press.
Mezzadra, S. (2011) The Gaze of Autonomy: Capitalism, Migration and Social Struggles. In: Squire, V. (ed.), *The Contested Politics of Mobility: Borderzones and Irregularity*. London: Routledge.
Millington, G. (2016a) Urbanization and the City Image in Lowry at Tate Britain: Towards a Critique of Cultural Cityism. *International Journal of Urban and Regional Research* 40(4): 717–35.
Millington, G. (2016b) The Cosmopolitan Contradictions of Planetary Urbanization. *British Journal of Sociology* 67(3): 476–96.
Millington, G. (2016c) *Urbanization and the Migrant in British Cinema: Spectres of the City*. Basingstoke: Palgave Macmillan.
Nail, T. (2015) *The Figure of the Migrant*. Palo Alto, CA: Stanford University Press.
Nancy, J.L. (2007) *The Creation of the World, Or, Globalization*. Albany, NY: SUNY Press.
Papadopoulos, D. and Tsianos, V.S. (2013) After Citizenship: Autonomy of Migration, Organisational Ontology and Mobile Commons. *Citizenship Studies* 17(2): 178–96.
Prakash, G. (2010) *Noir Urbanisms: Dystopic Images of the Modern City*. Princeton, NJ: Princeton University Press.
Purcell, M. (2008) *Recapturing Democracy: Neoliberalization and the Struggle for Alternative Urban Futures*. New York: Routledge.
Purcell, M. (2013) *The Down-Deep Delight of Democracy*. New York: John Wiley & Sons.
Rancière, J. (1999) *Dis-Agreement*. Minneapolis, MN: University of Minnesota Press.
Rancière, J. (2004) *The Politics of Aesthetics*. London: Bloomsbury.
Rancière, J. (2006) *Film Fables*. London: Bloomsbury.
Robins, K. (1996) *Into the Image: Culture and Politics in the Field of Vision*. London: Routledge.
Sennett, R. (1990) *The Conscience of the Eye: The Design and Social Life of Cities*. London: Faber and Faber.
Shapiro, H. (2010) *The Time of the City*. New York: Routledge.
Umland, A. (2001) Capturing Giacometti's Gaze. *MoMA* 4(8): 2–7.
Wachsmuth, D. (2014) City as Ideology: Reconciling the Explosion of the City Form with the Tenacity of the City Concept. *Environment and Planning D: Society and Space* 31: 75–90.
Watt, P. and Minton, A. (2016) London's Housing Crisis and Its Activisms: Introduction. *City* 20(2): 204–21.

23

Chicano Park's urban imaginary

Ethnic ties bonded to place and redistributive urban justice

Gerardo Francisco Sandoval

Introduction

Chicano Park is an important symbolic territory for Southern California's Chicano (Mexican American) imaginary. Chicano Park is a site of resistance, a public symbol of Chicano sovereignty, and a community-based activist struggle over the control of space and territory. Chicano Park became the site of contestation in 1978 when Chicanos took over the space under a freeway and bridge in San Diego, California, and declared it a public space (Rios, 2009). It is managed by the Chicano community and highlights the Chicano urban imaginary through its public arts – such as the murals painted on freeway pillars and public monuments depicting Chicano history and culture.

This chapter explores Chicano Park's urban imaginary and how neighborhood-based arts efforts relied on cultural representation to advocate for redistributive urban planning projects. I review Chicano Park's history as a symbolic space and a place for cultural performance and spiritual imaginaries. By engaging Anderson's notion of an imagined community, we see how Chicano Park's urban imaginary served to create an imagined political community that pressured the state to provide investments that would benefit the local arts milieu in the neighborhood. These political struggles are viewed by Castells as struggles over collective consumption where cultural identity serves as a form of territorial protection. I also engage Nancy Fraser's work on identity politics and understand Chicano Park's imaginary as a struggle between identity politics of representation and distribution.

The Chicano imaginary struggles over identity politics in Chicano Park can be seen as struggles over representation and redistribution. In terms of cultural identity representation, I analyze how the Chicano social movement served to cement the urban imaginary as a site of cultural struggle related to issues of progressive politics, Latin American revolutionary images, and workers' rights (such as farm work). I also analyze how the Chicano Park imaginary represents an anti-colonial imaginary based on indigenous (Mayan and Aztec) cultural semiology. This place of resistance becomes both a symbolic and spiritual space where Aztec Danza, sweat lodge, and indigenous rites of passage are practiced (see Figure 23.1). These representation struggles create spaces of resistance that

Figure 23.1 Mural of Aztec warrior at Chicano Park. Photograph by the author.

demand redistribution policies and projects to be implemented in the neighborhood, which include redistribution spatial policies as public investments in the arts, infrastructural investments (such as transit-oriented development linked to affordable housing), and public arts commissions linked to community improvement revitalization projects.

Hence, in this chapter, I argue that Chicano Park's urban imaginary served as symbolic territory for arts-based community revitalization grounded on cultural identity representation. The Chicano arts imagined community transformed urban planning investments into redistribution-oriented neighborhood revitalization efforts.

Chicano Park as urban imaginary

Chicano Park plays a central role in the Chicano urban imaginary. It is a public space near downtown San Diego, California, that encompasses the largest number of public murals in the United States. It is also now on the National Register of Historic Places, as it has a strong historical importance to the Chicano community of the United States. Most of the murals depict the historical struggles that Chicanos endured as their culture was colonized by the Spanish and then by the United States. The murals depict indigenous (Aztecs and Mayan) cultural symbols and images of Chicano political struggles. They also contain many religious (both indigenous and Catholic) images that serve as cultural reminders of the mestizaje (mixture of indigenous and Spanish blood, a product of the conquest). There are also statues of Mexican folk heroes such as Emilio Zapata, which add to the milieu of resistance and revolutionary struggle.

Chicano Park is in Logan Heights, the historical center of the Chicano community in San Diego. The park originated out of a struggle to save the neighborhood from further spatial segregation. When the California Department of Transportation built Interstate 5, it cut Logan Heights into two neighborhoods, which created Barrio Logan to the west of I-5 and left Logan Heights to the east. The neighborhood was further divided into four distinct spaces when a bridge was built to connect I-5 to Coronado Island (a higher income area of San Diego). When the California highway patrol decided to build a detention facility under the freeway (further separating the Logan Heights neighborhood) Chicanos in the neighborhood resisted. They organized marches and took over the space under the freeway. After a few days of protest, the state decided to end their plans for the detention facility and lease that land to the city for $1 a year. The community (led by the Chicano Park Steering Committee) took charge of the park and agreed to manage and protect it as a Chicano space. Chicano artists and activists from around the country transformed the space into a symbol of cultural resistance and Chicano sovereignty and self-determination; the public park evolved into a vibrant arts milieu that is protected by the Chicano community and serves as a central figure in the Chicano urban imaginary.

Chicano Park's urban imaginary is largely based on Aztlan mythology (Figure 23.2). Aztlan is the ancestral home of the Aztecs, one of the main indigenous civilizations in Mexico. Aztlan was said to encompass present day California, the southwestern United States and Mexico (until the Mayan territory of southern Mexico). This Aztlan imaginary provides Chicanos with an image of communion and a shared political community. Anderson's notion of an imagined community is useful to engage in understanding the type of shared political community that Chicanos have established in Chicano Park. Anderson argues,

Figure 23.2 Mural depicting Aztec symbolism that reads "All the Way to the Bay". Photograph by the author.

> Nations [are] an imagined political community... it is imagined because members of even the smallest nation will never know most of their fellow members, meet them, or even hear of them, yet in the minds of each lives the image of their communion.
> (*Anderson, 1983: 6*)

Although his book focuses on nations as imagined communities, Anderson's argument goes beyond that sphere as he argues that "imaginary as creation ... In fact, all communities larger than primordial villages of face-to-face contact are imagined. Communities are to be distinguished, not by their falsity/genuine, but by the style in which they are imagined" (Anderson, 1983: 6). Chicano's Aztlan imaginary provides Chicanos with a claim to territory. Chicanos are reclaiming the U.S. Southwest, which was taken from them first by Spanish colonists and later U.S. imperialism. This imaginary is based on spiritual and religious beliefs that pre-date colonialism and can further help Chicanos recreate their commonality. Finally, the Chicano imaginary provides a shared history to members, based on a sense of exploitation and unjust colonial power relations that are intermixed with racialized hierarchies (Pulido, 1996, 2006). Chicanos' claim to territory, a re-creation of their spiritual indigenous connections, and a shared history make them into a political community, an imagined community.

Neighborhood-based ethnic social movements grounded on an imagined political community were studied in-depth by Castells (Castells, 1997). Castells' classic work in *The City and the Grassroots* provides a good foundation for studying neighborhood social movements based on an imagined political community. Castells studied urban social movements throughout the world and came to the conclusion that social movements are a product of three key themes:

1. Demands focused on collective consumption, that is, goods and services directly or indirectly provided by the state.
2. Defense of cultural identity associated with and organized around a specific territory.
3. Political mobilization in relationship to the state, particularly emphasizing the role of local government.

(*Castells, 1983*)

Two of Castells' case studies in the book are particularly enlightening in understanding how neighborhood-based imagined communities became political projects for that neighborhood's territorial protection. Castells studied the Castro District in San Francisco and demonstrates how it became a neighborhood that identified as LGTBQ and how that identity served to unite the neighborhood and provide residents with political capital to restructure their community (Castells, 1983). Castells also studied San Francisco's Mission District, where he analyzed a neighborhood-based social movement that fought to stop urban renewal. Castells argues that the Mission District's progressive Latino community formed a coalition to politically mobilize against urban renewal and ultimately protect its ethnic identity and territory. This urban coalition was made up of urban-based churches, radical political groups, and traditional Latino organizations. This dense network of community organizations acted both as a deterrent against displacement and as a stimulus for individual rehabilitation and collective maintenance of the urban environment. In retrospect, it was an imagined community that relied on shared ethnic ties to create a political community that protected its territory.

The identity politics that emerged in these neighborhoods directly contributed to the increased political influence the struggles had on reshaping urban space. These struggles for urban justice can be viewed in terms of struggles over redistribution or recognition. Nancy Fraser argues that struggles over redistribution and recognition need to be viewed simultaneously and the intersectionalities between them understood. As Fraser explains it:

> The redistribution paradigm focuses on injustices it defines as socio-economic and presumes to be rooted in the economic structure of society. The recognition paradigm, in contrast, targets injustices it understands as cultural, which it presumes to be rooted in social patterns of representation, interpretation, and communication.
>
> (*Fraser and Honneth, 2003: 13*)

Hence, the urban imaginary struggles in these ethnic neighborhoods can be seen as struggles over identity politics. And these struggles over identity politics have real redistribution outcomes that need to be clearly identified. Another way of understanding the difference between redistribution and recognition is to understand their different forms of remedies to injustice. "In the redistribution paradigm, the remedy for injustice is economic restructuring of some sort. In the recognition paradigm, in contrast, the remedy for injustice is cultural or symbolic change" (Fraser and Honneth, 2003: 13). But Fraser argues that the distinction between both is a false antithesis: "One should roundly reject the construction of redistribution and recognition as mutually exclusive alternatives. The goal should be, rather, to develop an integrated approach that can encompass, and harmonize, both dimensions of social justice" (Fraser and Honneth, 2003: 26). Hence, to understand how the urban imaginary in Chicano Park changed the actual redistribution policies in the neighborhood, we have to understand how they shaped each other. That is, how demands for recognition of ethnic identity directly led to urban redistributive policies that included community benefits to these low-income neighborhoods (Young, 2011).

The identity politics that constitute the Chicano Park imaginary have roots in a history of racist urban planning policies targeting the barrio (Chicano neighborhood).

> Within the context of everyday life, el barrio is the reaffirmation of culture, a defense of space, an ethnically bounded sanctuary, and the spiritual zone of Chicana/o and Mexicana/o identity. It is a powerful, intense space that had defined the independence and resistance of a culture that predates Euro-American influences on city life and urban form.
>
> (*Diaz, 2005: 3*)

Chicano Park is a site where the Chicano community has been segregated and where environmental justice issues shaped by the shipyards that cater to the large naval base nearby have forced the community to resist and fight back. Eric Avila (2014) explains this type of ethnic identity and resistance in his book, *The Folklore of the Freeway.* Freeways destroyed communities of color but there was resistance, a type of regenerative community rebuilding. Culture played a role in neighborhood regeneration in the form of art and literature. Avila argues that Chicano Park is a space of resistance. It represents the underbelly of the modernist city:

> The folklore of the freeway thus includes the creative initiatives taken by people living in the border vacuums of barrios and ghettos to turn lemons into lemonade, making do with the injustices accompanying the postwar phase of urban modernization … The voices of dissent and the invocations of racial pride have been tattooed on the skin of public infrastructure, and they register the intent of local residents to accept the freeway on their own terms, not on those of transportation authorities.
>
> (*Avila, 2014: 178*)

The resistance in Chicano Park is the Chicano imaginary taking shape and as Avila states, "tattooed" on the pillars of the freeway. The Chicano urban imaginary is also an imaginary that is transforming all parts of the city, not just barrios:

> The words and images that define the folklore of the freeway illustrate the regenerative capacity of communities to clear the dust, regroup, and create anew. Their collective efforts enhance the cultural vitality of the entire city, not just of specific neighborhoods. Chicano Park, for example, became a state treasure in 1997, when the California legislature voted for its inclusion on the California Register of Historic Resources. In 2013, the National Park Service included Chicano Park on the National Register of Historic Places.
>
> (*Avila, 2014: 189*)

Hence, the Chicano imaginary is slowly making its mark in the Euro-centric imaginary that colonized the U.S. southwest: "The spatial Reconquista occurring in the Southwest will only accelerate in the next quarter century and is already socially and culturally transforming the urban environment" (Diaz and Torres, 2012: 34).

Representation and identity in Chicano Park

Chicano Park is a space of Chicano cultural identity representation. It is a public space created through political struggle and a physical materialization of the Chicano imaginary. As one of the original Chicano Park artists describes it, "Chicano Park is the umbilical cord of our Latin American culture. It gives us a sense of pride, like, we [Chicanos] are here. Whether immigrants stay here or move on north, it's a welcoming sign". The cultural representation in the park is seen through a social movement imaginary. The murals depict key events in the Chicano movement, like the high school walkouts in East Los Angeles and the farmworker struggles in the Central Valley of California. The murals also depict key radical Latin American Revolutionaries, like Ernesto "Che" Guevara. Other social issues depicted include the Chicana activist who struggles over both racism and sexism. As a community activist stated, "Without any historical or political context, unless you actually look at the murals and understand the character of the park, you don't understand that [the murals] have political connotations". And it is this political capital established via identity politics and cultural representation that makes this imagined space so powerful for Chicanos.

A key theme depicted in the artwork is an anti-colonial imaginary based on indigenous Aztec and Mayan culture and representation. This representation creates a spiritual connection to Latinos that celebrate their indigenous roots. As an artist explains, "The park is a sacred space for many people. There's a gel that brings us all together. Culture, history, traditions". The anti-colonial indigenous imaginary is not only depicted through the murals but through spiritual practices and ceremonies in the park. Danza, a form of Aztec dance and ceremony that has its origins hundreds of years ago in Central Mexico is performed twice a week in the kiosk at Chicano Park. There are three or four Aztec Danza groups that use the park and have been practicing their ceremonies at the park for forty years. Chicano Park is the place where danza originated in the United States and then spread to other locations in the country. The other form of indigenous spirituality practiced at Chicano Park is the sweat lodge, which takes place on the last Saturday of the month. These are indigenous practices that bring Chicanos together to meditate in a sweat lodge. That space is next to Interstate 5 and under the Coronado Bridge and is also surrounded by a community garden where indigenous herbs are cultivated (see Figure 23.3).

The new gateway sign in the neighborhood also depicts anti-colonial connotations. The sign was designed by one of the original Chicano Park artists, Armando Nunez, who helped to paint the first mural at the park, titled "Chicano Historical Heroes". Mr. Nunez has stayed very active in Chicano Park affairs and painted more murals. He had the idea of creating signage that would let people know they were entering Chicano Park in the 1980s. But his vision did not materialize until the recent revitalization of the neighborhood provided funds for the gateway. He designed the gateway sign using Aztec and Mayan semiology to celebrate the cultural identity of Chicanos and highlight their indigenous roots.

Figure 23.3 Cultural space in Chicano Park where "sweats" and other cultural and spiritual rituals take place. Photograph by the author.

> I did my research on Mayan, Combiack and Aztec cultures coming together ... Like the corn and beans being the staple food of our people. Everyone likes it, I haven't heard any negative comments, it means something. It has the sentiment of the community, people are very proud of that.

The gateway sign is also a symbol that depicts Chicano territory that has been institutionalized through the city. When people enter Chicano Park, they now know they are entering a space that has been historically imagined as a Chicano space. And community members have fought to depict that connotation for their neighborhood. As a Chicano Park artist explains,

> There is always a push from members of the community to say, no, it has to reflect our cultural history. If you see the Barrio Logan sign, you will see a corn in the center because for us Mexicans, it's very significant to us for our diet and history. There is always a push to have those architectural designs that reflect the community. There is a history of struggle of political combativeness and that has an impact on how the city designs things.

Representation of Chicano identity can also be viewed in the efforts to capture, reclaim and protect Chicano history in the park and neighborhood. Historical preservation in the

United States has traditionally focused on solidifying spaces of mainstream culture and making other "minority" cultures invisible in their efforts to highlight history (Arreola, 2004). But Chicano Park has been different. Chicano activists have embraced historical preservation and used it to protect their cultural identity in the park. A neighborhood activist conducted her own research of the history of Chicano Park and maneuvered the State of California's bureaucracy to have Chicano Park listed as a California State historical site. This same activist went on to solidify Chicano Park in the National Historical Registry. Hence, Chicano Park is a national historic site that, according to this Chicana historic preservation activist, "cannot be destroyed".

Another example of the anti-colonial imaginary shaping historical preservation efforts in the park is the story of the Aztec brewery. The Aztec brewery company moved from Mexicali to Barrio Logan in 1933 after prohibition. The brewery's tasting room was decorated with ceramic, glass, oil paintings, and wooden hand-carved images depicting Aztec and Mayan semiology. The artist, Jose Moya del Pino, was a Spanish muralist who had been influenced by Mexican muralist Diego Rivera. People describe walking into the tasting room as if walking into a temple. As a neighborhood activist describes it, the tasting room was, "Oh, beautiful... to die for". The Aztec Brewing Company ultimately closed and their building was demolished. But Chicano Park activists were able to lobby with the city to save some of the artwork in the brewery tasting room. Most of the indigenous symbolic artwork was preserved and taken to the Logan Heights library where it would now be on public display.

These efforts to reclaim and preserve Chicano cultural identity served to solidify representational claims to that contested space. Representation served as the means to both build political capital and to protect a territorial cultural space. The Chicano imaginary in Chicano Park was the glue that helped this community maintain its sense of place.

Redistributive claims reshaping Chicano Park

> I blessed the new [Chicano-themed] supermarket, the affordable apartments, all that new [El Mercado Del Barrio] complex.
>
> *Catholic priest in Barrio Logan*

Barrio Logan has gone through an urban revitalization process that has changed the physical character of the neighborhood. This neighborhood revitalization process is reminiscent of urban renewal struggles that displaced thousands of people of color in low-income neighborhoods (Ammon, 2016; Bullard and Johnson, 1997; Duneier, 2016; Gans, 1982; Mohl, 2008). Although contemporary, this displacement process is linked to market-based gentrification, which is transforming many inner cities (Angotti, 2008; Cahill, 2010; Dávila, 2004; Lees et al., 2010, 2015). The catalyst project that has begun to transform the neighborhood is called the Mercado Del Barrio development complex. This commercial and residential development now sits in the center of Barrio Logan and just adjacent to Chicano Park. It is comprised of 100 affordable housing units, local commercial businesses, and a large Chicano-themed supermarket as the development's centerpiece.

The Chicano-themed supermarket has a Latino façade with colorful artwork and sells products that cater to the Latino community. The apartments were initially slated to be market-rate housing, but neighborhood activists protested those and convinced the developer to make them affordable housing apartments. The complex is also walking distance from San Diego's trolley (light rail system), making it a transit-oriented development project. Next to the Mercado Del Barrio is a large parking structure that has a façade with pictures and art depicting key Chicano movement leaders and leaders from the Chicano Park takeover. There is also a new community college that was built next to the development. It is called the Cesar Chavez (after the Mexican-American farmworker activist) Community College and also has artwork depicting the Chicano movement and Chicano culture. Hence, all the new physical infrastructure, housing, and commercial developments being constructed in Barrio Logan serve as symbols of recognition of identity. But importantly, the efforts for identity recognition via Chicano symbols have directly led to redistribution policies such as affordable housing, locally owned businesses, and investments in educational and transportation facilities.

For example, the developers of the Mercado project (after getting political pushback from Chicano Park activists) saw the need to incorporate Chicano cultural symbols into their project. As the developer stated in an interview,

> It was an idea of the design team to integrate cultural murals into the project. It was important to do it ourselves and do something different in a retail center. We got the community involved in the process. We had kids involved in some of the mosaics.

Hence, pressures for recognition turned into a public participation mechanism that led to more redistribution outcomes. As one of the artists explains:

> The artists are a form of resistance; we have a lot of good muralists here in San Diego. They do some really nice artwork. The people who did original murals [at Chicano Park], some are still alive. Some of the new art in the rental housing, yeah, that art is not political, but it shows the cultural [attributes of our community]. The murals have baile folkórico or mariachis, so they have that cultural part. The parking lot even has a picture of Chunky (a Barrio Logan activist and musician) and Cesar Chavez.

Arts have become a tool for redistribution efforts in Barrio Logan. The city has started to invest in the neighborhood and is acknowledging the importance of supporting neighborhood-based artists. Some of these artists belong to urban planning advisory boards that play an important role in the city's investments in the community. For example, a Chicano-themed supermarket had been a community priority for many years, as there had not been a local supermarket in the neighborhood in thirty years. Hence, neighborhood activists pressured the city to ensure that a market would be located in the El Mercado Del Barrio development project. The city subsidized the construction of the market by providing the land for free. These are political redistribution decisions made by the city to ensure the Chicano milieu is maintained in the neighborhood. A Chicano artist reiterated the importance of the market: "A Latino-focused market has always been the hope, a Latino-themed grocer. That is something the community wanted".

Chicano artists transformed Chicano Park from an empty space under a freeway to an important cultural milieu. They are now doing the same for the entire neighborhood. They are, in essence, bringing Chicano Park from under the freeway and into the neighborhood. The new arts-based development in Chicano Park is led by local artists. Some are part of the older Chicano Park generation, but others are now Chicano Park's new generation of artists that were mentored by the older generation. This form of arts-based community regeneration is different than other artist-based gentrification processes that have attracted a lot of criticism for transforming neighborhoods and leading to business and housing displacement, as Meghan Rich's chapter in this volume demonstrates. What is occurring in Barrio Logan is different. It is neighborhood arts revitalization that is creating an arts milieu in the neighborhood and encouraging the city to invest more in the neighborhood but with a redistribution lens. As one of the original Chicano Park artists stated in an interview,

> The resurgence of the arts in the area is exciting. The extension of Chicano Park and the murals into the area is also exciting. The entire area will now be a thriving arts scene and community. That is the next phase of Barrio Logan.

This neighborhood arts-based revitalization has led to more businesses opening up on Logan Avenue, the main street that leads from Chicano Park to southern San Diego. Another well respected Chicano activist explains,

> small businesses opening up on Logan Avenue is a good thing. They bring in lots of youth and arts into the community. We have the Barrio Art Crawl that is putting Barrio Logan on the map in terms of artistic value. I firmly believe that it's a very good thing for the community. . ..it becomes part of the Barrio.

The Barrio Art Crawl is an effort to bring more attention to the artist milieu emerging in Barrio Logan. On one Friday evening a month, the local artists in the neighborhood come out to display their artwork in local galleries. This draws art enthusiasts from both the neighborhood community as well as from throughout the San Diego region. During the Arts Crawl, Logan Avenue becomes a very active street with hundreds of people visiting the art galleries, local restaurants, and bars. Art displayed at the galleries is not only limited to those of Chicano Park artists, but includes other artists from throughout the region. Hence, the Arts Crawl becomes a very exciting event that brings people from different cultures to experience Chicano Park and Logan Avenue.

The development of these businesses has been helped by the city's redistribution policies. The city has provided some low-interest loans, marketing help, and façade improvement assistance. There have also been efforts to start a business improvement district along Logan Avenue that is supported by the city. Neighborhood activists see the development of these small businesses and the arts district as positive change, as long as it contributes to the Chicano imaginary in the neighborhood. An important activist explains that the new business investments are "good in the context of our communities being able to have small businesses that are Chicano-owned, community run".

Discussion

> The park wants to maintain its autonomy ... you can't destroy the park. It is historical.
>
> *Chicano Park Historical Preservation Advocate*

The Chicano imaginary in Barrio Logan has helped the neighborhood revitalize while keeping the Chicano cultural arts milieu in the community. The Chicano imaginary emphasized the self-determination of its people, who were first colonized by the Spanish and later by the United States. One can view the continued racist urban planning interventions in the neighborhood as a contemporary form of continued colonization. Hence, Chicanos have been struggling to maintain their neighborhoods and communities for hundreds of years. These struggles are currently identity-based struggles over representation and redistribution. In this chapter, I have argued that Chicano Park's urban imaginary served as symbolic territory for arts-based community revitalization grounded on cultural identity representation. The Chicano arts imagined community transformed urban planning investments into redistribution neighborhood revitalization efforts.

Chicanos have relied on an urban imaginary to maintain their culture. They have used it as a form of empowerment as a subjugated ethnic underclass. Chicanos have created an understanding of their history linked to their "mestizaje", their combined Spanish and indigenous heritage and worldviews. Their urban imaginary has been a political tool to gain redistributive benefits from the state. And they have ultimately relied on their urban imaginary to make claims to space and territory. All of these functions related to the Chicano imaginary can be conceptualized as a political community à la Anderson's imagined communities concept. A Chicano Park historical preservation activist summarizes Chicano Park's experience as a form of imagined community, and speaks of the book *Barrio Logos,* by Villa (Villa, 2000):

> In the book *Barrio Logos*, Villa has the concept of Barrioization and Barriology. The history of all these communities you are really talking about a history of what he calls "Barrioization". Where you basically have created these urban barrios through racism. He argues that in response communities have created what he calls Barriologies. A form of wisdom. What is the thing in your life that is pushing you back and how do you react? So, Logan Heights, I-5, Coronado bridge helped create Barrioization. The maritime industry created this segregated, very isolated world where people were disempowered. And the response of the people was the creation of the murals. So, the idea was that here you had these huge pillars that we had to face daily, grey ugly, and no grass. And a guy by the name of Salvador Torre came and said, "Let's put murals here". That's the barriology. The reaction and the response. And now you have this incredible site in a very complicated relationship. It is the largest collection of outdoor murals in the world. And it all came out of this Barrioization and Barriology relationship.

The Chicano Park case clearly demonstrates that ethnic identity plays a key role in planning processes and reshapes urban redistributive policies. Political economic structures reshaping neighborhoods still create struggles over space (Fainstein, 2010; Harvey, 2005; Logan and Molotch, 2007; Mitchell, 2003), but ethnicity plays a key role in the resistance to neighborhood transformation. More research is needed to understand the role that ethnic identity plays for other ethnicities that have been discriminated against and marginalized. Do these communities have their own form of barriology based on their particular imagined community? For example, how would the Black urban imaginary play a role in contemporary urban redevelopment? We know how the Black community was segregated into ghettos, how they were displaced during urban renewal, and how the institutionalization via our criminal justice system has negatively impacted that community. But what is the urban imaginary that is pushing back on those systems of oppression for that ethnic group? The Chicano Park case demonstrates that we do not live in a post-racial society. It also demonstrates what the future possibilities can encompass if the urban imaginary is used as an imagined community to fight back and to create a cultural milieu that is based on empowerment and is authentic to Chicanos. An authenticity based on the historical struggles of those neighborhood residents and the ethnic ties that bond them to place. An arts-based regeneration that tells the story of that ethnic community, instead of erasing it.

References

Ammon, F.R. (2016) *Bulldozer: Demolition and Clearance of the Postwar Landscape*. New Haven, CT: Yale University Press.

Angotti, T. (2008) New York for Sale: Community Planning Confronts Global Real Estate. In: Angotti, T. (ed.), *Urban and Industrial Environments*. Cambridge, MA, London: MIT Press.

Arreola, D.D.. (ed.) (2004) *Hispanic Spaces, Latino Places: Community and Cultural Diversity in Contemporary America* (1st ed). Austin: University of Texas Press.

Avila, E. (2014) *The Folklore of the Freeway: Race and Revolt in the Modernist City*. Minneapolis: University of Minnesota Press.

Bullard, R.D. and Johnson, G.S.. (eds) (1997) *Just Transportation: Dismantling Race and Class Barriers*. Gabriola Island, B.C: New Society Publishers.

Cahill, C. (2010) Negotiating Grit and Glamour: Young Women of Color and the Gentrification of the Lower East Side. In: Lees, L., Slater, T. and Wyly, E.K. (eds.), *The Gentrification Reader*. London: Routledge.

Castells, M. (1983) *The City and the Grassroots: A Cross-Cultural Theory of Urban Social Movements*. California Series in Urban Development. Berkeley: University of California Press.

Castells, M. (1997) *The Power of Identity: The Information Age: Economy, Society and Culture*. Vol. 2. Oxford: Blackwell.

Dávila, A.M. (2004) *Barrio Dreams: Puerto Ricans, Latinos, and the Neoliberal City*. Berkeley: University of California Press.

Diaz, D.R. (2005) *Barrio Urbanism: Chicanos, Planning, and American Cities*. London: Routledge.

Diaz, D.R. and Torres, R.D.. (eds.) (2012) *Latino Urbanism: The Politics of Planning, Policy, and Redevelopment*. New York: New York University Press.

Duneier, M. (2016) *Ghetto: The Invention of a Place, the History of an Idea*. New York: Farrar, Straus, and Giroux.

Fainstein, S.S. (2010) *The Just City*. Ithaca: Cornell University Press.

Fraser, N. and Honneth, A. (2003) *Redistribution or Recognition? A Political-Philosophical Exchange*. Translated by Golb, J., Ingram, J. and Wilke, C.. London: Verso.

Gans, H.J. (1982) *The Urban Villagers: Group and Class in the Life of Italian-Americans*. Updated and expanded edition. New York: Free Press. London: Collier Macmillan Publishers.

Harvey, D. (2005) *A Brief History of Neoliberalism*. Oxford: Oxford University Press.

Lees, L., Shin, H.B., and López Morales, E.. (eds.) (2015) *Global Gentrifications: Uneven Development and Displacement*. Chicano, IL: Policy Press.

Lees, L., Slater, T., and Wyly, E.K.. (eds.) (2010) *The Gentrification Reader*. London: Routledge.

Logan, J.R. and Molotch, H.L. (2007) *Urban Fortunes: The Political Economy of Place*. 20th anniversary edition with a new preface. Berkeley: University of California Press.

Mitchell, D. (2003) *The Right to the City: Social Justice and the Fight for Public Space*. New York: Guilford Press.

Mohl, R.A. (2008) The Interstates and the Cities: The U.S. Department of Transportation and the Freeway Revolt. *Journal of Policy History* 20(2): 193–226.

Pulido, L. (1996) *Environmentalism and Economic Justice: Two Chicano Struggles in the Southwest*. Society, Environment, and Place. Tucson: University of Arizona Press.

Pulido, L. (2006) *Black, Brown, Yellow, and Left: Radical Activism in Los Angeles*. American Crossroads 19. Berkeley: University of California Press.
Rios, M. (2009) Public Space Praxis: Cultural Capacity and Political Efficacy in Latina/o Placemaking. *Berkeley Planning Journal* 22(1): 92–112.
Valle, V.M. and Torres, R.D. (2000) Latino Metropolis. In: *Globalization and Community*. Vol. 7. Minneapolis: University of Minnesota Press.
Villa, R. (2000) *Barrio-Logos: The Dialectic of Space and Place in Urban Chicano Literature and Culture* (1st ed). Austin: University of Texas Press.
Young, I.M. (2011) *Justice and the Politics of Difference*. New edition with foreword by Danielle Allen. Princeton: Princeton University Press.

24

Arts districts and the reimagining of neighborhood through arts and culture-based development

Meghan Ashlin Rich

Introduction

Post-industrial cities seeking to turn around declining neighborhoods have utilized arts and culture-based development in their revitalization strategies (Clark et al., 2002; Florida, 2002, 2004; Miles, 2005; Miles and Paddison, 2005; Strom, 2003). As such, state-sanctioned arts and entertainment districts have proliferated in many cities as part of a reimagining of urban neighborhoods through real estate revalorization and redrawing of neighborhood boundaries (Darchen, 2013; Noonan, 2013; Rich and Tsitsos, 2016; Stern and Seifert, 2007). This chapter explores the general processes and effects of themed arts cultural districts in urban neighborhoods through research from an ethnographic case study of the first state-sanctioned arts district in Baltimore, Maryland (USA), an amalgamation of three central neighborhoods renamed "Station North" in 2002. I begin this discussion with a general understanding of the processes of neighborhood theming and focus on "amenities", arts, culture, and entertainment.

Reimagining neighborhood through themed environments and the arts

In their now classic statement of cities as economic and political growth machines, *Urban Fortunes*, Logan and Molotch assert that "The reality of place is always open, making its determination an inherently social process" (Logan and Molotch, 1987: 47). Place is created and transformed by various political actors throughout the stratification system, from the local shopkeeper, to the neighborhood community association, to the political and economic machines in charge of urban development policy. Place is neither neutral nor independent of human thought and activity, but rather, it reflects, reaffirms, and reproduces ongoing social systems. As such, place exists in our imaginations as much as in concrete. To continue with Logan and Molotch's (1987) theoretical framework, the exchange value of place is wholly created by people, and thus as we imagine those neighborhoods, land value and rents fluctuate according to neighborhood reputation, formed in city dwellers' minds: "the ghetto", "the gayborhood", "the arts district", or "the university district". To raise the exchange value of a place, actors must transform its reputation through a rebranding and reimagining of neighborhood. As Greenberg's work on urban imaginaries and city branding affirms: "the space of the city is produced not only materially and geographically but also in the social imagination and through changing modes of cultural representation" (Greenberg, 2000: 228).

The reimagining of neighborhoods through branding and themes is a product of two overarching trends in real estate development and place revitalization in many cities throughout the world. The first is the result of globalization and the move from industrial production to individual consumption in post-industrial cities, heightening the importance of symbols and symbolic production to place (such as the proliferation of shopping malls/ centers) (Gottdiener, 1997). The second trend is the attempt by political forces to stave neighborhood and whole city decline through the theming of urban consumer-oriented districts, often investing millions of dollars of public funds in real estate re/development in hopes of attracting tourists, sports fans, and residents (such as the Guggenheim Bilbao in Spain and Camden Yards baseball stadium in Baltimore) (Evans, 2003; Miles, 2005). Since the 1990s, arts and culture theming of cities and neighborhoods has been a particularly popular tool for city/state governments to revitalize sluggish economies and reinvigorate tourist interest, a process that Strom terms "culture as development" (2003: 248). With urban growth, cities have become more dependent on amenities (entertainment and cultural offerings) to attract the highly educated so-called "creative class" (Clark et al., 2002; Florida, 2002, 2004). Strom argues that economic pressures have led to city leaders shifting from "selling" to "marketing", where "they seek to remake the city . . . to conform to the expectations of the affluent consumers they want to attract. Cultural institutions, associated with beauty, good taste, and higher purpose, become singularly important symbolic assets for image-conscious markets" (2003: 7). As the arts attract patrons, workers, volunteers, and visitors, they are seen as income generators by cities, as well as a boost to the economic and social fabric of the neighborhoods in which they are located.

Large arts projects/redevelopments are an obvious outcome of this urban development strategy, but what of arts and cultural districts? Are they always tied to a signature project or anchor, such as a museum? Are they a top-down process, with political and business leaders making decisions without regard to neighborhood stakeholders? What of the actual artists and production of art? Are artists already settled in arts districts or are they encouraged to do so through housing subsidies? Research on arts and cultural districts shows a wide-range of origination and implementation approaches, from "natural" districts in areas with existent high concentrations of artists and cultural amenities (Stern and Seifert, 2007, 2010) to the theming and aggressive redevelopment of areas with previously few cultural amenities (Noonan, 2013). Noonan's research (2013) on the effect of state-sanctioned arts and cultural districts on neighborhoods shows that there is a lack of clear economic growth, while, not surprisingly, there is some evidence of residential displacement over time in the districts included in his study. Darchen's investigation (2013) into Toronto's Master Plan to redevelop a downtown neighborhood as an "Entertainment District" shows how political and business entities use the creative city concept to revitalize an under-valorized area, all without a clear plan for how artists or other stakeholders will engage in this process.

The relationship between artists, neighborhoods, and gentrification is a now classic subject in urban sociology, from Zukin's investigation into the commodification of the lifestyles of New York's SoHo artist communities in *Loft Living* (1982) to the more recent work by Ocejo (2014) that describes the nostalgia of bohemian "early gentrifiers" of New York's Lower East Side, longing for the "bad old days" of the neighborhood drug addicts, squatters, and abandoned buildings. Artists are often viewed as the "shock troops" of gentrification, the first-wave of new settlers in industrial and/or declining neighborhoods, and yet they are often the victims of neighborhood "success" (Cameron and Coaffee, 2005; Currid, 2009; Markusen, 2006; Strom, 2010). As Lees argues (2000), once artists settle into derelict neighborhoods, capital investment and redevelopment are sure to follow on the backs of artists' sweat equity. Artists tend to have high levels of cultural capital and are an attraction for technology, media, and culture industry workers, what Ley (1996) calls "the new middle class" – or in other words, "the creative class" of Richard Florida's highly popularized thesis (2002). Working from Bourdieu's conception of the dialectic between middle-class taste, artistic cultural production, and cultural capital, Ley argues that artists are attracted to "authentic" places, poverty-stricken neighborhoods untouched by city planners and developers. Yet the professional class is attracted to the habitus of the artist, following their "aesthetic disposition":

> the aesthetic appropriation of place, with its valuation of the commonplace and off-center, appeals to other professionals, particularly those who are also higher in cultural capital than in economic capital and who share something of the artist's antipathy towards commerce.
>
> *(Ley, 2003: 2540)*

Built into the dialectical relationship between cultural and economic capital is the conflict between those who produce culture but are resistant to capital place "improvements" and those who are attracted to cultural production as consumers and wish to further the economic rise of neighborhoods through development and gentrification (Currid, 2009; Ley, 2003).

The outcome of this tension is that artists are often hostile to residential (through individual homeowners and community associations) and institutional (through community development corporations, universities, hospitals, philanthropic organizations, and political entities) attempts to revitalize neighborhoods, and for good reason. Because artists do not necessarily have middle-class access to networks of financial capital, without property ownership, artists cannot always choose to stay in gentrifying neighborhoods (Markusen, 2006; Rich and Tsitsos, 2016). Yet if a neighborhood is themed as an arts and cultural district, what can help stave off artist displacement when the neighborhood branding strategy stimulates real estate development and building rents rise? What of other neighborhood stakeholders, such as small business owners, renters, and poor "legacy" (longtime) residents, who, like the artists, are also threatened by gentrification? My research in the Arts and Entertainment District (A&E District) of Station North investigates these questions, finding that public-private partnerships and elite networks attempt to stave off displacement of the aforementioned groups, while at the same time, further theme the neighborhood as a center for artistic cultural production through arts-centered re/developments. I now turn to the establishment and purpose of A&E Districts in Maryland and specifically, Station North.

Creative placemaking and arts districts

Arts and entertainment-themed districts are inarguably tied to the trend in consumer-based cultural development strategies of urban revitalization and redevelopment, but are also integral to the creative placemaking trend in public policy and public arts funding. Markusen and Gadwa's white paper for the city of Philadelphia and the National

Endowment for the Arts (NEA) outlines the myriad ways creative placemaking can and has been implemented in cities throughout the United States, with arts districts being one of the most impactful, ongoing placemaking projects (2010b). They use Cleveland's Gordon Square Arts District and the city of Providence's revitalization through multiple arts districts as examples of successful urban creative placemaking projects that stimulated the economy and made neighborhoods more "livable" (Markusen and Gadwa, 2010b).

Definitions of creative placemaking are tough to pin down. Is any community project tied to cultural production "creative placemaking"? Markusen and Gadwa's white paper offers this definition:

> Creative placemaking ... partners from public, private, non-profit, and community sectors strategically shape the physical and social character of a neighborhood, town, city, or region around arts and cultural activities. Creative placemaking animates public and private spaces, rejuvenates structures and streetscapes, improves local business viability and public safety, and brings diverse people together to celebrate, inspire, and be inspired.
>
> *(Markusen and Gadwa, 2010b: 3)*

While they are aware of some downsides of creative placemaking projects, such as gentrification, residential displacement, lack of funding, and an absence of organizational cooperation, their unrelenting support of the power of artists, artistic production, and placemaking as an economic boon to communities has been challenged by skeptics. This includes Gadwa, who wrote in a subsequently published paper (Nicodemus, 2013) that creative placemaking is a malleable and "fuzzy" concept, wholeheartedly adopted by political elites, policymakers, and funding organizations in their development strategies. Others have criticized creative placemaking and arts districts as merely a branding strategy with little actual economic benefit for artists and neighborhoods (Evans, 2003, 2005; Rich and Tsitsos, 2016; Strom, 2010). Still others question the public inclusivity of placemaking activities (Darchen, 2013; Mathews, 2014; Montgomery, 2016; Sharpe et al., 2005; Stern and Seifert, 2007, 2008). What "place" is being made and by whom? Bedoya's now classic essay, "Placemaking and the Politics of Belonging and Dis-belonging" makes a forceful critique of the reimagining of cities through placemaking: "the blind love of Creative Placemaking that is tied to the allure of speculation culture and its economic thinking of 'build it and they will come' is suffocating and unethical, and supports a politics of dis-belonging employed to manufacture a 'place'" (Bedoya, 2013). He argues that for creative placemaking to be truly effective, it needs to be tied to the communities and the authentic spaces already in existence, rather than an effort to impose an outside aesthetic to "change" a community. This is a longstanding tension in public arts: what is truly public art and publicly informed artistic creation may not have any tangible economic value, meaning, it does not produce art that elites want to view in galleries or purchase (Gielen, 2011; Miles, 1997).

In my own investigation of the creative placemaking and revitalization strategies of the Station North Arts & Entertainment District (SNAED) in Baltimore, Maryland, I found tensions between different neighborhood stakeholders regarding the purpose of the district and its placemaking activities. To provide sociohistorical context, Baltimore has a population of 614,664 people (2016 U.S. Census estimate), down from almost a million residents in 1950. At its industrial peak, it was a hub of shipbuilding, textile, steel, and other types of manufacturing, and is still a major shipping port for the eastern U.S. While manufacturing has declined in the city, the technology, education, and health care sectors have grown. Like many cities in the U.S., Baltimore is a racially segregated city, a product of its Southern history of Jim Crow segregation laws, discriminatory real estate practices, and white/middle-class flight. There continues to be a large economic disparity between the majority population of African Americans and the minority population of whites (63.3 percent and 31.4 percent of residents, respectively), with African Americans' median incomes half that of whites. Whites in Baltimore are also more likely to have college degrees, own their own homes, and hold wealth than African Americans (Wells, 2017). Note: this research project was completed with William Tsitsos [Towson University] between 2013 and 2014 [with additional research in 2016–2017] and consisted of ethnographic participant observation and forty semi-structured interviews with neighborhood stakeholders, including residents, business owners, artists, and those who work for organizations involved in neighborhood revitalization.

"Station North" is a relatively recent renaming of an area in the center of Baltimore that consists of three distinct neighborhoods: Charles North, a largely commercial district north of Penn (Rail) Station; Greenmount West, a mixture of vintage manufacturing buildings and rowhouses; and Barclay, an economically distressed working class rowhouse neighborhood (the Station North A&E District only includes a few blocks of this neighborhood). According to the U.S. Census Bureau's 2015 American Community Survey, the main census tract in Station North has a population that is 62.3 percent African American, 29.9 percent white, 2.8 percent Asian, and 2.9 percent Latino/a. The white population has grown by 44.7 percent since 2010 and by 333 percent between 2000 and 2010 (U.S. Census Bureau, 2017).

When Station North won its bid to be the first A&E District in Baltimore in 2002, its application won on the strength of tying two quite disparate adjacent neighborhoods already chock full of artists, galleries, theaters, and DIY (do it yourself) creative spaces. "Station North" was thus born. Its current mission statement is:

> Station North Arts & Entertainment, Inc. employs an arts-based revitalization and placemaking strategy by managing quality public art projects, providing thought-provoking programming, and forging strong supportive relationships with local artists, designers, residents, businesses, and institutions to guide development in the Station North Arts & Entertainment District
>
> *(http://www.stationnorth.org/about/).*

One major placemaking project supported by SNAED is Open Walls, a mural project completed by internationally renowned artists over the springs of 2012 and 2014. It radically altered the landscape of the neighborhoods within Station North, with bright murals painted over homes, warehouses, parks, and office buildings. Not surprisingly, it has been viewed skeptically by some as a branding and development scheme with little residential input. A tavern owner commented:

> The [Open Walls] murals, people complain about . . . to me that's kind of like where the developers piss and by doing it in murals and marking their territory, because then they develop that – right around there. You know, [Open Walls] doesn't make any sense. It's not helping the economy . . . With gentrification, revitalization, you've got basically two options. One, *with* the people that are there or *without*, and they had a full plan of without.
>
> *(quoted in Rich and Tsitsos, 2016: 747–748)*

Even those I spoke to who represented pro-development and tourism organizations admitted that the Maryland Arts & Entertainment District program had little tax benefit for individual artists, but rather was a branding strategy for the neighborhood. The state-sanctioning of arts districts in Maryland has the purpose "to develop and promote community involvement, tourism and revitalization through tax-related incentives that attract artists, arts organizations and other creative enterprises" (Maryland State Arts Council, 2017). However, stakeholders report that most individual property owners and artists, even those who own their own warehouses with artists' live/work studios, do not receive tax benefits due to eligibility requirements and the complexity of the tax abatement applications.

When artists first started moving into the abandoned industrial warehouses of Greenmount West in the 1980s, there was little thought about state intervention for neighborhood revitalization and placemaking. According to the pioneering artists I spoke to who were able to access financial capital to buy their own buildings, the neighborhood was attractive simply because it held many enormous industrial buildings, ideal for large, noisy, or messy creative productions, and was in the geographic center of the city. As the derelict warehouse buildings started filling with artists, a community formed what Stern and Seifert (2007) call a "natural cultural district". In the nearby neighborhood of Charles North, galleries, independent theaters (both movie and performance theaters), and DIY spaces proliferated, reviving the major commercial thoroughfare of historic Charles Street.

Fast forwarding from the formation of the Station North A&E District in 2002 to today (2017): the pioneer artists may have been involved in the original bid for the designation, yet there is heavy ambivalence regarding neighborhood development that stems from the tangle of institutional interests and public-private partnerships that want the A&E District to economically "succeed". While the artists saw potential in the neighborhood to grow creatively and form a community, community developers and other institutional interests (universities, financial groups, philanthropic organizations, and city government) saw a chance to finally "fill the doughnut hole" in the center of the city, the "No Man's Land" undermined by white flight and financial disinvestment since the 1968 riots in the aftermath of Dr. Martin Luther King, Jr.'s assassination (Elfenbein, Hollowak, and Nix, 2011; Westbrook, 2014). And thus artists are used, as one artist put it, "as a tool of development" and gentrification, the first-wave that puts in the sweat equity to revive a neighborhood (Rich, 2017). The unintended consequence of artist communities is that they lay the groundwork for middle-class occupation en masse, allowing private and political interests in developing the neighborhood for middle-class tastes (Ley, 1996, 2003). Station North is similar to so many other gentrifying urban neighborhoods throughout the world, well documented by researchers in cities such as Barcelona (Martí-Costa and Miquel, 2011), New York (Mele, 2000; Smith, 1996; Zukin, 1982), Portland, Oregon (London, 2017; Shaw and Sullivan, 2011), Toronto (Bain, 2003; Catungal et al., 2009; Darchen, 2013; Grodach, 2013), and Washington, DC (Hyra, 2017).

Station North may depart from typical neighborhood gentrification narratives because there is a distinct awareness from community developers and their political and financial backers that the neighborhood has two separate populations at risk of displacement: 1) the long-term poor and working class African American residents (so-called "legacy" residents by neighborhood stakeholders) who replaced the white ethnic population in the 1960s as the factories shuttered and work became scarce; and 2) the aforementioned artists, most of whom rent their studio and living spaces and may live in buildings that are not compliant with the city's fire code. A representative from one of the most prominent community development corporations in Central Baltimore explained:

> One of the major challenges has also been in terms of affordable artist housing … building that relationship with the city's code enforcement department to not just blanket these neighborhoods with code enforcement, but be much more strategic about it, so we had to change the entire thinking of the code enforcement department on how they actually do what they do. Because if that were the case then all of those buildings along Federal and Oliver [Sts.] would be easily shut down, especially five years ago. None of them were up to code, and most of them aren't still, but then you lose the entire residential artist population for the most part in the neighborhood.
>
> *(quoted in Rich, 2017: 8)*

The conundrum for the city and community developers is complex. How do you revitalize an arts-themed neighborhood that will sustain itself economically through tourism, consumption, and commercial activity without displacing the very people that give it its authentic flavor? According to those working for Baltimore's most influential institutions, it is a delicate balance between supporting existing artists, residents, community associations, and businesses, and working behind the scenes to secure financial backing from major financial and foundational institutions, as well as tax credits from federal, state, and city entities. Through public-private partnerships, major redevelopments and new capital projects have been completed in the past ten years, including City Arts I and II, two new apartment buildings in Greenmount West, each with more than sixty subsidized, below-market-rate units for artists. Many of the other projects completed are tied to arts and cultural production,: the Fred Lazarus IV Center, a graduate studio center for Maryland Institute College of Art (MICA is one of the top-ranking arts and design colleges in the U.S.); the Motor House, a rehabilitated automobile storage warehouse that now houses artists' studio space and non-profit organizations; the Centre Theatre, a previously derelict arts deco theater that now is the home of the Johns Hopkins University/MICA joint film studies program; and the Baltimore Design School (grades 8–12) in the Lebow Building, formerly a Cork, Crown, and Seal Co. factory (see Rich, 2017 for a full list and description of major completed projects and business closings in Station North since 2002).

The theming of Station North as an arts and cultural district is a product of the "creative city" policy trend previously discussed in this chapter (Florida, 2002, 2004; Miles and Paddison, 2005). Even if there is little proof that arts and culture-based development helps revive economically distressed neighborhoods, there is no doubt that arts and culture-based revitalization makes neighborhoods more attractive to the middle class (Lees, 2000, 2003; Ley, 2003). However, if an arts-themed neighborhood displaces its longtime residents and its artists all in the name of making a "place", then it destroys the diverse cultural and economic basis that made it a locus of artists in the first place (Bain, 2003; Brown-Sarancino, 2009; Montgomery, 2016; Rich, 2015). If artists are not preserved through public and private intervention, the neighborhood will become an "arts district without artists" (Rich, 2015). To disrupt this gentrification cycle, public-private partnerships have reimagined the arts district as one tied to cultural production, rather than merely a place for the middle-class non-artist population to enjoy cultural events and, perhaps, to move into. With strong ties to the adjacent (and deeply financially endowed) universities, most of the new developments in Station North are tied to arts education and production.

Many community development leaders recognize that economic "success" of the neighborhood will inevitably drive out artists as rents rise. In response to these concerns, an experimental subsidization model of development is being trialed in some of the new/

rehabilitated buildings in the neighborhood. A director of a wealthy foundation explains the business model at work here:

> So, our model . . . which hasn't been done, but we're going to try to do it and figure out how to try and make it work is you have some commercial tenants who are paying market rates. That's the restaurant and maybe some of the office tenants and that helps subsidize some of the space that you can offer to artists at the affordable point that they can afford to pay, that six dollars a square foot or whatever it is. So, you know, that's what we're wrestling with now is how we can make the investment necessary, bring the building up to code and attract the commercial tenants so we can subsidize the artists and make it all sustainable without ongoing support from the foundation.
>
> *(quoted in Rich, 2017: 9)*

The question here is whether an artist subsidization model is sustainable for artists' economic stability (which is outside of the purview of this chapter) and artists' spaces for creative production and residency. Other concerns are whether the subsidization model creates a fabricated landscape of arts-themed developments, which may have very little to do with commerce and an economically robust, healthy neighborhood. A creative industry business owner complained about the artists subsidized developments:

> The neighborhood is all about the artists and it's not about creative industry . . . they talk a lot about the demographic changes that happened to the neighborhood, about what manufacturing jobs left and the neighborhood went down. But how they've chosen to solve that issue is basically by warehousing artists here. And all the activity seems to be geared towards putting artists . . . in "maker spaces", essentially. But there's nothing about any of these spaces that are conducive to commerce, which would keep them here. So basically they've created . . . a ghetto, of sorts, of artists and . . . it seems like they may feel like they're in the zoo . . . that they're zoo animals on display in the district and they're also resistant to the gentrification aspect of it.
>
> *(quoted in Rich and Tsitsos, 2016: 751–752)*

Many stakeholders in Station North noted the conflict between real estate development and artist sustainability, as property values rise due to renewed interest in this centrally located neighborhood. Artists helped to make the arts district possible through their sweat equity, yet (as noted in the quote above) artists and other neighborhood stakeholders are well aware that they are at risk of displacement as residential and commercial rents rise. In this sense, artists are "tools of gentrification" and developers, there to generate capital interest in economically depressed, abandoned neighborhoods, but ultimately unable to sustain their presence without ongoing financial backing (Rich, 2017). We have already witnessed a number of long-running independent arts collectives shuttered in the past few years due to real estate pressures. These include the Bell Foundry, a DIY (do it yourself) collective and artists' warehouse (condemned in 2016 a few days after the Oakland, CA warehouse fire that tragically killed thirty-six people) and the Hour Haus, a recording studio and music rehearsal space that had been active for 25 years. Are there other models, beyond subsidization, that can sustain an artistic presence in the neighborhood?

Figure 24.1 The condemned Bell Foundry arts collective warehouse next to a newly built office complex in Greenmount West.

An alternative model is to shift artists from a renter class to an owner class. Before Station North became an official A&E District, artist collectives were established in two giant shuttered factories in Greenmount West: The Cork Factory and Area 405. The financing for artist-owned buildings can be complicated because many artists lack the capital resources needed for large loan approvals of industrial buildings, where most require hundreds of thousands of (U.S.) dollars' worth of construction to make them legally inhabitable for humans (for instance, a Limited Liability Company [LLC] had to be formed in order for the financing to be approved for the purchase of the 66,000 square foot building that became Area 405). The two collectives mentioned here are at much less risk of displacement because the buildings and units within the buildings continue to be artist owned. However, there are many historic single-family rowhomes surrounding these buildings, which since 2002 have been rehabilitated by mostly middle-class families, rather than artists. While there have been a number of city-run programs to encourage homeownership and rehabilitation of vacant houses in Greenmount West, most artists I talked to maintained that they could not get loans to purchase homes in the neighborhood. Rather, the perception is that "speculators" with deep pockets are capitalizing on the gentrification of the neighborhood and buying up properties they never plan to live in, some using city-financed anti-vacancy incentives to do so.

From the perspective of non-artist residents and the community associations in the neighborhood, it is imperative to protect the neighborhood from overdevelopment. Maintaining artists is not a priority, instead, the emphasis is on neighborhood quality-of-life issues such as "crime and grime" and attracting middle-class families through green spaces and strong schools. Hyperaware that there are mounting pressures on longtime residents, the New Greenmount West Community Association worked with Baltimore City planners to establish a neighborhood plan, one that includes a diverse set of interests:

> Our vision for Greenmount West is that our community continues to be welcoming to emerging and successful working people, professionals, as well as artists and artisans. We believe the neighborhood population will grow substantially and become a more diverse community, in both age and income, while maintaining our ethnic, racial, and lifestyle diversity. We also envision that Greenmount West will be an attractive low- to medium-density housing alternative situated next to the higher-density residential and entertainment development envisioned for Charles North, encouraging more family oriented and long-term residential households.
>
> *(Greenmount West Community Association, 2010: 7)*

The vision statement above juxtaposes the very divergent visions for the two main neighborhoods that make up Station North, with Greenmount West's wish for growth through residential diversity in comparison to the commercial and entertainment developments in Charles North. In contrast, the Charles North master plan's executive statement is: "Charles North is Baltimore's next possible 'impossible' project. Similar to the Inner Harbor of the 1970s, it will create a major 'world class' and multi-ethnic destination in the heart of the city" (Charles North Vision Plan, 2008: 4). This includes themed reenvisioned sub-neighborhoods (such as "20th Street Asia Town Neighborhood", "North Avenue Market", "Cultural Center", and "Design Center") with revitalized commercial districts and newly built high-rise buildings, all tied to the highly anticipated multi-million dollar redevelopment of Penn Station (Rich, 2017). As a resident active in the Charles North neighborhood association explained,

> When we did the urban renewal on this plan we wanted to encourage development . . . So we decided . . . that we wanted as much [building] diversity and height as we could get . . .we could have buildings of the same scale as Manhattan. [Laughs] Now it seems absurd but we needed people to have vision because what we had here is a train station and empty parking lots and very few people. There's not that many residents and we need to encourage residents to come in, it's not just an artist's community . . . I think artists get confused. This is a commercial district.
>
> *(quoted in Rich, 2017: 11)*

The Charles North Vision Plan does recognize artists as vulnerable to displacement, but it is unclear how this will be avoided without artist subsidization models as currently used in the Motor House, which offers artists' studios at under-market rates. At the time of writing, the first new high-rise development in Station North is nearing completion: the Nelson Kohl Apartment Building, with 103 market-rate units, partially financed by state tax credits and an investment from Wendell Pierce (the actor who played Detective Bunk Moreland in *The Wire*, the beloved HBO television show set in Baltimore). It is not clear if any of the units will be set aside as "affordable housing", but the rhetoric surrounding the groundbreaking of the building constructs a vision in which this artsy development supports multiple communities "without displacement" (Simmons, 2016). Because the Kohl Building is steps away from the train station and in the heart of the A&E District, it is unlikely to offer cheap rents and not incentivize further development in the neighborhood. Homeowners in the neighborhood who can

Figure 24.2 The Copycat Building in Greenmount West. It has offered artists live/work rental studios since the 1990s.

afford higher tax rates welcome this type of development, and thus the neighborhood associations are generally supportive. Renters without means, including legacy (long-time) residents, artists, and other marginalized groups, may be displaced in the end. These types of developments do nothing to ease housing shortages, but rather attract a more white-collar, well-heeled population to the area. While "empty parking lots" start getting filled with people (who have jobs and pay taxes) the re-population of the neighborhood may be viewed by those artists and bohemians who previously claimed the neighborhood as a recolonization.

Reimagining Baltimore

In this chapter, I have considered the use of arts and culture-based development to spur revitalization in declining urban neighborhoods. Since the 1990s, use of arts and entertainment (A&E) districts and creative placemaking have been popular redevelopment strategies of states and policymakers. Although there is little evidence that A&E districts and creative placemaking activities produce real economic gains, there is much research that shows that these revitalization strategies leave a lasting mark on city landscapes, populations, and in how we collectively imagine and experience existing neighborhoods. Neighborhoods are living things that shift as populations and cultures shift. And yet as policymakers fell in love with the "creative city" thesis, neighborhoods are subjected to the whims of developers, state revitalization programs, and anchor institutions (such as universities and hospitals), all with "good intentions" to "lift all boats". Is it possible to maintain class and race diversity in arts-themed neighborhoods when any redevelopment to attract consumers and residents will inevitably raise land rents?

That said, the reimagining of Central Baltimore from a "No Man's Land" to the Station North Arts & Entertainment District does not fit into any neat gentrification trope of the top-down corporate development model. The stakeholders involved in neighborhood change represent many interests, including community development companies, a prestigious arts and design college, and a diverse set of homeowners and renters, artists, musicians, and workshop/gallery owners. The community associations that represent the two major neighborhoods in Station North were politically savvy in getting ahead of the major development decisions so that they have input into neighborhood change. Still, as one legacy resident and café owner bluntly explained, people who do not work or live in the neighborhood are "rearranging the chess board" (Rich and Tsitsos, 2016). There is a general feeling from residents that not all voices are heard equally and that there is a conflict between the "regular" non-artist population (who care more about "crime and grime" than preserving artists' spaces) and artists who are under threat of displacement as real estate is gobbled up by public-private partnership projects (ironically, many of them arts themed). Add two major anchor institutions (MICA and Johns Hopkins University) into the mix of stakeholder interests, and the picture becomes even more complex; these institutions are private, not democratically elected, and do not have to answer to constituents, as in the case of the Mayor's Office (Rich and Tsitsos, 2017).

The push-back from residents and neighborhood activists is palpable. For months, prominent graffiti on an empty billboard on North Avenue in 2014 asked, first, "Who Is Creative Placemaking?" and subsequently, "Who is Land Banking?" After the 2015 protests and uprising in response to the death of Freddie Gray (a young African American man from West Baltimore who died in police custody) and continued conversations within the Baltimore arts communities regarding the privileging of white artists over artists of color, LGBTQ artists, and other marginalized communities, institutions involved in continued revitalization must be more than "thoughtful" in their efforts. To return to Greenberg's work on city branding, "multiple urban imagineers may coexist and compete, but only those with the requisite political, economic, and cultural capital will have the power to brand an imaginary that will have enduring impact on the social life of the city" (2000: 255). Heads of institutions that influence development in Baltimore, as well as city leaders, must balance the wish for economic success and the need for social justice in this transforming city.

References

Bain, A. (2003) Constructing Contemporary Artistic Identities in Toronto Neighborhoods. *Canadian Geographer* 47(3): 303–17.

Bedoya, R. (2013) Placemaking and the Politics of Belonging and Dis-Belonging. *GIA Reader* 24 (1). Available at: http://www.giarts.org/article/placemaking-and-politics-belonging-and-dis-belonging

Brown-Sarancino, J. (2009) *A Neighborhood that Never Changes: Gentrification, Social Preservation, and the Search for Authenticity*. Chicago: University of Chicago Press.

Cameron, S. and Coaffee, J. (2005) Art, Gentrification, & Regeneration: From Artist as Pioneer to Public Arts. *European Journal of Housing Policy* 5(1): 39–58.

Catungal, J.P., Leslie, D., and Hii, Y. (2009) Geographies of Displacement in the Creative City: The Case of Liberty Village, Toronto. *Urban Studies* 46(5/6): 1095–114.

Clark, T., Lloyd, R., Wong, K., and Jain, P. (2002) Amenities Drive Urban Growth. *Journal of Urban Affairs* 24(5): 493–515.

Currid, E. (2009) Bohemia as Subculture; "Bohemia" as Industry: Art, Culture and Economic Development. *Journal of Planning Literature* 23(4): 368–82.
Darchen, S. (2013) The Creative City and the Redevelopment of the Toronto Entertainment District: A BIA-Led Regeneration Process. *International Planning Studies* 18(2): 188–203.
Elfenbein, J., Hollowak, T., and Nix, E. (eds.) (2011) *Baltimore '68: Riots and Rebirth in an American City*. Philadelphia, PA: Temple University Press.
Evans, G. (2003) Hard-Branding the Cultural City – From Prado to Prada. *International Journal of Urban and Regional Research* 27(2): 417–40.
Evans, G. (2005) Measure for Measure: Evaluating the Evidence of Culture's Contribution to Regeneration. *Urban Studies* 42(5/6): 959–83.
Florida, R. (2002) *The Rise of the Creative Class...And How It's Transforming Work, Leisure, Community, and Everyday Life*. New York: Basic Books.
Florida, R. (2004) *Cities and the Creative Class*. New York: Routledge.
Gielen, P. (2011) Mapping Community Art. In: De Bruyne, P. and Gielen, P. (eds.), *Community Art: The Politics of Trespassing*. Amsterdam: Valiz/Antennae Series, 15–33.
Gottdiener, M. (1997) *The Theming of America: Dreams, Media Fantasies, and Themed Environments*. Boulder, CO: Westview Press.
Greenberg, M. (2000) Branding Cities: A Social History of the Urban Lifestyle Magazine. *Urban Affairs Review* 36(2): 228–62.
Grodach, C. (2013) Cultural Economy Planning in Creative Cities: Discourse and Practice. *International Journal of Urban and Regional Research* 37(5): 1747–65.
Hyra, D. (2017) *Race, Class, and Politics in the Cappuccino City*. Chicago: University of Chicago Press.
Lees, L. (2000) A Reappraisal of Gentrification: Towards A "Geography of Gentrification." *Progress in Human Geography* 24(3): 389–408.
Lees, L. (2003) Super-Gentrification: The Case of Brooklyn Heights, New York City. *Urban Studies* 40(12): 2487–509.
Ley, D. (1996) *The New Middle Class and the Remaking of the Central City*. Oxford: Oxford University Press.
Ley, D. (2003) Artists, Aestheticisation and the Field of Gentrification. *Urban Studies* 40(12): 2527–44.
Logan, J.R. and Molotch, H.L. (1987) *Urban Fortunes: The Political Economy of Place*. Berkeley, CA: University of California Press.
London, J. (2017) Portland, Oregon, Music Scenes and Change: A Cultural Approach to Collective Strategies of Empowerment. *City & Community* 16(1): 47–65.
Markusen, A. (2006) Urban Development and the Politics of a Creative Class: Evidence from a Study of Artists. *Environment and Planning* 38(10): 1921–40.
Markusen, A. and Gadwa, A. (2010a) Arts and Culture in Urban and Regional Planning: A Review and Research Agenda. *Journal of Planning Education and Research* 29(3): 379–91.
Markusen, A. and Gadwa, A. (2010b) *Creative Placemaking* White Paper. Washington, DC.
Martí-Costa, M. and Miquel, M.P. (2011) The Knowledge City against Urban Creativity? Artists' Workshops and Urban Regeneration in Barcelona. *European Urban and Regional Studies* 19(1): 92–108.
Maryland State Arts Council. (2017) *Arts & Entertainment Districts*. Available at: https://www.msac.org/programs/arts-entertainment-districts.
Mathews, V. (2014) Incoherence and Tension in Culture-Led Redevelopment. *International Journal of Urban and Regional Research* 38(3): 1019–36.
Mayors' Institute on City Design, National Endowment for the Arts. Available at: http://www.nea.gov/pub/creativeplacemaking-paper.pdf.
Mele, C. (2000) *Selling the Lower East Side: Real Estate, Culture and Resistance in New York City*. Minneapolis: University Of Minnesota Press.
Miles, M. (1997) *Art, Space and the City: Public Art and Urban Futures*. London: Routledge.

Miles, M. (2005) Interruptions: Testing the Rhetoric of Culturally Led Urban Development. *Urban Studies* 42(5/6): 889–911.
Miles, S. and Paddison, R. (2005) The Rise and Rise of Culture-Led Urban Regeneration. *Urban Studies* 42(5/6): 833–39.
Montgomery, A. (2016) Reappearance of the Public: Placemaking, Minoritization, and Resistance in Detroit. *International Journal of Urban and Regional Research* 40(4): 776–99.
Nicodemus, A.G. (2013) Fuzzy Vibrancy: Creative Placemaking as Ascendant U.S. Cultural Policy. *Cultural Trends* 22(3/4): 213–22.
Noonan, D.S. (2013) How U.S. Cultural Districts Reshape Neighborhoods. *Cultural Trends* 22(3/4): 203–12.
Ocejo, R. (2014) *Upscaling Downtown: From Bowery Saloons to Cocktail Bars in New York City*. Princeton, NJ: Princeton University Press.
Rich, M.A. (2015) Arts Districts Without Artists *Metropolitics*, 23 March. Available at: http://www.metropolitiques.eu/arts-districts-without-artists.html.
Rich, M.A. (2017) "Artists are a Tool for Gentrification": Maintaining Artists and Creative Production in Arts Districts. *International Journal of Cultural Policy* 80(1): 1–16.
Rich, M.A. and Tsitsos, W. (2016) Avoiding the "Soho Effect" in Baltimore: Neighborhood Revitalization and Arts and Entertainment Districts. *International Journal of Urban and Regional Research* 40(4): 736–56.
Rich, M.A. and Tsitsos, W. (2017) New Urban Regimes in Baltimore: Higher Education Anchor Institutions and Arts and Culture-Based Neighborhood Revitalization. *Education and Urban Society* 50(6): 1–24.
Sharpe, J., Pollock, V., and Paddison, R. (2005) Just Art for a Just City: Social Art and Social Inclusion in Urban Regeneration. *Urban Studies* 42(5/6): 1001–23.
Shaw, S. and Sullivan, D.M. (2011) White Night: Gentrification, Racial Exclusion and Perceptions and Participation in the Arts. *City & Community* 10(3): 241–64.
Simmons, M. (2016) Colorful, Artsy 103-Unit Market-Rate Apartment Complex Underway in Station North. *Baltimore Business Journal*, 26 October. Available at: https://www.bizjournals.com/baltimore/news/2016/10/26/colorful-artsy-103-unit-market-rate-apartment.html.
Smith, N. (1996) *The New Urban Frontier: Gentrification and the Revanchist City*. New York: Routledge.
Stern, M. and Seifert, S. (2007) *Cultivating "Natural" Cultural Districts Brief*. Philadelphia, PA: The Reinvestment Fund.
Stern, M. and Seifert, S. (2008) *From Creative Economy to Creative Society*. Philadelphia, PA: The Reinvestment Fund.
Stern, M. and Seifert, S. (2010) Cultural Clusters: The Implications of Cultural Assets Agglomeration for Neighborhood Revitalization. *Journal of Planning Education and Research* 29(3): 262–79.
Strom, E. (2003) Cultural Policy as Development Policy: Evidence from the United States. *International Journal of Cultural Policy* 9(3): 247–63.
Strom, E. (2010) Artist Garret as Growth Machine? Local Policy and Artist Housing in U.S. Cities. *Journal of Planning Education and Research* 29(3): 367–78.
U.S. Census Bureau. (2017) *2000 and 2010 Population Summary Files, 2015 American Community Survey*. Available at: www.census.gov.
Wells, C. (2017) Report Highlights Economic Disparities between Races in Baltimore. *The Baltimore Sun*, 30 January. Available at: http://www.baltimoresun.com/news/maryland/baltimore-city/bs-md-racial-wealth-divide-20170130-story.html.
Westbrook, H. (2014) Baltimore's Station North: Art, Alchemy, and Creative Placemaking in Action. *Europe in Baltimore*, 15 January. Available at: http://europeinbaltimore.org/station-north/.
Zukin, S. (1982) *Loft Living*. Baltimore, MD: Johns Hopkins University Press.

25

Jia Zhangke's cinematic vision of urban dystopia in contemporary China

Yingjin Zhang

Urban dystopia in contemporary China

We may start our examination of dystopia with Fredric Jameson's somber observation:

> [T]he postmodern city seems to be in permanent crisis and is to be thought of, if at all, as a catastrophe rather than an opportunity. As far as space is concerned, the rich are withdrawing ever more urgently into their gated communities and their fortified enclosures; the middle classes are tirelessly engaged in covering the last vestiges of nature with acres of identical development homes; and the poor, pouring in from the former countryside, swell the makeshift outskirts with a population explosion.
>
> (*Jameson, 2010: 22*)

Jameson may have had Los Angeles in mind when he theorizes the postmodern city as dystopia, but recent urban studies has brought attention to similar global developments in "other cities, other worlds", such as Beijing, Cairo, Istanbul, Johannesburg, Mumbai (or Bombay), and São Paulo (Huyssen, 2008). If we modify Jameson's description of the middle classes, who have filled identical high-rise apartments in urban China instead of encroaching into nature as their counterparts do in the United States, his observation above fits the typical cityscape of contemporary China, where gated communities have appeared in ever-larger numbers and a floating population of urban migrant workers from the countryside – estimated at 282 million by the end of 2016 (CLB) – has labored far away from home to sustain the miracle of China's unprecedented economic growth since the early 1990s.

Like its Western counterpart, the postmodern city in China – although "postsocialist" is often preferred (McGrath, 2008) – evokes disturbing images associated with dystopia:

> We see landscapes defined by ruin, death, destruction. We see ... derelict buildings, submerged monuments, decaying cities, wastelands, the rubble of collapsed civilizations. We see cataclysm, war, lawlessness, disorder, pain, and suffering. Mountains of uncollected rubbish tower over abandoned cars.
>
> (*Claeys, 2017: 3*)

Such dystopian images have found a powerful expression in Jia Zhangke's cinematic vision of urban dystopia in contemporary China, which has persisted from his college student short *Xiao Shan Going Home* (Xiao Shan huijia, 1995) and his debut underground feature *Xiao Wu* (Xiao Wu, 1997) to his recent historical reflection *Mountains May Depart* (Shanhe guren, 2015). As an internationally celebrated film director (Berry, 2009; McGrath, 2008: 129–64), Jia has expanded his urban imaginaries, through an intricate web of images, songs, and dialog, from his small hometown city of Fenyang in Shanxi province to China's capital city of Beijing, and further to foreign cities such as Paris and Ulaanbaatar (Ulan Bator), all referenced in *The World* (Shijie, 2004), his first feature officially approved for production and exhibition in China.

This chapter investigates dystopia as Jia's distinctive contribution to urban imaginaries in contemporary China. Historically, dystopia in contemporary China emerged in reaction to the past utopia of socialist nation-building and the present euphoria of market capitalism in which the official slogan is, unambiguously, "to get rich is glorious". If we accept that dystopia, seen as "utopia's twentieth-century doppelgänger", "is a utopia that has gone wrong, or a utopia that functions only for a particular segment of society" (Gordin et al., 2010: 1; for other terms related to utopia, see Claeys, 2017: 273; Vieira, 2010: 4), we recognize that urban dystopia in postmodern, postsocialist China has emerged in reaction to two kinds of utopianism that swept over China in quick succession – socialism of the 1950s–1970s and consumerism from the 1990s to the present. First, the socialist utopia projected an idealized society of equality and sufficiency for all, but periodic nationwide political campaigns in the name of a permanent revolution fundamentally destroyed social equilibrium and left the population in constant fears of persecution and deprived of basic needs to sustain their idealism. Second, the consumerist utopia promises abundant private wealth and fast-speed development, but dreams of affluence and mobility are glorified to cover up an ever-widening gap of economic inequality, regional unevenness, environmental damages, emotional wounds, and other widespread problems in China's integration into globalization.

As an artistic vision in contemporary China, dystopia has found expression not only in cinema but also in other forms of arts and literature. Jeffrey Kinkley demonstrates that China's new historical novels from the 1980s onward prefer social satires on the dark side of human nature and that, in literary dystopia, "it is movement through time, not space, which takes the novels' pent-up characters into different worlds" (2014: 1). By contrast, contemporary Chinese visual arts often concentrate on space and place in projecting dystopic urban imaginaries, although such imaginaries are informed by historical references (Visser, 2004; Wu, 2005: 165–233). Similarly, recent Chinese cinema grounds its dystopian vision on urban spaces and places, especially those associated with underprivileged people and endangered by rapid development. Just as the "eutopia" (the good place) of socialism and the "euchronia" (the good place in the future) of communism are both subject to interrogation in literary dystopia (Yan, 2009, 2012), the utopia of globalization in which China celebrates its rise as a global superpower is refigured as urban dystopia in contemporary Chinese cinema from the perspective of the marginal, the liminal, and the estranged.

In what follows, I examine Jia Zhangke's cinematic vision of urban dystopia in contemporary China in three sections – devoted respectively to his rendition of conceptual dystopia in terms of mobility, liminality, and spectrality, his projection of visual dystopia onto debris, ruins, and walls, as well as his intensification of emotional dystopia by way of estrangement, escape, and exile. Rather than taking dystopia merely as cautionary tale or pessimistic scenario, I conclude in favor of conceptualizing critical dystopia as artistic intervention that demands justice and preserves hope.

Conceptual dystopia: mobility, liminality, spectrality

For dystopia to take shape, an external view is necessary to conceptualize something that has gone wrong or something that works only for a segment of society. In Jia Zhangke's cinematic oeuvres to date, this external view is typically obtained from a mobile spectatorship in a specific type of China's urban imaginaries: *xiancheng*, the county-level city, administratively above a township (*zhen*) and a village (*xiang*) but below a district (*diqu*)-level city (*shi*) and a province (*shen*). Such mobile spectatorship enables Jia to explore China's changing postmodern, postsocialist cityscapes, and *xiancheng* functions as a space of liminality that carries profound contradictions in China's urban imaginaries in the era of globalization.

Jia's cinema has been described as "an attempt at cognitive mapping" that has produced "a particular typology" of *xiancheng* as an increasingly fragmented urban sphere in contemporary China, while Jia is seen as "more engaged in finding a legitimate perspective from which to capture a reality that is simultaneously slipping away from experience and coming back to haunt and overwhelm it at an abstract, mythological level" (Zhang, 2010a: 76). The perspective Jia has found, I suggest, is that of dystopia, which foregrounds a mobile spectatorship vis-à-vis an alienating cityscape and captures images of liminality that verges on imminent disappearance and ghostly returns, thereby haunting the present in visual, sonic, and conceptual realms. Interpreting *xiancheng* as "the in-between, generic area where the daily reality of contemporary China is laid bare", Xudong Zhang elaborates on its ambivalence: "With no clear-cut boundaries or sharp distinctions between rural and urban, between industrial and agricultural, between high and low cultures, *xiancheng* becomes a meeting place for all kinds of forces and currents, whether contemporary or anachronistic" (2010: 77). *Xiancheng* becomes liminal in both spatial and temporal terms, both here and elsewhere, both present and absent, both grounded in the past and yearning for the future. As such, it slips away from utopia and threatens to vanish into dystopia.

From the start of his career, Jia has relied on the wanderer to capture the liminality of his hometown Fenyang as a typical dystopic *xiancheng* in China. Like his prototype Xiao Shan (played by Wang Hongwei), who is stranded in Beijing before the Chinese New Year, the titular pickpocket (Wang Hongwei) in *Xiao Wu* finds himself alienated in his hometown Fenyang as "a wanderer in a cityscape where he is unable to establish roots or a foundation" (Lu, 2014: 324). Wearing glasses and oversized Western suit and pants, Xiao Wu roams Fenyang streets, sometimes alone, sometimes with his underage male protégés, and one time with his temporary love interest Hongmei, a karaoke club hostess who subsequently abandons him for a client out of town without saying good-bye. In his aimless promenade with Hongmei, the camera captures – in a documentary style – random street scenes of closed store fronts, half-demolished buildings, debris-scattered alleyways, and congested intersections. Flaunting her fashionable outfits, hairdo, and high-heeled shoes in an unappreciative urban environment, Hongmei soon gets bored, stops by a rundown store and places a long-distant call to her mother, pretending she is auditioning for a film role in Beijing. Xiao Wu stands nearby, smokes a cigarette, and listens amidst street noise. There is nothing glamorous about Fenyang: streets are nondescript, buildings dilapidated, and brick walls worn-out and filthy. Fenyang is devoid of any spectacular signs of globalization that showcase metropolitan cities like Beijing and Shanghai as symbols of the consumerist utopia.

As Michel de Certeau theorizes:

> To walk is to lack a place. It is the indefinite process of being absent and in search of a proper. The moving about that the city multiplies and concentrates makes the city itself an immense social experience of lacking a place.
>
> (*1984: 103*)

Through acts of walking, urban places are rearticulated as a series of lacunae and anti-texts, as that which has always already escaped or contradicted the "proper" text – the utopian dream proffered by the original urban planning. Understood in this light, Xiao Wu's walking in Fenyang composes a migratory city text marked by a kind of liminality that gestures toward illegitimacy, right under the watchful eye of the authorities who insist on the law and order and announce on loud speakers along the streets a serious crackdown on crimes (a practice reminiscent of the socialist past). Indeed, Xiao Wu's illegitimacy – or the illegitimacy of his migratory "anti-texts" in China's *xiancheng* – is highlighted time and again by people around him: by a former pickpocket friend Xiao Yong who has transformed himself into a private entrepreneur and savors the spotlight of the local television coverage; by Xiao Wu's father who wields a wooden stick and drives Xiao Wu out of their shabby rural house for being unfilial and useless; and by policeman Hao who discovers Xiao Wu's sympathetic acts of returning personal ID cards after keeping the stolen money from his victims' wallets. Interestingly, in the memorable final scene, in which Xiao Wu is handcuffed by policeman Hao to an electricity pole on the street side, the camera unexpectedly adopts Xiao Wu's point of view and shows a crowd of onlookers who interrupt their daily routines and gather around him. In this moment of apparent stasis, the mobile camera acts as Xiao Wu's returned gaze and opens a new space of liminality that recognizes the participation of the onlookers, and by extension the film's viewers, in conceptualizing Fenyang as a sure icon of urban dystopia.

Example of such mobile spectatorship abounds in Jia's subsequent films when his motif of wandering (Holtmeier, 2014) is extended to various vehicles of transportation. In *Unknown Pleasures* (2002), Binbin and Xiao Ji, two unemployed young drifters, are often seen riding a motorcycle across Datong, another inland city not far from Fenyang, while their frustrated desires are articulated by their random acts of moving around town, as rendered in the following three scenes. In one, Yuanyuan, Binbin's girlfriend, who is leaving Datong for college education in Beijing, rides a bicycle in an empty waiting hall in the local railway station. Slowly, she circles around, waiting for Binbin to join her, but Binbin, who has just given her a Motorola cellphone as a farewell gift, sits still and watches in silence as Yuanyuan rides her bicycle to an adjacent waiting hall and disappears out of view. In another scene, Xiao Ji rides the motorcycle across a dried-up river bed and tries repeatedly to race up the steep river bank, while the viewer is shown a depressing panoramic view of identical old gray residential buildings under a gloomy sky in the background. A similar dystopic view of this residential neighborhood is shown earlier when Qiaoqiao, Xiao Ji's love interest, walks across the debris-filled river bed, her hands holding a black shawl over her head to block the sunlight. In the third scene, after Binbin is arrested in an attempted bank robbery in imitation of *Pulp Fiction* (Quentin Tarantino, 1994), with fake explosives around his waist instead of a handgun, Xiao Ji escapes on his motorcycle and races along the newly completed Beijing-Datong Highway. He runs out of fuel and abandons his motorcycle in the rain. The long shot of the unattended solitary motorcycle on the highway symbolizes the inaccessibility of the under-developed *xiancheng* to China's affluent capital city, at least from the perspective of under-privileged inland urban youths.

Contrary to the utopian vision of the human mastery of technology showcased in the arrival at the city by train in *Berlin, the Symphony of a Great City* (Walter Ruttman, 1927) and *Man with the Movie Camera* (Dziga Vertov, 1929), in Jia's films "the train becomes a symbol of both mobility and entrapment" (Lovatt, 2012: 422). In *Unknown Pleasures*, the trains' departures and arrivals are announced offscreen. In *The World*, many migrant workers from Shanxi are confined to The World Park, a theme park in Beijing offering scaled-down structures of global tourist landmarks, such as the Eiffel Tower in Paris and the Twin Towers in New York, as architectural attractions (Mello, 2016). The elevated monorail train circles around the theme park through designated areas such as India and Mongolia, but it practically leads to nowhere. A similar scene of concealed entrapment is performed at the beginning of Jia's *Platform* (Zhantai, 2004), in which actors of the Fenyang county cultural and performing troupe deliver a song-and-dance piece titled "The Train Running to Shaoshan" while sitting on wooden stools in a line to simulate the passenger train on stage. The imagined mobility of traveling in homage to Shaoshan in South China, the birthplace of the supreme leader Chairman Mao Zedong, pays lip service to the bygone socialist utopia and dramatizes the dilemma of *xiancheng* artists who struggle, over ten years (1979–1989), to transition from the socialist to the postsocialist era.

Another vehicle deployed to foreground the contradiction of mobility and entrapment is the airplane. *The World* features the female protagonist Tao (Zhao Tao) dressed as a flight attendant sitting in a grounded airplane in the theme park, "a visual sign of the mobility-immobility dichotomy" (Mello, 2014: 193). The imagined freedom of migrant workers is further depicted in two other scenes. In one, two Shanxi migrants watch an airplane fly over the sky on top of a high-rise building under construction, as neither of them has ever taken an airplane trip, and neither will live in a high-rise apartment any time soon. In another scene, a flash sequence animates a drawn figure of Tao, again dressed as a flight attendant, soaring in the sky over Beijing's cityscape – an example of desirable freedom otherwise unattainable in her regimented daily life (Zhang, 2010b: 86–9).

The contrast between mobility and entrapment heightens the sense of liminality, and Jia tactfully introduces enigmatic figures in *The World* to allude to other perspectives and other lives. In the opening credit sequence, a ragpicker appears in a long shot as an unexpected intruder into an otherwise serene panoramic view of the distant theme park and high-rise buildings, and this "vagrant scavenger with debris-filled rucksack" (Wagner, 2013: 366) serves as a ghostly reminder of the dystopic vision that hovers above the film's image-track and sound-track. Indeed, Jia's construction of "spectral soundscapes" in *The World* augments conceptual dystopia in the sonic realm, and silent appearances of human figures conjure up phantom spirits: "the migrant's experience of liminality represents a kind of ghostliness" (Lovatt, 2012: 423). After a migrant worker dies in an incident on the construction site, his aged parents arrive from Shanxi and are shown in utter silence in the hospital, almost like ghosts. Later, the parents remain silent, without tears, on the construction site where the deadly accident occurred and where other migrant workers burn paper money to mourn the death of their fellow Shanxi folk in the dark night.

Integral to his dystopian vison, "Jia's disenfranchised characters occupy a space of liminality", where their experience "is fractured, nonsynchronous and discordant", and their silence creates "a temporal pause – a symbolic space of resistance to the relentless linear trajectory of the grand narrative of postsocialist China" (Lovatt, 2012: 435). Adopting the perspective of *xiancheng* and foregrounding a mobile spectatorship associated with the under-privileged, Jia's vision of urban dystopia has produced "a universe of rented spaces haunted by a nowhere or by dreamed-of places" (De Certeau, 1984: 103). It is not a surprise that Jia ends *The World* in a pitch-dark screen, where Tao and her migrant worker boyfriend, both presumed to have died in a gas leak in a friend's suburban apartment, are heard in a phantom dialog: "Are we dead?" "No, this is only the beginning". It is a beginning of their afterlives as ghosts that belong to nowhere – not even in their hometown – and that haunt spaces of liminality such as life and death, light and darkness, utopia and dystopia.

Visual dystopia: debris, ruins, walls

The motifs of mobility, liminality, and spectrality similarly find resonances in Jia Zhangke's documentary, *I Wish I Knew* (Haishang chuanqi, 2010), which was commissioned for the 2010 Shanghai World Expo to illustrate its optimistic official slogan, "better city, better life". However, in symbolic resistance to the teleological narrative of an ever-progressing China, Jia skillfully inserts liminal spaces of dystopia between his key segments of interviews by featuring the phantom appearance of an unnamed fictional character played by Zhan Tao. Dressed in white (typical of the ghost in the Chinese imagination), she roams the contemporary global city of Shanghai and appears unexpectedly here and there like "an angel of history", a witness to generations of people in Shanghai who have suffered in times of wars, revolutions, persecutions, and migrations (Chiu and Zhang, 2015: 135–52). One such liminal scene is captured by Jia's camera in a long shot when Zhao, her back to the viewer, picks her way along debris-scattered, water-filled pavements in the Shanghai Bund – the debris generated temporarily in 2009 by the renovation of this popular tourist area before the 2010 Shanghai Expo.

Jia's obsession with urban debris can be traced back to *Xiao Shan Going Home*, in which his camera captures Beijing's unsavory scenes of grimy alleyways, shabby houses, and cramped rooms frequented by migrant workers in the mid-1990s. By the time he shot *Xiao Wu*, he found in urban demolition a perfect example of visual dystopia. As Sheldon Lu explains, "*Chai* (demolition) is the very theme of much contemporary Chinese visual art" as it "points not only to the physical demolition of the old cityscape but more profoundly to the symbolic and psychological destruction of the social fabric of families and neighborhoods" (2007: 167). Yomi Braester concurs, "Demolition sites are seen as timekeepers of urban history and cinema – that is, as an endeavor to retain the spatial repositories of personal and collective memory in effigy" (2007: 162). It is in demolition sites that "Jia's poetics of vanishing" (Zhang, 2010a: 87), which paradoxically works to retain the visible evidence of such vanishing, finds a perfect visual expression.

The best film to represent demolition as visual dystopia is Jia's *Still Life* (Sanxia haoren, 2006). This is set in Fengjie, another inland *xiancheng*, but this time an ancient town in central China near the picturesque Kuimen Gorge, which has now vanished under the rising Yangtze River water after the completion of the world's largest engineering project, the Three Gorges Dam. The film follows Han Sanming, a male miner from Shanxi, who travels to Fengjie to look for his long-estranged ex-wife and their daughter. Piles of debris are shown in what is left of the old town where Fengjie, and migrant workers are hired to demolish buildings and salvage reusable construction materials.

Sounds of primitive human acts of hammering are heard against the eerie sight of half-demolished structures. Indeed, in *Still Life*, demolition is no longer a mere metaphor but a real act caught in camera, as we see in a crucial scene described below:

> Han and his wife (whom he has finally found), have wandered off into the ruins of Fengjie at dusk ... The couple eventually wanders to an empty room, its walls streaked with burn marks and broken by yet another giant opening, even on the top but jagged and blown out at the sides near the bottom... The two face each other in silence until a large crashing sound draws them to the giant window. Through this opening they look into the distance at a multistory building that implodes and swiftly disappears in a cloud of dust.
>
> (*Byrnes, 2012: 79*)

As this scene illustrates, the ruin in Jia's films is as concrete as it is metaphorical and affective (Schultz, 2016). Based on his reading of the same scene, Pheng Cheah asserts that "*Still Life* exposes how globalization makes a world through ruination, or better yet, makes a ruined world" (2013: 196). Images of ruins punctuate the entire film, from a riverside junkyard to half-demolished buildings to the tall structure imploding to the ground mentioned above. By highlighting a migrant worker walking on a ruined map in *Still Life*, Jia foregrounds demolition and ruins and extends his vision of urban dystopia to a larger scale of globalization.

The power of ruins as visual dystopia has been acknowledged in scholarship. As if to answer the question, "Why has contemporary Chinese art since the 1990s developed such strong interest in urban ruins, as shown in endless examples of photographs, installations, performances and films that document demolished residential houses and dilapidated industrial sites?" (Wu, 2012: 9), Tze-lan Sang views ruins as "a rich reservoir of symbolic and allegorical possibilities, often mined by artists and writers to signify moral decay, the crumbling of value and belief systems, spiritual desolation, and nostalgia" (2017: 233). In Jia's case, ruins are also visualized as a site of spreading disease, as public health workers dressed in protective gears are seen spraying amidst urban wreckages, as well as a site of death, as a migrant worker is shown buried dead under a pile of debris and his body is carried away – these similar scenes are inserted in both *Still Life* and *Dong* (2006), Jia's documentary shot on location in Fengjie in the same year as *Still Life*, to highlight the liminality of fiction and fact, imagination and reality, art and history.

In Jia's cinematic oeuvre, the walls, like the ruins, often serve as an icon of visual dystopia, especially when they have stood intact despite historical changes. In *Platform*, Jia presents several impressive scenes of the ancient city walls and contrasts them with ubiquitous worn-out, dirt-covered gray brick walls of derelict houses in Fenyang. The city walls are frequented by Cui Mingliang (Wang Hongwei) and Yin Ruijuan (Zhao Tao), whose dating scenes are often rendered in medium and long shots. In one scene, on top of the city walls, the motionless camera captures Cui and Yin taking turns coming in and out of the frame, each facing the other whose view is blocked by a corner of the wall. This visual game of hide and seek foreshadows the ups and downs of their tenuous romantic relationship.

In Cecília Mello's reading, "the splendid city walls ... suggest not only immobility and entrapment but also the past and the weight of history which is the necessary 'slow' or 'still' counterpart to the fastness of the present", and Jia's recurring images of the walls thus shift the viewer's focus from time toward space, from slowness toward stillness

(2016: 140). Just as stillness is ingrained in the materiality of the walls, *Platform* works slowly to preserve public and private memories in the mnemonic walls, as well as in music and songs. In this sense, the walls in Jia's films function as what Laura Mulvey theorizes of cinema in general – capturing a "fossilized trace of the past ... the past as loss, as a jumbled accumulation of ruin and trace that survives the inexorable process of time's passing and human forgetting" (2004: 144).

The ancient city walls reappear, like "a phantom from the past" (Xiao, 2015: 31), at the end of Jia's *A Touch of Sin* (Tian zhuding, 2013): traveling to Shanxi to find a new job, Xiaoyu (Zhao Tao) walks by giant city walls against dusty winds, runs into a crowd of seniors who head in the other direction, and sees an open-air regional opera performance in a distance surrounded by a stretch of city walls with a magnificent tower in the middle. She joins the crowd and hears the magistrate on stage questioning the wronged woman Su San three times in a row: "Do you understand your sin?" The camera first cuts to Xiaoyu's despondent look (for she previously killed a sexual predator in self-defense in the film's third part and reported the killing by calling the police herself) and then shows, in a frontal medium shot, the silent crowd of onlookers who display no emotions on their faces. The magistrate's unanswered question regarding sin, guilt, justice, and violence is thus "hurled at the audience both on and off screen" as the film concludes with an imaginary court session where "everyone stands trial" (Xiao, 2015: 31). Evocative of visual dystopia, the city walls stand as a witness to historical injustice and human tribulations.

Emotional dystopia: estrangement, escape, exile

In comparison with conceptual and visual dystopia, emotional dystopia in Jia Zhangke's films may be less visible but proves to be equally unsettling or even lethal. In *Xiao Wu*, after his father drives him away from home, treating him as an outcast like the rest of society, the pickpocket looks around, and the rotating point-of-view shot reveals his childhood environment to be alienating to him and dystopic to the audience. Near the end of *Platform*, Mingliang's father has moved out to live with his mistress in another part of Fenyang, and his parents' emotional gap is symbolized by a deep ditch stretching along an entire street and separating houses on both sides, a conspicuous sight of urban renovation in contemporary China. In *A Touch of Sin*, the last straw that breaks a young factory worker's will to live is his unsympathetic mother demanding for money over a long-distant phone call: he walks into the factory compound where identical residential buildings tower above him on both sides, returns to his cramped dormitory room, jumps out of the balcony with his arms spread out as in flight, and hits the ground with his face down.

From his initial irony of Xiao Shan's failure to buy a train ticket to go home in *Xiao Shan Going Home* to his latest "decades-spanning, continent-hopping epic" (Foundas, 2015) centered on a separated family in *Mountains May Depart*, Jia has tracked the radical transformation of family, friendship, and intimacy in globalization and has consistently projected emotional dystopia through the tropes of estrangement, escape, and exile.

Set mostly in Fenyang, *Mountains May Depart* is divided into three parts, each announced by a specific year on the screen: 1999, 2014, and 2025. The first part begins with an exuberant group dance to the tune of the Pet Shop Boys' 1993 rendition of "Go West", a 1977 song by Village People:

Together, we will go our way
Together, we will leave someday
Together, your hand in my hands
Together, we will make our plans

The rest of the film, however, dramatizes the deterioration and termination of such "togetherness" among friends and families in postsocialist Fenyang and futuristic Melbourne. A love triangle among Shen Tao (Zhao Tao) and her two suitors, Jinsheng (a flamboyant gas-station owner who purchases a coalmine when coal prices hit the bottom) and Liangzi (a taciturn worker in a coalmine) is resolved when Liangzi decides to leave town, in an escape from losing Tao to Jinsheng. Liangzi refuses to accept the wedding invitation Tao has delivered to him; instead, he throws his house key over the roof in front of Tao and vows never to return home again. By the time the first part ends, Jinsheng is overjoyed at the birth of his son, whom he names Daole, a homophone for Dollar in Mandarin Chinese.

In the second part, Jinsheng and Tao have parted ways in divorce, and Daole/Dollar returns on a passenger flight from Shanghai, where he attends an international school and is indulged in cosmopolitan spoils like horse-riding and ski-boating, to join Tao in Fenyang for the funeral service of Tao's father, who died unexpectedly away from home in a railway station. The motif of house keys returns in two scenes in this middle part. In one scene, Tao gives Daole/Dollar her house keys as she accompanies him to Shanghai, intentionally taking the train to buy more time to bond with her son, who is fast slipping away from her as it has been arranged for him to emigrate to Australia by his father, now an investment capitalist in Shanghai named Peter. In another scene, Tao gives Liangzi's house key, which she retrieved from the roof top, to Liangzi, who has quietly returned to Fenyang with a wife and a baby, after working as a coalminer in the nearby Hebei province for fifteen years, and who now suffers from a serious lung disease. When Tao leaves Liangzi in his neglected house without electricity with 20,000 yuan for medical treatment and walks away with her dust-filled wedding invitation, she stops by a gray brick wall and has difficulty controlling her emotions.

The third part, set in the Melbourne of 2025 and conducted mostly in English dialog, projects a different vision of urban dystopia than those in Jia's earlier films. Now a twenty-something bored with everything, Dollar quits college and serves as a waiter at a restaurant in revolt against his father, Peter/Jinsheng, who lives in a luxury beachside apartment with a panoramic ocean view but never enjoys his affluence in Australia. With repeated shots of open landscapes and distant traffic on the highway, the Melbourne setting is more allegorical than futuristic, because Dollar now does not speak Chinese and has no memory of his mother, calling himself a test-tube baby in a Chinese class. Triggered by a Cantonese song his middle-age Chinese teacher Mia plays, which reminds him vaguely of what Tao let him listen to on their train journey in 2014, Dollar experiences moments of *déjà vu* and murmurs "tao" (which means waves in Chinese and is his mother's given name) to pounding sea waves. The film then cuts to Tao preparing dumplings in her Fenyang house, alone, as if she instinctively overhears Dollar's murmuring of "tao" faraway in Melbourne. She goes out for a walk in the snow. Against a familiar scene of a walled temple and a tall pagoda, Tao starts dancing by herself, as the Pet Shop Boys' cheerful tune of "Go West" increases in volume on the soundtrack.

For Elena Pollacchi, "The cacophony of images, languages, screen formats, and timeframes that converge in *Mountains May Depart* bring to mind Jia's postmodern cinematic trajectory", which "reveals the weaknesses of the utopian project" that is now officially promoted as the China Dream (2018: 220). The film's emotional dystopia thus conjures up "cacotopia", a place of widespread unpleasant feelings (Vieira, 2010: 4). Similar to critics who have found the Melbourne segment "awkward" (Foundas, 2015) or "the weakest" part of the film (Cheshire, 2016), Peter Bradshaw is unimpressed with "Jia's almost sci-fi sketch of the future" and surmises: "Did Tao, in fact, dream the movie's entire final Australian section? Jia allows us, fleetingly, to suspect this" (2015). Regardless of whose dystopian vision it is, Jia's futuristic Melbourne delineates what may be called "carcerotopia", "the most extreme form of prison state" (Claeys, 2017: 56), which proves profoundly ironic in *Mountains May Depart* because, for both Peter/Jinsheng and Dollar/Daole, it is a matter of self-incarceration or self-exile in a land of freedom (albeit historically a land of incarceration and exile for the British empire). Through estrangement, escape, and exile, *Mountains May Depart* compels its viewer to recognize emotional dystopia as a disturbing force and illustrates this insight: "Dystopia thus describes negative pasts and places we reject as deeply inhuman and oppressive, and projects negative futures we do not want but may get anyway" (Claeys, 2017: 498).

Critical dystopia in cinema

As suggested by carcerotopia, visions of dystopia tend to dwell on negativity, and scholarship on dystopia sometimes ends in pessimism, if not cynicism. Kinkley, for instance, soberly concludes his study of dystopia in China's new historical novels:

> Precisely where in life and art to look for justice, goodness, and creativity, when in life one sees rampant unpunished evil, ecological catastrophe, ... unfortunate outcomes in the grand sweep of history ... is a question with no resolution on the horizon. Utopia lacks wonder; dystopia lacks justice.
>
> (*2014: 207*)

Yet, even though the age of globalization may no longer evoke wonder and inspire utopia, I would argue that critical dystopia still demands justice and preserves hope. As demonstrated in this chapter, Jia Zhangke's vision of urban dystopia in contemporary China demands justice not only in his exposure of "rampant unpunished evil" in *A Touch of Sin*, "ecological catastrophe" in *Still Life*, and "unfortunate outcomes in the grand sweep of history" in *Mountains May Depart*, but also in his glimpses of hope, however momentary, in renewed human bonding. It is illuminating, therefore, to juxtapose two readings of the scene in *Still Life*, where Han Sanming places his right arm around his ex-wife's shoulder after they watch a high-rise building implode on the distant hill. On the one hand, Cheah realizes that, weighed heavily by dystopia, *Still Life* "can guarantee nothing" to the viewer because what it promises in a gesture of community "is also a principle of ruination, perhaps ruination itself" (2013: 203). On the other hand, as Corey Byrnes emphasizes, the scene conveys that "their sense of mutual responsibility and affection is revived by their shared vulnerability; the eerie collapse is also a synecdoche of the larger menacing and transforming external world that is best navigated communally" (2012: 79).

We may further expound dystopia in relation to "catastrophe" used by Kinkley above and by Jameson in the opening quotation of this chapter. In Jameson's case, "catastrophe" follows "crisis", which is translated into Chinese as *weiji*, "made of two characters that, read independently, mean 'jeopardy' and 'opportunity'" (Li, 2008: 265). Rather than Jameson's binary of the postmodern city as a "permanent crisis" or "catastrophe rather than an opportunity", urban dystopia at once recognizes "jeopardy" (*wei*) and preserves "opportunity" (*ji*). In this sense, dystopia is not all consumed by negativity. It is not always anti-utopia because dystopia does not abandon the opportunity of imagining hopeful alternatives. Peter Fitting thus differentiates dystopia from anti-utopia: "The critique of contemporary society expressed in the dystopia implies (or asserts) the need for change; the anti-utopia is, on the other hand, explicitly or implicitly a defense of the status quo" (2010: 141).

Reconceptualized this way, dystopia may be pessimistic but is not always cynical, certainly not entirely nihilistic. After all, "dystopia places us directly in a dark and depressing reality, conjuring up a terrifying future if we do not recognize and treat its symptoms in the here and now" (Gordin et al., 2010: 2). An imperative behind dystopia is therefore to preserve the potential for critical intervention. "Dystopias that leave no room for hope do in fact fail in their mission", Fátima Vieira admits, yet critical dystopia believes that there is "still a chance for humanity to escape, normally offering a glimmer of hope at the very end of the narrative" (2010: 17). Critical dystopia negates the negation of earlier reactions against utopia because it includes "at least one eutopian enclave or holds out hope that the dystopia can be overcome" (Claeys, 2017: 281–2).

In this positive light, I contend that Jia Zhangke's critical dystopia of contemporary urban China functions in the same way as literary dystopia does in general: "to warn us against and educate us about real-life dystopias" (Claeys, 2017: 501). Despite his pessimism, Cheah holds out hope in his comment on *Still Life*,

> To bear witness and to make a celluloid record of disappearing forms of human life or communities that are being destroyed by global capitalist modernization so that such cultures can be preserved in posterity for the gaze of the world at large: this is one of the main aims of world cinema.
>
> (*2013: 192*)

Motivated by a distinctive aesthetic of "neorealist preservation against neoliberal creative destruction" (Li, 2016: 169), Jia Zhangke's cinema of urban dystopia contributes directly to such a project of preserving cultures in world cinema.

References

Berry, M. (2009) *Xiao Wu, Platform, Unknown Pleasures: Jia Zhangke's "Hometown Trilogy"*. New York: Palgrave Macmillan.

Bradshaw, P. (2015) Mountains May Depart Review: Jia Zhang-ke Scales New Heights with Futurist Drama. *The Guardian*. Available at: https://www.theguardian.com/film/2015/may/20/mountains-may-depart-review-jia-zhang-ke-scales-new-heights-with-futurist-drama.

Braester, Y. (2007) Tracing the City's Scars: Demolition and the Limits of the Documentary Impulse in the New Urban Cinema. In: Zhang, Z. (ed.), *Urban Generation: Chinese Cinema and Society at the Turn of the Twenty-First Century*. Durham, NC: Duke University Press, 161–80.

Byrnes, C. (2012) Specters of Realism and the Painter's Gaze in Jia Zhangke's *Still Life*. *Modern Chinese Literature and Culture* 24(2): 52–93.

Cheah, P. (2013) World as Picture and Ruination: On Jia Zhangke's *Still Life* as World Cinema. In: Rojas, C. and Chow, E. (eds.), *The Oxford Handbook of Chinese Cinemas*. New York: Oxford University Press, 190–206.

Cheshire, G. (2016) Film Review: Mountains May Depart. *Roger Ebert*, 12 February. Available at: https://www.rogerebert.com/reviews/mountains-may-depart-2016.

Chiu, K.F. and Zhang, Y. (2015) *New Chinese-Language Documentaries: Ethics, Subject and Place*. London: Routledge.

Claeys, G. (2017) *Dystopia: A Natural History*. Oxford: Oxford University Press.

CLB (no date) Migrant Workers and Their Children. *China Labor Bulletin*. Available at: http://www.clb.org.hk/content/migrant-workers-and-their-children.

De Certeau, M. (1984) *The Practice of Everyday Life*. Trans. Rendall, S. Berkeley: University of California Press.

Fitting, P. (2010) Utopia, Dystopia and Science Fiction. In: Claeys, G. (ed.), *The Cambridge Companion to Utopian Literature*. New York: Cambridge University Press, 135–53.

Foundas, S. (2015) Film Review: Mountains May Depart. *Variety*. Available at: http://variety.com/2015/film/reviews/mountains-may-depart-cannes-film-review-1201501026/.

Gordin, M.D., Tilley, H., and Prakash, G. (eds.) (2010) *Utopia/Dystopia: Conditions of Historical Possibility*. Princeton, NJ: Princeton University Press.

Holtmeier, M.A. (2014) The Wanderings of Jia Zhangke: Pre-Hodological Space and Aimless Youths in *Xiao Wu* and *Unknown Pleasures*. *Journal of Chinese Cinemas* 8(2): 148–59.

Huyssen, A. (ed.) (2008) *Other Cities, Other Worlds: Urban Imaginaries in a Globalizing Age*. Durham, NC: Duke University Press.

Jameson, F. (2010) Utopia as Method, or the Uses of the Future. In: Gordin, M., Tilley, H., and Prakash, G. (eds.), *Utopia/Dystopia: Conditions of Historical Possibility*. Princeton: Princeton University Press, 21–44.

Kinkley, J. (2014) *Visions of Dystopia in China's New Historical Novels*. New York: Columbia University Press.

Li, D.L. (2008) *Yi Yi*: Reflections on Reflexive Modernity in Taiwan. In: Berry, C. (ed.), *Chinese Films in Focus II*. New York: Palgrave Macmillan, 265–72.

Li, D.L. (2016) *Economy, Emotion, and Ethics in Chinese Cinema: Globalization on Speed*. London: Routledge.

Lovatt, P. (2012) The Spectral Soundscapes of Postsocialist China in the Films of Jia Zhangke. *Screen* 53(4): 418–35.

Lu, J. (2014) Walking on the Margins: From Italian Neorealism to Contemporary Chinese Sixth Generation. *Journal of Italian Cinema & Media Studies* 2(3): 317–33.

Lu, S. (2007) *Chinese Modernity and Global Biopolitics: Studies in Literature and Visual Culture*. Honolulu: University of Hawaii Press.

McGrath, J. (2008) *Postsocialist Modernity: Chinese Cinema, Literature, and Criticism in the Market Age*. Stanford: Stanford University Press.

Mello, C. (2014) Jia Zhangke's Cinema and Chinese Garden Architecture. In: Lucia Nagib, L. and Jerslev, A. (eds.), *Impure Cinema: Intermedial and Intercultural Approaches to Film*. London: I.B. Tauris, 183–202.

Mello, C. (2016) If These Walls Could Speak: From Slowness to Stillness in the Cinema of Jia Zhangke. In: De Luca, T. and Jorge, N.B. (eds.), *Slow Cinema*. Edinburgh: Edinburgh University Press, 137–49.

Mulvey, L. (2004) Passing Time: Reflections on Cinema from a New Technological Age. *Screen* 45(2): 142–55.

Pollacchi, E. (2018) Jia Zhangke's *Mountains May Depart* (2015) and the China Dream, or, How Chinese Art Cinema Learned to Stop Worrying and Love Chinese Soft Power. In: Voci, P. and Luo, H. (eds.), *Screening China's Soft Power*. London: Routledge, 212–28.

Sang, T. (2017) Ruinscapes in Modern Chinese Literature and Visual Culture. *Frontiers of Literary Studies in China* 11(2): 233–8.
Schultz, C. (2016) Ruin in the Films of Jia Zhangke. *Visual Communication* 15(4): 439–60.
Vieira, F. (2010) The Concept of Utopia. In: Claeys, G. (ed.), *The Cambridge Companion to Utopian Literature*. New York: Cambridge University Press, 3–27.
Visser, R. (2004) Spaces of Disappearance: Aesthetic Responses to Contemporary Beijing City Planning. *Journal of Contemporary China* 13(39): 277–310.
Wagner, K.B. (2013) Jia Zhangke's Neoliberal China: The Commodification and Dissipation of the Proletarian in *The World* (Shijie, 2004). *Inter-Asia Cultural Studies* 14(3): 361–77.
Wu, H. (2005) *Remaking Beijing: Tiananmen Square and the Creation of a Political Space*. Chicago: University of Chicago Press.
Wu, H. (2012) *A Story of Ruins: Presence and Absence in Chinese Art and Visual Culture*. London: Reaktion Books.
Yan, L. (2009) *Dream of Ding Village*. Trans. Carter, C. London: Corsair.
Yan, L. (2012) *Lenin's Kisses*. Trans. Rojas, C. New York: Grove Press.
Zhang, X. (2010a) Poetics of Vanishing: The Cinema of Jia Zhangke. *New Left Review* 63: 71–88.
Zhang, Y. (2010b) *Cinema, Space, and Polylocality in a Globalizing China*. Honolulu: University of Hawaii Press.

26

ICONi©Cities

Global imaginaries of urban dispossession

Uli Linke

Introduction

The idea of globalization holds an often-invoked promise for a better world, a way of life in which the violent barriers of racial consciousness have been pushed aside, even obliterated, by a universal quest for justice, human rights, and prosperity. National histories of genocide and racial terror, seen as symptoms of a failed modernity, are located in the past, a temporal order that is opposed to an emergent future, a world of possibilities, a terrain without frontiers. But such a turn to desired or projected states of being, as I argue here, tends to divert critical attention from the deeply entrenched and persistent project of mapping global disparities by race. Globalization has not produced a singular or unified world-order. Global capitalism does not operate as a single-space economy, "as a non-contradictory, uncontested space" (Hall, 2000: 32). In the new millennium, the structural order of things, including "the relationship among residence, race, and rights" (Appadurai, 2003: 346), has been destabilized. Political forms, cultural imaginaries, social lives, and economic interests may engage global possibilities along different and sometimes contradictory trajectories.

Under these volatile conditions, global processes not only "open up the possibility for the (re)articulation of racial, sexual, and national identities" (Thomas and Clarke, 2006: 27), but also tend to "revivify long-established colonial and Orientalist tropes" (Graham, 2006: 256). Patterns of racial ordering persist in the endeavors of international organizations, worldwide policy-making, and globalized culture industries. They are also evident in "the vast gaps between North and South" and the corresponding "planetary correlation of darkness and poverty" (Winant, 2001: 305). Yet racialization, I suggest, also persists in the ways in which European nationals continue to imagine themselves against "Others" across continental divides – across the Atlantic, from Africa to the Americas and Asia. Such entrenched systems of racial signification have real-world consequences for the allocation of power and privilege on a global scale. This is especially true, as Paul Gilroy observed, when a corresponding representational schema "assigns differential value to lives lost according to their locations and supposed racial origins or considers that some human bodies are more easily and appropriately humiliated, imprisoned, shackled, starved, and destroyed than others" (2003: 263). The racialist impetus for these tension-fraught and sometimes violent realities of a global world has been rendered opaque by an emphasis on the "redistributive, democratizing, and empowering effects" (Winant, 2001: 301). The push to globalization can thereby be (mis)perceived as a beneficent, liberatory project.

My chapter examines how racial hegemonies are sustained and perpetuated by the ways in which urban spaces inhabited by peoples on the margins of the world economy are imagined, represented, and brought to public visibility. As globalization "repudiates fixed territories, sacred spaces, and hard boundaries in favor of unstable flows" (Luke, 2004: 121–2), the expansive reach of commodity capitalism has transformed areas of urban misery into tourist destinations. The inventories of global travel culture are, in turn, stocked with ever-new venture tours into the urban netherworlds of dispossession: the mega-slums of Asia, Africa, and the Americas (Frenzel et al., 2012). Branded as memorable travel destinations by media industries and commercial enterprise, ghettos, favelas, barrios, slums, and townships have become international tourist attractions: living (open-air) museums or human zoos. Unfettered by the legacies of colonialism, slavery, and empire, corresponding representations of slum-dwellings are circulated as iconic artifacts of poverty for transnational consumption.

Prevailing cultural fantasies about distant urban worlds are amplified by digital communication technologies that rely on "a globally linked network through which images, media forms, cultural products, and texts circulate throughout the world" (Sturken and Cartwright, 2009: 390). The upsurge in action-adventure films and science fiction movies, including *City of God* (2002), *Banlieue 13* (2004), *District 9* (2009), *Slum Dog Millionaire* (2010), or *Black* (2014), reveals how cultural fantasies of ghetto-worlds of squalor and ethnopolitical terror are transformed into popular forms of entertainment. The pivotal conditions for domination and exploitation, which Michael Omi and Howard Winant (1994) so eloquently termed "racialization", are thus no longer produced solely from within the ideological territory of nation-states. The assignation of racial meanings to concrete bodies, practices, and places has been globalized. In other words, the expansive capitalist world system is at once also a "global racial formation" (Winant, 2001: 3–6). Although the codex of race has not entirely lost its local or national imprints, racial formations are evident on a planetary scale, as a worldwide phenomenon. Propelled by the phantasms of "racial neoliberalism" (Goldberg, 2009), these transglobal logics propagate ever-new imaginaries of spaces, maps, and peoples.

The manner whereby these mobile slum-imaginaries are manufactured for transnational (and specifically European) attention is the subject of this chapter. Rather than scrutinizing image-making encounters at specific tourist destinations, I examine how distant places of urban poverty are visualized, suffused with racial signs, and inserted into global circuits of commodity capitalism. My examination builds on the bourgeoning literature of slum tourism, a term that refers to sightseeing endeavors and venture capitalism in areas of urban poverty. This travel-genre belongs to a broader phenomenon variously described as "dark tourism" (Lennon and Foley, 2000). Such touring practices focus attention on places of human tragedy that can be branded, showcased, and meaningfully reimagined by travelers. While promising to deliver memorable experiences, these emergent urban tourist sites also appeal to a growing public desire to retrieve a sense of justice or agency from a world perceived as inhumane or uncaring. Such endeavors, as Kevin F. Gotham asserts, rely on "a process that involves the deployment of symbols and imagery to neutralize negative publicity, counter stigma, and project globally a coherent and transparent image of urban rebirth and vitality" (2007: 843–4). Propelled by efforts to forge alternative imaginaries and visions of disenfranchised urban life, the expansive industry of slum tourism relies on local and international partnerships to attract consumer attention, but also engages the broader enterprise of image-making. While navigating competing fields of vision, transglobal tourist fantasies of urban poverty are shaped by a "new politics of visibility" (Freire-Medeiros, 2013: 128–34). Slums are promoted as destination sites by visual attention. The tourist gaze is enticed by signs, symbols, and images that aim to inspire investment and secure commitments to travel and sightseeing. My chapter expands on these insights by scrutinizing how mobile images of urban poverty garner public attention by appealing to racial aesthetics of space.

Building on this notion of the "traveling city" (Freire-Medeiros, 2013: 56), I examine how the world's enclaves of urban poverty are imagined, trafficked, branded, and consumed along transnational media circuits as signifying practices in a global panopticon of race. Central to my inquiry is the seductive appeal of those images of "shantytowns" or the "black ghetto" that circulate as popular commodity forms throughout Europe's metropolitan centers. Given the multitude of image-industries and image-makers, competing representations of urban poverty are manufactured for European public attention by aesthetic, symbolic, and affective means, ranging from the romance of despair or humanitarian compassion to a nostalgic longing for premodern signs of a deprived but simpler life. Yet, despite the range of messages conveyed to different publics, my research suggests that the repertoire of slum-images draws on a shared set of iconic signs. While the symbolic mediation of the visual material permits multivocal ambiguities, the cultural apprehension of what is "seen" is guided by racializing codes. When retrieved from across the globe, and transported into Europe, slum-iconicities are not merely consumed as cultural exemplars or artifacts of faraway urban worlds, but also interpreted as visual evidence of racial alterities. The radical otherness of slum-inhabitants is *rediscovered* in the sensuous depictions of exotic naturalism, the metaphoric renderings of cultural essences, and the iconic representations of primal bodies and premodern vitalities. In tracing the manner whereby the logics of spectacle and entertainment have come to organize the circulation of slum-images, I show how signifiers of race are variously produced and encoded in the iconicities of urban dispossession. The visual mytho-logics I investigate are integral to what Omi and Winant termed "racial projects": slum-iconicities encode "simultaneously an interpretation, representation or explanation of racial dynamics and an effort to reorganize and redistribute resources along particular racial lines" (1994: 56). In this chapter, however, my focus is somewhat more limited in scope. While attentive to the economic implications of racialization, my analytic concern is directed toward uncovering the symbolic language of slum-representations.

Africanizing the US ghetto: signs of the strange, monstrous, and intimate

While the world system of growing inequalities has been described in terms of a "planet of slums" (Davis, 2007) or as a "ghetto formation" (Winant, 2001), the geopolitical dispersal of urban poverty invites further inquiry. Linked to the expansive growth of mega-cities, a closer reading of these sites reveals that contemporary landscapes of marginalization are not confined to the great metropolises of the Global South. Poverty may occupy a variety of urban orbits. People without means are forced to inhabit shelters or build houses on the outer margins of world cities. In the United States, trailer parks and tent cities are extensions of this neoliberal pattern of urban exclusion, which relocates the disenfranchised in natural settings: parks, forests, fields. In other parts of the world, informal urbanization lays claim to land outside of cities, generally in "hazardous, health-threatening locations" (Davis, 2007: 121). Such emergent shanty-settlements consist of large numbers of self-built housing, often with few infrastructural provisions. In contrast to these metropolitan fringe expansions, poverty in North America is spatially confined to an urban core. Such "disparate sociospatial formations" (Wacquant, 2008: 3) have become competing templates in a global imaginary of race.

In the United States, the "inner-city slum" or urban "ghetto" stands as an ideogram of salient non-white others: disenfranchised "poor people concentrated in derelict cores and inner suburbs" (Davis, 2007: 31). Constructed as a carceral space, the ghetto emerges as a "proxy-prison" or "poverty warehouse" in urban centers (Mendieta, 2007). Confined to "the racialized core of the US metropolis", the ghetto is envisioned as a template of violence and illegality, as a negative social space that breeds and reproduces criminality (Wacquant, 2008: 2). Human beings, who find themselves entrapped within these spatial confines, are imagined as part of a "multitude" of undesirable others, as signs of a dangerous, teaming humanity that needs to be contained (Mbembe, 2017: 11). Framed by these imaginaries, inner-city residents are treated as societal surplus – unwanted "waste" products: valueless, useless, and disposable (Giroux, 2007). It is revealing that these meanings of disposability and confinement have been encoded in the very lexicon as "ghetto". The term has been traced to Renaissance Venice, where it referred to things and people "cast out" or "discarded" (from vulgar Latin **jectare*, to throw, cast), meanings that bring into focus the linguistic imprint of the contemporary world of iron manufacturing. While *ghetto* denoted "foundry" in standard Italian (Sennett, 1996: 231), the Venetian phrase c*ampo ghèto* referred to an insular place at the city's margins, which was used for cooling and storing slag – the waste produced by smelting ore. This site became a "prophylactic space" for ethnic segregation and social exclusion, the *Venetian Ghetto* – the very place where Jews were compelled to live after 1516 (Sennett, 1996: 228–41). The semantics of the "ghetto" traveled across the Atlantic to North America (via the routes of slave-trade), where the Venetian loan-word came to refer to analogous practices of socio-spatial segregation. In the early twentieth century, when city neighborhoods in the United States were racialized as "black" and African-American, the urban *ghetto* acquired its modern meanings: a carceral space inhabited by "those who have been left behind", those destined to endure "segregated lives" in "zones of poverty" and "common failure" (Sennett, 1996: 368). From its inception, the "ghetto" was conceived as a location for ethnoracial segregation, the designated space for "urban outcasts" (Wacquant, 2008). The camp, the ghetto, and the prison thus emerge as historically interconnected sites of population control.

Propelled into the present, the construct of the "ghetto" fuses two political moments of late capitalism: a regime of warehousing poverty (by dehumanization) and controlling surplus labor (by criminalization). Depictions of the ghetto as a penal containment-space are founded on racial fantasies of threat. In the United States, as Stephen Steinberg asserts, "we speak euphemistically of 'the urban jungle'" and "portray the inner city as a haven of pathology, disorder, and immorality", which comes close to "declaring its inhabitants 'uncivilized'" (2010: 217). The euphemism of the inner-city ghetto as a "jungle" implies wilderness, a dangerous place, the habitat of untamed nature, "where the wild things are" (Mexal, 2004: 238): the primal, violent, and sexually unrestrained. As Stephen J. Mexal elaborates:

> The urban wild is a fraught, dangerous space . . . a symbolic space charged with racial and economic tensions, a space outside space. As such, the ideology of wilderness dictates that it be fenced off, that it be restrained as a site of geographic and political domination.
>
> *(2004: 238)*

Popular representations of the ghetto in turn mobilize signs that confirm the negated humanity of those imagined to inhabit the inner city: the figure of the "black" other, whose existence is defined by the spatial and temporal coordinates of the "urban jungle", which is a construct inflected by the racial logic of the white gaze and the controlling exercise of state power (Wacquant, 2008). Entrapped within this metaphorical field, as Allen Feldman notes, the body of the black other is marked by "the stigma of animality", a "beastial imagery" that simultaneously prescribes police responses of "taming and caging" (1994: 409). Such racial archetypes are sustained by additional semiotic frames. The "urban jungle" metaphor can be linked to an *africanization* of the ghetto. I hereby argue that the discourse of Africa, as analyzed by Achille Mbembe, operates as a template for the iconicity of the US inner-city "ghetto".

> The African human experience constantly appears in the discourse of our times as an experience that can only be understood through a *negative interpretation*. Africa is never seen as possessing things and attributes properly part of "human nature"... At another level, discourse on Africa is almost always deployed in the framework (or on the fringes) of a meta-text about the *animal* – to be exact, about the *beast*: its experience, its world, and its spectacle.
>
> *(Mbembe, 2001: 1)*

Following this meta-frame, the imagined life of the "ghetto" unfolds under several related signs: "the strange and monstrous" and the "intimate" (Mbembe, 2001: 1–3). Such notions are encoded in signifiers of the ghetto-look, whereby the ascribed animalistic or presocial disposition of the salient ghetto-other is reified and recycled in transglobal imaginaries. The televised spectacle of hurricane Katrina's destructive force along the Gulf Coast in 2005 provides a pertinent example. In the wake of the storm, the "Africanization" of US inner cities was rendered most visible when global audiences saw the total neglect of stranded poor black people in New Orleans, with media reports of violence and looting wildly exaggerated, revealing the long-standing racial stereotypes of black savagery (Giroux, 2007). Fantasies of the lawless, Africanized city, renamed "Little Somalia" by rescue workers, suggested the need to restore order by force and to liberate the urban "enterprise zone" with military combat tactics from the "insurgency" of black resident-survivors (Lipsitz, 2006: 454). Such notions of racial threat resurfaced during the 2008 US presidential campaign, when the candidacy of Hawaiian-born Barack Obama was contested by denying the plausibility of his American citizenship: as a black man, he was cast as a Kenyan immigrant, thereby imagined as no more than a "displaced African savage", and no less than a potential "Black-Arab-Muslim terrorist" (Linke, 2010). These racial machinations were documented by mainstream media in the same manner as the image-feeds from post-Katrina New Orleans. Reports of the inundated city, with its broken levees, ignited public anxieties about black violence out of bounds, a savagery unleashed by the terrifying deluge that had flushed out the "wild" from a forgotten urban habitat in the southern United States. As the city lay in ruins, venture capitalists began to organize sightseeing tours to the sites of destruction, accompanied by the sale of Katrina souvenirs for ghetto-tourists. Thereby the dead zones of urban immiseration and the demodernized spaces of blackness were not merely commoditized, but presented as "a collage of fixed and static images that [were] marketed and interpreted for tourists" (Gotham, 2007: 838). But Africanized imaginaries of the US inner city have been shaped by additional means, and by other forms of branding.

Visions of the black ghetto, including tropes of the urban wild, have been given tangible form as portable commodities, as exchange objects that can be traded, owned, bought, and sold. Forged from the traits of racialized identities, these commodity artifacts are mass-produced as styling accessories for mainstream consumers. Cultural machinations of the ghetto toy with notions of the "monstrous" black body, an untamed beast, whose raw animal nature hints at the possession of claws or talons. By purchasing fake acrylic "ghetto fab nails" at a store, interested buyers can adorn themselves with the desired look to project an edgy, non-conforming identity. The same holds true for other urban fashion gear, such as "ghetto fab wigs" or "afro visor hats" that are shaped from the hair of black urban residents. As a hairstyle and head covering made by the urban poor, this form of body art is synthetically reproduced and sold as a wardrobe supplement. Yet streetwear designs from the black ghetto do not merely seek to evoke images of the dangerous, animalistic, and strange. Performances of the abject body draw on racialized typifications of the inner city and turn them into expressions of contempt. For instance, when enacting the figure of the "uncivilized savage", black youth engage fashion styling practices such as "low-riding" or "sagging" to show disrespect: fastened below the hip, jeans are worn to slide down the buttocks and expose the wearer's naked skin or undergarments. This urban black style originated in prison culture, where the clothing issued to inmates was often too large and belts were prohibited. Initially popularized by hip-hop and gangsta artists in the late 1990s to bring attention to the ghetto-prison complex, the practice was subsequently adopted by young urban blacks as a gesture of solidarity with the disproportionately black prison population. For urban youth, the practice of "sagging" has become an "act of defiance", "a way to be radical and savage", and "show some disrespect" (Blake, 2017). Representations of the Africanized ghetto as an uncivilized space are thereby not only challenged but transmuted into anti-establishment performances. These urban street wear styles, with their unruly, disruptive, and rebellious qualities, have intense consumer appeal. Black street-fashion accessories are commoditized, mass-produced, and branded as ghetto chic, thereby transforming dominant notions of a contested black civility into white countercultural fashion statements.

Ghetto chic or slum fashion is embedded in a discourse of mimesis and performance that transforms the bare life of poverty into domesticated, aestheticized objects – free-floating signifiers for forging identities that are performed on bodies uncoupled from the symbolic space of the black ghetto. Commoditization unlinks and uncouples signs by metonymic fragmentation and dissection – procedures that are frequently cited in feminist literature as demeaning, objectifying, and fetishistic. Global appropriations and enactments of the ghetto-look may be experienced as liberatory by mainstream consumers. Such identity practices and forms of mimicry are particularly effective when the appropriated styling elements are either derived from the voices of the "ghetto" or self-produced by the dispossessed (music, body art, clothing). In these instances, the imagined ability to "tap into reservoirs of cultural or historical traditions of resistance" (Rao, 2012: 675) intensifies the signifying capacity of purchased proxy-objects, magnifying the sense of intimacy with black cultural styles that contest normative assumptions within prevailing identity regimes. Symbolically charged, and commodified as fashion, the ghetto-look thus provides a safe (prosthetic) entry into a rebellious stance on consumer culture while inviting participation in its identity conferring promises. Conceptions of the US inner-city ghetto as a negative space and its Africanized inhabitants as "violent", "uncivilized", and "premodern" are thereby continuously recycled in tangible form. Images of the failed city and signs of the abject body are offered up for purchase to global consumers as embattled identity accessories. In this process, whereby the cultural tropes of the Africanized "ghetto" are sold for profit, the politics of spatial segregation and unequal opportunity recede from public visibility.

Tropicalizing the terrain of urban frontier: signs of nature, art, and beauty

How do these racializing machinations of the "inner-city" poor in North America compare to views of urban deprivation in the Global South? What types of fantasies are propagated by images of those informal mega-settlements called "shantytowns"? While the portable signs of the Africanized black ghetto are encoded by racial histories and assertions of a failed modernity, shantytowns are seen as part of a "new world" order (Neuwirth, 2006). In contrast to the racially homogenized lifeworld of the "black ghetto", shantytowns are typified as settlements with utterly diverse inhabitants: refugees, poor families, displaced farmers, inter/national labor migrants, returning combat veterans, unemployed workers, and low-income professionals (Perlman, 2010). Although this very diversity of urban immigrant population is often experienced as threatening by nationalist regimes, local city elites, and middle-class urbanites, who see flux, mobility, and difference as signs of a destabilizing national and global economic order, the translocal imaginary paints a different picture. Shantytowns are imagined as a new urban frontier situated on the edge of wilderness, where makeshift houses are built in undesirable locations. Urban squatters are described as pioneer settlers who transform uninhabitable spaces: "unstable hillsides, rubbish mountains, chemical dumps" (Davis, 2007: 121). Shantytown occupants are regarded as positive agents, as industrious workers, who produce value and transform environments while patiently enduring the repressive apparatus of state and municipal governments. The figure of the shanty-resident is idealized as "spirited, intelligent, and hard-working", with "family values", and who "invests in community" (Neuwirth, 2007: 71, 73). Seen as energetic, creative, and vibrant, "with the motivation and willingness to work" (Perlman, 2010: vxiii), shantytown inhabitants are imagined as urban frontier settlers, as builders of houses, families, and communities, who strive to overcome conditions of displacement, impermanence, and poverty. The visual codes whereby shantytowns enter into global commodity culture are centered on the products of labor: the homes and living spaces built by squatters. By drawing on long-term research in Rio de Janeiro (Brazil), Janice Perlman suggests that Latin American shantytowns (favelas) can "be seen as the precursors to the 'new urbanism' with their high-density, low-rise architecture. [The houses are] owner-designed, owner-built, and owner-occupied" (2010: vxiii). In contrast to popular machinations of the "black ghetto", which spectacularize the abject body to reify race, place, and difference, shantytowns are typified by pictorial representations of the built environment. Visual attention, I argue, shifts from the racialized ghetto-body to signifiers of landscapes of urban architecture and material culture. The *tropicalized* world of shantytowns, in opposition to the iconicity of the "black ghetto", unfolds under the signs of nature, art, and beauty.

This representational focus is exemplified by the work of German photographer Peter Bialobrzeski (University of the Arts, Bremen), whose stunning visual interpretation of urban dispossession have garnered international recognition. Distinguished by the World Press Photo Award in 2010 and the Erich Salomon Prize in 2012, the photographer's shanty-images have been featured in galleries, museums, and art shows worldwide, including New York, Shanghai, Paris, Hamburg, Berlin, and Frankfurt. In one acclaimed exhibit, *Case Study Homes*, Peter Bialobrzeski (2010) shows a series of photos of makeshift buildings from Baseco, a squatter settlement at the edge of Manila, which is home to approximately 70,000 people – mostly labor migrants from the Philippine countryside. The shantyhouses are concentrated near the port of Manila, along the banks of the Pasig River Bay, which is a hazardous landscape that stretches across the grounds of a former landfill and chemical dumpsite. The Case Study photographs reveal provisional structures, small dwellings made from drift wood, bits and pieces of cardboard, corrugated metal, and other cast-off materials. As revealed by the photographer's portrait shots of single houses, each unit is uniquely fashioned. Assembled from scavenged materials, as distinct architectural forms, "these buildings bear an anarchical, piratic, improvised appearance" (Schmitt, 2009). The images capture the material conditions of rural displacement and transience by showing an urban enclave composed of homes with tattered walls and stick-branched roofs – a series of makeshift constructs hesitantly anchored to the ground by uneven wooden poles.

The images magnify architectural eccentricities in a naturalized urban landscape. Public spaces are shown devoid of people, history, and culture. Signs of social life are missing. The photos "provide no commentary; the scenes are marked by both an exaggerated lack of identity and a paradoxical self-referentiality" (Glasmeier, 2008). Poverty artifacts are shown as auto-iconic art objects. The shacks are displayed as curiosities, strange but fascinating pebbles on a beach. Nearly all traces of the inhabitants have been removed. Even the sandy-ground bears no evidence of human presence: sign of domestic activity, including scraps of building material or scatters of garbage, have been expunged from the visual field. Glimpses of freshly washed clothing hung out to dry in the sun allude to the occupants' presence. In the photographs, shantyhouses are shown as edifices of architectural distinction and conjure attention by vibrant colors: "From a Westerner's perspective, the makeshift dwellings with their colorful tarpaulins and converted advertising billboards turn into works of art: they are collages of color and diversity" (Schmitt, 2009: 2). The Baseco settlement is not only beautified but staged as a world of wondrous and exotic houses, an urban dreamscape in which the allure of color functions as "the vehicle of spectacle" and as "the language of consumption" (Lutz and Collins, 1993: 94).

Color, art, and nature are combined as signs of a *tropical*ized landscape, in which the shantyhouses, much like the figures in Gauguin's South Sea paintings, suggest vibrancy, a love of life, and a surplus of creative energy. This is articulated by the following pastiche of commentaries by exhibit spectators: the inhabitants "try to make something beautiful out of whatever little they have" (Nandy, 2010); "These are creative people who . . . find ways to design ANYTHING to be desirable" by using the "'styling' elements of poverty" (Karsten (PolluteLessDotCom), 2007). Such readings disclose a preoccupation with shantytowns as works of art, an interpretation that imagines the house-builders as creative agents. As Barbara Kirshenblatt-Gimblett suggests, the transmutation of material artifacts into art forms refashions such objects in terms of a "universalizing rhetoric" that grants such cultural edifices civilizational status: the discourse of "art transcends" the classification of "ethnic, folk, or primitive" material worlds by drawing on a "humanizing" language of "legitimation" and "inclusiveness" (1998: 25, 242–3). In other words, art objects defy ethnoracial classification because they are by definition unique. Whereas the bodies and lifestyle practices of US "ghetto" inhabitants are imagined and commodified as "quintessential attributes" – as racialized traits of a negated humanity (Kirshenblatt-Gimblett, 1998:12) – the photographic images of Baseco shantyhouses incite attention through a turn to artistic appreciation. By using the universalizing code of art as a frame of reference, European spectators cannot merely mediate cultural difference but can engage in racializing practices of looking without having to abandon their elite subject position.

The photographed houses are rendered visually devoid of contextualizing signifiers. There are no identifying markers of locality, country, and continent. Situated in an aestheticized natural setting, the shanty-homes have been deterritorialized and dehistoricized. Since the inhabitants are absent from the visual frame, the images are rendered globally portable without racializing cues. The viewing public can conjure variable interpretations about the identities of the homebuilders or their life circumstances. As articulated by one exhibit visitor, "The developing world shantytowns are the new global grassroots of style. They inspire aesthetically, displaying amazing inventiveness and endless formal variation. Shantytowns are exemplary recyclers and have extremely modest environmental footprints" (Imomus, 2009). As suggested here, urban poverty (as gleaned from the photos) is deemed a positive "practice" that promotes creative survival skills and enhances environmental sustainability. But when conceived "as models of utopian sustainability", as Janice Perlman argues, conditions of poverty are depoliticized by a global imaginary that romanticizes deprivation "as the solution to environmental problems, overpopulation, and housing shortages" (2010: 334). In the photos, the alter-realities of global capitalism are expunged from the visual field by transforming poverty artifacts into aestheticized art objects: clean, colorful, and uniquely spectacular. In an emancipatory move that simultaneously obscures the human faces of the inhabitants, the photographer's images "let 'art' speak for itself" (Kirshenblatt-Gimblett, 1998: 245). Such representational strategies toy with a problematic postulate: "life is art" in the urban tropics. Aestheticized and naturalized, the global conditions of urban inequality have been overtly de-racialized. Practices of looking can thereby proceed undistracted by the political realities of Philippine militarization, US army bases, and the resultant ruination of life worlds.

Slum architecture as commodity fetish

The desire to bring the "colorful" (race-based) alter-realities of urban life from other continents to Europe is puzzling. Why do white publics long to see shantytowns as iconic artworks of the poor? While the showing of "slums" invites visual intimacy with the signposts of poverty, such exhibits also reify machinations of race and difference. When shown as beautified but defleshed architectural forms, as I have argued, the shanty-artifacts sustain Europe's imperialist machinations of tropicalized urban worlds. But these patterns of looking also entail a self-civilizing or whitening procedure by deflecting attention from the dark underground spaces of European history. Whereas the urban life worlds of the Global South are brought to visibility as ubiquitous slum-formations, the existence of the urban "ghetto" in contemporary Europe is officially denied. In the aftermath of National Socialism, the Holocaust, and World War II, the state-sanctioned concentration of select population groups in urban space has become a contentious issue. The image of the ghetto as a carceral space and a "political machine of 'exclusionary enclosure'" (Mendieta, 2007: 384) evokes memories of state violence: anti-Semitism, the *Warsaw Ghetto*, concentration camps, and genocide. In Europe, the regime of the ghetto signifies an unspeakable reality – racial mass murder (Goldberg, 2006). Excised from public discourse, confined to an imaginary exile (in the past or across continents), spatial agglomerations of urban alterities in Europe are rendered unseen. Ethnic enclaves are treated as a cultural affront. This is suggested by popular references to "the Turkish ghetto" in Berlin (Agabeyoglu, 2014: 6–7; Hadatty, 2017; The New York Times, 1977), which became a racialized space in the late 1960s. Initially populated by Turkish migrants, and later by Syrian refugees, this urban enclave retained its stigma as a Muslim-Arab-African neighborhood in German media and public discourse. Yet even this urban space was to be commercially branded. The Berlin ghetto has become a tourist destination (Hadatty, 2017; Rayasam, 2016), with guided sightseeing tours into the urban zones of ethnic difference.

In other European cities, the non-white poor and racially dispossessed are pushed out of the urban centers, out of public sight, into the metropolitan peripheries, "with immigrant and unemployed populations marooned in highrise housing on the urban outskirts" (Davis, 2007: 31). Exemplary are the *banlieues* of Paris or Stockholm's "Little Baghdad", where, "as in other European urban centers, the racial becomes the spatial" (Pred, 2000: 98). The peripheralization of the immigrant poor or disenfranchised is not confined to Europe. The displacement of select population groups to the outer fringe of metropolitan centers is a worldwide phenomenon. European cities, however, may be unique in their effort to gain visual access to the life worlds of the "urban outcasts" from the Global South: by "bringing the exotic" back home (MacCannell, 1976: 13) and by facilitating spectacular exhibits of shantytowns from the global elsewhere-places of "darkness" and poverty. The optics of these expansive shanty-communities of the Global South have gained attention among various actors, including European city-builders. Based on a paradigmatic case study of Lagos, Dutch architect Rem Koolhaas, for instance, has appropriated the spatial contours of southern mega-slums as an aestheticized design-form or "meta-model" for future first-world building practices (Hecker, 2010; Rao, 2012: 674–80). Although promoted as acts of solidarity with the world's dispossessed and underprivileged, such projects rest on commonsense postulates of white entitlement: the prerogative to reduce salient others to a repertoire of signs for publicity and shock effect; the ability to appropriate the lives of others as an imaginative stage for self-expression; and the power to reconstruct inhabitable shantyhouses in Europe – much like prosthetic skins that are fitted and draped over white bodies as the mask of "black-face" to enable a desired performance.

Such appropriations of the existential life worlds from the Global South have become commonplace in urban Europe. Departing from past practices, racial subjugation has shifted from taking possession of bodies and land to an extraction of imaginative cultural labor – as a form of symbolic capital and as a countercultural resource. European slum tourism has entered a new era. In addition to the commoditization of shantytown architecture by consumer-travelers, political activists, and architects, efforts are underway to mainstream these slum-habitats as fetishized theme parks. Clubbing faux-shantytowns as ethnicized entertainment spaces has become a popular European pastime. This is suggested by the construction of an imitation-shantytown as a corporate theme park in Germany. During the Johannesburg World Cup soccer games in the summer of 2010, a private enterprise in Berlin sought to increase its profit margins by building a South African "slum" complex in the center of the city. The idea was to bring Johannesburg to Berlin by creating "a slum-look-alike fun park" (Mösken, 2010). The project commenced by scavenging urban junkyards for the building materials. Cardboard, old tires, rusty containers, and corrugated metal sheets were hauled to a designated site on the Spree riverbank next to *Bar 25*, a popular Berlin club. The scavenged items were assembled on club grounds to create an imitation-shanty complex, a phantasmatic slum city accessorized with a shaman-shack bar, paintings of stereotyped black people on the fence surrounding the fake "squatter" compound, and a gaming space with fake laundry hanging from clotheslines. The club's shantytown theme park included additional entertainment areas: "a mini-stadium to watch the games on the big screen, a skateboard park, beach volleyball, a 'Kapstadt' [Capetown] grill restaurant" (BPM Bella, 2010). Berlin's residents were invited to enter this simulated South African "township" construct to watch live soccer games, listen to techno back-beat music, drink champagne, eat the open-fire grilled steaks, and enjoy the realism of the experience (Mösken, 2010). The shanty "township" in Berlin was built as a total entertainment experience and eccentric lifestyle choice, a full immersion fantasy-experience in a neoliberal consumer paradise. This venture could be read as a postmodern enactment of German whiteness (Linke, 1999): spectacularized in urban public space, white body performances are here staged in a make-believe slum-setting against the decivilized black (African) others, who as allusive onlookers (painted figures on the fence) watch white consumers act out their secret longing for the authentically primitive.

Global imaginaries of urban poverty

In a global world, conceived as a space of possibilities, a terrain without frontiers, the vast disparities between wealthy power elites and the disenfranchised poor have not diminished. Globalization has intensified existing inequalities. Under these conditions, propelled by shifting imaginaries of race, the iconic vestiges of "slum" life from the Global South are circulated as popular commodity forms throughout Europe's metropolitan centers. Depictions of urban poverty are branded for white consumer publics that long to refashion their social identities by physical or symbolic contact with racialized icons of difference. But these visual encounters and representational practices, as I have argued, are not necessarily static or predictable. Capitalist imperatives and the manufacture of elusive authenticities may intersect to produce "counterintuitive" results (Comaroff and Comaroff, 2009: 1). Under globalization, ethnospatial and georacial logics can be reclaimed for profit. Likewise, the imaginative geographies of privilege and disadvantage may be reconfigured: forged by recourse to space, signifiers of race, and by new consumer practices.

From this perspective, we can further decipher the negation of the Africanized US "ghetto". In the global imaginary, the "black ghetto" is the antinomy of an open-ended future. Stigmatized by machinations of race, its inhabitants "are vanishing into the sinkhole of poverty in desolate and abandoned enclaves of decaying cities" (Giroux, 2007: 309). The US "ghetto" holds no promise but presents the ruins of modernity, an archive of the failure of the industrial city, its crumbling infrastructure, and its pathologized residents, who must be spatially contained or imaginatively harnessed – "caught in a forever petrified freeze-frame" (Lafleur, 2010: 208). These imaginings are reified by culture industries through the mass production of racialized body-traits (nails, hair, garments) that can be sold to white consumer publics as countercultural identity accessories. The oppositional status of the "black ghetto" as a "savage", "uncivilized", and "violent" space is thus crucial for sustaining white fantasies about distinction and distance. The shantytown, by contrast, engenders a more optimistic interpretation. As an artifact of globalization, it embodies the "hope of history": each foot of non-arable urban frontier-land transformed by squatter settlers, each house built, however incomplete, is perceived as a "relaunch of an ongoing promise, a 'not yet', a 'what is coming', which – always – separates hope from utopia" (Mbembe, 2001: 206). While the end of modernity is conceived through temporal metaphors of urban ruination and traces of the abject body (the *Africanized* ghetto), the future of globalization itself remains yet undetermined. Located "at the nexus of global forces of transnational flows and networks of activity" (Gotham, 2007: 827), urban squatters and their transformative labor are seen as iconic manifestations of this unconsummated future "hope of history". Shantytowns, identified as productive works of transglobal (postmodern) agents, engender desire and embody longing, which can be manufactured and encoded in consumable signs. The artworks, photos, and dioramas of shanty-architecture transport these fantasies across Europe, in exhibits of tropicalized landscapes of urban worlds that are apprehended as "naturally" and "essentially" colorful, vital, and playful (childlike). These tropicalized machinations of shantytowns, while enticingly appealing, are nevertheless a form of racialization.

Among European publics, critical awareness of these portable imaginaries of urban poverty as racial representations of people and space remains unarticulated. The use of "slums" as globally mobile spectacles requires that core images be detached from local life and history to produce a repertoire of free-floating emblems and signs that can be variously deployed, assembled, and discarded, depending on shifting cultural desires in a capitalist market. When pushed into the global realm, the iconicities of urban poverty are "sucked into the timeless present of the all-pervasive virtual space of consumer culture" (Huyssen, 2003: 10, 28). As portable commodity objects, which are trafficked by stripping away historical realities and concrete societal contexts, slum-artifacts move in visual fields that enact relations of inequality and European privilege: by the manner of representation, by exhibition practices, and the acts of visual consumption. In this process, which likewise relies on the mediating intervention of photography, diorama art, and sculpture, the formative conditions of global urban inequalities are reified, even normalized. Urban alterities are imagined, exhibited, and seen through racializing visual codes that promote, following Frantz Fanon, "an aesthetic of respect for the status quo" (1994: 3). Such a visual habitus envelops urban poverty to be experienced as "pleasing, ultimately even beautiful" (Mirzoeff, 2011: 3–4). In this process, the racial imaginaries of empire are revivified to regain a foothold in the white metropolis, traversing European exhibit grounds through the conduit of "tropicalized" or "Africanized" representations of global urban dispossession, which are commodified, decolonized, and thereby rendered "safe" for white consumer publics.

Acknowledgements

This chapter is a revised, edited, and much shortened version of Linke (2014) "Racializing Cities, Naturalizing Space", published by *Antipode*, vol. 46, issue 5, pp. 1222–39 (www.blackwell-synergy.com). Copyright @ Uli Linke. I gratefully acknowledge the journal, its editorial board, and Blackwell Publishing.

References

Agabeyoglu, N.Y. (2014) *Deutsch-Türkische Rapmusik in Berlin*. Berlin: LIT Verlag.
Appadurai, A. (2003) Sovereignty without Territoriality. In: Low, S.M. and Lawrence-Zúñiga, D. (eds.), *The Anthropology of Place and Space*. Oxford: Blackwell, 337–49.
Bialobrzeski, P. (2010) *Case Study Homes*. Ostfildern: Hatje Cantz.
Blake, J. (2017) The Fall of the Sagging Pants Era Is upon Us. *CNN*, 20 October 2017. Available at: http://www.cnn.com/2017/10/20/us/sagging-pants-era-over/index.html. Accessed 30 December 2017.
BPM Bella. (2010) Berlin's Infamous Bar 25 Does World Cup Joburg Style. *BPM Life*, 22 June 2010. Available at: http://bpmlife.co.za/?p=1725. Accessed 10 October 2014.
Comaroff, J.L. and Comaroff, J. (2009) *Ethnicity, Inc*. Chicago: University of Chicago Press.
Davis, M. (2007) *Planet of Slums*. London: Verso.
Fanon, F. (1994) *The Wretched of the Earth*. Trans. Philcox, R. New York: Grove.
Feldman, A. (1994) On Cultural Anesthesia. *American Ethnologist* 21(2): 404–18.
Freire-Medeiros, B. (2013) *Touring Poverty*. New York: Routledge.
Frenzel, F., Koens, K., and Steinbrink, M. (eds.) (2012) *Slum Tourism*. London: Routledge.
Gilroy, P. (2003) Where Ignorant Armies Clash by Night. *International Journal of Cultural Studies* 6: 261–76.
Giroux, H.A. (2007) Violence, Katrina and the Biopolitics of Disposability. *Theory, Culture and Society* 24(7–8): 305–9.
Glasmeier, M. (2008) The Presentness of the Unpredetermined. *LA Galerie Lothar Albrecht*. Available at: http://lagallery-frankfurt.de/PBTransitionUK.pdf. Accessed 10 October 2017.
Goldberg, D.T. (2006) Racial Europeanization. *Ethnic and Racial Studies* 29(2): 331–64.
Goldberg, D.T. (2009) *The Threat of Race*. Oxford: Blackwell.
Gotham, K.F. (2007) (Re)Branding the Big Easy. *Urban Affairs Review* 42(6): 823–50.
Graham, S. (2006) Cities and the War on Terror. *International Journal of Urban and Regional Research* 30(2): 255–76.
Hadatty, M. (2017) Vom Türken-Ghetto Zur Touristenattraktion. *RBB Fernsehen/Kowalski & Schmidt*, 9 April 2017. Available at: https://www.rbb-online.de/kowalskiundschmidt/archiv/kowalski-schmidt-09-04-2017/vom-tuerkenghetto-zur-touristenattraktion.html. Accessed 22 December 2017.
Hall, S. (2000) The Local and the Global. In: King, A.D. (ed.), *Culture, Globalization and the World-System*. Minneapolis: University of Minneapolis Press, 19–39.
Hecker, T. (2010) Slum Pastoral. *Space and Culture* 13(3): 256–69.
Huyssen, A. (2003) *Present Pasts*. Stanford: Stanford University Press.
Imomus [Nick Currie] (2009) Click Opera – Case Study Homes. *LiveJournal*, 29 April 2009. Available at: http://imomus.livejournal.com/453897.html. Accessed 10 October 2017.
Karsten (PolluteLessDotCom). (2007) Any Change Is Welcome. *Grist*, 1 October. Available at: http://www.grist.org/article/fashion-victim-of-climate-change#comments. Accessed 10 October 2017.
Kirshenblatt-Gimblett, B. (1998) *Destination Culture*. Berkeley: University of California Press.
Lafleur, M. (2010) Tracing the Absent-Present. *TOPIA: Canadian Journal of Cultural Studies* 23–24: 203–26.
Lennon, J. and Foley, M. (2000) *Dark Tourism*. London: Continuum.
Linke, U. (1999) *German Bodies*. New York: Routledge.
Linke, U. (2010) Fortress Europe. *TOPIA: Canadian Journal of Cultural Studies* 23–24: 100–20.
Linke, U. (2011) Technologies of Othering. In: Ribeiro Sanches, M., Clara, F., Ferreira Duarte, J., and Pires Martins, L. (eds.), *Europe in Black and White*. Chicago: University of Chicago Press/Intellect Books, 123–42.
Linke, U. (2014) Racializing Cities, Naturalizing Space. *Antipode* 46(5): 1222–39.
Lipsitz, G. (2006) Learning from New Orleans. *Cultural Anthropology* 21(3): 451–68.
Luke, T. (2004) Everyday Techniques as Extraordinary Threats. In: Graham, S. (ed.), *Cities, War and Terrorism*. Oxford: Blackwell, 120–36.

Lutz, C.A. and Collins, J.L. (1993) *Reading the National Geographic*. Chicago: University of Chicago Press.
MacCannell, D. (1976) *The Tourist*. Berkeley: University of California Press.
Mbembe, A. (2001) *On the Postcolony*. Berkeley: University of California Press.
Mbembe, A. (2017) *Critique of Black Reason*. Trans. Dubois, L. Durham: Duke University Press.
Mendieta, E. (2007) Penalized Spaces. *City* 11(3): 384–90.
Mexal, S.J. (2004) Consuming Cities. In: Allister, M. (ed.), *Eco-Man*. Charlottesville: University of Virginia Press, 235–47.
Mirzoeff, N. (2011) *The Right to Look*. Durham: Duke University Press.
Mösken, A.L. (2010) Johannesburg 24: Corporate Slumworld. *Exberliner*, 30 June 2010. Available at: http://www.exberliner.com/reviews/johannesburg-24-corporate-slumworld. Accessed 10 October 2017.
Nandy, A. (2010) Planning Cannot Eliminate Slums. *OneWorld South Asia*, 14 April 2010. Available at: http://southasia.oneworld.net/opinioncomment/planning-cannot-eliminate-slums. Accessed 10 October 2017.
Neuwirth, R. (2006) *Shadow Cities*. New York: Routledge.
Neuwirth, R. (2007) Squatters and Cities of Tomorrow. *City* 11(1): 71–80.
Omi, M. and Winant, H. (1994) *Racial Formation in the United States* (2nd edn.). New York: Routledge.
Perlman, J. (2010) *Favela*. New York: Oxford University Press.
Pred, A. (2000) *Even in Sweden*. Berkeley: University of California Press.
Rao, V. (2012) Slum as Theory. In: Crysler, C.G., Cairns, S., and Heynen, H. (eds.), *The Sage Handbook of Architectural Theory*. London: Sage, 671–86.
Rayasam, R. (2016) Berlin's Hipster Ghetto. *Foreign Policy*, 18 January 2016. Available at: http://foreignpolicy.com/2016/01/18/berlins-hipster-ghetto-neukolln-refugees-migration-immigrants/. Accessed 4 January 2018.
Schmitt, B. (2009) *Case Studies*. Trans. Schede, S. L.A. Galerie Lothar Albrecht, 27 March 2009, 1–4.
Sennett, R. (1996) *Flesh and Stone*. New York: W.W. Norton.
Steinberg, S. (2010) The Myth of Concentrated Poverty. In: Hartman, C. and Squires, G.D. (eds.), *The Integration Debate*. New York: Routledge, 213–27.
Sturken, M. and Cartwright, L. (2009) *Practices of Looking* (2nd edn.). New York: Oxford University Press.
The New York Times. (1977) Turkish Ghetto in West Berlin Plagued by Unemployment. *The New York Times Archives*, 4 March 1977. Available at: https://nyti.ms/1RE7Zfh. Accessed 4 January 2018.
Thomas, D.H. and Clarke, K.M. (2006) Introduction. In: Clarke, K.M. and Thomas, D.H. (eds.), *Globalization and Race*. Durham: Duke University Press, 1–34.
Wacquant, L. (2008) *Urban Outcasts*. Cambridge: Polity Press.
Winant, H. (2001) *The World Is a Ghetto*. New York: Basic Books.

27

Imagining the entitled middle-class self in the global city

Tiny Times, small-town youth, and the New Shanghainese

Tsung-yi Michelle Huang and Muzi Dong

Introduction

Since the era of reform and opening up, China has made proactive moves to encourage the country's cities on to the path toward modernity, paving the way with ideologies of developmentalism and urbanism. 2011 marked a watershed moment for the country, as it was the point at which the urban population outnumbered the rural population in China for the first time in history. By the end of 2013, there were estimated to be 658 cities in China: the four first-tier metropolises, Beijing, Shanghai, Guangzhou, and Shenzhen, dozens of second-tier cities, as well as over 500 medium and small cities (third tier and below). First-tier cities are metropolises that assume prominent positions in China's social activities such as politics and economics. The term usually refers to Beijing, Shanghai, Guangzhou, and Shenzhen, but in some variations can include cities such as Tianjin and Chongqing. Such stratification of China's cities is not determined by the government or the authorities. Instead, it is a concept initially based on the real estate industry's all-round assessment of mainland cities' comprehensive strength.

One of the main tactics Mao employed that enabled him to lead the Communist forces to victory during the Chinese Civil War was "encircling the cities by holding rural areas". Now, however, urban areas are sprawling relentlessly, "encircling" rural areas (Visser, 2010). The cities' phenomenal rate of expansion suggests that China has entered an urban age and that there has been a radical change to, if not a reversal of, the Communist Party of China's (CPC) anti-urbanism policy. As Chen Yingfang argues, "[i]f we cannot fully grasp the meanings of cities to China and their significance to the state, it would be impossible for us to account for the logic of change of contemporary Chinese society" (Chen, 2012: 18). While in recent years a growing number of scholars have concentrated on the "the urban" as the spatial anchoring to address the social, political and economic transformation of contemporary Chinese society, as Chen proposes, less effort has been made to consider the significance of the imaginary construct of the urban as a dominant trope and landscape of China.

Shanghainese Gu Zheng, the associate director of Jia Zhangke's *Xiao Wu*, maintains that it is the small town Fenyang that is demonstrative of "the real China":

> Jia keeps telling me that Shanghai and Beijing are just showcases for China. If I want to see the real China, I have to follow him to his hometown Fenyang. But it is so disappointing to see the real thing...
>
> *(Zheng, 2010: 198)*

Gu suggests that urban China is often only partially represented. While "the real China", a small town like Fenyang, is barely visible in the rhetoric and representations of contemporary urban China, it is, rather, the Manhattanized showcases, such as Shanghai and Beijing, that embody the modern city *par excellence* in the collective imagination. When the global city becomes the blueprint of development for cities of all tiers across China, a generic cityscape ensues and metropolitan lifestyles inevitably eclipse other forms of urban experience. One thereby witnesses the disappearance of many counties, towns, and cities at or below the third-tier level in political, academic, and everyday discourse. A large number of small-towners, in turn, aspire to cross the rural-urban divide, eager to kiss their humble hometowns goodbye and pursue a new identity as middle-class citizens in the "big city". Against this context of the skewed imagination of urban space and experience, this chapter seeks to unpack the cultural politics and imagination of classed selves in the metropolis, exploring the relationship between the aspirant subject – the "small-towner" – and the hierarchical formation of cities in the context of China's rapid urbanization that began from the 1980s onwards. As the issue is being approached from a cultural studies perspective, this chapter starts with a brief overview of China's urbanization and intra-city competition. The effects of this hierarchized space on subject formation will then be approached by examining a few representative urban writings by small-towners to tease out the implications of the binary imaginary of the small town and the global city. Having established the context of small-towners' urban imaginary, subsequent sections turn to the works of Guo Jingming, a small-town-boy-cum-Shanghainese writer. Both Guo's early writings on Shanghai fever and his novels of the city's second-generation *nouveau riche* (*Tiny Times* 小時代) will be examined in an effort to reach a critical understanding of the methods by which an entitled middle-class self-image – marked by the accumulation of cultural capital, conspicuous consumption, and the aestheticization of one's lifestyle – is formed in a first-tier global city and why the success stories of this Shanghai newcomer are always haunted by the specter of his abandoned small-town identity.

Urban China and its small towns

Since the 1980s, place-making strategies that aim to push cities to "go global" by creating urban glamour zones in the image of Manhattan have been *en vogue*. The rapid construction of Special Economic Zones (SEZs) in the 1980s, as exemplified by Shenzhen, and the transformation of Pudong from farmland to the glittering central business district of Shanghai in the 1990s, became developmental milestones that encouraged other cities to experiment with and pursue their own Manhattan dreams. Promoted by local governments and hyped by the media, discourses of joining the "global-city" club to become "bigger, taller and faster" soon formed the basis for China's urban planning development, celebrated as the *sine qua non* of Chinese modernity in the global era (Huang, 2004; Dirlik, 2005: 37–9; Xu and Yeh, 2005; Wei and Leung, 2005; Han, 2014:

52–62; Jing, 2012: 68). The previous two decades not only saw megaprojects, including the construction of sky scrapers, undertaken in first- and third-tier cities alike, but also the sharp increase of "fast cities", rapidly growing cities and those built on sand (Datta and Shaban, 2016: 23), both of which dramatically transformed China's urban landscape (Chen, 2009; Ma and Wu, 2005; Zacharias and Tang, 2010; Zhang and Zhao, 2009).

The overall development of urban China in recent decades has been very uneven: while coastal cities have enjoyed great advantages, cities in the interior and small towns tend to be marginalized under the central government's urbanization master plan. The discrepancy gives rise to a hierarchical relationship between cities and towns. It has been observed that "the income and development gap between cities and the countryside is now being replaced by an equally fundamental gap between small, particularly inland cities and the giant coastal metropolises" (Davis, 2006: 7). As cities become the strategic site and measure of development, intense intra-city competition follows. The unprecedented urban competition takes Chinese cities into an era of rankings (Ma, 2013). In addition to categorizing cities in an administrative hierarchy, i.e. city-size ranked from first- to fifth-tier cities, ranking lists of all types permeate both official and media competitiveness reports. Although the highest ranked cities such as Beijing, Shanghai, and Guangzhou are always the center of attention, this chapter argues that it is those seemingly lackluster small towns at the bottom of these rankings that shed light on a much less engaged aspect of China's urbanization and the representational power of imagery and rhetoric of great metropolises.

Under the influence of the myth of global cities as the epitome of China's modernization, which tempts with the promise of the good life that great metropolises offer, small towns and villages are driven to see themselves as incomplete forms of metropolis, and urbanization is, as a consequence, viewed as the key to local economic development and to a new way of life. Unfortunately, when desires for fast development are coupled with their lack of any kind of resources (including talent, touristic appeal, and aesthetic value), it is not unusual to find that the *in situ* urban transformation of the small towns, which revolves around the superficial imitation of global city landscapes, not only falls short of their primary goal of building a grandiose city but renders them homogeneous, dehistoricized, and kitschy. As pale copies of authentic metropolises, small inland towns can easily become the opposite of global cities, economic backwaters and outcasts from modernity.

Contemporary studies of urban modernity, under the auspices of postcolonial theories, question first and foremost the mainstream discourse related to urban modernity. Dibazar, Lindner, Meissner, and Naeff stress that our mainstream definition of modern cities is often based on the historical and geographical particularity of a limited few western metropolises, which makes it hard for us to look beyond their specific experience with industrialization, social space transformation, and cultural aesthetic development; correspondingly, the establishment of the urban modernity paradigm is accompanied by the othering of the so-called "backward", "underdeveloped" regions. Those non-western city centers are always frequently compared with their modern western "originals" that they are "pretending" to be like – no doubt, this thus presumes that the so-called "modern city" actually has an established archetype (2013: 648–50). Robinson also proposes the concept of "ordinary city" to critique the dominant global/world city discourse, referring to a cosmopolitan approach to cities in contrast to the narrowly focused global world city paradigm: "ordinary cities bring together a vast array of networks and circulations of varying spatial reach and assemble many different kinds of social, economic and political processes". In other words, ordinary cities are "diverse, complex and internally differentiated" (2006: 109). De Kloet and Scheen's (2013) study of Pudong in Shanghai as a *shanzhai* (山寨) global city is another case in point. They point out that the global city is merely an imaginative construct, an abstract concept made up of numerous skyscrapers (and perhaps water fountains). The connotation of "*shanzhai*" is telling – its original meaning being the bandit stronghold hidden illicitly in the mountains and away from state power (2013: 699–700). A *shanzhai* global city is thus a unique copy of the so-called original which in fact no longer exists. Its existence and experience challenge the set of mainstream thinking dominating contemporary analyses of China's urbanization based on originality and fakeness (2013: 706).

As metropolitanization in the direction of the global/world city becomes the contemporary form and upgraded version of Chinese urbanization (Liu, 2006, 2013), metropolises as a prevalent ideology have shifted to become the concrete space of future lifestyles (Chen, 2012: 19). The new spatial order of urban spaces (first-tier cities vs. small towns) concomitantly gives rise to what Hoffman calls a "hierarchy of desire" (2001: 45, 2010: 70), referring to the way the younger generation maps their aspirations of social mobility (choice of where to work and live) on to the hierarchies of desirable cities in the official ranking. The glittering Beijing, Shanghai, and Guangzhou lend themselves as the concrete spaces aligned with young people's career ambitions. To some extent, when great metropolises become the defining national landscape, urbanites naturally assume a privileged identity in contemporary China.

The yearning to "qualify" for a cosmopolitan urban identity finds one of its best expressions in contemporary writings by "small-towners" since the 1990s. The rise of urban literature in the 1990s, replacing rural writings, was enabled by the emergence of a process of urbanization in China that was unprecedented in its scale and speed. The rapid expansion of cities has allowed a new generation of small-towners, along with migrant workers, to seek out opportunities and resources in these metropolises. Driven by aspirations to urban life in China, many small-town writers have recast their life goals and the subject of their writings – now the bright lights of the big cities are deemed a worthy subject for their creative works rather than their boring kitschy hometowns. Two types of narratives dominate the writings of these small-towners with aspirations to urban life. Some narrate the bitter and overwhelming experience of the urban by people from other provinces, such as the novels of Qiu Huadong and Xu Zechen on drifters in Beijing. Others choose to shy away from their own small-town memories, instead positing themselves as middle-class "new urbanites" leading a charmed life in the urban glamour zone, as seen in the semi-autobiographical works by Wei Hui and Zhao Po, the so-called "beauty writers".

In contrast to the trendy texts on sexy urban glamour, a salient feature of writing on hometowns since the 1990s is a key transformation from recounting one's pining and affection for one's hometown to an overwhelming desire to escape the countryside and the small-town that one calls home. For example, in the works of the "drifter generation" (writers from small towns or counties) such as Ah Yi and Lu Yao, being labeled as a small-towner is nothing less than trauma, "a long-lasting self-abasement and self-denial" in their own words (Lu Yao, quoted in Zhao, 2015). The frustration and angst of being born as a small-towner, interpreted as a genealogy of misfortune, appears as a recurring theme in the narratives of small-town experiences. Before judging these small-town writers as overly sentimental, one needs to take a closer look at how these feelings are structured against the backdrop of the above-mentioned urbanization of China. There are two reasons that account for small-towners' melancholy. First, most of these hometowns have "disappeared" in their attempts to imitate first-tier cities. When homogeneous, mass-produced scenes of modernity erase a landscape of local history and the sense of place in the small towns where one grew up, personal memories lose their ability to anchor identity. Second, the hierarchy of desire speaks to the sense of crisis felt by those cities at the bottom of the rankings. One must leave the backward small towns so as to be granted access to the variety of opportunities and profusion of pleasures that only metropolises can provide. The dream of climbing the ladder of the urban hierarchy is nowhere more poignantly expressed than by Ah Yi:

> I cried to myself and swore to God that I have to depart now, to the town, the county. No, it is not good enough. I have to go to the city, the capital city of the province, the coastal city, the municipality, the capital city of the nation, to New York.
>
> *(2012: 11)*

Torn up by the anxiety of getting stuck at the bottom of the social/spatial order, as well as by the longing to experience urban life, small-town youth imagine New York as the final destination of their mobility, and global cities serve as the new hometown in which they wish to settle down for a rich and refined life. It is in this context that bestselling author Guo Jingming, originally a small-towner from Sichuan, and his works on Shanghai, emerged to represent a new urban imaginary.

Guo Jingming: imaginaries of a global city and narratives of outsiders

For Guo Jingming the fatal defect of small towns lies in their lack of sensitivity to urban life and style; he thereby refuses to call such an unremarkable town his hometown. The writer's disdain at the backwardness of small towns suggests one can only find "authentic" modernity in the mega-cities. This ideology accounts for why in much of his early prose Guo's discontent with his hometown is disclosed through his obsession with Shanghai:

> It seems that my roots were put down in Shanghai. It is like a wandering nerve; I have wandered all the way here. Unbelievable ... Why were those who do not love Shanghai born here? God must have been mistaken...
>
> *(2014a: 234–35)*

For Guo, as well as many other small-towners, there is no paradise superior to Shanghai. Like Hong Kong, Shanghai benefits from the legacy of modernity brought about by the colonial, or semi-colonial, history of the previous century (Abbas, 2000; Lee, 1999). In the global era, under the banner of developmentalism, one of the recurrent themes in the discourse of Shanghai's city-marketing is to graft the image of today's New Shanghai on to the reputation of Old Shanghai as the "Paris of the East". Shanghai plays a leading role in shaping both urban modernity and a new metropolitan social subject. Put another way, Shanghai is the "global" counterpart to cities all over the country. Shanghai has quickly become a city in which millions of Chinese pursue material interests and social mobility. In this light, it is no wonder that Guo, a melancholy, sensitive, and conceited small-town young man, sees Shanghai as his "real home" and makes the decision to "carry his little backpack" to start the journey home "despite his family's objection and people's sneering" (Guo, Li and Mao, 2013: 8).

Shanghai functions as an irreplaceable symbol central to Guo Jingming's writings from the outset of his career, creating an aura and fitting in with a form of rhetoric that allows him to steep both his works and himself in a self-important glow. In other words, the meaning of Shanghai as the starting point of Guo's upward trajectory differs from those Shanghainese writers from other provinces in the 1930s, but resembles Honoré de Balzac's portrait of outsiders who came to Paris. Guo worships Shanghai as the city of all Chinese cities, just like the Paris of nineteenth century France, a city that held the dreams, ambitions and futures of all those who come from other provinces (Harvey, 2003: 29–32). In Guo's portrayal, countless small-town youths take Shanghai as a focus for their aspirations:

> Shanghai! O Shanghai! It is a name of innumerous fancies and innumerous pains. Like a colossal ultimate fantasia, it is also the infinite wasteland of ashes. Day by day, how many walk out of the railway station underpass with youthful ambition. As they look up to see the high rises surrounding the station's north square, they are awed by the magnificent urban structure. Little do they know that this is only the plainest and the most ordinary district of Shanghai, that there are even more gilded places, even more loci of extravagance. All these and always these are daggers hidden beneath the boulevard of tomorrow, ready to penetrate through the surface at all times, rupturing the fantasy of your youth. You will hear the blast.
>
> *(Guo, Li, and Mao, 2013: 8)*

The perception of the metropolis of these provincial youths comes first of all from the vivid, raw stimulation of landscape, which offers a visual experience drastically different from small-town scenery. Hyperbolizing the modern spectacle of the metropolis is the manifest feature of Guo Jingming's Shanghai writing as well as his persistent form of representation. This metropolitan experience is indeed similar to Guo's own when he first arrived in Shanghai – obsession with the city rapidly turned into an intense account of the sensory impact of the majestic modern architecture with the exclusionist local culture bewildering the outsiders. He describes in his prose the world he saw when "emerging from the People's Square Metro Station" for the first time: "Immense. Spinning. Luminous. Mesmerizing. Cold. Stiff. Vogue. Contemptuous. Shanghai" (2013a: 306). By throwing out successive adjectives and periods, the writer's idiosyncratic wording freeze-frames a small town boy's experience of shock and trauma when first arriving in the busy cold metropolis.

Guo Jingming's ambivalence toward Shanghai echoes the experiences of Parisians of provincial origins. Harvey points out in his research on Balzac's Paris that "Parisians of all classes lived in a state of denial and distrust of their rural origins"; the provincial migrants have to go through "complex rituals of integration" into the metropolis, and "[o]nce incorporated, they never look back" (2003: 30–2). Like numerous provincial characters in Balzac's works (or like Balzac himself), Guo underwent the difficult lifestyle change from provincial town to the mega metropolis. After achieving fame and becoming a registered Shanghai resident, Guo self-identifies as Shanghainese and sets out from this perspective when commenting on contemporary Shanghainese literature. His writing has also transformed, shifting from a provincial admiration for and obsession with Shanghai to an appreciation for the metropolis from the viewpoint of a native. His masterpiece *Tiny Times* can be seen as the product of this process.

Shanghai in literature, Shanghai spirit, and *Tiny Times*

Shanghai has been a popular subject of urban literature for a long time: it was the Janus-faced city of darkness and light for the "New Sensation School" in the 1930s, a bleak dystopia of sin and evil in left-wing literature, and the legendary city of love in wartime in Eileen Chang's stories, to name but a few. As Shanghai has experienced unprecedented changes over recent decades, new images and representations of the city proliferate. Concomitant with its transformation at full speed to become East Asia's global city is an image-based marketing campaign that posits Shanghai as a charming and cosmopolitan city *par excellence* by the city government (Berg and Yuan, 2014; Dynonj, 2011; Huang, 2006). At the same time, a rich body of urban writing presents Shanghai here-and-now and New Shanghainese in distinctive ways. On observing the "Shanghai fever" of recent decades and the types of writings produced along with this trend, one finds that central to mainstream writings of Shanghai since the reform and opening up period is a proclivity for recognizing and representing only certain types of Shanghainese people and their lived space, which largely corresponds to the "modern side" of the city. For writers like Wei Hui and Mian Mian, New Shanghai is a sexualized metropolis that facilitates young people in embracing their carnal desires. In nostalgic stories of Old Shanghai, on the other hand, Shanghai is a city with a legacy of good taste, most vividly represented as the urbanites' subtle, exquisite, and sensuous experience of everyday life in Wang Anyi's novels of Shanghai alleyway houses (*longtang* 弄堂) or, as Cheng Naishan and Chen Danyan remind us, representing a perception of "true nobility" in China and their upper/middle-class aesthetics. Most of these urban writers leave behind "the world 'outside of the neon light,'" which also undergoes the modernization process of the city but is only faintly visible in Shanghai writings, as Luo Gang (2007: 91) observes.

This propensity of representing the modern side of Shanghai finds its most recent and extreme version in Guo Jingming's *Tiny Times*, a coming-of-age trilogy that takes place from the year 2008 onward, the thirtieth anniversary of China's reform and opening up. The trilogy is mostly concerned with the life of four young women as they make their way out of university into today's careerist, consumerist Shanghai, going through betrayals and breakups, misunderstandings and forgiveness, disease and death, boardroom politics and corporate takeovers. Guo claims that these novels which are composed of almost a million characters, are a "*golden hymn dedicated to Shanghai*" (emphasis added). In an interview he reveals the motives behind the writing of *Tiny Times*:

> Whenever we talk about Shanghai, we always think of Eileen Chang's passionate romance or the heroine's turbulent life in Wang Anyi's *The Song of Everlasting Sorrow*. Unfortunately, there is no Shanghai story of here and now ... *The New Shanghai of 2008 has never been written*. I hope people will think of my *Tiny Times* when they talk about New Shanghai in the future.
>
> *(Hu, 2008: 23, emphasis added)*

Guo's New Shanghai story, as we will discuss below, is one that passionately embraces the dazzling modern cityscape, the consumer economy, the brand-name luxuries, and sumptuous lifestyle. It is not only an attempt to capture the look and feel of Shanghai in his own era, but a manifesto for successful New Shanghainese and a manual for those who aspire to middle-class life in the emerging and expanding global city.

A best-seller for three consecutive years from 2010, the *Tiny Times* trilogy has been the most commercially successful novel ever written on Shanghai. The night before its film adaptation was released in 2013, 6.7 million copies of the trilogy had already been sold. In addition to the film series, which made 2 billion RMB at the box office, related manga, musical, and TV renditions have drawn the attention of different audiences. Guo's huge success won him praise as "the practitioner of Chinese Dreams" in 2013 by *Southern Weekly*. Despite the unparalleled popularity of the *Tiny Times* series, Guo Jingming and his Shanghai stories have provoked skepticism and unleashed controversy. The author's legitimacy in writing about Shanghai has been questioned: many critics were bitter about Guo's pretentious "Shanghai native" tone in telling the urban romance (Tubinggen Carpenter, 2013) and how he "keeps trying to erase his Sichuan small-town background to act as a true Shanghainese" (Huang, 2011: 7). *Tiny Times* is also dismissed simply as Guo's "effort to make up for being trapped in a small-town and thus denied the chance to enjoy glitzy days in the metropolis", touting "a tutorial of the glamour and luxury of Shanghai" to other "small-town youth" from urban-rural fringe areas (Han, quoted in Wu, 2008: B11). Along a similar line of argument, his representations of Shanghai have been criticized as hollow and flat, a one-dimensional drama full of "materialistic spectacle without cultural substance" (Fang et al., 2013: 87). Shanghai's "diversity and heterogeneity", as one critic argues, "are flattened by Jin Mao Tower, World Financial Center, and Hang Lung Plaza (Plaza 66)", it is reduced to "a city without history, interchangeable with New York, London and Tokyo" (Huang, 2011: 6). In short, the critics of the franchise insist that it is a fake Shanghai story written by an outsider who pretends to be a native and therefore it lacks an authentic "Shanghai spirit".

The stark contradiction between the vehement criticism of *Tiny Times* and the series' immense popularity, we argue, begs the question of not only how to understand the alleged absence of Shanghai spirit in *Tiny Times*, but what social and psychic implications "authentic Shanghainess" could raise today. In other words, Guo's writings about Shanghai deserve serious attention since his works cast light on the idealized images of a New Shanghai and a new middle-class generation after the city reached its peak of development in the 2000s. As a self-made immigrant from another province, i.e. a New Shanghainese, the author endeavors to open up new urban experiences and sensibilities for the young generation. Interestingly, this New Shanghainese always assumes the role of a native rather than a stranger from a faraway small town. In what follows there will be a close analysis of Guo's narrative strategies in representing class distinctions so as to explain the hierarchical relationship between Old and New Shanghainese and the logic of exclusion inherent in these urban identities.

Distinction, New Shanghainese, and the middle-class self in *Tiny Times*

Before delving into a detailed explanation of how *Tiny Times* contributes to an understanding of the emerging entitled middle-class self and particular class relations in Shanghai, it is important to contextualize the term "New Shanghainese", which gained its popularity in official discourse and academic circles around the year 2000. In a 2001 interview, Huang Ju, the former leader of the Shanghai Municipal People's Government, defined "New Shanghainese" as "those who foray into the world in the spirit of adventure, citizens of the world and also Chinese" (Xiong, 2002: 129). The same year New Shanghainese was adopted as the theme of a big academic conference, which identified

"New Shanghai" as the spirit of hospitality to welcome foreign cultures and people as an ocean embraces all rivers (Shanghai Zhengda Institute, 2002). The vision of Shanghai's cosmopolitan future is renewed in this very idea of New Shanghainese.

In lauding the city he admires enormously, Guo dances to the same tune as that of the official discourse of New Shanghainese, which defines cosmopolitanism as the essence of its culture and keeps its assumption of class divisions from sight. The novel paints the New Shanghai in 2008 as a global city open to all, one that sees "numerous people come and go" (2013b: 311). In presenting the fluidity and diversity of the urban scene, Guo shows a parade of passers-by, providing a panoramic view of the New Shanghainese with different social roles:

> young white-collar workers with Marc Jacobs handbags, young graduates trying their luck in the job market, those numerous Oriental-looking customers at Starbucks, the poker face of a salesperson at a luxury store in the Bund, a distressed and resentful woman from an alleyway house, and the upper-class women waiting for their afternoon tea.
>
> *(2013b: 10–1)*

Yet a closer look at Guo's writing reveals that the secret of modern life in Shanghai lies in its spectacular public space: the city may welcome all, but class division is the repressed reality that constantly seeps through the cracks of the façade of cosmopolitanism. This contradiction is clearly illustrated by the juxtaposition of brand-name high-heels and working-class boots:

> At People's Square, one gets to see the crystal hollow-cutting Jimmy Choo high-heels that any woman would trade her soul for and at the same time the green People's Liberation rain boots that remind us of the good old days of laboring.
>
> *(2014a: 156)*

In an attempt to express the cosmopolitan spirit of Shanghai, Guo places people from different classes in one scene, but in so doing, he suggests that the outsider and the underclass can easily be identified by their taste, even if they share the same urban space with those who wear Jimmy Choos.

It is no accident that one of the persistent themes of the trilogy is the contradiction between the tasteless lower-status non-locals and the tasteful middle-class Shanghainese. The scene of tourists flocking into Shanghai during holidays is a typical urban spectacle in Guo's works, which demarcates the visitors:

> On the Riverside Avenue across from the Bund, countless visitors come and gasp in admiration, holding their cameras and jostling each other to get the best shooting location. They wear the same cheap clothes from large chain-stores and yell out-loud, "Here! Here!" The luxury world across the street is only twenty meters away.
>
> *(2013b: 11)*

The noisy and tasteless tourists are obviously out of place – even though physically the world of luxury brands is right in front of their noses, they have no access to the world of taste. Unlike those visitors who are incompatible with Shanghai's metropolitan landscape, the subtle, well-groomed professional class blends in perfectly into the background of the urban glamour zone.

The representation of class distinction is highly spatialized in *Tiny Times*. Guo characterizes his "100 percent real rich second-generation" as someone who lives in the "Tiny Shanghai", spending her daily life only in the central district and the previous concession regions, the Huangpu and Jingan Districts. The rest of city, the newly expanded area, and even Pudong are categorized as the distasteful "non-Shanghai". Gu Li never sets foot in places outside the inner ring of the city. She would "vomit when her car passed out of the central ring", and she always angrily hangs up her friend's "phone call from the outer circle" (2013b: 306). As far as this super-rich girl is concerned, Pudong is still the remote suburbs that cannot be reached by subway; it is so remote that extra roaming fees are charged. She even refuses to believe that the "Shanghai International Finance Centre Mall in Pudong has more fashion brands than Plaza 66 in Jingan District" (2014c: 139). Shanghai's urban landscape is flattened, partial, and skewed since for the novelist only the urban glamour zone suits his protagonist's profile as a "true noble". What is noteworthy about Guo's privileged city-user is that despite the fact that Lujiazui, Pudong, has served as the landmark of New Shanghai's urban development and a symbol of the city's global modernity, for some Shanghainese, Pudong cannot cast off its stigma as an outlying area that has just gained a few fancy buildings. For them, only the prime districts in Old Shanghai across the Huangpu River are "truly" representative of the glamour of the city.

To shape the new elite personality, Guo is almost obsessed with portraying people newly arrived to the city as lacking in taste. The way he represents social distinction dramatizes Pierre Bourdieu's seminal account of tastes as "the practical affirmation of an inevitable difference" (1987: 56). A pertinent example in the novel is super-rich girl Gu Li's vicious comments on the northern Chinese long-johns (*qiuku* 秋褲) as a fashion no-no and a symbol of vulgarity and bad taste. "People running in *qiuku*" in the streets baffles her, not to mention that "some women wear *qiuku* to the Hermes shop at Yintai Center". The ultra-chic Shanghai girl claims in her histrionic tone that this scene is "the most terrifying scene I have ever seen since the 1998 Japanese horror film *Ring*" (2014c: 16–7). In a sense, hinging the image of authentic Shanghainese on their differences from the lower class renders *Tiny Times* a guidebook to urban life in contemporary China, not unlike the etiquette manuals of the nineteenth century. The function of etiquette manuals was not only to inform latecomers to the city how to be respectable urbanites but "also provide the knowledge to spot imposters, because the newly arrived, the autodidact, will unavoidably give away signs of the burden of attainment and incompleteness of his/her cultural competence" as Beverly Skeggs rightly notes (2004: 136).

Gu Li's sarcasm has a dual function in Guo's class bifurcation: while mocking and spotting the tasteless outlanders, the author employs this ventriloquist's dummy to brag about his newly found fame, wealth, and distinction in Shanghai, i.e. showing his transformation from a small-town boy to a successful urban writer/entrepreneur. At one point in the novel, when the girls talk about the well-known novelist Guo Jingming, Gu Li grumbles that:

> [T]he only thing I know about him [Guo] is that once I saw a men's dress shirt in Dior, but the salesmen refused to sell it to me. They said that Guo had ordered it, so it is reserved for him. And they never restock that one. What a bitch!
>
> *(2013b: 185)*

With the knowledge of new goods and lifestyle as cultural capital, Guo demonstrates that he has not only acquired authority in matters of taste but gets to legitimize himself through his own literary production.

Class in its dual senses (as cultural taste and social hierarchy) thus played a dominant role in the construction of the self in *Tiny Times*. The desire to return to "Old Shanghai" always underlies the novels, despite its seeming appreciation of "New Shanghai" charm. Whether in diminishing the ever expanding metropolis to the former concession zone during the Republican Era in its portrayal of new Shanghai, or in fabricating an ahistorical upper-class family to uphold a "rich second-generation" protagonist (inasmuch as a "rich first generation" self), *Tiny Times* shows the author's eagerness to legitimize his status as a "new aristocrat" by posing as "old money". Guo's identification of "Shanghai-ness" therein is clear. *Tiny Times* reformulates the contemporary affluent metropolitan area with its Old Shanghai city heritage as the essence of "New Shanghai" and idealizes "New Shanghainese" as the "real aristocrats" of this "vintage" global city in the twenty-first century. In doing so, *Tiny Times* posits the new urban middle-class culture as the contemporary "Shanghai spirit".

Ironically, it seems that no matter how Guo Jingming attempts to justify his extravagant world of luxuries, his identity as a spokesperson for his beloved city and the image of Shanghai/Shanghainese in *Tiny Times* is held suspect by literary and cultural critics and many Shanghainese readers. This paradox can be addressed by examining the identity of New Shanghainese. Since its appearance in 2000, the identity of New Shanghainese tends to beget ambivalence. At first it mainly catered to the rising elite class in the global city as mentioned above. Despite Shanghai's privileged status as a global city in the making, when it comes to tastes and distinction there exists an obvious hierarchy: the Old Shanghainese are viewed as the custodians of Shanghai culture, whereas, as latecomers in Shanghai, New Shanghainese have had to acquire taste to ensure their class position and access to the city as a resource. Here the "new" points to some essential lack, the difference between the local and the outsider based on their symbolic capital. The term has been used in recent times in media and the internet in a pejorative sense, usually referring to those who acquire household registration (*hukou* 戶口) in the city or generally all newcomers, from high and low social ranks alike. In its worst form, the discourse on New Shanghainese expresses a strong sentiment of localism and identifies the newcomers as "locusts" in the city or predators feeding on Shanghai's resources. In light of the negative implications of New Shanghainese in these contexts, it is not surprising that Guo's huge commercial success has inspired some controversy. The self-designated identity as a native Shanghainese, a signature of Guo's Shanghai stories, is a double-edged sword. On the one hand, playing the role of a local enables Guo to structure his *Tiny Times* around the binary between Shanghainese and outlanders so as to present his stories of status distinction with displays of "good taste", conspicuous consumption, and the aestheticization of everyday life. On the other hand, while Guo is having fun identifying country-bumpkins in his Shanghai stories, he has been subject to harsh criticism of being an imposter by native Shanghai critics and readers. In other words, when a New Shanghainese attempts to pass as an authentic local resident and spokesperson for the city, the privileged identity, i.e. the Old Shanghainese, come to the foreground to show that the immigrant is always an outlander and the new rich should not presume to be old money. Beneath the highly glossed cosmopolitan spirit of New Shanghainese as a legacy of Old Shanghai's openness lies the excluding logic of class distinction: In the world of *Tiny Times*, no tourists or immigrants from other provinces escape Guo's scathing judgment as vulgar and lacking in discernment. By the same token, in the real world, if Guo was not born a Shanghainese but a small-towner, there is indeed no chance for him to become one.

Conclusion

Through a textual analysis of *Tiny Times*, this chapter argues that the status of native Shanghai "aristocrat" to which Guo Jingming aspires is actually representative of the newly elevated middle-class pouring into the global city from all over the country (i.e. the New Shanghainese). The Shanghai of the 1990s is, no doubt, the representative space of the society's collective imagination of "Chinese metropolitan middle-class life". Along with social transformation and urban regeneration, the aestheticization of everyday life via symbolic consumption has since become the quintessential identity performative of the contemporary urban middle-class. Guo's success thus comes face to face with the historical context and epochal problematic as such: through social change, as "great metropolises" replace "small towns" as the defining landscape of China, "the rich second-generation" surpasses "the rich first generation" as the hip and desirable social subject. Authenticity of origin and status become the identity myth of contemporary Chinese cities and individuals. However, the end of the previous stage of drastic economic reform and its redistribution of capital have solidified social re-stratification, making identity transformation from "the lower class" to "the upper class" today much more difficult than during the previous decades of the reform period. Accordingly, when "small-town boy" Guo Jingming attempts to play the "new cultural intermediaries" in Bourdieu's terms (1984: 325–6) and brokers his charismatic lifestyle and aesthetics to the public, their favor cannot be easily won. Guo's "famous brands extravaganza" in *Tiny Times*, in which he parades luxury brands from Prada sweaters and Dior gowns to Armani sofas, is instead deemed symptomatic of status anxiety. This anxiety prompts him to seize symbolic capital while betraying the learned performativity of a provincial outsider, if not a country bumpkin. As a result, public opinion has dismissed his materialistic Shanghai writing as the "*nouveau riche* mentality" of a small-town boy.

Examining how Guo Jingming reconstructs the identity of a social subject from urban experience in the current hierarchized China, and how he showcases and justifies his logic in the novels *Tiny Times*, we find that however much he has flourished, the writer seems to be constantly haunted by the specter of his small-town origins. If Wang Xiaoming has implied that behind "the half-face myth" of "successful people", in fact, lies the unspeakable means of wealth accumulation (2000: 29–36), Guo's Shanghai writing, in turn, unveils "the other half a face" of the "new middle class" – the front man of this metropolitan myth – to be the genealogy of "small-town boys". When "the Guo Jingmings" show off their aesthetic cultivation while instructing people on how they can become metropolitan middle-class elites, they have to forsake their identification with their hometown and small-town experience to achieve this shift in identity, so as to establish their authority as experts in "metropolitan sensibilities". Meanwhile, the strong performativity and pretentiousness of Guo's Shanghai writing is also, to a certain extent, catering to market demand. Despite its flawed narratives, the *Tiny Times* series has provided numerous Chinese teenagers with precisely the imaginative romantic metropolitan experience they seek. The prosperous metropolis in Guo's writing is not only the "real" global city but "China's" global city. Shanghai's specific locale allows it to carry the imagination of a transnational middle-class identity in the native context.

References

Abbas, A. (2000) Cosmopolitan De-Scriptions: Shanghai and Hong Kong. *Public Culture* 12(3): 769–86.

Ah Yi. (2012) *Mofan Qingnian* [Model Youth]. Beijing: Delphin.

Berg, P.O. and Yuan, R. (2014) Developing and Branding a Polycentric Mega-City: The Case of Shanghai. In: Berg, P.O. and Björner, E. (eds), *Branding Chinese Mega-Cities: Policies, Practices and Positioning*. Cheltenham: Edward Elgar, 21–41.
Bourdieu, P. (1987) *Distinction: A Social Critique of the Judgement of Taste*. Cambridge: Harvard University Press.
Chen, X.M. (ed.) (2009) *Shanghai Rising: State Power and Local Transformations in a Global Megacity*. Minneapolis: University of Minnesota Press.
Chen, Y.F. (2012) *Chengshi Zhongguo De Luoji* [City: Chinese Logic]. Beijing: Joint Publishing.
Datta, A. and Shaban, A. (eds.) (2016) *Mega-Urbanization in the Global South: Fast Cities and New Urban Utopias of the Postcolonial State*. New York: Routledge.
Davis, M. (2006) *Planet of Slums*. New York: Verso.
De Kloet, J. and Scheen, L. (2013) Pudong: The Shanzhai Global City. *European Journal of Cultural Studies* 16(6): 692–709.
Dibazar, P., Lindner, C., Meissner, M., and Naeff, J. (2013) Questioning Urban Modernity. *European Journal of Cultural Studies* 16(6): 643–58.
Dirlik, A. (2005) Architectures of Global Modernity, Colonialism, and Places. *Modern Chinese Literature and Culture* 17(1): 33–61.
Dynonj, N. (2011) Better City, Better Life? the Ethics of Branding the Model City at the 2010 Shanghai World Expo. *Place Branding and Public Diplomacy* 7(3): 185–96.
Fang, W., Song, S., Guo, S., Ji, Y., Long, H., and Hong, X.M. (2013) 'Dashidai' Haishi 'Xiaoshidai'? – Guanyu Xiaoshidai De Zhenghouxing Jiedu ['Great Times' or 'Tiny Times'? Interpreting *Tiny Times* Symptoms]. *Creation and Criticism* 24: 83–90.
Gu, Z. (2010) Women Yiqi 'Shua' Dianying [Let's 'Play' Films Together]. In: Z, J. (ed.), *Jiazhangke Guxiang Sanbuqu – Xiaowu* [Jiazhangke's Hometown Trilogy: Xiaowu]. Jinan: Shandong Pictorial Agency, 196–206.
Guo, J.M. (2013a) *Yuan Feng Cai Chen* [May the Wind Cut the Dust]. Wuhan: Chang Jiang Literature and Art Press.
Guo, J.M. (2013b) *Xiao Shidai 1.0 Zhezhi Shidai* [Tiny Times 1.0 Dirty Secrets Make Friends]. Wuhan: Chang Jiang Literature and Art Press.
Guo, J.M. (2014a) *Shousui Baiju* [Remembrance of Things Past Like a White Horse]. Shanghai: Oriental Publishing Center.
Guo, J.M. (2014b) *Xiao Shidai 2.0 Xutong Shidai* [Tiny Times 2.0 To Be Blind, To Be Loved]. Wuhan: Chang Jiang Literature and Art Press.
Guo, J.M. (2014c) *Xiao Shidai 3.0 Cijin Shidai* [Tiny Times 3.0 I Will Always Miss You Like a Darling]. Wuhan: Chang Jiang Literature and Art Press.
Guo, J.M., Li, Q., and Mao, M. (2013) *Xiao Shidai Dianying Quanjilu* [Tiny Times Film Full Record]. Wuhan: Chang Jiang Literature and Art Press.
Han, S.M. (2014) *Chuanmei Zhi Cheng: Meijie Shiye Xia de Chengshi Zhongguo* [City of Communication: Chinese Cities on the Horizon of Media]. Beijing: China Social Sciences Press.
Harvey, D. (2003) *Paris, Capital of Modernity*. New York: Routledge.
Hoffman, L.M. (2001) Guiding College Graduates to Work: Social Constructions of Labor Markets in Dalian. In: Chen, N.N., Clark, C.D., Gottschang, S.Z. and Jeffery, L. (eds.), *China Urban: Ethnographies of Contemporary Culture*. Durham, NC: Duke University Press, 43–66.
Hoffman, L.M. (2010) *Patriotic Professionalism in Urban China: Fostering Talent*. Philadelphia, PA: Temple University Press.
Hu, X. (2008) Guo Jingming: Weisheme Buneng You Wenxuefuhao [Guo Jingming: Why Can't We Have Rich Writers]. *Huaxi Metropolis Daily*, 26 September, 23.
Huang, P. (2011) 'Dashidai' Yu 'Xiaoshidai' – Han Han, Guo Jingming Yu '80hou' Xiezuo ['Great Times' and 'Tiny Times' – Han Han, Guo Jingming and 'Post-80s' Writing]. *Southern Cultural Forum* 3: 5–10.

Huang, T.Y. (2004) Dushi Kongjian De Shengchan: Quanqiuhuade Shanghai [The Production of Urban Spaces: Shanghai as a Global City in the Making]. *Taiwan: A Radical Quarterly in Social Studies* 53: 61–83.

Huang, T.Y. (2006) The Cosmopolitan Imaginary and Flexible Identities of Global City-Regions: Articulating New Cultural Identities in Taipei and Shanghai. *Inter-Asia Cultural Studies* 7(3): 472–91.

Jing, C.M. (2012) Zhongguo Chengshihua Jingcheng Zhong De Zhengfu Xingwei Yanjiu [A Study of Governmental Behaviors during Chinese Urbanization]. *China Market* 20: 64–70.

Lee, L.O. (1999) *Shanghai Modern: The Flowering of a New Urban Culture in China, 1930–1945*. Cambridge: Harvard University Press.

Liu, S.L. (2006) Dushihua Jinchenglun [On the Process of Building Metropolitan]. *Academic Monthly* 12: 5–12.

Liu, S.L. (2013) Guanyu Woguo Chengzhenhua Wenti De Ruogansikao [Thinking about Urbanization and Townization in China]. *Academics* 3: 5–13.

Luo, G. (2007) Kongjian De Shengchan Yu Kongjian De Zhuanyi – Shanghai Gongren Xincun Yu Shehui Zhuiyi Chengshijingyan [Production and Transformation of Space: New Residential Areas of Workers in Shanghai and the Experience of a Socialist City]. *Journal of East China Normal University* 6: 91–6.

Ma, L. (2013) Chengshi Paihangbang: Liuxing, Wenti Yu Zhanwang [City Ranking: Trends, Problems and Prospects]. *Journal of Gansu Institute of Public Administration* 3: 24–35.

Ma, L.J.C. and Wu, F.L. (eds.) (2005) *Restructuring the Chinese City: Changing Society, Economy and Space*. New York: Routledge.

Robinson, J. (2006) *Ordinary Cities: Between Modernity and Development*. London: Routledge.

Shanghai Zhengda Institute. (2002) *Xin Shanghairen* [New Shanghainese]. Beijing: Oriental Press.

Skeggs, B. (2004) *Class, Self, Culture*. London: Routledge.

Southern Weekly. (2013) 2013 Zhongguomeng Jianxingzhezhijngci [2013 Salute to Practitioners of Chinese Dreams], 19 December. Available at: http://www.infzm.com/content/96851.

Tubinggen Carpenter. (2013) Xiaoshidai: Yingpanmen De Shanghaitan [*Tiny Times*: Shanghai of 'Hardware'"]. *Mtime.com*, 3 July. Available at: http://i.mtime.com/t193244/blog/7637800/.

Visser, R. (2010) *Cities Surround the Countryside: Urban Aesthetics in Postsocialist China*. Durham, NC: Duke University Press.

Wang, X.M. (2000) Banzhanglian De Shenhua [Myth of Half Revealed Face]. In: Wang, X.M. (ed.), *Zai Xinyishi Xingtai De Longzhao Xia: 90 Niandai De Wenhua He Wenxue Fenxi* [Under the New Ideology: 1990s Cultural and Literary Analysis]. Nanjing: Jiangsu Peoples Press, 29–36.

Wei, Y. and Leung, C.K. (2005) Development Zones, Foreign Investment, and Global City Formation in Shanghai. *Growth and Change* 36(1): 16–40.

Wu, B. (2008) Mawanjia Han Han 'Xiashui' Wan PK [After Quarrelling, Han Han Decides to PK with Guo Jingming]. *Guangzhou Daily*, 22 November, B11.

Xiong, Y.Z. (2002) Shanghairen De Guoqu, Xianzai He Weilai [Shanghainese: Past, Present and Future]. In: Institute, S.Z. (ed.), *Xin Shanghairen* [New Shanghainese]. Beijing: Oriental Press, 117–30.

Xu, J. and Yeh, G.O. (2005) City Repositioning and Competitiveness Building in Regional Development: New Development Strategies in Guangzhou, China. *International Journal of Urban and Regional Research* 29(2): 283–308.

Zacharias, J. and Tang, Y. (2010) Restructuring and Repositioning Shenzhen, China's New Mega City. *Progress in Planning* 73: 209–49.

Zhang, L. and Zhao, X.B. (2009) City Branding and the Olympic Effect: A Case Study of Beijing. *Cities* 26: 245–54.

Zhao, Z.J. (2015) 'Xiancheng Qingnian' Lu Yao [Small-Town Youth: Lu Yao]. *Jiangnan Time*, 1 April, B01.

Part V
Speculative and transformative

28

Urban imaginaries and the palimpsest of the future

Nick Dunn

Introduction

Since the publication of the UNFPA's 2007 report, we have been repeatedly informed that our society will be an urban one, yet within the complexity of our cities there is a paradox. Cities are championed (Glaeser, 2011; Hollis, 2013) as the places within which dense dynamics of economy and technology will enable us to flourish and where we can transform our lives and be happy (Montgomery, 2013). However, it is also because of the very same forces that many problems are occurring within urban society, which makes the question of how we form visions for urban futures critical. Imagination is key to processes of creating visions, which in turn directly informs those which we support in processes of becoming and the delivery of futures.

Formulating visions for urban development has often been marginalized or dismissed as being inconsequential. However, if the Neuman and Hull (2009) statement "if we cannot imagine, then we cannot manage" holds true, the practices of conceptualization, envisioning and performing urban futures is vital to our ability to deal with increasing urban complexity. Numerous writers, including Hall (2002) and Pinder (2005), have emphasized that imagining cities is a normative practice within the disciplines of urban design and planning, and the value of doing so is fundamental to our ability to understand alternative visions for collective life. Despite this, it is increasingly apparent that such professions seem unable to transfer this skill toward the production of new visionary imaginaries of the urban that illustrate a significant diversity of ideas. This is odd, even contradictory. At a time when discourse concerning the role of cities as *the* object of our times is perhaps at its greatest momentum (Graham, 2010; Robinson, 2006; Sassen, 2001), even to the point of what we might understand as "planetary urbanism" (Brenner and Schmid, 2015), it is remarkable that proposals for radical alternatives are few. Why should this be? Two matters present themselves as possible explanations. The first is our inability to define what we are discussing when we explore and examine the "urban" as critiqued by Gleeson (2012). The second is constituted by the concomitant forces that have, particularly over the last fifty years or so, gradually converged our urban imaginaries to its current narrow bandwidth of possibilities, notable for its convergence of future scenarios rather than divergence (Hunt et al., 2012).

This latter point is perhaps best illustrated by the dominance of the "smart cities" paradigm. This paradigm, itself diverse and with multiple narratives, is often perceived as one of singularity, in which technology supports optimal efficiency in many, if not all, aspects of urban life. Although the visions for smart cities hint at a brave new world that will eradicate urban ills in the name of progress, with futures that are always clean, smooth, green, and vibrant, they also necessarily open up serious questions concerning such apparently frictionless and perfect environments and who operates them (Greenfield, 2013; Hollands, 2008; Thomas et al., 2016). Why would we not want to live in a world depicted like this, especially when the antonyms of "smart" plunge us into an unhelpful dialectic when imagining radical alternatives? We will return to the urban imaginary of smart cities later but it is a useful reminder of path dependency when we try to conceive of different futures. Perhaps this is why now it is particularly salient to examine how we may better enable the reworking or vision processes so that rather than closing down futures we find ways of opening them up. Key to such processes is the role of imagination.

The power of imagination

So can imagining the future change it? Do the ideas we have for urban futures build up over time and echo throughout history? In this section, I am going to further explore the idea that perhaps the problem regarding urban imaginaries in the present time is not necessarily that we have unattainable ideas, but that we have a lack of them. Many of our imaginative ideas for urban futures are born in the past. More specifically, we can conceive of the history of urban imaginaries as similar to a palimpsest, i.e. it contains a layering of ideas, some of which have been reused or significantly altered but still bear visible traces of its earlier form. Indeed, despite the claims foregrounded by technological innovation, it is also possible to argue that we have not traveled very far at all when it comes to radical visions for urban life. One of the explanations for this is what we might term "imaginative lock-in", i.e. our inability to think beyond relatively normative trajectories when conceiving urban futures and their lifestyles. This may not necessarily be the result of an immediate dearth of extreme or challenging urban imaginaries, but might be the consequence of those dominant narratives that, as they make certain futures visible, they occlude or even discredit others. In this regard the history of urban imaginaries is not unlike many other histories. Simultaneous processes of globalization and pervasive digital technologies have enabled the production and dissemination of urban imaginaries through interlinked global ideoscapes and mediascapes (Appadurai, 1990). However, despite both the cultural and geographic diversity of their origin, these interconnected ideo-/mediascapes typically congregate into a surprisingly limited number of types. Which stories we tell and how they are shared to form collective visions for urban life, therefore seems more pertinent than ever.

The role of imagination is fundamental to processes of conceptualization, envisioning, and performing urban futures. The importance of such creativity in forming urban imaginaries extends in other ways to their questioning of reality, reshaping our spatial conceptions, or providing expressions of alternatives. In his last book, *What is the Future?* (2016), John Urry explained that the various methods for envisaging futures, visions, and the role of imagination can have powerful consequences and are a major way of bringing the state and civil society back into the collective dialog about futures, especially if the focus is upon social and not just technological futures. As he concludes, "a planned future may not be possible, but a coordinated one may be the best show in town" (Urry, 2016: 191). This is our value as architects, designers, urbanists, and other agents for positive change. It is also where imagination comes in, by enabling designers, planners, stakeholders, and the public to develop suitable ideas to help guide the forces and complex situations or urban development, while also keeping alternative options as open as possible. Ache provides further emphasis: "vision-making processes become very important in such a context, in the best case creating open political horizons interested in becoming and the 'midwifing of futures'" (2017: 1).

When architects, planners, and urban designers design, they typically create something that does not exist yet; and we tend to lose sight of this most visionary aspect of these professions. It is in moments of great optimism or great crisis that architects and urbanists have returned to the more visionary side of their practices. At their best, these practices are willing to look at the potential future as a way to explain and address present dilemmas. In this way they respond to actual needs. The most challenging answers, however, have only one thing in common: the power of imagination to transport us to another reality. Most importantly, they also act as conduits for ideas and are able to share and explore pluralistic possibilities to reconsider the world we live in.

In *Envisioning Real Utopias*, Erik Olin Wright points toward the role of the social imaginary in constructing possibility: "what is possible pragmatically is not fixed independently of our imaginations, but is itself shaped by our visions" (2010: 21). Wright favors real utopias because he considers them more useful for future options than imagined totalities. However, I suggest that this misses an important point with regard to the instrumentality of design in future-forming visions. Furthermore, it is also evident we need to better understand the way imaginative ideas travel and influence us from a variety of different sources.

World-building and the reflexivity between fact and fiction

Urban imaginaries have been produced and disseminated throughout history, with arguably the first holistic vision for collective life being that of Thomas More's *Utopia* of 1516. The last 200 years, however, have been particularly fruitful because of the degree to which imaginaries developed in tandem with the growth of urbanization through the industrialization of processes, especially in the West. Whether Engels in Manchester, Hausmann in Paris, Leonidov in Moscow, Ferris in New York, or Sant'Elia in Milan, spectacular visions for urban life were offered and (largely) disappeared. Despite the variances in their stylistic tendencies and features, the morphological similarities are notable, yet it would only be after the Second World War, with the rise of consumer culture – specifically the automobile – that a new form of urbanization would be unleashed. Deeply connected to this was the widespread infiltration of television into people's homes so that consequently, as Sanford Kwinter has observed, "the American city began to explode spatially, but only as a quilted interlock of increasingly confined and abstract synthetic environments" (2000: 509). A key component of this transformation was the visual dimension of television, in being able to vividly communicate the mid-twentieth century preoccupations with the atomic age, space travel, and various counter cultures, bringing them into the homes of millions. Here, many imagined future worlds involving spectacular technologies, time travel, roads and trains in the sky, robots, off-earth communities, and so on were richly detailed and appeared alongside factual programs and news bulletins, seemingly describing similarly wonderful advances in technology and human progress.

This brings us to consider the complex ways that fiction explores plausible near-future urban contexts that, in addition to speculations about the future, can also provide powerful commentaries on, and critiques of, the nature of contemporary social life. Importantly, writers too have begun to argue that epistemological boundaries separating fiction from non-fiction are far more porous that often recognized. This holds relevance for urban imaginaries because they typically work with futurity, projecting ideas about how our urban landscape might be. Numerous authors (Abbott, 2016; Clear, 2009; Hewitt and Graham, 2015) have discussed how these boundaries are crossed, especially in the case of urban planning, architectural design, and science fiction, since the visionary element of the architectural and planning disciplines is a strong, integral part of traditional activity. Bassett et al. (2013) have argued that the relationship between innovation in the real world and science fiction is one of mutual engagement and perhaps even co-constitution. Kitchin and Kneale (2005) have identified this reciprocity as contemporary urbanism shaping science fiction, which then works in complex ways to affect the imagination, experience, and construction of contemporary urbanism. They also emphasize the interplay between science fictions writers, its readers, and the development of space.

Futurama was one of the most influential and enduring attempts to imagine how things might be. This giant model of an urban future, designed by Norman Bel Geddes and exhibited in 1939 as part of the General Motors pavilion at the New York World's Fair, was a significant example of an urban imaginary directly influencing the future. It was also at this international exhibition that designers understood the opportunities to be found in urban, private, and public spaces beyond industrial objects. Presenting a preview of what the city would be in 1960, astounded visitors circulated the huge model via a conveyor belt, observing skyscrapers, expressways, and automated farms. Importantly an example such as this reminds us that visible futures are branded, shaped as arguments replete with different registers of information, and as specific options are promoted this may mean alternatives are concealed or even discredited. While with the benefit of hindsight it is possible to detect in Futurama some of the subsequent developments in the American urban landscape, it is less evident how these ideas endure, perish, or evolve to influence future trajectories.

This enables us to respond to one of the principal criticisms of Joseph Voros' (2003) futures cone, which is often used to illustrate different types of alternative futures, which is that it is unidirectional. So, although it is able to show all the imaginable futures, i.e. inside the cone, we might be able to conceive, it takes no account of the past starting as it does from "now". Coulton and Lindley (2017) remind us that, perhaps counter-intuitively, the past has a significant role in understanding the intentionality of an idea as much, if not considerably more so, than the future. The cone also implies an accepted view of the past and the present. Indeed, as Law and Urry (2004) observed, there is a plurality of pasts and futures that are individually constructed to assemble a collective reality. At this juncture, it is worth considering what such history may be able to tell us.

A visual history of the future

So let us rewind a little here. Urban imaginaries have long been dreamed up and shared by a wide range of artists, architects, filmmakers, and visual designers, among others. Images of urban futures are crucial, as they enable a future-orientated society to have a conversation across different communities and with the public. More specifically, while they all have relevance to the context of urban representation or future scenarios, they are also culturally and socially important because they are reflecting points of time historically, thus reflecting a society's attitudes. Perhaps most importantly, they are a way for the public, communities, and experts from various professions and academic disciplines to access ideas about how our cities and our futures may potentially be. How valuable is this? Why are such visions significant? What do they tell us?

In 2014, I was commissioned by the UK Government's Office for Science to write a report specifically on this subject (Dunn et al., 2014). In our research we examined nearly 1000 different future cities and urban imaginaries. Given the sheer volume of material available, we quickly realized that we would have to try to understand how they could be classified and related to one another in order to make sense of what we were studying and be able to provide insights. In the end, for the purpose of the report we chose ninety-four future cities that were prominent types to give as large an overview as possible to the reader. We classified the materials surveyed to identify primary elements within each image and then recorded these. This was a dual process: on the one hand we organized the visualizations in relation to categories and the way in which they have been produced; while on the other hand we analyzed the depictions for their thematic content and which dominant elements of urban or rural life they portrayed. Once we had collated this information, we then examined potential clusters and groupings of visualizations to understand patterns and trends. Our attempt to visualize the connections between these examples, what type of media was used to produce them, and what themes they communicated, as can be seen in Figure 28.1, perhaps ended up just a messy and complex as a city itself!

The scope of the report meant that the period under study spanned from the beginning of the twentieth century to the present day (i.e. 2014 when the report was produced). Clearly, this is a huge topic and we had to be discerning in our approach. Our attempt to categorize the different ways in which urban imaginaries have been depicted enabled us to draw out overarching narratives and thematic patterns for how urban life has been envisaged. From this we established a series of future city categories and dominant paradigms. Why should this matter? The value of re-examining the histories and visual materials of the past, and the manner in which they sought to extrapolate or project urban imaginaries, is evident in the increasing body of work related to this endeavor (for example: Brook, 2013; De Wit and Alexander, 2013; Mami et al., 2011). In addition, as Murphy (2016) has argued, the importance of built work that has been subsequently demolished, the apparent failure of technology to deliver upon its promise, or even imagined projects that were never realized upon our reading of the future of the urban should not be underestimated. This is a point worth emphasizing for the evaluation of where we can be, as it gives scope to competing city forms and their respective challenges. In addition, our relationship with, and, understanding of, the built environment has also radically altered during the timeframe under scrutiny. The burgeoning interest in cities in recent years has cast them to the forefront of public consciousness through direct, everyday experience and numerous mainstream and social media platforms. Thus the impact of how urban imaginaries are conceptualized, disseminated, and performed across various media, means that we cannot and should not underestimate their power upon our thinking and future strategies.

This work accounted for the proliferation and impact of urban imaginaries across a spectrum of media. In order to contribute to the understanding of such varied materials, it also sought to identify and establish connections between different characteristics within

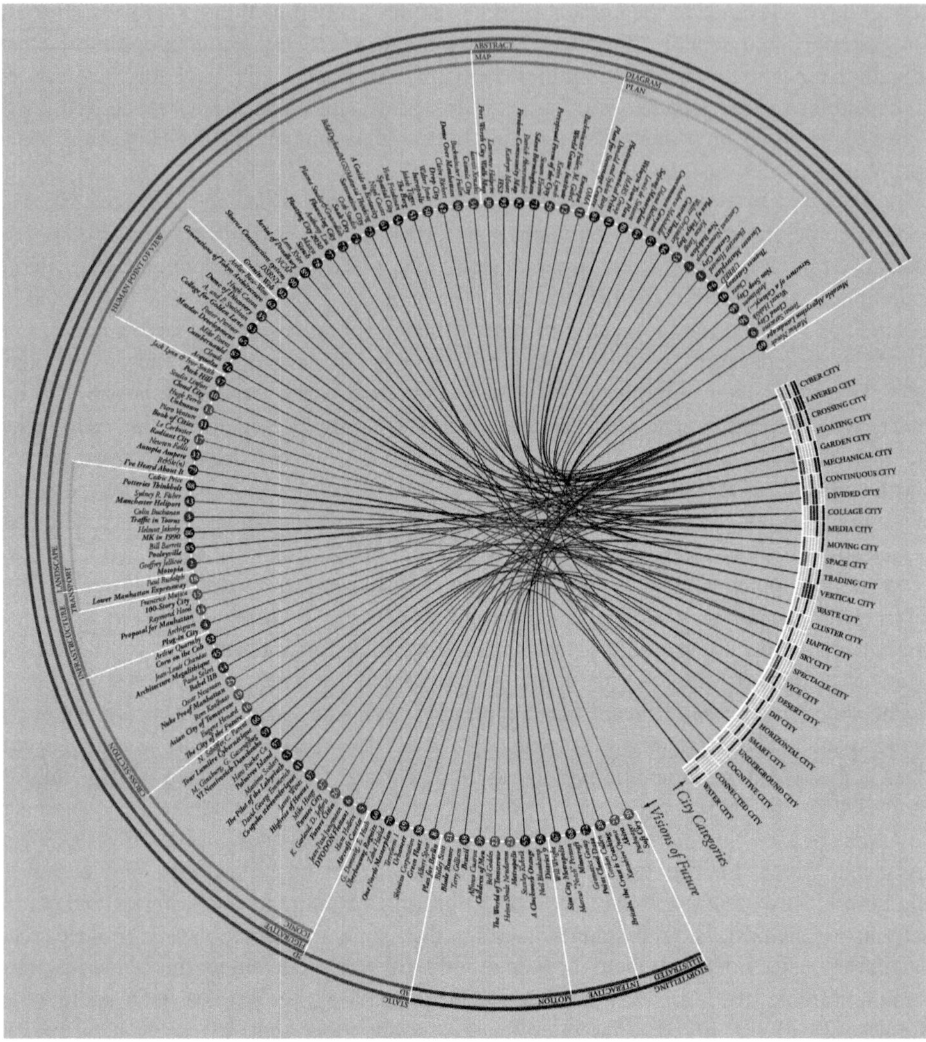

Figure 28.1 Taxonomy for visualization of future cities, 2014.

these visualizations, including method of production, contexts, technologies, socialities, digital features, and data.

This led to the setting out of six dominant visual paradigms of urban imaginaries:

- *Regulated Cities* – urban imaginaries that integrate aspects of rural/country/green living.
- *Layered Cities* – portrayals that have multiple explicit but fixed levels typically associated with different types of mobilities.
- *Flexible Cities* – urban imaginaries that allow for plug-in and changes but are still fixed in some manner to context.

- *Informal Cities* – present urban imaginaries that suggest much more itinerant and temporary situations and include walking, nomadic, and non-permanent sites for inhabitation.
- *Ecological Cities* – depictions of urban imaginaries that demonstrate explicit ecological concerns, renewable energies, and low or zero carbon ambitions.
- *Hybrid Cities* – urban imaginaries that deliberately explore the blurring between physical place and digital space, including augmented reality and "smart" cities.

These six principal paradigms were subsequently checked for their integrity and flexibility. The survey of further visualizations of urban imaginaries, initially collated during the early stages of research for the report, facilitated a cross-checking process to see if they could all be accommodated within one or more of the six paradigms above. Admittedly, the nomenclature of these paradigms is open to critical debate and further scrutiny. This is expected and welcomed in order to further such research. However, for the purpose of the present discussion they represent comparatively discrete and robust types that provide overarching, primary classes within which the twenty-eight categories of future cities could be identified.

When arranged in relation to a timeline (Figure 28.2) the different themes can be appreciated from a conventional historical perspective. This enables six visual dominant paradigms to be understood as flows throughout the time period examined, illustrating connectivity and reoccurrence, where applicable.

The intention of this work is to provide a useful resource for catalyzing and rethinking the potential of perspectives on urban imaginaries more widely. It enables dominant trends and patterns, such as the apparent recurrence and growth of more socially engaged future urban visions in the early twenty-first century. However, we need to be circumspect when trying to draw neat conclusions from this kind of survey. On the one hand, this identifiable trend may be representative of greater societal and global ambitions of ecological and social sustainability for urban life. But an alternative reading might suggest that the branding of contemporary visions is tailored to align with political and economic agendas, so that these ecological urban imaginaries are likely to represent a wide spectrum from legitimate and innovative strategies to deliver low or zero carbon development to proposals that have been subject to "green-washing". While this demonstrates the agency and plasticity of visualizations of urban imaginaries, it also raises complicated issues concerning the communication and interpretation of them. Environmental concerns and technological possibilities point toward the two most dominant paradigms that, since compiling the original report, have only increased in their prominence: Ecological Cities and Hybrid Cities. However, these are often bound together in a compelling urban imaginary narrative, that of the Smart City.

Is there an art to being "smart"?

The Smart City discourse is highly pervasive, offering an optimistic view on what can be achieved in digitally enabled cities that utilize data extensively to address and improve the operation of various urban issues and management systems (Ratti and Claudel, 2016; Townsend, 2013). While many of the ambitions and goals behind smart cities are positive and potentially beneficial for collective life, the over-reliance on software that typically features in their concept has led to their visions largely being promoted by major IT

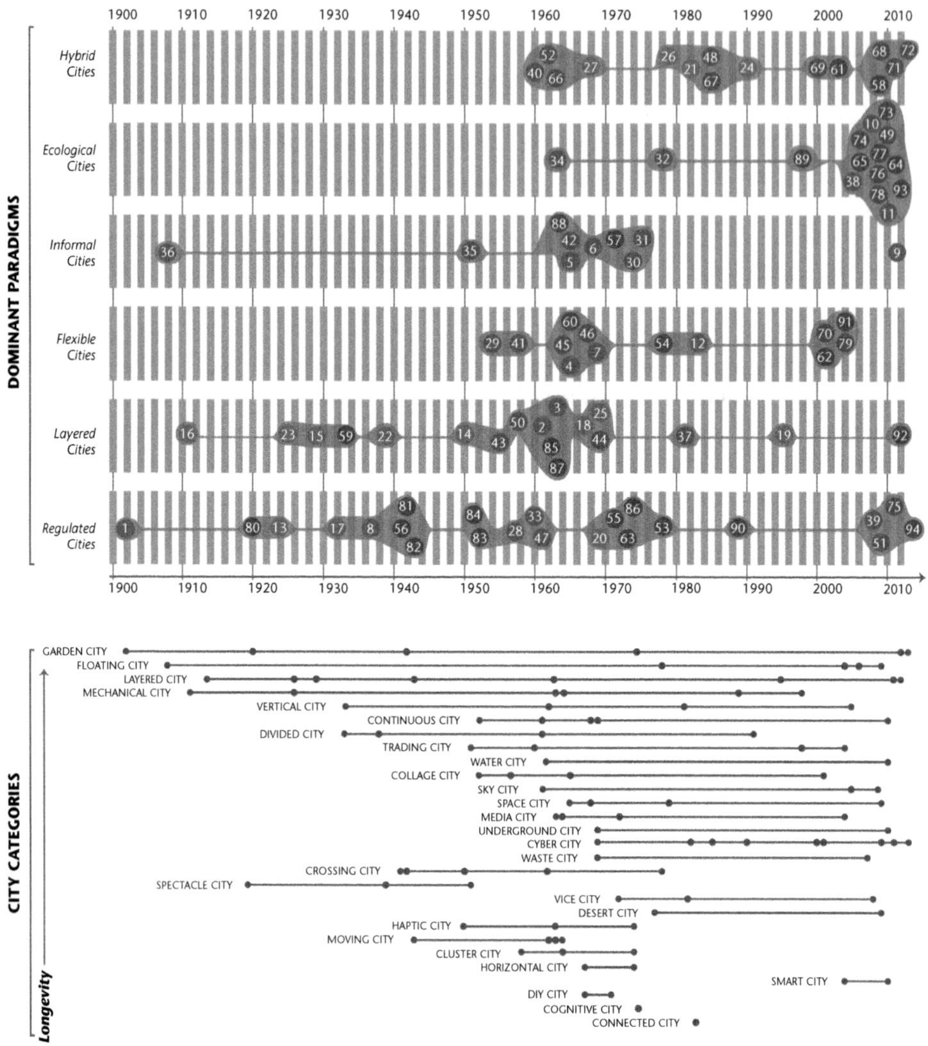

Figure 28.2 Timeline of the six principal paradigms and twenty-eight future city categories between 1900 and 2014.

corporations who have a vested interest in the deployment of technical solutions for city development and management. As Gillian Rose (2017) has observed when discussing the Future Life project by Siemens, it presents a pleasurable, albeit smooth and untethered view, replicating digital visuality rather than actual spatial experience, and it is here that we may detect some problematic issues. This example is synonymous with other corporate visions, which despite their innocuous display contain powerful agency in how we conceive of future urban life. By representing the future from a technocratic position, these urban imaginaries consolidate the role of corporations in providing a dominant view "where data and software seem to suffice and where, as a consequence, knowledge, interpretation and specific thematic expertise appear as superfluous" (Söderström et al., 2014: 308). Indeed, despite their diversity of approach and features, the imaginary of most smart cities is one of conspicuously bland, generic, ahistorical, apolitical spaces whose identity is characterized by information technologies that could be applied anywhere. So why are they so successful?

As Lars Lerup noted in *After the City*, memes help affect our values. They do not exactly determine our values, but they direct and constrain them. A variety of memes is thus somatically and socially produced and disseminated. A good example of this is the Smart City – a great meme: easy to remember and difficult to contest on some levels – after all do we really want to live in a stupid city? Even if we resist the technocratic determinism that such a paradigm is largely predicated upon, we find ourselves struggling, grasping for radical alternatives for urban imaginaries or resuscitating the ghosts of previous ideas, the Garden City being an enduring example in the context of the UK since its original conception in 1898.

Right now, as cities face new challenges and are reexamined for their potential to provide radical systemic revision and social restructuring, it is essential to reclaim urban imaginaries as a means of questioning the present and demonstrating that what we think may be impossible is possible (Levitas, 2013). It may be that a major barrier in this reworking is the scope of the object in view. Brenner and Schmid's (2015) planetary urbanism suggests that modern societies are necessarily underpinned by the urban and that it has become a worldwide condition in which all aspects of life are intertwined and inseparable from it. Importantly for our discussion here, they also state "the urban is a collective project in which the potentials generated through urbanization are appropriated and contested" with the urban being "produced through collective action, negotiation, imagination, experimentation, and struggle" (Brenner and Schmid, 2015). It is here that we can identify some of the hopes and beliefs in the Smart City idea. By absorbing other variations on urban futures, the Smart City imaginary is powerful: its very intangibility directly correlating to its appeal. This has also facilitated its position as the site for urban experimentation *par excellence*, since numerous future pathways for cities and urbanism are keen to develop a technology-based, specifically data-driven, version. In the contemporary climate and particularly within the Smart City movement, dissonance is rarely encountered and, as a result, no real challenging alternative is able to emerge. How can we respond and provide a pluralistic and collective account of urban imaginaries to better enable us to identify radical alternatives?

False dichotomies and addressing complexity

One of the imperatives going forward may be to understand the false dichotomy of urban imaginaries presented by the Smart City paradigm, as it suggests anything alternative to it is inherently regressive or inefficient. In order to do so, it is important to recognize many urban imaginaries for how they are presented, i.e. they are exactly that: stories broadcast *at* people rather than being more open, suggestive, and enabling, perhaps empowering, the viewer to respond and actively shape their own formulation. Therefore, it appears vital that we are able to create urban imaginaries that include and valorize plurality and agonism (DiSalvo, 2010) especially in the visioning process (Pløger, 2004). Such an approach, as Pollastri et al. note, not only challenges analytical and rational methods of examining the future by enabling controversies and diversities that characterize the urban to emerge, but also stimulating that we may "question assumptions, propose unthinkable alternatives and highlight unforeseen conflicts" (Pollastri et al., 2017: S4375).

While in times of rapid change and uncertainty there may be considerable comfort in finding clarity and agreement, it is suggested here that articulating divergence is a vital step to exploring truly radical alternatives. Stepping back from a solution-orientated approach affords us to visualize and better understand latent tensions, and to critically question assumptions about what futures are or should be desirable. This also recalls Wood's (2007) argument for "micro-utopias" and its plea for all citizens to imagine alternative futures. Therefore, the argument for socially constructed urban imaginaries alongside and combined with technological ones is fundamental to our ability to develop compelling visions that reflect the competing and pluralistic complexities of contemporary urbanization, its processes, and the lived experiences within it.

There are three key issues here that warrant further research. First, how such visioning as creative agonism may be explored more fully, to ensure it provides robust urban imaginaries that are able to capture and articulate the complexity of situations that dominant paradigms currently omit (indeed, the belief in smart cities as a technical system that can manage such contexts is part of their considerable appeal for governments, city leaders, and other stakeholders). Second, having been able to formulate new urban imaginaries, is the examination of potential delivery mechanisms for the translation of vision into actions to achieve these futures. Third is an acknowledgement of the amounts of time, energy, and commitment that such a process requires, which is essential to its success, not only within the practices of the associated design disciplines, policymakers, and governance organizations, but also of the need for it to be taken seriously academically.

In trying to avoid path dependency, urban imaginaries can only provide transformative capacity if they are able to account for elements that explicitly produce friction toward a radically alternative future (Albrechts, 2015). Different ways of thinking about cities and our fundamental relationships and experiences within them are essential. Urban imaginaries require us to explore improbabilities, paradoxes, and risks against which we can question the consensual and rationalistic narratives that currently dominate our purview. Pioneers of the future may need to identify apparent voids in existing places and narratives through which we can reimagine the urban. This inquiry therefore demands that we find new spaces for creative rethinking (Dunn, 2016) or articulations of alternative futures (Porritt, 2013), so that we may better conceive different trajectories. Perhaps then we might be able to create and share urban imaginaries that offer an escape from the paths we are currently converging along.

References

Abbott, C. (2016) *Imagining Urban Futures: Cities in Science Fiction and What We Might Learn from Them*. Middletown, CT: Wesleyan University Press.

Ache, P. (2017) Vision Making in Large Urban Settings: Unleashing Anticipation? In: Poli, R. (ed.), *Handbook of Anticipation: Theoretical and Applied Aspects of the Use of Future in Decision Making*. Cham, Switzerland: Springer, 1–21.

Albrechts, L. (2015) Ingredients for a More Radical Strategic Spatial Planning. *Environment and Planning B: Urban Analytics and City Science* 42(3): 510–25.

Appadurai, A. (1990) Disjuncture and Difference in the Global Cultural Economy. *Theory, Culture & Society* 7: 295–310.

Bassett, C., Steinmueller, E., and Voss, G. (2013) *Better Made Up: The Mutual Influence of Science Fiction and Innovation*. Nesta Working Paper 13(7).
Brenner, N. and Schmid, C. (2015) Towards a New Epistemology of the Urban? *City* 19(2–3): 151–82.
Brook, D. (2013) *A History of Future Cities*. New York: W.W. Norton & Company Inc.
Clear, N. (2009) A Near Future. *Architectural Design* 79(5): 6–11.
Coulton, P. and Lindley, J. (2017) Vapourworlds and Design Fiction: The Role of Intentionality. *The Design Journal* 20(Suppl. 1): S4632–42.
De Wit, W. and Alexander, C.J. (2013) *Overdrive: L.A. Constructs the Future 1940–1990*. Los Angeles: Getty Research Institute.
DiSalvo, C. (2010) Design, Democracy, and Agonistic Pluralism. In: *Design & Complexity*. Montreal. Available at: http://www.drs2010.umontreal.ca/data/PDF/031.pdf.
Dunn, N. (2016) *Dark Matters: A Manifesto for the Nocturnal City*. Winchester: Zero Books.
Dunn, N., Cureton, P., and Pollastri, S. (2014) *A Visual History of the Future*. London: Foresight Government Office for Science, Department of Business Innovation and Skills, HMSO.
Glaeser, E. (2011) *Triumph of the City: How Our Greatest Invention Makes Us Richer, Smarter, Greener, Healthier, and Happier*. London: Penguin.
Gleeson, B. (2012) Critical Commentary. The Urban Age: Paradox and Prospect. *Urban Studies* 49(5): 931–43.
Graham, S. (2010) *Cities under Siege*. London: Verso.
Greenfield, A. (2013) *Against the Smart City: The City Is Here for You to Use* (1.3 edn.). New York, NY: Do Projects.
Hall, P. (2002) *Cities of Tomorrow: An Intellectual History of Urban Planning and Design in the Twentieth Century* (3rd edn.). Oxford: Blackwell Publishing.
Hewitt, L. and Graham, S. (2015) Vertical Cities: Representations of Urban Verticality in 20th-Century Science Fiction Literature. *Urban Studies* 52(5): 923–37.
Hollands, R.G. (2008) Will the Real Smart City Please Stand Up? *City* 12: 303–20.
Hollis, L. (2013) *Cities Are Good for You: The Genius of the Metropolis*. London: Bloomsbury.
Hunt, D.V.L., Lombardi, R.D., Atkinson, S., et al. (2012) Scenario Archetypes: Converging Rather than Diverging Themes. *Sustainability* 4(4): 740–72.
Kitchin, R. and Kneale, J. (eds.) (2005) *Lost in Space. Geographies of Science Fiction*. London: Continuum.
Kwinter, S. (2000) Television: The Infrastructural Revolution. In: Koolhaas, R., Boeri, S., Kwinter, S. et al. (eds.), *Mutations*. Barcelona: Actar, 508–23.
Law, J. and Urry, J. (2004) Enacting the Social. *Economy and Society* 33(3): 390–410.
Lerup, L. (2000) *After the City*. Cambridge, MA: The MIT Press.
Levitas, R. (2013) *Utopia as Method: The Imaginary Reconstitution of Society*. New York, NY: Palgrave Macmillan.
Mami, K., Miho, T., Naotake, M., et al. (2011) *Metabolism, the City of the Future*. Tokyo: Mori Art Museum.
Montgomery, C. (2013) *Happy City: Transforming Our Lives through Urban Design*. London: Penguin.
Murphy, D. (2016) *Last Futures: Nature, Technology and the End of Architecture*. London: Verso.
Neuman, M. and Hull, A. (2009) The Futures of the City Region. *Regional Studies* 43(6): 777–87.
Pinder, D. (2005) *Visions of the City: Utopianism, Power and Politics in Twentieth-Century Urbanism*. Edinburgh: Edinburgh University Press.
Pløger, J. (2004) Strife: Urban Planning and Agonism. *Planning Theory* 3(1): 71–92.
Pollastri, S., Boyko, C.T., Cooper, R., et al. (2017) Envisioning Urban Futures: From Narratives to Composites. *The Design Journal* 20(Suppl. 1): S4365–77.
Porritt, J. (2013) *The World We Made*. London: Phaidon.
Ratti, C. and Claudel, M. (2016) *The City of Tomorrow: Sensors, Networks, Hackers, and the Future of Urban Life*. New Haven, CT: Yale University Press.
Robinson, J. (2006) *Ordinary Cities: Between Modernity and Development*. New York: Routledge.

Rose, G. (2017) Screening Smart Cities: Managing Data, Views and Vertigo. In: Hesselberth, P. and Poulaki, M. (eds.), *Compact Cinematics: The Moving Image in the Age of Bit-Sized Media*. London: Bloomsbury Academic, 177–84.
Sassen, S. (2001) *The Global City: New York, London, Tokyo*. Princeton: Princeton University Press.
Söderström, O., Paasche, T., and Klauser, F. (2014) Smart Cities as Corporate Storytelling. *City* 18(3): 307–20.
Thomas, V., Wang, D., Mullagh, L., and Dunn, N. (2016) Where's Wally? In Search of Citizen Perspectives on the Smart City. *Sustainability* 8(3): 207. 1–13.
Townsend, A.M. (2013) *Smart Cities: Big Data, Civic Hackers, and the Quest for a New Utopia*. New York, NY: WW Norton & Company Inc.
UNFPA. (2007) *State of World Population 2007: Unleashing the Potential of Urban Growth*. New York: United Nations Population Fund.
Urry, J. (2016) *What Is the Future?* Cambridge: Polity Press.
Voros, J. (2003) A Generic Foresight Process Framework. *Foresight* 5(3): 10–21.
Wood, J. (2007) *Design for Micro-Utopias: Making the Unthinkable Possible*. Aldershot: Gower.
Wright, E.O. (2010) *Envisioning Real Utopias*. London: Verso.

29

Emergent imaginaries

Place, struggle, and survival

Andrea Gibbons

Introduction

In recent decades, a growing convergence around a new vision for the transformation of both space and society seems to be emerging from the subaltern positionality found at the intersection of racial and spatial injustice in the United States. This chapter explores this convergence by undertaking a thematic analysis of the founding documents of three different, but in many ways interlocking, national alliances of grassroots, subaltern organizations undertaking community and labor organizing: the Environmental Justice Movement (EJM), the Right to the City Alliance (RTTC), and the Movement for Black Lives (M4BL). In their shared expressions of both a new ideal and the pathways of struggle to achieve it, the potential outlines of a new urban imaginary can be drawn, rooted in the knowledges of growing subaltern movements that are articulating with critical theory, but refusing to be bound by it.

To meaningfully come together, each of the three national alliances has worked collectively to establish a set of core values and a vision of the world while prioritizing the laying out of principles that encompass the breadth of their work. This process generated and documented collective goals and priorities for social and spatial change, and clearly indicated the process and method by which they believe such change could be achieved. As the M4BL states: "We want this platform to be both a visionary agenda for our people and a resource for us" (M4BL, n.d.). They have emerged over a period of twenty-four years and through multiple connections between individuals and organizations, from the emergence of the first National People of Color Environmental Leadership Summit in 1991 (which established the Principles of Environmental Justice), the foundation of the Right to the City Alliance in 2007, to the first convening of the Movement for Black Lives in 2015. There is no straightforward developmental trajectory between these movements, rather organizations and community organizers have come together in different ways to respond to changing circumstances and challenges, building on what has come before while incorporating new people and ideas. It does signal, however, a process of ongoing development of spaces and national conversations between subaltern organizations allowing them to more fully develop solidarities, and draw on each other's experience and long histories of activism and struggle. In looking at their hard-forged principles, four common fundamental understandings emerge: (1) the need for struggle around not single but interconnected issues, cultural as well as material, at various scales as they are experienced in everyday subaltern life; (2) an increasingly sophisticated understanding of the intersectionality of oppression and unity around the principle that those oppressed must lead the struggle for liberation; (3) a focus on processes of direct democracy and transparency within each organization that facilitate this liberation, as well as demands for a transformation of formal structures of democracy and participation to give people real power over the circumstances and environments which shape their lives; (4) a vision of complete social and political transformation, but one whose momentum is capable of being built via intermediate demands for immediate reform.

If accepted that those most oppressed lead the struggle for liberation, there follows very particular ramifications for critical thought. It demands a "knowing with, understanding, facilitating, sharing, and walking alongside" rather than a "knowing about" and "guiding", rare in the more vanguard-oriented Western-centric tradition (De Sousa Santos, 2016: ix). A number of academics have been key partners in each of these movements through accepting the role of walking alongside, and in this spirit this chapter is presented primarily as an exercise in listening to what is being said and building an understanding of what is happening on the ground. In this, it references a body of theory already found to resonate with these movements and the questions they are asking, along with the questions raised for critical theory in service to transformative social change toward our rights to the city and creating a better world.

Subaltern space and struggle

While the concept of the subaltern is not particularly common in the context of the United States, this chapter draws on the work of Laura Pulido – who builds on the term's more traditional use within postcolonial studies of the Indian subcontinent (see Guha, 1988; Pulido, 1996) – while also noting the renewed life of Gramsci's work in emergent Latin American theorizations of counter-hegemony and struggle (see De Sousa Santos, 2016; Mignolo, 2012; Modonesi, 2010). "Subaltern" is "a name for the general attribute of subordination … whether this is expressed in terms of class, caste, age, gender and office or in any other way" (Guha, 1988: 35). It is theoretically useful in the way that it encapsulates the structural positionality that these movements have claimed, and which serves as a basis for unity among a variety of groups connected by one or more such intersectional attributes. Member organizations of the three alliances studied define their membership through diverse intersectional understandings of oppression, each prioritizing the leadership role of their members in struggle. A few examples – the first two from members of the M4BL's United Front, the second two taken from an RTTC brochure in the year after its founding – illustrate this diversity:

> Freedom, Inc. engages low- to no-income communities of color in Dane County, WI. We work to end violence against people of color, women, those that non-traditionally gender identify, youth, and our elders…
>
> (*Freedom Inc., 2017*)

> We affirm the lives of Black queer and trans folks, disabled folks, undocumented folks, folks with records, women, and all Black lives along the gender spectrum.

> Our network centers those who have been marginalized within Black liberation movements
>
> (*Black Lives Matter, 2017*)

> CAAAV builds the power of Asian immigrant and refugee communities in NYC ... guided by a global analysis of migration, labor, and poverty and how these experiences are shaped in the U.S. by nationality, immigration status, gender, race, and class. CAAV, 2017

> [The Miami Workers Center] work[s] to achieve this by initiating and supporting grassroots organizations that are led by the people most affected by the major issues of our time: poverty, racism, and gender and sexual oppression.
>
> (*A Funder's Guide: Right to the City Alliance, 2008*)

Pulido's work on two rural Chicano environmental justice struggles describes how activism emerging from a subaltern positionality differs from mainstream activism, even when organizing around the same issues and sharing a broadly similar politics:

> subaltern environmentalism is embedded in material and power struggles, as well as questions of identity and quality of life. Dominated communities engaged in environmental struggles do not disaggregate their various identities and needs. Although they may engage in strategic essentialism, the practice of reifying aspects of one's identity for political purposes, they recognize the multiple identities and the various lines of domination and power that need to be resisted and challenged. They build complex movements which simultaneously address issues of identity as well as a wide range of economic issues (production, distribution, and uneven development), thereby defying the various models and paradigms social scientists have created to impose meaning on collective action, in particular, environmentalism.
>
> (*Pulido, 1996: xv*)

This is quoted at length, because it reflects fairly comprehensively the three alliances analyzed here and the ways that community organizers themselves think about them, in terms of both philosophy and strategy. For example, Gihan Perera, Executive Director of the Miami Workers Center and one of the founding members of the RTTC, has said:

> one of our most basic understandings is that we organize those who are most directly impacted by oppression to directly confront the powers which deny them of their rights. This is not just a reflection of an organizing method but an indication of a political principle. It's a question of leadership of the oppressed, of the working class, and people of color in particular. We're not just all humans. We are people, classes, races, ethnicities, genders with distinct and varied relationships to power. We believe that those whose power and rights are most crushed must be central to leading the fights for their own liberation.
>
> (*Perera, 2008: 12*)

The category of subaltern encompasses individual positionalities in a flexible, relational way based around a constellation of mediating characteristics. As Pulido writes, "It is important to realize that positionality does not refer to a specific person or group per se but is rather a position that can be filled by any individual" (1996: 28). What they share is a "counterhegemonic, or subaltern, location – they exist in opposition to prevailing powers" (Pulido, 1996: 4). Explorations of how different characteristics intersect within hierarchies of dominance build on the foundational work on intersectionality by Kimberlé Crenshaw (1989) and of course bell hooks' theorization of interlocking oppressions and the ways in which the margins can become spaces of possibility and resistance for the oppressed (1984). Hooks (1989; also drawing on Freire, 2000) acknowledges that while positionality opens up possibilities, movement must be founded on a collective decision to name and confront interlocking structures of oppression. It is in this way that the oppressed can free themselves of their oppression, and they are the only ones who can do so. Katherine McKittrick's work on this theory emerging from Black experience "calls attention to the ways in which the subaltern self attends to and creates workable material and imaginary geographies" (2006, 56), even as she draws on Patricia Hill Collins' critique of the imagery of marginalization as a concept that can flatten out hierarchies and deterritorialize struggle. A more robust, geographical understanding of the counterhegemonic position is needed. Each of the movements looked at here is rooted in struggle over the physical spaces of both communities and bodies, and it is from precisely this position that they are working to define and confront the sources of their interlocking oppressions.

Thus, in the United States, the subaltern position is at once economic, social, political, and geographic, but always structured through race. Ruth Gilmore calls racism "a death-dealing displacement of difference within hierarchies" (2002: 16), as is evident not only in the spectacular and open violence of slavery, genocide, and police brutality and murder, but also the violences of segregation and structured discrimination deeply rooted in the history of planning, law, regulation, professional practice, and white homeowner organizing. Segregation's violence is implicit in the vast racialized differences in assets, wealth, health, and life-expectancy stemming from (but not reducible to) the wholesale removal of resources from communities as part of white flight and disinvestment shaping the broader movements of financial capital. The results include a lack of adequate jobs, decent housing, good schooling, adequate infrastructure, supermarkets and quality produce, banks (replaced by check cashers and payday loans), greenspace and parks (yet an overabundance of polluting industry and landfills), and public transportation (Lipsitz, 2011; Sharkey, 2008). While this has perhaps been most visible and most studied in urban areas, the multiple grassroots movements around the siting of toxic factories, nuclear waste, and landfills have shown similar dynamics – the preservation of the best land for whites while dumping unsightly and unhealthy industries in communities of color – operating in rural areas and on reservations (Cole and Foster, 2000). These historic patterns, and the ways that white supremacy has been built into definitions of property value, continue to structure the wider movement of capital and financialization, driving continued disinvestment in some communities and the mass displacement of working class and people of color from others as capital returns to develop city centers (Gibbons, 2014).

Segregation also works to spatially preserve white privilege through safeguarding amenities for specific communities, while removing from sight and mind the poverty and despair of asset-stripped communities of color. The resulting ignorance serves to absolve whites of racism as defined by individual acts of violence and aggression, allowing them to benefit from a racist system without necessarily ever confronting that fact (Pulido, 2000). It also limits the power of analysis emerging from outside these spaces, that at best continues to liberally treat such communities as problems to be solved. At worst it allows the blame for poverty, discrimination, and death at the hands of the police to be shifted onto the very people upon whom these violences are inflicted (Pulido, 2000, 2002; Taylor, 2016). The argument here is that these communities are in fact the sources of the liberatory praxis necessary to transform our cities and our world.

This category of subaltern refines, and challenges, some of the theorizing around urban struggle and the right to the city in particular. Harvey (2012), highlights the work of both Lefebvre and Castells in broadening the Left's understanding of revolutionary agency from a strict class definition of the proletariat to the urban dweller (see also Marcuse, 2009). What these alliances highlight – particularly the M4BL's push back against the seemingly endless murderous violence against Black men and women that proceeds with impunity – is that this is not enough. The virulent racism and violence that has emerged from the shadows since Trump's election to the presidency only underscores the importance of a deeper understanding of the fractured identities encompassed by "urban dweller", and the necessity of actively privileging collective subaltern leadership with support from allies in the struggle for liberation.

This chapter uses the words of the movements themselves to explore what *would* be enough. It highlights the vibrancy of strategic thought and power of today's subaltern struggles that must be made central to the broader fight for liberation. It begins with a brief history of each of the three alliances, before looking at their convergence around the interconnected issues they each prioritize, the necessary centrality of subaltern positionality and intersectionality to struggle, their focus on democracy and horizontal organizing, and their drive toward a new transformative politics.

The environmental justice (EJ) movement

The oldest and most complex of these groups, the EJ Movement differs from the others in that it does not have a formal mechanism or process for "joining" or membership, although an index of people of color environmental organizations was developed. It emerged, according to Cole and Foster, "organically out of dozens, even hundreds, of local struggles and events and out of a variety of other social movements" (2000: 19). They name the civil rights movement, the anti-toxics movement, Native American struggles, the labor movement, the involvement of academics, and to a much smaller degree the traditional environmental movement, as part of its genesis, while situating this moment of movement within a history of action around environmental justice dating to the landing of Columbus. These broad roots, along with the breadth of its guiding principles, have inspired a wealth of literature struggling with its definitions and working toward clarity around what is and is not "Environmental Justice" (Agyeman, 2005). Key moments are recognized however, such as the 1982 fight in Warren County against PCB-contaminated dirt, the 1987 report "Toxic Wastes and Race In the United States: A National Report on the Racial and Socio-Economic Characteristics of Communities with Hazardous Waste Sites", and the first National People of Color Environmental Leadership Summit held in Washington, DC in 1991, which brought hundreds of people and many different strands of the movement together over two days. The preamble to its principles states:

> WE, THE PEOPLE OF COLOR, gathered together at this multinational People of Color Environmental Leadership Summit to begin to build a national and international movement of all peoples of color to fight the destruction and taking of our lands and communities, do hereby re-establish our spiritual interdependence to the

> sacredness of our Mother Earth; to respect and celebrate each of our cultures, languages and beliefs about the natural world and our roles in healing ourselves; to ensure environmental justice; to promote economic alternatives, which would contribute to the development of environmentally safe livelihoods; and to secure our political, economic and cultural liberation that has been denied for over 500 years of colonization and oppression, resulting in the poisoning of our communities and land and the genocide of our peoples, do affirm and adopt these Principles of Environmental Justice.
>
> (*Alston, 1991*)

The seventeen principles of the EJM formulated during the summit are explored below, as well as the broad range of issues they encompass.

Cole and Foster believe that three things further unite EJM activists together as a movement: motives, background, and perspective. Most grassroots activists are motivated by the need to fight for both their health and homes – an "immediate" and "personal stake" in the outcome of struggle absent from those with more privilege (2001: 33) – and they "are largely, though not entirely, poor or working-class people. Many are people of color that come from communities that are disenfranchised from most major societal institutions" (2001: 33). Both motives and background here could serve as a fair description of the subaltern, particularly in the way that these drive their third characteristic: radical tactics and strategy toward a structural transformation of society as the only possible long-term solution to the many problems their communities face.

The right to the city alliance (RTTC)

The Right to the City Alliance in the U.S. emerged from conversations between the heads of three social and economic justice organizations: Miami Workers' Center, Strategic Actions for a Just Economy (SAJE, Los Angeles) and Tenants and Workers United (Alexandria, VA). An initial conference hosted in LA in 2007 (which I attended as an organizer and researcher with SAJE) brought together over twenty groups engaged in grassroots community organizing, along with a handful of academics and allies. As the website states:

> Right To The City Alliance (RTC) emerged in 2007 as a unified response to gentrification and a call to halt the displacement of low-income people, people of color, marginalized LGBTQ communities, and youths of color from their historic urban neighborhoods. We are a national alliance of racial, economic and environmental justice organizations.
>
> (*Right to the City Alliance, n.d.*)

The Right to the City, is, of course, a broader idea put forward by Lefebvre in 1968 as "a cry and a demand" for a just city shaped to the needs and desires of *all* of its residents by these residents themselves (Lefebvre, 1996). While the slogan of a Right to the City has been taken up in multiple ways across the world, it resonated strongly with the founding organizations, who used it as a starting point for twelve principles that could unify their vision (Goldberg, 2010). At present, the RTTC involves thirty-three organizations in sixteen cities – this number has fluctuated over the years, with the number of cities represented expanding while the number of organizations themselves has dropped from a 2009 high of over forty in nine major cities (Perera, 2008; Right to the City Alliance, n.d.). While each organization continues to undertake community organizing at the local level, the alliance has succeeded in coordinating a number of actions at the national level and continues to develop (Fisher et al., 2013; Right to the City Alliance, n.d.).

The movement for black lives (M4BL)

Resistance to multiple violences has always been present in African American communities. As the webpage for the Movement for Black Lives states:

> Our resistance and rebellion are not new, but like other times in history have come to a critical mass, and the bravery of those in Ferguson and across the country captured the attention of the world.
>
> (*M4BL, n.d.*)

What is new, then, is the recent explosion of community organizing, unlike the problems that it seeks to address located in centuries of discrimination, segregation, and oppression. The M4BL has sought to focus this vast energy into a platform similar in form and content to both the EJM and RTTC but much greater in scale: a national convening of over 2000 people in multiple organizations created a broad platform and plan for action beginning in 2015. However, the M4BL emerges from an even broader national outpouring of anger and action, in some places galvanized and almost everywhere focused by the #BlackLivesMatter tag.

Patrisse Cullors, Opal Tometi, and Alicia Garza founded #BlackLivesMatter after the murder of Trayvon Martin and the acquittal of George Zimmerman in 2012, and it now forms the basis of a national network of chapters. For them, #blm is "a tactic to (re)build the Black liberation movement" (Black Lives Matter, 2017). In his depiction of the movement from its explosion after the murder of Michael Brown in Ferguson, Wesley Lowery writes:

> Black Lives Matter is best thought of as an ideology. Its tenets have matured and expanded over time, and not all of its adherents subscribe to them in exactly the same manner – much the way an Episcopalian and a Baptist . . . could both be described as a Christian.
>
> (*Lowery, 2017: 87*)

Garza also talks about the very different ways that the hashtag has been taken up, too often discounting the labor, creativity, and love of queer Black women (Black Lives Matter, 2017). The M4BL sought to honor these origins and intersectional identities while bringing together the wide diversity of groups, organizations, and networks to begin the process of creating a national platform and the policy priorities for action (M4BL, n.d.). Endorsing organizations that are also founding members of the RTTC include FIERCE, FUREE, and Causa Justa/Just Cause. Right to the City itself as a broader alliance has also signed on to the M4BL platform.

In looking thematically across these three movements, four main areas serve as the basis for exploring a potential convergence toward a liberatory imaginary: the interconnectedness of the issues faced; the necessity of subaltern leadership; the centrality of collective analysis and theory along with direct democracy in struggle; and the underlying transformational vision emerging out of this subaltern positionality.

Interconnectedness of issues and scales

Breadth of issues: three rallying calls to national alliances

Despite the difference in central focus, each of the three movements shares a concern with the principle issues articulated by the others, along with many of the same particular demands. Each has also attempted, through the formation of a platform, to proactively define their struggle. In the words of Gihan Perera:

> Those of us most affected by this have been trying to fight back as best as we can – by fighting against developments, by trying to hold onto the neighborhoods. But we end up taking on fights on multiple fronts: around housing, around education, around transportation. And all of those fights become separate and often reactive.
>
> The Right to the City Alliance and frame is an attempt to say, "Can we determine our own agenda?"
>
> (*Heller and Perera, 2007: 8*)

The understanding of the interconnectedness of the issues faced is even more explicit in the description of the founding summit of the Environment Justice Movement by one of its conveners, Dana Alston:

> For people of color, the environment is woven into an overall framework and understanding of social, racial, and economic justice. The definitions that emerge from the environmental justice movement led by people of color are deeply rooted in culture and spirituality, and encompass all aspects of daily life – where we live, work, and play. This broad understanding of the environment is a context within which to address a variety of questions about militarism and defense, religious freedom and cultural survival, energy and sustainable development, transportation and housing, land and sovereignty rights, self-determination, and employment.
>
> (*Alston, 1991*)

Within this self-defined context, the first principle of the EJM proclaims "the sacredness of Mother Earth, ecological unity and the interdependence of all species, and the right to be free from ecological destruction", while the remaining sixteen principles demand democracy, sustainability, and give detailed actions required around specific environmental dangers (The Principles of Environmental Justice, 1991).

The struggle for environmental justice has been incorporated into both the RTTC and the M4BL in very similar ways. Both use a framework of rights but define these very broadly outside of a narrowly legal framework. For the RTTC, it is encompassed as one of their principles:

> **Environmental Justice:** The right to sustainable and healthy neighborhoods and workplaces, healing, quality health care, and reparations for the legacy of toxic abuses such as brownfields, cancer clusters, and superfund sites.

The M4BL incorporates it within economic justice, speaking in even broader terms of

> A right to restored land, clean air, clean water and housing and an end to the exploitative privatization of natural resources – including land and water. We seek democratic control over how resources are preserved, used and distributed and do so while honoring and respecting the rights of our Indigenous family.
>
> (*M4BL, n.d.*)

The M4BL also incorporates demands for reparations to undo the multiple kinds of environmental damage already inflicted, their use of the term "environmental racism" consciously invokes the centrality of white supremacy in patterns of environmental degradation and control over land and resources.

> Reparations for the wealth extracted from our communities through environmental racism, slavery, food apartheid, housing discrimination and racialized capitalism in the form of corporate and government reparations focused on healing ongoing physical and mental trauma, and ensuring our access and control of food sources, housing and land.
>
> (*M4BL, n.d.*)

Both alliances also honor the particular struggle of Indigenous peoples in the U.S. For the RTTC, this is expressed in a specific principal of "Indigenous Justice", which also highlights the often urban nature of native struggles:

> The right of First Nation indigenous people to their ancestral lands that have historical or spiritual significance, regardless of state borders and urban or rural settings.

Member organizations of the two alliances encompass these broad definitions of environmental justice as it also articulates with economic and social justice:

> ...[to] build broad movements for social, economic and restorative environmental change.
>
> (*Highlander Center, M4BL*)

> ...believe in the inherent right of all peoples to clean air, water, land and other resources necessary to meet their basic needs and live with dignity;
>
> *(Environmental Justice Advocates of Minnesota, M4BL)*

> ACE builds the power of communities of color and lower income communities in New England to eradicate environmental racism and classism and achieve environmental justice.
>
> (*ACE, RTTC*)

The second rallying call is around ownership and control over land, and the right of communities to the city. It is the primary focus of the RTTC, made explicit in their first two principles of unity:

> **Land for People vs. Land for Speculation:** The right to land and housing that is free from market speculation and that serves the interests of community building, sustainable economies, and cultural and political space.

> **Land Ownership:** The right to permanent public ownership of urban territories for public use.
>
> (*Perera, 2008*)

A demand is also included around full investment in communities, particularly in infrastructure, a growing concern among environmental justice organizations and activists (Agyeman, 2005):

> **Services and Community Institutions:** The right of working class communities of color to transportation, infrastructure, and services that reflect and support their cultural and social integrity.
>
> (*Perera, 2008*)

This is also a clear call within the other two movements in claims around land ownership, control and investment. This thread runs throughout the various principles of the Environmental Justice movement, beginning with the preamble: "…to fight the destruction and taking of our lands and communities". In many ways, the EJM principles go the furthest, seeking to radically reframe the nature of the human relationship with the planet and the other organisms with whom human beings share it:

> 1. Environmental Justice affirms the sacredness of Mother Earth, ecological unity and the interdependence of all species, and the right to be free from ecological destruction.
>
> (*Alston, 1991*)

It is from that basis that it incorporates a discourse of ethical land use and sustainability.

> 3. Environmental Justice mandates the right to ethical, balanced and responsible uses of land and renewable resources in the interest of a sustainable planet for humans and other living things.
>
> (*Alston, 1991*)

The M4BL's demand for "A right to restored land, clean air, clean water and housing", seems to echo this broader framework. Their platform incorporates struggles around land and development under the rubric of economic justice:

> We demand economic justice for all and a reconstruction of the economy to ensure Black communities have collective ownership, not merely access.
>
> (*M4BL, n.d.*)

The M4BL also explicitly acknowledges the long history of disinvestment in Black communities, and explores the need to address this under the platform principles of both *economic justice*, which includes restructuring of taxes for "a radical and sustainable redistribution of wealth", and *invest-divest*, which demands investment in the improvement of communities through the divestment in oppressive systems of incarceration and control.

Among member organizations' mission statements, this is expressed in various ways that all acknowledge the centrality of struggle over land rights in their work, while connecting that to broader economic and social struggles in different ways:

> ... engaging in broad based coalition organizing for black food and land, increasing visibility of Black led narratives and work, advancing Black led visions for just and sustainable communities, and building capacity for self-determination.
>
> (*National Black Food and Justice Alliance, M4BL*)

> ONA helps neighbors connect day-to-day crises like skyrocketing property taxes and immigration raids with broader causes, including gentrification and the criminalization of entire peoples ...
>
> (*Olneyville Neighborhood Association, RTTC*)

> ... organizing Oakland residents to advocate for housing and jobs as human rights ...
>
> (*Just Cause Oakland, RTTC & M4BL*)

The final call for alliance is the focus of the Movement for Black Lives:

> We demand an **end to the war against Black people**. Since this country's inception there have been named and unnamed wars on our communities. We demand an end to the criminalization, incarceration, and killing of our people.
>
> (*M4BL, n.d.*)

This encompasses a multitude of demands: "[a]n immediate end to the criminalization and dehumanization of Black youth across all areas of society", an end to capital punishment and money bail; the repeal of crime and immigration bills targeting Black immigrants; an end to violence against and active support for the right of Black trans, queer, and gender nonconforming people; an end to mass surveillance of Black communities; a demilitarization of law enforcement; an end to the privatization of police, prisons and all criminal justice related services; an end to jails, detention centers, youth facilities and prisons among others (M4BL, n.d.).

Police brutality and state violence have long been part of the struggle highlighted within the other two national alliances, while also central to the work of a number of their member organizations. This is described in its broadest possible terms within the original principles of the EJM, perhaps not expressing clearly the connections between police brutality and occupation within the U.S. and military occupations and U.S. imperialism of other nations, but still unequivocal in its denunciations:

> 15. Environmental Justice opposes military occupation, repression and exploitation of lands, peoples and cultures, and other life forms.
>
> (*Alston, 1991*)

For the RTTC, this is expressed slightly more specifically in two principles:

> **Freedom from Police and State Harassment:** The right to safe neighborhoods and protection from police, immigration, and vigilante repression.
>
> **Immigrant Justice:** The right of equal access to housing, employment, and public services regardless of race, ethnicity, and immigration status and without threat of deportation by Immigration and Customs Enforcement or employers.

Unsurprisingly given the historic and continuous nature of police brutality against communities of color, this is also central to many member organizations.

> Originally named "Committee Against Anti-Asian Violence", CAAAV was...one of the first groups in the U.S. to mobilize Asian communities to fight police brutality and other forms of racially motivated violence...
>
> (*CAAV, RTTC*)

> Safe Streets defines its constituency as communities most impacted by violence, police harassment and incarceration.
>
> (*Safe Streets, RTTC*)

> Dignity and Power Now (DPN) ... fights for the dignity and power of incarcerated people, their families, and communities. In doing so DPN wages a fight for everyone because the prison industrial complex forms an imaginative limit on everyone's capacity to envision freedom and liberation.
>
> (*DPN, M4BL*)

In terms of additional concrete demands, reparations, jobs and education also form part of each alliance's work. The RTTC demands reparations both for toxic abuses as above, as well as principle 10: "**Reparations:** The right of working-class communities of color to economic reciprocity and restoration from all local, national, and transnational institutions that have exploited or displaced the local economy" (Perera, 2008). For EJM, it is described in principle 9 as "the right of victims of environmental injustice to receive full compensation and reparations for damages as well as quality health care" (The Principles of Environmental Justice (EJ), 1991). The idea of reparations is explored much more deeply by the M4BL as a key platform demand, expanded to address past and current inequities through full and free access to education at all levels, guaranteed minimum income, school curricula that explore colonialism and slavery while celebrating the struggle of Black communities, and the "immediate passage of H.R.40, the 'Commission to Study Reparation Proposals for African-Americans Act'" (M4BL, n.d.). In its struggle around work:

> 8. Environmental Justice affirms the right of all workers to a safe and healthy work environment without being forced to choose between an unsafe livelihood and unemployment. It also affirms the right of those who work at home to be free from environmental hazards
>
> (*M4BL, n.d.*)

M4BL demands the right for workers to organize, a renegotiation of the Trans-Pacific Partnership and other free trade agreements to prioritize rights of workers and communities, development of cooperatives and "social economy networks", job programs that "provide a living wage and encourage support for local workers centers, unions", and protections for workers in industries not currently "appropriately regulated including domestic workers, farm workers, and tipped workers, and for workers" (M4BL, n.d.) . In terms of education, both EJM and M4BL demand free access to an education that both celebrates the diversity of culture, heritage and struggle within communities in the U.S., while also recognizing the murderous history of colonialism, slavery, appropriation and discrimination.

Breadth of scale: organizing at the local, regional, national and international level

In coming together, all three of the alliances connect local community organizing and activism to larger scales. As described above, the EJM reframes everything in terms of the "the sacredness of Mother Earth", and the ecological understanding of the interconnectedness of all things. They also explicitly acknowledge the connections between city and country:

> 12. Environmental Justice affirms the need for urban and rural ecological policies to clean up and rebuild our cities and rural areas in balance with nature, honoring the cultural integrity of all our communities, and provided fair access for all to the full range of resources.
>
> (*Alston, 1991*)

This is echoed with the principles of the RTTC

> Rural Justice: The right of rural people to economically healthy and stable communities that are protected from environmental degradation and economic pressures that force migration to urban areas.
>
> (*Right to the City Alliance, n.d.*)

Both the EJM and M4BL have member organizations working in rural areas and small towns, while the RTTC is urban-centered but organized into regional groupings that recognize rural-urban connections (Right to the City Alliance, n.d.). The RTTC's focus on the city drives the form of its own international outlook:

> Internationalism: The right to support and build solidarity between cities across national boundaries, without state intervention.

The M4BL also acknowledges its U.S.-centric focus, while acknowledging how the issues it confronts are much larger in scope.

> While this platform is focused on domestic policies, we know that patriarchy, exploitative capitalism, militarism, and white supremacy know no borders. We stand in solidarity with our international family against the ravages of global capitalism and anti-Black racism, human-made climate change, war, and exploitation.
>
> (*M4BL, n.d.*)

It also contains specific references to international polices

> An end to the Trans-Pacific Partnership and a renegotiation of all trade agreements to prioritize the interests of workers and communities.
>
> (*M4BL, n.d.*)

Just as the EJM references specific corporate forms:

> 13. Environmental Justice opposes the destructive operations of multinational corporations.
>
> (*Alston, 1991*)

Subaltern communities freeing themselves (and by extension everyone else)

In each of the discourses of these three movements, the priority is placed on self-liberation. While this is not set forth clearly within the Principles of Environmental Justice themselves, it is implicit in the preamble (as quoted above), which contains a clear understanding of a continued colonial and oppressive relationship between people of color and a white establishment responsible not just for environmental destruction, but for an unjust share of that burden (poisoning, destroying) falling on communities of color. Just as clear is the recognition that this destruction has been hitherto unrecognized and unacknowledged by the larger (white) environmental movement, that communities of color have a central role in rebuilding and reimagining a better way of life, and that the struggle needs to be conducted on economic, political, and cultural fronts. The idea that communities should speak for themselves remains central to EJM as a whole, and the movement brings together poetry, interviews, roundtables, and storytelling with academic articles as equal formats for building theory and movement (Adamson, Evans & Stein, 2002; Agyeman, 2005; Cole and Foster, 2000; Pulido, 1996).

The RTTC picks up on some of these threads in the preamble to their principles:

> Right to the City was born out of desire and need by organizers and allies around the country to have a stronger movement for urban justice. But it was also born out of the power of an idea of a new kind of urban politics that asserts that everyone, particularly the disenfranchised, not only has a right to the city, but as inhabitants, have a right to shape it, design it, and operationalize an urban human rights agenda. Right to the City Alliance, n.d..

Again, it is the subaltern demanding the right to live in and shape the city, taking power through collective organization. One academic present at the founding convening, Harmony Goldberg, compares the RTTC to the National Domestic Workers Alliance:

> these organizations have decided to take on the task of re-building the organized power of the people who are the front-lines of neoliberalism. This is based on the belief that oppressed people have the most interest in changing the system and – if organized – the most power to actually win change. These organizations' commitment to the methodology of grassroots organizing is also based on a commitment to the self-determination of oppressed people … building the collective power and leadership of working class people and people of color to win real changes in the daily lives of their communities.
>
> (*Goldberg, 2010: 103–104*)

The M4BL makes the leadership of their coalition even more explicit, writing:

> We believe in elevating the experiences and leadership of the most marginalized Black people, including but not limited to those who are women, queer, trans, femmes, gender nonconforming, Muslim, formerly and currently incarcerated, cash poor and working class, disabled, undocumented, and immigrant. We are intentional about amplifying the particular experience of state and gendered violence that Black queer, trans, gender nonconforming, women and intersex people face. There can be no liberation for all Black people if we do not center and fight for those who have been marginalized.
>
> (*M4BL, n.d.*)

All three recognize the intersectionalities of oppression while prioritizing subaltern voices. This does not diminish the need for broad alliances but broadcasts a clear message around the different roles for allies in these growing movements. Jelani Cobb's article on the rise of Black Lives Matter contains an eloquent quote from Alicia Garza: "San Francisco broke my heart over and over. White progressives would actually argue with us about their right to determine what was best for communities they never had to live in" (Cobb, 2016). All sixty-three of the RTTCs core members and the M4BL's united front declare their commitment to such bottom-up change through grassroots organizing, and a commitment to being driven by their members. For the M4BL there is a vital and clear understanding that it is not just Black voices that must elevated, but those Black voices that are most marginalized.

This marks another type of change from past histories of struggle. A great deal has been written, particularly in relation to the civil rights movement, about the prioritizing of causes and cases that highlight and battle one aspect of oppression while presenting an individual in all other ways identical (or superior in the ways that Rosa Parks was superior) to a mainstream norm. Michelle Alexander is eloquent about the ways that this has undercut resistance to what she calls the New Jim Crow, while it is also described by Aldon Morris, Barbara Ransby and others (Alexander, 2011; Morris, 1984; Ransby, 2005). In response to criticism from Reverend Al Sharpton among others, a press release from Ferguson Action, stated

> We are decentralized, but coordinated. Most importantly, we are organized. Yet we are likely not respectable negroes. We stand beside each other, not in front of one another. We do not cast any one of ours to the side in order to gain proximity to perceived power. Because this is the only way we will win. We can't breathe. And we won't stop until Freedom.
>
> (*About this Movement, 15 December 2014, quoted in Taylor, 2016*)

This jettisoning of the hierarchies of respectability and formal leadership, often found within oppressed communities, usefully complicates the idea of the subaltern as it links to understandings of intersectionality and works to dismantle various hierarchies of power. This highlights a need for individual activists to be flexible – sometimes in the forefront, sometimes working as allies – given the complex nature of identity and the various shifting patterns of privilege and discrimination as they relate to class, race, gender, sexuality, age, and disability. Within this, there is also a commitment to a horizontal alliance of groups accountable to their grassroots bases and positionality in each of these national groupings. As Barbara Ransby highlights, this is an emerging model for social movement and organizing very different from the hierarchies and charismatic leadership visible in earlier generations (Ransby, 2017).

Direct democracy

This horizontal structure of organizing mirrors demands around the democracy as both process and outcome of struggle central to all three alliances. For the EJM:

> 7. Environmental Justice demands the right to participate as equal partners at every level of decision-making, including needs assessment, planning, implementation, enforcement and evaluation.
>
> *The Principles of Environmental Justice (1991)*

> Environmental Justice affirms the fundamental right to political, economic, cultural and environmental self-determination of all peoples.
>
> *(Alston, 1991)*

For the RTTC:

> **Democracy and Participation:** The right of community control and decision making over the planning and governance of the cities where we live and work, with full transparency and accountability, including the right to public information without interrogation.
>
> *Right to the City Alliance (n.d.)*

And for the M4BL:

> We demand a world where those most impacted in our **communities control the laws, institutions, and policies that are meant to serve us** – from our schools to our local budgets, economies, police departments, and our land – while recognizing that the rights and histories of our Indigenous family must also be respected.
>
> *M4BL, n.d.*

> We demand **independent Black political power and Black self-determination** in all areas of society. We envision a remaking of the current U.S. political system in order to create a real democracy where Black people and all marginalized people can effectively exercise full political power.
>
> *(M4BL, n.d.)*

In this way, all three movements claim the necessity of a radical transformation.

Transformation and reform

None of these organizations has a party line or dogma, yet each contains a stringent analysis of capitalism, and each challenges different aspects of its fundamental nature in placing profit over other human values. In combining such critique with community-organizing methods that have traditionally eschewed ideological framings, these subaltern grassroots organizations represent something (relatively) new (Goldberg, 2010; Sen, 2003). For the EJM, critique focuses on a understanding of the sacredness of earth, the interdependence of all species, and above all in a new frame demanding that we "make the conscious decision to challenge and reprioritize our lifestyles to ensure the health of the natural world for present and future generations" (The Principles of Environmental Justice, 1991). This is echoed by the RTTC's first principle, "Land for People vs. Land for Speculation". In reorganizing life and economy around the acceptance of such principles, it is questionable how much would be left that is recognizably capitalist. The M4BL (n.d.) states:

> We reject false solutions and believe we can achieve a complete transformation of the current systems, which place profit over people and make it impossible for many of us to breathe.
>
> (*M4BL, n.d.*)

These demands are not couched in the traditional language of the intellectual Left, but chosen to resonate with the grassroots bases of these subaltern organizations. The M4BL marks the difference between stopping immediate harm and greater transformation, but notes how one can lead to the other:

> This document articulates our vision of a fundamentally different world. However, we recognize the need to include policies that address the immediate suffering of Black people. These policies, while less transformational, are necessary to address the current material conditions of our people and will better equip us to win the world we demand and deserve.
>
> (*M4BL, n.d.*)

The need for a larger vision continues to be a priority however. As Patrisse Cullors (one of the founders of #BlackLivesMatter) says

> When our political activism isn't rooted in a theory about transforming the world, it becomes narrow; when it is focused only on individual actions instead of larger systemic problems, it becomes short-sighted.
>
> (*quoted in Heatherton and Camp, 2016: 37*)

The question, thus, becomes how to develop theory through praxis that helps achieve full transformation. Each of the three platforms presented here, if pursued to their full and complete realization, would have a revolutionary effect on the world as we know it, while each also presents more incremental steps toward this kind of transformational change. Harmony Goldberg describes the importance of this:

> these organizations make a distinction between "fighting for reforms" and "reformism", that is, the belief that reforms can meet the fundamental needs of oppressed people. The fight for reforms can be a part of the process of building power for a longer-term transformative struggle.
>
> (*Goldberg, 2010: 106*)

Conclusion

This chapter looks at the imaginaries of radical hope and struggle emerging from subaltern communities of color; an antidote to the imaginaries of despair so typically projected onto to them, whether they be inner city neighborhoods or reservations. It is of necessity an all-too-brief exploration of the theoretical potential of the subaltern in the U.S. context, and what implications this might have for critical theorizations of agency and movement building. It outlines (also all too briefly) a growing convergence around a liberatory subaltern imaginary able to encompass complex constellations of issues and scales, prioritize intersectional understandings in demanding the leadership of subaltern communities in their own liberation, develop direct democracy and horizontal organizing, and provide transformative visions of social change along with a roadmap of intermediate steps in building power to achieve it.

What is the urban imaginary here, then, what is to be won? A transformed body of urban residents secure in their homes and in their collective power over land and resources, with an ability to transform space to support a fullness of life not profit. A shared humility about humanity's place on the planet. An understanding of how cities connect to regional, national and international scales, with the needs of future generations put first in every personal, political and economic decision. Cities where all are free from violence and oppression. A government of the people that not only provides for its people, but that prioritizes the work and reparations necessary to undo the multiple injustices of the past and involves everyone in the creation of a better future. Beginning to summarize such a list already fails to do justice to the praxis of the three movements discussed here. Yet it also begins to show how their vision points to a world as far beyond capitalism as it is a world necessary to build if we wish to survive past the point of destruction to which capitalism has brought us. Above all, it is a vision of a city and a world to be won collectively through a movement for liberation already in motion, led by subaltern communities in the fight to preserve life itself. It is time to join in making the road by walking, to seek out the cracks in the oppressive edifice, to help open up the way forward and keep it open.

References

A Funder's Guide: Right to the City Alliance. (2008) Available at: http://righttothecity.org/wp-content/uploads/2014/07/FundersGuide.pdf. Accessed 20 September 2017.

Agyeman, J. (2005) *Sustainable Communities and the Challenge of Environmental Justice*. New York: New York University Press.

Alexander, M. (2011) *The New Jim Crow: Mass Incarceration in the Age of Colorblindness*. New York: New Press.

Alston, D. (1991) The Summit: Transforming a Movement. *RP&E Journal* 2(3/4): 14–17.

Black Lives Matter (2017) *Black Lives Matter*. Available at: http://blacklivesmatter.com/. Accessed 27 September 2017.

CAAV. (2017) *CAAV Organizing Asian Communities*. Available at: http://ccheonline.org/node/332. Accessed 30 September 2017.

Cobb, J. (2016) The Matter of Black Lives. *The New Yorker*, March. Available at: https://www.newyorker.com/magazine/2016/03/14/where-is-black-lives-matter-headed. Accessed 1 October 2017.

Cole, L.W. and Foster, S.R. (2000) *From the Ground up Book Subtitle: Environmental Racism and the Rise of the Environmental Justice Movement*. New York: New York University Press.

Collins, P. (2009 [1990]) *Black Feminist Thought: Knowledge, Consciousness, and the Politics of Empowerment*. New York: Routledge.

Crenshaw, K. (1989) *Demarginalizing the Intersection of Race and Sex: A Black Feminist Critique of Antidiscrimination Doctrine, Feminist Theory and Antiracist Politics*. University of Chicago Legal Forum. Special issue *Feminism in the Law: Theory, Practice and Criticism*. University of Chicago Law School, 139–68.

De Sousa Santos, B. (2016) *Epistemologies of the South: Justice against Epistemicide*. London: Routledge.
Fisher, R. et al. (2013) "We are Radical": The Right to the City Alliance and the Future of Community Organizing. *The Journal of Sociology & Social Welfare* 40(1): 157–82.
Freire, P. (2000) *Pedagogy of the Oppressed*. London: Bloomsbury.
Gibbons, A. (2014) *Segregation in Search of Ideology? Hegemony and Contestation in the Spatial and Racial Configuration of Los Angeles*. London: The London School of Economics and Political Science.
Gilmore, R.W. (2002) Fatal Couplings of Power and Difference: Notes on Racism and Geography. *Professional Geographer* 54(1): 15–24.
Goldberg, H. (2010) Building Power in the City: Reflections on the Emergence of the Right to the City Alliance and the National Domestic Worker's Alliance. In: Team Colors Collective (ed.), *Uses of a Whirlwind: Movement, Movements, and Contemporary Radical Currents in the United States*. Oakland: AK Press, 97–108.
Guha, R. (1988) On Some Aspects of the Historiography of Colonial India. In: Guha, R. and Spivak, G. (eds.) *Selected Subaltern Studies*. Oxford: Oxford University Press, 35–44.
Harvey, D. (2008) The Right to the City. *New Left Review* 53: 23–40.
Harvey, D. (2012) *Rebel Cities: From the Right to the City to the Urban Revolution*. London: Verso.
Heatherton, C. and Camp, J.T. (eds.) (2016) *Policing the Planet: Why the Policing Crisis Led to Black Lives Matter*. New York: Verso.
Heller, C.C. and Perera, G. (2007) *The Right to the City: Reclaiming Our Urban Centers, Reframing Human Rights, and Redefining Citizenship*. Available at https://www.tides.org/fileadmin/user/pdf/TheRightToTheCity.pdf. Accessed 20 September 2017.
Hooks, B. (1989) *Talking Back: Thinking Feminism, Thinking Black*. Boston: South End Press.
Hooks, B. (2014 [1984]) *Feminist Theory: From Margin to Center*. 3rd ed. New York: Routledge.
Lefebvre, H. (1996) *Writings on Cities*. Oxford: Blackwell Publishers.
Lipsitz, G. (2011) *How Racism Takes Place*. Philadelphia: Temple University Press.
Lowery, W. (2017) *'They Can't Kill Us All': The Story of Black Lives Matter*. New York: Penguin Books.
Marcuse, P. (2009) From Critical Urban Theory to the Right to the City. *City* 13(2–3): 185–97.
McKittrick, K. (2006) *Demonic Grounds: Black Women and the Cartographies of Struggle*. Minneapolis: University of Minnesota Press.
Mignolo, W.D. (2012) *Local Histories/Global Designs: Coloniality, Subaltern Knowledges, and Border Thinking*. Princeton: Princeton University Press.
Modonesi, M. (2010) *Subalternidad, Antagonismo, Autonomia: Marxismo Y Subjetivación Política*. Buenos Aires: Clacso.
Morris, A.D. (1984) *The Origins of the Civil Rights Movement: Black Communities Organizing for Change*. New York: New Press.
Perera, G. (2008) Claiming the Right to the City: A Question of Power. Available at: http://www.urbanhabitat.org/files/15.Perera.pdf. Accessed 20 September 2017.
Pulido, L. (1996) *Environmentalism and Economic Justice: Two Chicano Struggles in the Southwest*. Tucson: University of Arizona Press.
Pulido, L. (2000) Rethinking Environmental Racism White Privilege and Urban Development in Southern California. *Annals of the Association of American Geographers* 90(1): 12–40.
Pulido, L. (2002) Reflections on a White Discipline. *The Professional Geographer* 54(1): 42–9.
Ransby, B. (2005) *Ella Baker and the Black Freedom Movement: A Radical Democratic Vision*. Chapel Hill, NC: University of North Carolina.
Ransby, B. (2017) Black Lives Matter Is Democracy in Action. *New York Times*, 21 October. Available at https://www.nytimes.com/2017/10/21/opinion/sunday/black-lives-matter-leadership.html?smprod=nytcore-iphone&smid=nytcore-iphone-share. Accessed 19 November 2017.
Right to the City Alliance (n.d.) *Right to the City: Member Organizations*. Available at http://righttothecity.org/about/member-organizations/. Accessed 1 October 2017.

Sen, R. (2003) *Stir It Up: Lessons in Community Organizing and Advocacy.* San Francisco: John Wiley & Sons, Inc.
Sharkey, P. (2008) The Intergenerational Transmission of Context. *American Journal of Sociology* 113(4): 931–69.
Taylor, K.Y. (2016) *From #Blacklivesmatter to Black Liberation*. Chicago: Haymarket.
The Movement for Black Lives (2017a) *About Us*. Available at https://policy.m4bl.org/about/. Accessed 1 October 2017.
The Movement for Black Lives (2017b) *Platform*. Available at https://policy.m4bl.org/platform/. Accessed 1 October 2017.
The Movement for Black Lives (n.d.) *Reparations*. Available at https://policy.m4bl.org/reparations/. Accessed 27 September 2017.
The Principles of Environmental Justice (EJ). (1991) Available at: http://www.ejnet.org/ej/principles.pdf. Accessed 20 September 2017.

30
Queer urban imaginaries

Ben Campkin

What roles do urban policies and practitioners play in supporting LGBTQ+ communities? This question arose in the finalization of the United Nations *New Urban Agenda* (United Nations, 2016) – a key international policy framework, adopted in December 2016, designed to promote "a new model of urban development that is able to integrate all facets of sustainable development to promote equity, welfare, and shared prosperity". This influential document, which centers on inclusive, participatory cities as a cornerstone of sustainable urban development, identifies specific groups vulnerable to multiple forms of discrimination and violence. Controversy arose when LGBTQ+ people were omitted. This was the result of lobbying by a group of seventeen countries, led by Belarus, with some of the worst records of violence and intimidation towards those communities: human rights abuses, criminalization including capital punishment and other forms, and climates of oppression, discrimination and homo- and transphobia. The move is perhaps unsurprising, given that legal recognitions and social acceptance vary so greatly, and that there are still seventy-three countries with laws criminalizing homosexuality.

Governments and campaigners marshalling arguments *for* the inclusion of LGBTQ+ people, those who were part of the consultation process, recalled earlier UN policies where these groups had been specifically named. They also pointed to the particular forms of violence and overt and covert discrimination enacted against these individuals, and evidence of increasing levels of risk towards them. By noting that "urban modern cities" are "where the majority of LGBTI2S [Lesbian, Gay, Bisexual, Trans, Intersex and Two-spirited] youth, organizations and individuals live and operate", they suggested a particular need to consider the right to the city from the perspective of people with these identities (Vancouver Queer Consultation Group, 2018). The inclusion of LGBTQ+ minorities as a vulnerable group would have widened the *New Urban Agenda's* attention to the rights of girls and women to more broadly consider vulnerabilities through the spectrum of sexual and gender diversity. "Cities are where LGBT people go to find community but in many nations they don't just face silencing but extreme violence", commented Ellen Woodsworth, chair of Women Transforming Cities International Society, Vancouver, who campaigned for the document to specifically include LGBTQ+ rights (Totaro, 2016). However, the Belarus-led coalition was successful: there are no specific references to these vulnerable groups. The document instead reaffirms heteronormative family values.

This incident gives new impetus to consider queer urban imaginaries, and the ways that LGBTQ+ communities have shaped cities and will shape urban futures. The moment of policy erasure surrounding the *New Urban Agenda* can be read alongside critiques of the exclusion of queer scholarship from some of the dominant debates in urban theory – an arena in which, as in the UN Habitat III process, Henri Lefebvre's "right to the city" is invoked. Focusing attention on the recent "planetary urbanization" debate inspired by Lefebvre's work, Natalie Oswin, a geographer who has used postcolonial and queer approaches to study sexuality and space in different settings internationally, has critiqued a neglect of theoretical contributions rooted in identity politics (Oswin, 2016). She problematizes urban analyses that focus on capitalist logics at the expense of – rather than in parallel with – consideration of the inequalities wrought by heteronormative and patriarchal structures (Oswin, 2016); those structures, that is, which have been illuminated by feminist, queer, and race critical scholars, and those who work through the lens of intersectionality in their analyses. Oswin wishes to

> write against a radical approach to urbanization that focuses predominantly on capitalist exploitation and class dynamics, and for a radical approach to urbanization inspired by Marxist insights as well as insights gleaned from other critical approaches such as queer, feminist, postcolonial and critical race theories.
>
> *(ibid: 3)*

Oswin also points out that

> many scholars over many decades have produced work that shows that heteronormative and capitalist logics are intertwined with processes of urbanization all over the globe, that the regulation of intimacy is a cornerstone of efforts to limit public spheres everywhere, and that queer thought and politics have much to contribute to any critical project to produce an emancipatory commons.
>
> *(ibid: 5)*

For Oswin, a wide-reaching project of attending to subaltern urban populations is necessary, and a radical queer urban imaginary would be one that works across empirical, theoretical, and activist work to achieve this. This chimes with recent work at the interface of planning and architectural practice, queer activism, and urban theory in different contexts internationally (Doan, 2011, 2015; Goh, 2017; Pilkey et al., 2017).

Feminist and queer scholars from many urban disciplines, including geography, urban studies, and architectural history have, since the 1990s, paid attention to gender, sexuality, and space, and the reciprocal productions of urban space and minority gender and sexual identities (Bell et al., 1994). The urban imaginaries considered by these scholars have been diverse and have focused on a range of public and private spaces. They have understood explicitly that sexual orientation is constructed and performed spatially and phenomenologically (Ahmed, 2006; Reed and Castiglia, 2011). They have worked to disrupt the dominant heteronormative ideological production of space and to understood how models of "gay space" have interacted with wider dynamics of place-making, practices of consumerism, processes of gentrification and entrepreneurial urban imaginaries (Andersson, 2011; Brown, 2008; Collins, 2004; Doan, 2015; Rushbrook, 2002). Recently, there has been increasing interest in the ways that LGBTQ+ heritage is embedded in, or can be narrated through, the built environment (Andersson, 2012; Campkin and Marshall, 2016, 2017; Goh, 2017; Historic England, 2016; Oram, 2011, 2012; Schulman, 2013).

In architecture, there was an initial interest in queer space (or more accurately gay male space) in the 1990s, represented by North American scholars and curators such as Joel Sanders (1996), Aaron Betsky (1997), and Henry Urbach (1996). These writers and practitioners developed identity-based agendas for architecture and queered the architectural canon in multiple ways – by reading queerness in canonical figures and built works, and by introducing new objects of enquiry, for example. There was a lag in these architectural theoretical debates and design and curatorial practices in the 2000s, although, in the meantime, a significant body of work continued to develop in geography. Such work, like that of the architectural debates of the 1990s, has been critiqued for being limited to certain dominant forms of gay identities and enclaves, and a limited number of metropolitan centers in the global north (Brown, 2008).

In addition to geographers, historians and cultural and literary scholars have investigated London's past through attention to queer urban imaginaries (Cook, 2014; Cook et al., 2007; Hornsey, 2010; Houlbrook, 2005; Jennings, 2007). In part, this has involved the study of queer visual and spatial practices, such as cruising (Turner, 2003). The city provides the preconditions for the emergence of homosexual identities in the modern period (Hornsey, 2010). Similarly, it provides one of the contexts for the broad array of expressions of gender and sexuality, including trans and non-binary identities, in the present moment (Campkin and Marshall, 2016). There is now a notable resurgence of interest in queer architectural imaginaries, more broadly understood than Betsky's (1997) earlier conceptualization. Examples include the recent TU Delft "Queering Architecture" seminar which focused on "queering the discipline of architecture" towards approaches that are "more inclusive from an LGBT perspective"; or design practices engaging with spaces and activism related to LGBTQ+ people of color communities in New York City (Goh, 2017).

LGBTQ+ night-spaces in London

Since the summer of 2012, the elimination of queer space in London – as safe venues for LGBTQ and queer-identifying people – has galvanized debate and collective action through performance, research, and activism, and has emerged as an issue of wider media, public, and policy discourse. This has focused, in particular, on commercial nightlife space, which, for these communities, often enwraps civic and welfare space (Campkin and Marshall, 2016; 2017). Dominant modes of urban development have been protested and stalled, and alternative visions of change have been articulated. These have highlighted sites of heritage, through queer place-making, such as the guided tours by Queer Tours of London. These thematic tours draw attention to, and performatively narrate, LGBTQ+ built and social heritage. They are also future-facing – building capacity, including human and financial resources, to sustain queer activist campaigns. The organizers use the tagline "a mince through time" to evoke the subversive power of defiantly queer ways of being in and appropriating the urban environment. The tours focus on important social and legal histories and embed these within specific places and buildings.

Another example would be the weekly vigils held outside the iconic cabaret venue, pub, and nightclub, The Black Cap, since it closed and was put up for sale. These campaigners have created the Black Cap Foundation, under which banner they have produced an unauthorized estate agent's sales brochure for the venue. They also produced numerous testimonials in support of the Black Cap's value to the community and to individuals, with a survey of nearly 200 respondents. Milestones in their campaign so far include winning the legal status of Asset of Community Value, and subsequently winning the refusal of an appeal by the owners against that designation. They also achieved a moratorium on the sale of the venue, and have won and maintained the support of other local businesses, politicians, and the local borough authority.

In these and other cases, community-valued spaces are being recognized, celebrated, and protected, insofar as they can be through the planning system. Purely commercial understandings of "success" or "failure" in urban development have been critiqued as inadequate, and the asymmetrical power relations between landowners and investors and their agents, and venue operators and clients, has been exposed. In the Black Cap case and elsewhere, new structures for community ownership of venues have been proposed and piloted. These conflicts over closures of LGBTQ+ venues produce modes of urbanization based on queer modes of kinship and through attending to queer histories that are embedded in the spaces themselves and in the folk practices associated with them, such as specific forms of drag, performance, and traditions of activism.

In these situations, contemporary discussions of sexuality, subjectivity, and relationality in queer theory can be connected to theoretical debates in urban studies. To understand queer space and urban imaginaries, urban theory can be read through queer studies, and gender and sexuality studies, and vice versa. For example, in order to formulate a critical and queer approach to urban heritage, reference to recent insights about failure, dispossession, futurity, and relationality in queer theory provide multiple avenues for reflection and action (Campkin and Hunt, 2017).

The immediate and pressing practical question for communities, campaigners, built environment practitioners, and policy-makers, has been, "What structures and practices are in place to protect LGBTQ+ spaces and built heritage, and what challenges and potentials do these raise?" Many cultural and nightlife venues, deemed "safer spaces" in different ways, have been closed, sold, and/or converted to other uses, even when apparently vital and commercially viable. The closures have taken place rapidly, in the shadows, serving to maximize developers' profits at the expense of established communities, and frequently underpinned – typically for property-led regeneration – by opaque and offshore financial transactions. Each venue closure requires attention to specific circumstances, and these are not always readily available as public information, but general trends are evident, including negative impacts of large-scale infrastructure developments, a lack of implementation of safeguarding measures in the existing planning system; the sale and change of use of property by landlords whereby venue owners, operators, and clients have severely limited negotiating power compared with large organizations; as well as disproportionate rent increases being imposed in contexts of redevelopment and/or gentrification (Campkin and Marshall, 2017).

Higher-profile spaces, such as The Black Cap or the iconic cabaret and performance venue the Royal Vauxhall Tavern, have become symbolic battlegrounds and have drawn attention to the wider and longer-term elimination of queer space (Alwakeel, 2015; Brown, 2015). In the latter case, a powerful campaign arose when the venue was sold to a developer, and the campaigners, RVT Futures, are working towards a community buy-out after successfully winning Asset of Community Value status and an architectural heritage Grade II listing – the first to be awarded based primarily on social value (Walters, 2015).

Media attention has emphasized the closure of commercial LGBTQ+ entertainment venues, but civil society organizations and charities are closing at an alarming rate as a result of the combined pressures of government and local government cuts and the costs of space (Campkin and Hunt, 2017: 247). Such organizations have often formed part of clusters with commercial and nightlife venues and so are equally if not more vulnerable to waves of gentrification. LGBTQ+ night-time spaces also form important roles within their neighborhoods, providing specific welfare services to their clients, while boosting business for local shops, integrating with other elements of localized cultural and creative industries.

One angle on the history of queer space in London, as in other Western metropolitan centers, suggests that it has flourished most readily in failed, ruined, marginal, and stigmatized space (Andersson, 2011; Andersson and Campkin, 2009; Binnie, 1997). This is as true of radical utopian queer art as it is of middle-class gay pioneer gentrifiers or everyday cruisers. As such, it is notable that a contemporary moment of apparent failure – when multiple venues have closed or been forced to close – has witnessed, rather than the vanishing of queer space, a rejuvenated queering of London through the actions of campaigners and other queer space producers. In this context, artists, activists, venue owners and customers, entrepreneurs, academics, and others have been propelled to debate and assess the histories, qualities of, and present and future need for queer spaces and urban imaginaries – and to act individually and in networks to protect it. In response, there has been an urgent direction of energy – first reactive, but now more proactive and strategic – towards the protection of certain spaces. There are also parallel discussions in other cities about the effects of gentrification on queer space, the assimilation of former LGBTQ neighborhoods, and/or the desexualization of spaces formerly associated with sex (Bell and Binnie, 2004; Doan, 2011). In London, and in other cities such as New York, this situation has coincided with, and has begun to capitalize on, a new prioritization of minority heritage by institutions. In the UK context there has been a new attention to built heritage associated with LGBTQ-identified people (Historic England, 2016), as well as a wider inclusion of LGBTQ heritage narratives in museum discourse and the public sphere (Mills, 2006). A city-wide Queer Space Network has also been formed across different activist campaigns, lobbying local government and producing a *Queer Vision for London*. This document argues that "The GLA must make queer culture a priority for the city, to support and promote a vibrant queer culture, and the empowerment of all queer people, with all the social and economic benefits this brings" (Queer Spaces Network, 2016). In order to achieve this, it sets out practical ways forward in the protection of queer spaces, such as listing all LGBTQ+ spaces pre-dating 1986 as "legacy venues" in parallel with the system implemented in San Francisco.

The responses of London's queer communities to recent threats to close iconic performance venues, where queer utopias have been staged and imagined, has involved systematic use of legal and policy frameworks for protecting architectural and community heritage and media interventions, in conjunction with performative activism, and performance as activism, in which queer history has been mobilized towards the future. In this, there is a refusal of conservation, per se, as well as a refusal to let go. Thinking through the possible tensions between heritage and queerness, or heritage-led urbanization and the queering of space, seems necessary as a longer-term project. Now, as in past moments of synergy between queer and urban activism, there is also a need for affinities across minority groups whose rights to heritage are being infringed, and for a critical awareness of inequalities wrought by contemporary neoliberal urbanization and gentrification (Schulman, 2013).

Figures collated by UCL Urban Laboratory show that 58 percent of licensed LGBTQ+ nightlife premises closed in the past decade (Campkin and Marshall, 2017). Yet this figure only refers to licensed premises, excludes certain kinds of spaces such as saunas, and excludes events that may occupy multiple venues which may or may not be primarily designated for LGBTQ+ use. Looking only at licensed premises creates a bias towards spaces owned and operated by and for white, cis-gendered, gay-identifying men (ibid). Women, trans, non-binary and BAME communities, who otherwise lack dedicated venues (and face barriers to opening them), have traditionally been better served by events, whether longstanding or ephemeral. Longstanding events have important social outreach functions and value to LGBTQ+ communities. Recent research also illuminates new events aimed at groups excluded from mainstream LGBTQ+ commercial scenes, which addresses perceived problems with those scenes, as well as a resurgence of spaces in south-east London, where LGBTQ+ nightlife existed historically but had declined until recently (ibid). This geography is constantly shifting and requires attention to the contexts behind and beyond the headlines and dominant media discussions.

Through the work of Queer Spaces Network, the community response to venue closures has been paralleled by – and linked in to – activities in City Hall. The recently appointed London Mayor, Sadiq Khan, has positioned himself as an outspoken advocate of LGBTQ+ rights and of the value of LGBTQ+ venues. He has done so in a number of ways: hosting an LGBTQ+ reception at City Hall to coincide with London Pride and highlighting the absence of such receptions in the preceding administration, under Conservative Mayor Boris Johnson; commissioning research to ascertain the extent of licensed venue closures and the reasons for closure; using the research findings to inform his update of the city's main strategic plan (*Draft London Plan,* 2017) and to produce a *Cultural Infrastructure Strategy* to "identify what we need in order to sustain London's future as a cultural capital" – a manifesto commitment framed around property-related threats to the city's cultural competitiveness, including high rents (Khan, 2016). LGBTQ+-friendly policies have their equivalents in other cities, internationally, but the Mayor's strong positioning on the issue of closures, and on the value of LGBTQ+ night-spaces for the capital in terms of culture and diversity, as well as economy, is notable. So too is the attempt to engage organizations representing queer, trans and intersex people of colour communities (QTIPOC), such as UK Black Pride.

In November 2016, Mayor Khan appointed Amy Lamé as the capital's first "night czar", echoing other global city leaders who have their sights set on developing the night-time economy. He is not alone in attending to the management – and economic potential – of the urban night. In September 2017, New York City Mayor Bill de Blasio instituted an Office of Nightlife, with comparable objectives of both regulation and protection of night-spaces and activities. London's Night Czar, Lamé, originally from New Jersey, recently authored the first children's book on LGBTQ+ rights, *From Prejudice to Pride* (Lamé, 2017). She combines grassroots knowledge with experience in mainstream city politics, having served as Mayoress of Camden, one of the London local boroughs, between 2010 and 2011. Lamé understands first-hand the radical social and cultural potential of night-space. She has worked in London's LGBTQ+ night-spaces since 1992, and co-founded and runs the award-winning artistic and activist club night "Duckie". She also co-founded and chaired RVT Futures, set up to fight the sale, closure, and redevelopment of the Royal Vauxhall Tavern, as the venue that hosts Duckie. The campaign has been highly successful to date, thanks to a skilled and dedicated group of core members.

Things are now at an interesting turning point. Backed by supportive policies and statements from the Mayor and Night Czar, successful campaigns to protect or re-open iconic night-spaces are gaining momentum and publicity. Another case is that of the popular East London late-night LGBTQ+ venue, the Joiners Arms, which closed prior to a large mixed-use development. Here, the local council recently made an unprecedented move, voting unanimously to insist that the site's developers grant a 25-year lease for an LGBTQ+ venue within the reconfigured site. Additionally, pre-requisites for planning permission included that the developer contribute "£130,000 towards fit-out costs of the venue, a 12-month rent-free period, a 22 percent extension of the floor-space designated for the venue, and additional commitments to enhancing the sound-proofing of the venue, including for the smoking area" (Friends of the Joiners Arms, 2017). The main campaign group are now in the process of forming a community cooperative to bid for the venue when it goes to tender. Key to this, and a number of other campaigns, has been an openness to "evolve" the venue, within a changing context, rather than preserve it.

The recent research on London's LGBTQ+ night scenes and spaces provides opportunities to understand how queer urban imaginaries are bound into the design and operation of particular spaces, and the entanglements of those spaces within, or their disruption of, wider processes of urbanization. The newly LGBT-supportive policy context has been formed in part through engagements with a radical queer imaginary of urban change, represented by Queer Space Network's *Queer Vision for London* and other activists' statements. Yet the Mayor and City Hall's vision is formulated through well-trodden neoliberal planning tropes of the "global" or "creative" city, and underpinned by nigh-time economy boosterism. In summary, it is driven by forms of international competitiveness between cities that exacerbate the financialization of the built environment at the expense of cultural and social infrastructure. To what extent can radical queer imaginaries sit comfortably within or alongside this mainstream acceptance of the status quo? At this moment, and as a campaign to create a new LGBTQ+ community center in London gathers momentum, it is worth reflecting on an earlier period of interaction between London's LGBTQ+ communities and urban policy. In this way, contemporary interactions between activists, civil society organizations, policy-makers and built environment professionals in London can be placed within a longer history of the changing imaginaries of LGBTQ+ inclusion.

The municipal lesbian and gay imaginary of the London Lesbian and Gay Centre

The ambitious multi-use London Lesbian and Gay Centre was conceived in 1982. It was supported as part of a radical social agenda, including an array of ground-breaking "equal opportunities" policies and community funding strategies, developed under Mayor Ken Livingstone's leadership of the Greater London Council (GLC). On 9 September 1982, the GLC's Grant's Sub-Committee awarded the London and Lesbian and Gay Centre Steering Committee, a group of volunteers convened under the GLC's Gay Working Group, £15,500 to employ two development workers to advise on the establishment of the Centre, which "would offer recreational, cultural and educational activities, a meeting place, and an information and gay rights service" (GLC, 1982). The GLC's wider commitment was to "help all sections of the community, especially those suffering hardship, handicap, disadvantage and discrimination" (GLC, 1982). The idea was developed through Gay Working Party discussions and, later, the Centre's Steering

Figure 30.1 The front elevation of the London Lesbian and Gay Centre, information leaflet, 1985. Source: Hall-Carpenter Archives.

Committee, including open public meetings, and with the support of legal and architectural professionals the Committee were able to galvanize.

The original proposal, developed through the GLC's Gay Working Party in 1982, states that:

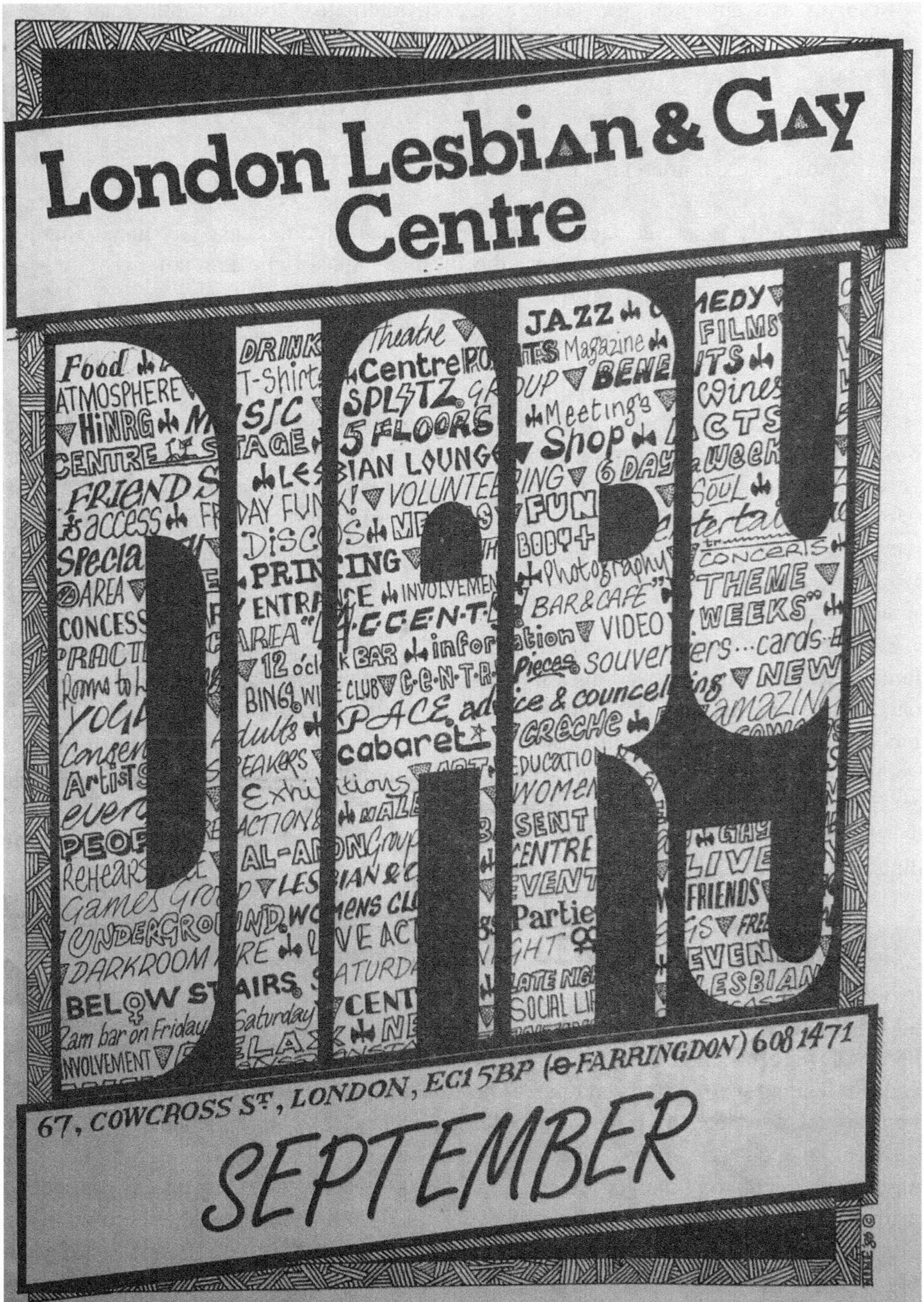

Figure 30.2 *London Lesbian* and *Gay Centre Diary*, September 1986, cover design by NINE. Source: Hall-Carpenter Archives.

> Our concept is of a central community based centre run by lesbians and gay men for lesbians and gay men, providing a relaxed alternative to the commercial "scene", which often excludes women, older and younger people, and those without much money.
>
> *(GLC, Gay Working Party, 1982a)*

The critique of the commercial gay scene was explicit, with it being understood as:

> predominantly a service facility for fairly affluent middle-class gay men. There are still very few places for lesbians, the disabled, those who are not rich enough to afford pubs and clubs etc. to meet let alone participate.
>
> *(London Lesbian and Gay Centre, 1983a)*

The ambitious imaginary for lesbian and gay human rights extended beyond London, with the Centre as accompaniment to the GLC's Gay Working Party charter, *Changing the World: the London Charter for Lesbian and Gay Rights* (1985). This is an important historical precedent of proactive support for the lesbian and gay community. It outlined a vision to address the needs of the time, such as providing counter-narratives to aggressively negative media stereotyping, and providing information on housing and employment rights. As a visioning exercise it was very attuned to the international context of human rights.

Early in discussions about the Centre, Griffith Vaughan Williams of the Campaign for Homosexual Equality, the leading lesbian and gay rights organization at the time, wrote a concept note summarizing previous explorations of the potential of a gay center. This also emphasized the international dimensions of the Centre, such as providing a focal point for overseas visitors to find information. Part of the argument was that Britain was one of the few European capitals not to have such a Centre, and that London needed to catch up with Manchester, which already had one, funded under the Urban Aid Programme of Prime Minister James Callaghan's Labour Government in 1978.

The building that was purchased in 1983, 67–69 Cowcross Street, comprised five-storeys of a poultry-processing facility that had previously served nearby Smithfield Market. The premises were found rapidly in order to maximize the potential contribution from the GLC's capital budget, with £751,000 being provided for the building, conversion and repair, with further funding for furniture and equipment. The architects eventually appointed were McLean Ditlef-Nielsen Quinlan, with Fiona McLean as the lead. The available funding was partly used towards accessibility features including a disability access ramp, disabled toilets, and induction looping. More than an architectural conversion transforming industrial space to mixed cultural and community uses, this was a self-consciously articulated statement of the GLC's Equal Opportunities agenda. It continued to be adapted incrementally following the opening, and was flexible enough to accommodate a wide range of uses by day and into the night, being open every day except Monday, and until midnight most nights. The building included a disco, bars, café, bookshop, a women-only floor and coffee bar (The Orchid), crèche, meeting rooms and workshop spaces, a shop (Centre Pieces), printing and typesetting workshop, and a photographic darkroom. The ground and lower floors were the more public areas, including a large reception, with more specialist spaces on higher storeys.

The Centre was constituted "for the advancement of lesbians and gay men, regardless of age, sex, race, ethnic origin, creed or disability" (LLGC, 1983b). As well as providing spaces for women, the Steering and later Management Committees had a proactive approach to reaching out to older and younger peoples' groups and providing space for their meetings. Yet even in the initial discussions there were anxieties about "how one centre could cope with the differing interests and needs of the lesbian and gay community" (GLC, Gay Working Party, 1983). At an open meeting organized by the Centre's Steering Committee at County Hall (which functioned as City Hall under the GLC) a note asks: "What specific provisions should be made about who can use the centre? Can it exclude certain groups e.g. Tories, religious groups, TV/TS [Transvestite/Transexual], Paedophile", revealing some of the preoccupations and biases of the day (London Lesbian and Gay Centre, 1983b). Plans for the center chimed with broader debates about the possibilities and pitfalls of coalitions across different identity-based political agendas, with some lesbians arguing for women to put their efforts into developing a separatist center under a parallel "Women's City" project. However, others won the argument that compartmentalizing and over-stretching resources was too risky, and instead successfully secured a women-only space within the plans (GLC, 1982b). From the desire for inclusion, the agreed Management Committee structure was inevitably complex, with one pamphlet noting an Electoral Reform Society worker's comment "that it was one of the most complicated he had ever seen" (London Lesbian and Gay Centre, 1984).

The Centre accommodated and fostered a range of autonomous organizations that were serving hugely important social services to gay and lesbian communities dealing with the stresses and traumas of queer life in the 1980s and early 1990s – for example, organizing and campaigning on health and HIV/AIDs, employment rights, housing, police monitoring, and providing counter-narratives to homophobia evident in the mainstream media. This included provision of specific spaces for certain organizations. The Hall-Carpenter Archives, the largest archive for the study of gay activism in Britain since the 1950s, were housed here, along with the Project for Advocacy, Counselling and Education (PACE), the capital's main charity promoting LGBTQ+ mental health from 1985 until it closed in 2016. So too were important telephone-based information and support services. There were also meeting rooms and facilities for a wider range of voluntary sector initiatives promoting collective and individual self-care, as well as cultural and leisure organizations and creative practitioners, such as artists wanting to exhibit their work and music and drama production and performance.

The Centre had a strong visual identity throughout its existence and from the outset there was an explicit focus on the dissemination of accessible information for diverse audiences. The logo comprised an appropriation of the pink triangle that Nazi concentration camp prisoners identified as homosexuals were forced to wear. This triangle was adapted into a range of graphic forms during the Centre's lifetime. In early publications the triangle appeared as a kite; later being incorporated into a London skyline, placing the building and the communities it represented within a spatio-temporal imaginary of modern and historic London constructed through reference to some of the capital's most iconic buildings. The *Centre News Team* and its publications, like the Steering and Management Committees, were explicit in aiming at a wide audience: the team "shall endeavour not to discriminate against any person/group because of race, creed, colour, sexual preference, religious belief or political viewpoint" (LLGC, 1989). The strong visual identity expressed in publications was accompanied by events that brought together the autonomous organizations operating under the Centre's umbrella. Pride was organized from the Centre, and in this it provided the symbolic heart of an annual ritual to commemorate the Stonewall Riots through marching the streets, rallies, and an ambitious

Figure 30.3 London Lesbian and Gay Centre, *Some useful information*, pamphlet, 1985. Source: Hall-Carpenter Archives.

program of events and activities attracting thousands of people from London and elsewhere. At such moments, the GLC-supported vision for Lesbian and Gay equality was projected within and beyond the city.

At the point of opening, on 9 April 1985, the voluntary Steering Committee was formalized as a Board of Directors and an elected Management Committee. The Centre was constituted as a "Company Limited by Guarantee Not Having Share Capital", that is, with no distribution of shares to members (GLC, 1982a). There were twelve full-time staff when the Board took possession in January 1985, in preparation for the official opening in April 1985. There were 1400 registered members one year after opening, and an estimated 500,000 annual visitors. Membership comprised both individuals and representatives nominated by lesbian and gay groups. Alongside these structures, the GLC's ethos was to support lesbian and gay minorities with "self-help" initiatives. As it evolved, the organizational structure set out to widely represent different user groups and to be inclusive of women and ethnic minorities.

However, during the planning and lifetime of the Centre there were criticisms of a lack of transparency in decision-making. As early as 1984, the management felt the need to defend their emphasis on "doing" over talking – that is, consulting with members (LLGC Press Statement, 1984). Such conflicts and complex internal dynamics between the different groups the Centre encompassed are perhaps inevitable in the development of a building and project that were so socially ambitious and incorporated so many people and activities. At the end of the 1980s, an incoming General Manager, Oscar Watson, referred to the Centre as "a strange beast, a monster which needs constant attention and feeding to keep it alive", and wrote: "It's devoured people, money, time and energy and every conceivable resource imaginable and still, all it takes is a letter in a paper or a tiff in the bar and it needs major surgery" (London Lesbian and Gay Centre, 1989). In the archives, criticisms of the management of the Centre, and statements about its financial health, are often stressed in communications from the management themselves, acknowledging issues that have been raised and presenting changes of direction. For example, in 1986, an Extraordinary General Meeting was called in response to a controversy that the Centre was not doing enough to reach black and minority ethnic lesbians and gays. At this point, a newly appointed Management Committee took the opportunity to define themselves as agents of change, to the extent of labelling the Centre "racist" and tabling a motion to employ black outreach workers and take other actions in response (LLGC, 1986a; LLGC, 1986b).

Another conflict, which broadened out to wider debates in feminist politics and was written up at the time in *Feminist Review*, focused on whether SM (sadomasochist) groups should be able to meet at the Centre (Ardill and O'Sullivan, 1986). In this way, the Centre became a focal point and literal meeting place for a range of intersecting political imaginaries. It also became an emblem in mainstream political debates in the House of Commons and media to signify all that was wrong with Mayor Livingstone and the GLC's radical left-wing political agenda. In parliament, the use of public funds was controversial. Thatcher referred to such GLC initiatives for sex workers, lesbians, and teens as "a disgraceful waste of money and a disgraceful imposition of increases on the tax burden" (Hansard, 1982). Apart from a narrative of the squandering of directly allocated public resources, the management had to navigate a bleak landscape of funding cuts more generally.

Conceived as a "social enterprise", a term that had emerged in the 1970s, the Centre fell into a debt crisis in 1988, as well as suffering cuts to the grants schemes it and the organizations it accommodated relied upon. After the Thatcher government eventually closed the GLC, the Centre was given the opportunity to purchase the building for £510,000, from the London Residuary Body set up to manage the GLC's assets. This led, in 1989, to a funding campaign. However, the Centre's financial problems worsened over time and it closed in 1992. The buiding now houses a non-LGBTQ+ specific wine bar/restaurant and an advertising agency.

Figure 30.4 67–69 Cowcross Street today. Photography by the author, 2018.

In a *Vice* article from 2016, "Remembering the 1980s Lesbian and Gay Centre That Didn't Last a Decade" (Hastings, 2016), the Centre is presented as a failure, with quotes from those who were there and involved in a number of ways conveying a strong sense of disappointment. There are references to political infighting, critiques of staff being chosen for specific political affiliations, being over-generously paid, practicing poor financial management and even theft and fraud; of the policing of the use of certain spaces for those with the "correct" identities; and of programming that provoked tensions between groups (Hastings, 2016). The reality, those quoted suggest, directly contrasts with the "symbol of lesbians and gay men working together on such diverse issues as women-only space, AIDS, childcare and services for gays" celebrated in a pamphlet on the Centre's first birthday. Taking seriously the negative evaluations evident in the testimonies the article collates, and avoiding nostalgia, such criticisms are counter-balanced by recognition of the pioneering, experimental, and radical nature of the underpinning vision.

A number of characteristics of the urban imaginary embodied within the GLC's policies and the Centre are worth noting. Firstly, the inclusion of diverse groups with overlapping but distinct agendas and interests, and attempts to address what might now be discussed as intersectional differences in terms of access to and need for the space and associated resources. Significant privilege was given to minorities within the diverse London Lesbian and Gay community, with attempts to direct funding to understand their experiences and needs through research and meaningful consultation. Secondly, the ethos of cooperation underpinning the conceptualization, evidenced in both a municipal operational structure and accommodation of informal or DIY approaches. Thirdly, the co-location of services, research, and communications in order to gather and disseminate evidence and counter the misinformation and homophobia of the time. Fourthly, the strong connections to borough-level local government in Islington and, through proactive seeking of resources and access to other local government funding streams, to other administrative districts. Acknowledging the difficulties and complexity that engulfed the Centre and led eventually to its closure, it is important to see these in the context of the time, in order that this "failed" experiment can fertilize new initiatives and infrastructures to support LGBTQ+ communities to enact radical queer imaginaries. That is to say, imaginaries that protect the rights and spaces of minority communities, and support forms of queer kinship that are locally embedded but which also extend beyond city and national borders. Working in utopian ways against the dominant heteronormative and patriarchal structures of capitalist urbanization, there is bound to be difficulty and disappointment, but that makes a grounded politics of hope ever more important.

References

Ahmed, S. (2006) *Queer Phenomenology: Orientations, Objects, Others*. Durham, NC: Duke University Press.

Alwakeel, R. (2015) Royal Vauxhall Tavern: ANOTHER London Gay Venue Threatened. *Evening Standard*, 16 April 2015. Available at: http://www.standard.co.uk/news/london/royal-vauxhall-tavern-another-london-gay-venue-threatened-by-developers-10181665.html.

Andersson, J. (2011) Vauxhall's Post-Industrial Pleasure Gardens: "Death Wish" and Hedonism in 21st-Century London. *Urban Studies* 48(1): 85–100. Available at: https://doi.org/10.1177/0042098009360238.

Andersson, J. (2012) Heritage Discourse and the Desexualisation of Public Space: The "Historical Restorations" of Bloomsbury's Squares. *Antipode* 44(4): 1081–98.

Andersson, J. and Campkin, B. (2009) Literary Representations of Cottaging in London. In: Gershenson, O. and Penner, B. (eds.), *Ladies and Gents: Public Toilets and Gender*. Philadelphia: Temple University Press.

Ardill, S. and O'Sullivan, S. (1986) Upsetting an Applecart: Difference, Desire and Lesbian Sadomasochism. *Feminist Review* 23: 31–57. Available at: https://doi.org/10.1057/fr.1986.19.

Bell, D. and Binnie, J. (2004) Authenticating Queer Space: Citizenship, Urbanism and Governance. *Urban Studies* 41(9): 1807–20. Available at: https://doi.org/10.1080/0042098042000243165.

Bell, D., Binnie, J., Cream, J., and Valentine, G. (1994) All Hyped Up and No Place to Go. *Gender, Place & Culture* 1(1): 31–47. Available at: http://www.tandfonline.com/doi/abs/10.1080/09663699408721199.

Betsky, A. (1997) *Queer Space: Architecture and Same Sex Desire*. London: HarperCollins.

Binnie, J. (1997) *A Geography of Urban Desires: Sexual Culture in the City*. Unpublished PhD thesis, University of London.

Brown, G. (2008) Urban (Homo)Sexualities: Ordinary Cities and Ordinary Sexualities. *Geography Compass* 2(4): 1215–31. Available at: https://doi.org/10.1111/j.1749-8198.2008.00127.x.

Brown, M. (2015) London Gay Pub the Royal Vauxhall Tavern Is Given Grade II Listing. *The Guardian*, 9 September 2015. Available at http://www.theguardian.com/uk-news/2015/sep/09/london-gay-pub-royal-vauxhall-tavern-given-grade-ii-listing.

Campkin, B. and Hunt, J. (2017) Letters Home. In: Pilkey, B., Scicluna, R.M., Campkin, B. and Penner, B. (eds.), *Sexuality and Gender at Home: Experience, Politics, Transgression*. London: Bloomsbury Publishing, 232–50.

Campkin, B. and Marshall, L. (2016) *LGBTQ+ Nightlife in London, from 1986 to the Present: Interim Findings*. London: UCL Urban Laboratory. Available at: https://www.ucl.ac.uk/urbanlab/docs/LGBTQI_nightlife_in_London_from_1986_to_the_present_-_interim_findings.pdf.

Campkin, B. and Marshall, L. (2017) *LGBTQ+ Cultural Infrastructure in London, 2006–Present*. London: UCL Urban Laboratory. Available at: http://www.ucl.ac.uk/urbanlab/research/lgbtqi-space.

Collins, A (2004) Sexual Dissidence, Enterprise, and Assimilation: Bedfellows in Urban Regeneration. *Urban Studies* 41(9): 1789-806.

Cook, M. (2014) *Queer Domesticities: Homosexuality and Home Life in Twentieth-Century London*. Basingstoke: Palgrave Macmillan.

Cook, M., Cocks, H.G., Mills, R., and Trumbach, R. (eds.) (2007) *A Gay History of Britain. Love and Sex between Men since the Middle Ages*. Oxford: Greenwood World Publishing.

Doan, P.L. (ed.) (2011) *Queerying Planning: Challenging Heteronormative Assumptions and Reframing Planning Practice*. Farnham: Ashgate Publishing.

Doan, P.L. (ed.) (2015) *Planning and LGBTQ Communities: The Need for Inclusive Queer Spaces*. London: Routledge.

Friends of the Joiners Arms. (2017). Available at https://thejoinersliveson.wordpress.com/. Accessed 7 January 2018.

Goh, K. (2017) Safe Cities and Queer Spaces: The Urban Politics of Radical LGBT Activism. *Annals of the American Association of Geographers* 18 December 2017: 1–15. Available at: https://doi.org/10.1080/24694452.2017.1392286.

Greater London Authority. (2018) *Draft New London Plan*. Available at https://www.london.gov.uk/what-we-do/planning/london-plan/new-london-plan/draft-new-london-plan/. Accessed 7 January 2018.

Greater London Council. (1982) *GLC Gives More Community Grants*. Public Relations Branch News Service, No. 395.

Greater London Council. (1982a) *Gay Working Party*. London Lesbian and Gay Centre. proposal and concept statement.

Greater London Council. (1982b) Gay Working Party, Note of the meeting of Lesbian Organisations held on 24 August at 7 PM, undated, c.1982.

Greater London Council. (1983) Gay Working Party. *Minutes*, 27 July 1983.

Hansard (1982) *House of Commons Prime Minister's Questions*, 25 November 1982. Available at https://www.margaretthatcher.org/document/105059. Accessed 22 August 2018.

Hastings, C. (2016) Remembering the London Lesbian and Gay Centre that Didn't Last a Decade. *Vice*, 27 August 2016.

Historic England. (2016) *Pride of Place: England's LGBTQ Heritage*. Historic England. Available at: https://historicengland.org.uk/research/inclusive-heritage/lgbtq-heritage-project/. Accessed 7 January 2018.

Hornsey, R. (2010) *The Spiv and the Architect: Unruly Life in Post-War London*. Minneapolis: University of Minnesota Press.

Houlbrook, M. (2005) *Queer London: Perils and Pleasures in the Sexual Metropolis, 1918–1957*. Chicago: University of Chicago Press.

Jennings, R. (2007) *A Lesbian History of Britain. Love and Sex between Women Since 1500*. Oxford: Greenwood World Publishing.

Khan, S. (2016) Making the Most of Arts, Culture and Creativity – Sadiq Khan – A Mayor for All Londoners. Available at: http://www.sadiq.london/making_the_most_of_arts_culture_and_creativity. Accessed 3 January 2018.

Lamé, A. (2017) *From Pride to Prejudice: A History of the LGBTQ+ Movement*. London: Wayland.

London Lesbian and Gay Centre. (1983a) *What Is the London Lesbian and Gay Centre?* Pamphlet.

London Lesbian and Gay Centre. (1983b) Steering Committee, Open Meeting at County Hall, 12 December 1983.

London Lesbian and Gay Centre. (1984) Cowcross 2. Pamphlet.

London Lesbian and Gay Centre Press Statement. (1984) 24 January.

London Lesbian and Gay Centre. (1986a) Extraordinary General Meeting. *Minutes*, 16 November 1986.
London Lesbian and Gay Centre. (1986b) *Centre Tackles Racism.*
London Lesbian and Gay Centre. (1989) *Centre News*. December–January.
Mayor of London. (2018) *Draft London Plan*. August.
Mills, R. (2006) Queer Is Here? Lesbian, Gay, Bisexual and Transgender Histories and Public Culture. *History Workshop Journal* 62(1): 253–63. Available at: https://doi.org/10.1093/hwj/dbl006.
Oram, A. (2011) Going on an Outing: The Historic House and Queer Public History. *Rethinking History* 15(2): 189–207. Available at: https://doi.org/10.1080/13642529.2011.564816.
Oram, A. (2012) Sexuality in Heterotopia: Time, Space and Love between Women in the Historic House. *Women's History Review* 21(4): 533–51. Available at: https://doi.org/10.1080/09612025.2012.658178.
Oswin, N. (2016) Planetary Urbanization: A View from Outside. *Environment and Planning D: Society and Space*, 3 November: 540–6. Available at: https://doi.org/10.1177/0263775816675963.
Perry, F. (2016) Why Won't the Global Agenda for Inclusive Cities Recognise LGBTQ Citizens. *The Guardian*, 19 October 2016. Available at: https://www.theguardian.com/cities/2016/oct/19/un-new-urban-agenda-inclusive-cities-lgbtq-rights-habitat-3.
Pilkey, B., Scicluna, R.M., Campkin, B., and Penner, B. (eds.) (2017) *Sexuality and Gender at Home: Experience, Politics, Transgression*. London: Bloomsbury Academic.
Reed, C. and Castiglia, C. (2011) *If Memory Serves: Gay Men, Aids, and the Promise of the Queer Past*. Minneapolis: University of Minnesota Press.
Rushbrook, D. (2002) Cities, Queer Space, and the Cosmopolitan Tourist. *GLQ: A Journal of Lesbian and Gay Studies* 8(1): 183–206.
Sanders, J. (1996) *Stud: Architectures of Masculinity*. New York: Princeton Architectural Press.
Schulman, S. (2013) *The Gentrification of the Mind: Witness to a Lost Imagination*. Reprint edition. Berkeley: University of California Press.
Totaro, P. (2016) Belarus Leads Group of about 17 Nations to Block LGBT Rights in U.N. Cities Plan – Sources. Available at: https://uk.reuters.com/article/uk-un-habitat-lgbt/belarus-leads-group-of-about-17-nations-to-block-lgbt-rights-in-u-n-cities-plan-sources-idUKKCN12B2TC?il=0. Accessed 3 January 2018.
Turner, M. (2003) *Backward Glances: Cruising Queer Spaces in London and New York*. London: Reaktion.
United Nations. (2016) The New Urban Agenda. Habitat III. Available at: http://habitat3.org/the-new-urban-agenda/. Accessed 3 January 2018.
Urbach, H. (1996) Closets, Clothes, Disclosure. *Assemblage* 30: 63–73. Available at: https://doi.org/10.2307/3171458.
Vancouver Queer Consultation Group. (2018) Endorse Queer Declaration Inclusion in Implementation of UN Habitat 3 New Urban Agenda. Available at: https://www.change.org/p/endorse-queer-declaration-calling-for-inclusion-in-the-implementation-of-the-new-urban-agenda-of-un-habitat-3. Accessed 7 January 2018.
Walters, B. (2015) Read Our Application to Make the RVT a Listed Building. *Not Television*. 8 September 2015. Available at: http://www.nottelevision.net/read-our-application-to-make-the-rvt-a-listed-building/.

31

Crafted imagination

Future-builders and the contemporary logic of experimentalism

Federico Savini

Planning, institutionalized imaginaries, and imagination

This chapter discusses emerging approaches of imaginary production. To do so, it conceptualizes the tense relationship between planning, imaginaries, and imagination. It defines planning as a practice of organizing the production of new urban imaginaries, that is, the process of imagination. Secondly, it provides a critique of an emergent approach to organized imagination, which is becoming widely used in today's urban policy, namely the urban laboratory and the role of future-building agencies.

> Planning (or "development") also highlights a developmental movement from past to future. It implies that it is possible to decide between appropriate actions now in terms of their potential impact in shaping future socio-spatial relations. This future imagination is not merely a matter of short-term political expediency, but is expected to be able to project a transgenerational temporal scale, especially in relation to infrastructure investment, environmental management and quality of life.
>
> (*Healey, 2004: 46*)

In this text, planning theorist Patsy Healey defines the practice of planning and the constitutive role that the future plays in it. Planning is understood as an organized collective practice of imagination about a future state of a place, namely the constellation of social relations and physical artifacts. Urban imaginaries are both enabling framings and products of these practices. On the one hand, actors engaging in planning activities take part in a process of cooperation, deliberation, imagination, and collective sense-making that may result in a novel meaning of place. On the other, the same actors are always situated in context, where existent imaginaries shape their views, aspirations, and narratives. In this process, an imaginary works as a "symbolic matrix" (Gaonkar, 2002: 1) or structuring component (Castoradis, 1976: 145) that provides actions with meaning. Each of these actors carries an already pre-constituted vision of what a particular place or city is (or will be), namely an understanding of its present and the expectations of its possible futures.

To disentangle the relationship between planning processes, imagination, imaginaries, and the institutional-physical structures of cities, in this chapter I conceptualize planning as any attempt of organizing a collective process of imagination of place. Imagination is the production of new imaginaries, which differ from existent ones. In doing so, planning works in tension with existent imaginaries. Following Arjun Appadurai,

> the imagination has become an organized field of social practices, a form of work (in the sense of both labor and culturally organized practice), and a form of negotiation between sites of agency (individuals) and globally defined fields of possibility
>
> (*Appadurai, 1996: 31*)

Planning is one particular field of imagination among many. Existent imaginaries are the framings that build the context within which imagination and planning take place. As context, these imaginaries are institutionalized into all kinds of rules, procedures, and physical artifacts that constrain individual and collective imagination in cities.

The relationship between planning, imaginaries, and imagination is always problematic because planning inevitably reduces social complexity, providing biases to the imaginative process, which eventually frustrates the range of imaginable possible futures. Planning is a "killer of alternatives" (Metzger, 2015) because it reduces the range of possibilities through its spatial selectiveness. It has an inherently discriminatory capacity in its epistemologies (Madanipour, 2013; Savini, 2016). Its tendency to frustrate new imaginaries springs from the institutional frameworks that imagination has to cope with, the financial, legal, and physical elements through which urban form is regulated (Savini et al., 2014). Yet, because it sets boundaries, planning also enables those processes that allow for a redefinition and reconstitution of these boundaries. In other words, planning aims at the *re*-production of institutions through a process of imagination (Salet, 2017). These institutions can be material, such as zoning procedures and standards of environmental qualities of financing mechanisms, but also immaterial, such as established idioms, icons, and dictionaries of equality or justice. Planning institutions, such as laws, procedures, and instruments of governments, are substantiated by images and codes, such as zoning maps, narratives, and languages that are long-lasting and fully internalized by actors (Yanow, 1993). Institutionalized, "social imaginaries are ways of understanding the social that become social entities themselves, mediating collective life" (Gaonkar, 2002: 4). Architectural artifacts, the most visible product of planning practices, can be understood as not "only a means of *expressing/signifying* existing elite power, but also as one of the most effective means for *instituting* new social relations" (Kaika, 2011: 970). Symbols, understood as immaterial frames of action, do underlie institutional structures and procedures of decision-making, and planners are actively redefining and questioning these meanings in the carrying out of urban interventions (Dembski, 2014).

The history of planning ideas can be understood as a dynamic tension between those imaginaries that have institutionalized into rules, laws, procedures, and physical artifacts and the search for novel approaches to produce new imaginaries. Jane Jacobs' activist projects proposed a different imaginary of neighbourhood living by actively contesting the institutionalized imaginaries of the modern city, such as the top-down zoning regulations and the established norms of car mobility. Many of the most innovative planning ideas have struggled to safeguard processes of imagination against institutionalized imaginaries. The vision of a regional network of garden cities was at odds with the imaginary of the industrialized city, based on the centrality of the factory and the administrative institutions of municipal governments. The vision of a hyper-diverse urbanism problematizes the institutionalized practice of zoning and rational governmentality inherited from modernism. The idea of a just city undermines the regulatory and financial foundations of commodified housing (Uitermark, 2009). Finally, a different future of the relationship between cities, bodies, and nature radically questions the institutions that regulate human-nature metabolisms within contemporary capitalism (Kaika and Swyngedouw, 2014). Planning history is marked by continuous struggle between existing institutions and the search for spaces of imagination.

This tension between imagination and institutions can be addressed in different ways, through different modalities and planning approaches. An approach to imagination is yet never neutral but always symptomizes the balances of powers in the process of urban change. In discussing modalities of future-making, Groves has specifically categorized practices of building futures into different "styles of anticipation" (Groves, 2016). Styles differ according to the type of rules and actors that are mobilized in the imaginative process. Each approach involves an assemblage of performative instruments, such as laws, architectural artifacts, financial means, narratives, plans, concepts, and other semiotic elements, that allow a particular vision of the future to gain "salience" *vis á vis* other possible ones. It is therefore crucial to explicitly question why certain styles gain particular legitimacy and political power at one point in time. This means to question the politics of future-building, uncovering the modalities, actors, and approaches through which particular imaginative practices are given capacity to act legitimately as producers of future.

Not all planning practices are given legitimacy in contemporary cities. Politicians, planners, citizens, and corporate actors recognize some practices as *capable of* generating new imaginaries, while they may disqualify others as irrelevant or even dysfunctional to prosperity or urban growth (Purcell, 2016). At the base of this argument lies the assumption that cities are dynamic social entities where imagination can occur in different forms any time a particular group of agents begins to reflectively problematize their present condition and its future change (Graham and Healey, 1999). Yet, the capacity of these imaginative practices to institutionalize new imaginaries by changing physical and regulatory frameworks is not evenly distributed. Some practices may be marginalized and eventually left to die while others may be nurtured and sustained actively through concrete monetary and legal means by governments and urban development corporations. The act of recognizing a particular process as potentially imaginative of urban futures represents the first discriminatory act of an institutionalized imaginary: a particular actor or group of actors is given the "power to make the future" (Urry, 2016: 11) and it is identified as a legitimate organizer or enabler of the future-building process. This means that, while being highly imaginative, some practices of imagination may have no impact on the change of consolidated institutions and their underlying imaginaries.

The main challenge for organizing imaginative practices is thus, firstly, to capture the imaginative potential of social practices to address urban problems and, secondly, to provide the conditions, instruments, and settings that enable emerging imaginaries to change institutional frameworks. A decade ago, Sandercock (2004) urged planners to be more audacious, creative, and therapeutic in order to be open to radical imagination in the context of a rapidly changing urban society. In her words, imagination needs actors that are able to reflect on their needs while at the same time engaging with new roles, acts, languages, and meanings. Today, planning scholars agree that planning needs open communication, deliberation, contextual interaction, serendipity, improvisation, and gaming to produce new futures (Healey, 2009). Building on these insights, planners have tried to diversify the repertoire of modalities to enable imagination. Since the 1990s, strategic thinking has been one of the most diffuse ways to navigate the dilemma between institutional boundaries and the urgency to enable imagination. The search for "strategies" rather than for "blueprint plans" became a powerful approach to guide processes of city-regional development without frustrating the possible imaginaries of place (Albrechts et al., 2017). Strategies have been understood as the 'drawing out of a sense of potentialities and possibilities from multiple unfolding relations, within which to set actions that will intervene in these unfolding relations in the hope of further particular objectives and qualities' (Healey, 2007: 30). Spatial visions, rather than zoning, have been used to mobilize public, private, and civic actors to set a shared idea of the future, an idea that sets a common frame for subsequent actions and interventions in space.

This legacy of strategic spatial planning is in permanent transformation as planners are in search of new approaches to cope with increasing socio-environmental complexity and to protect public values from corporate interests. In the aftermath of the Global Financial Crisis, planners have been building a new repertoire of imaginative practices, motivated by the need to enable imagination in times of urban complexity and environmental-economic change. Despite the positive energy behind the search of novelty in the setting-up of imaginative processes, these new approaches, as I discuss below, show a problematic attitude to urban futures and imaginary production. Today's approaches to urban imagination appear to bypass the inherent tension between institutions and imagination by commodifying and objectifying the process of imagination. They turn imagination into an output to be crafted and delivered by an emerging constellation of agencies, specialized in the building of urban futures. Through these approaches, imagination is objectified as a goal, rather than a process, and the future is commodified as a service, rather than as an emergent result of collective action.

In the following, I firstly discuss the emerging profile of the future-building agency, a driving agent of imaginative processes today. Secondly, I critically reflect on one of the institutional set-ups within which future-builders operate, the so-called "urban living laboratory": these are crafted arenas of imagination guided by logics of experimentalism, instrumental design, and spatio-temporal isolation. I conclude by discussing the inherent problems in these two set-ups and commenting on the risks and potentials of these practices in light of the urgency to rethink contemporary processes of urban change.

The future as service: the emerging landscape of future-builders

Today some may retrieve the modern term of "futurologists" or "futurist" to distinguish experts of the future from experts of the present. For the sake of this chapter, I will use the term "future-builder" or "future-building agencies" to refer to those agencies or set of agents that are specifically concerned with the building and designing of possible urban futures. In the aftermath of the Global Financial Crisis, the future has been increasingly "corporatized" (Urry, 2016: 11), and turned into a commodity to be produced or into a service to be commissioned by companies, public agencies, and civic actors. This development results from an increasing crisis of legitimacy of public governments, a rising public awareness over the unknown impacts of the economy on the future of cities, and the widespread belief that progress depends on expert technological advancement. Within this context, the constellation of the planning actors recognized as legitimate by governments, civic groups, and corporations has expanded to include a broad set of experts, from data scientists, ICT programmers, private accountants, and experts that are given the legitimate task to organize action oriented to trace, if not even predict and realize, the changing future of urban spaces.

The emergent profile of *future-building* agencies differs from the classic role of the planning technocrat or of imaginative political activists. Futurologists are not necessarily planning experts but they do influence planning processes by taking a direct stance on those processes. Not rarely, agencies advising on the future are spin-offs of architectural firms, combining the use of planning methodologies, such as a scenario-making and stakeholder analysis, with methods of visual narration, marketing, and dramaturgy, art, and serious games (Poplin, 2012). Future-builders claim an expertise that allows actors to catalyze energy around the building of a collective and comprehensive image of the future. They do not define themselves as political activists but offer support for building an "objective" and yet inclusive view of the future. Future-builders are not technicians of policy-making regulations and administration, but instead demonize mechanical bureaucracies. On the one hand, they often claim that regulations are boundaries to the imagination, yet, on the other, they try to instrumentalize these regulations selectively to enact or constrain particular processes of interaction. They practice what Peck has defined as "urbanology" (Peck et al., 2013), an original combination of celebratory narratives of urban growth and careful use of statistical tools to simulate scientific arguments (Peck, 2016). Future-building agencies often provide an attitude, a mindset, or a guiding style for institutional change. They advocate free imagination, emancipated from what Max Weber had epitomized as the "iron cage" of bureaucracy. The claim for a radical imagination is accompanied with the ambition to move planning processes into the private sphere of the public opinion-making. The makers of the future are experts that supposedly bridge the gap between the public, private, and civic spheres of planning.

Future-building agencies encapsulate a celebration of imaginative-utopian activism and an attitude of demonization of the public planner as bureaucrat. In the exercise of this expertise, the future of cities is often intended in an idealized form, while the future-maker is conceived as visionary, engaging in risky ideas, and realistically pragmatic, applying existing technologies in context. Following this ethos, future-makers often attempt to mobilize narratives of radical change, remarkably optimistic of the opportunities entailed in the use of emergent technologies and entrepreneurially innovative urban practices. In this sense, the makers of the future engender a seemingly "populist" (Swyngedouw, 2009) attitude towards planning imagination: the value of citizens' inclusion is celebrated as key while the role of corporate actors and governments are held to be necessary. The future is depicted as catastrophic while being simultaneously envisaged as a positive "haven" (Law and Urry, 2004) through a dramaturgy that aims at motivating the audience. At the same time, however, there is a lack of critique of the exclusionary potential that a particular vision may entail. In general, conflicts are strategically avoided and carefully organized through methodologies of negotiation and conflict management.

The current popularity of urban future-building is exemplified in the way smart-city policies have been supported in the city of Amsterdam. In 2016, the European Commission elected Amsterdam as the European capital of innovation (i-Capital). This was the result of a long-term process of coalition-building around the image of a city that is "connected", "sustainable", "green" and "prosperous". In 2014, the newly elected government of the city promised to take seriously the challenge of urban sustainability, instituting an alderman for this and directing political and financial resources to the development of a smart-city policy. This idea draws on principles of technological progress in the field of urban development and mobility, the synergy between major economic players in the city, in the fields of telecommunication, energy, mobility, and security, and small-scale innovative ideas emerging from the creative industrial sector.

Several planning regulations, regarding mobility, security, and energy management, had to be adapted in order to facilitate this synergy in practice. This was not only a technical endeavour. It also entailed a deeply symbolic and discursive exercise of building a coherent idea of the city as a powerhouse of a technological future, and these narratives are crafted, spread, and carried by a constellation of professional agencies (Kitchin et al., 2017; Söderström et al., 2014).

Today's practitioners of urban future-building include a handful of players at the interface between the public administration, the corporate sector, and the cultural industry of the city. As early as 2008, the first "Amsterdam Smart City" platform was initiated by the city of Amsterdam, in cooperation with energy and telecommunication infrastructure providers in order to create a show-case for investments in the field of ICT. Among the proponents of these initiatives there is a constellation of agencies, specifically active in stimulating the use of ICT in the urban context. One of the key players is the chief technology officer (CTO) of the city of Amsterdam, directly working under the mayor and therefore independent from specific aldermen's political mandates. He is tasked with creating links between the business community, the government, and the civic society around all that regards technology and urbanism. The CTO has the mandate to build synergies and networks around a policy of a smart future, and to connect imaginaries of urban change with institutional resources to carry out concrete projects of urban innovation. A wide range of future-building agencies revolve around the work of these institutional subjects, especially in the field of design and architecture. They include visionary studios of socio-technological and spatial innovation working on urban development projects. ARUP, based in London, is a prime example of an international consultancy firm, and is involved in the visionary design of the island of Zeeburg, East of Amsterdam, and of the Amstelkwartier, south-east of the centre. These firms build futuristic imaginaries of place and at the same time advise on the architectural solutions for urban projects (Figure 31.1).

The future-building sector usually involves a heterogeneous constellation of firms, agents, and individuals, ranging from consultants, decision-making analysts, cultural

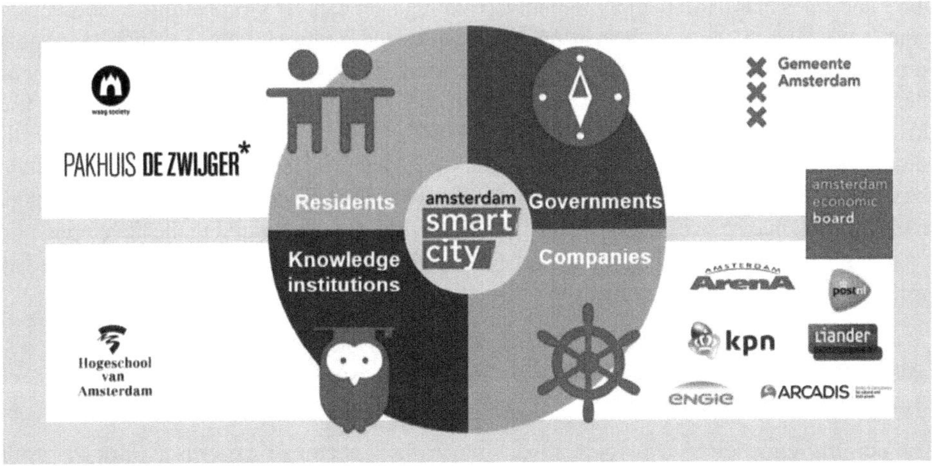

Figure 31.1 Model illustration of the Amsterdam Smart City Platform. Image courtesy of the City of Amsterdam.

activists, and independent research agencies. Despite their differences, these actors share the same working logic: to support governments and corporations in designing innovative and realizable futures. International future-building companies working in Amsterdam include *Changeist, Bits and Atoms* (by Townsend, author of the book *Smart Cities*) and *Sustainism,* a one-man project of envisioning techno-driven minimalist urban living. Cultural institutions in Amsterdam are also active producers of future imaginaries of society-city interaction. A remarkable example of this tendency is the *Pakhuis de Zwijger,* which has played a key role in the sketching of a charismatic narrative of political activism for eco-living in times of technological progress.

Today, this cultural hub of Amsterdam has become a key place where new discourses are showcased and where international thought-leaders are inspiring the broader public. The work of these institutions is mostly related to the building of a city-wide network of urban innovators and actively supporting existent urban projects. Building on ideas of virtual democracy and social data, these actors build a narrative that connects technological progress with democratic engagement at the neighbourhood and regional level (Franke et al., 2015). In the field of socio-technical research, *The Waag Society* is another crucial actor in the building of Amsterdam as an innovative hub of ideas and future imaginaries. Adopting a broader perspective, it is possible to trace a much wider network of agencies who act as experts in providing educational support for future-builders, such as the *School of Foresight*, the *Future Consult* advisory firm, and the network of Dutch futurologists, the Dutch Future Society (2016). All of these actors are networkers in the process of future-making and they are recognized by local institutions as legitimate experts in the process of reimagining the city. It is possible to see the leaders of all these agencies holding the award of the European Capital of Europe together on the event's website (Municipality of Amsterdam, 2017).

The emergence of future-builders as a particular form of imaginative agency reveals a double distortion of the idea of imagination in contemporary planning. Firstly, the process of imagination is associated with a specific task or service, something that can be provided, carried out, and dispensed depending on the specific demands of a particular context. Future-building agencies are neither public nor corporate actors in the strict sense. Their (apparent) freedom of foresight, audacity, and creativity springs from their autonomy from the procedures and rules that instead apply to public planning agencies. They constitute a landscape of experts working at the interface of public, private, and civic projects. Through systems of formal partnership or through public opinion-making, the visions that these actors convey are likely to have an important impact on how both corporate and public funding is allocated for urban development projects. Secondly, by crystallizing future-building as an expertise, the future becomes an actual object of assessment, the actual product against which the quality of the service can be assessed. As I show below, the growing salience of future-building agencies in the promotion of processes of imagination of the future is embedded in the development of a particular repertoire of planning approaches, guided by a logic of urban experimentalism, instrumental design, and spatio-temporal isolation.

Laboratories of the future: experimentalism, instrumental design, and institutional isolation

The current tendency to constitute urban futures as a sector of expertise, I argue, results from and simultaneously legitimizes a particular approach to urbanism oriented to *crafted imagination*. The term "living laboratory" has gained widespread international currency in recent years and it is largely used to group a broad set of practices in the field of smart urbanism, business development, urban creativity, and transition management (Mensink et al., 2010). There is no single definition of the term, especially in its urban use, but it is usually deployed to refer to both a methodology and a space of decision-making that is oriented to innovation. It is depicted as an "arena" or space for interaction, oriented towards the (co)production of the future as well as a specific approach of stimulating socio-technical innovation through place-based interaction between humans and technologies (see Evans et al., 2016). Its first uses in an urban context dates back to 2006, when the Finnish president of the EU instituted the European Network of Living Labs, with the aim to stimulate a wave of European, national, and local policies to test new technologies in tackling urban problems, and to match aims of business development with policy-making. The EC defines living labs as instruments that:

> ... stimulate new ideas, provide concrete research challenges and allow for continuous validation of research results. At a pan-European level, a large-scale network of living labs could become a strong tool for making the innovation process of industry more efficient and dynamic by stimulating the involvement of citizens of differing cultures and societal backgrounds who can provide rich feedback in context on the use and impact of the technologies being researched.
>
> (*EC, 2006*)

This particular definition of living labs encapsulates a peculiar mix of assumptions underlying the making of labs: the goals of digital and technological prototyping are combined with the ambitions of industry-driven and citizen-supported innovation. Despite its origins in the field of the industrial R&D sector, the term "laboratory" has increasingly appealed to a wider audience of urban activists, governments, and professionals. To capture the reasons for this trend, it is useful to trace the genealogy of living labs across the recent history of urban innovation and creative development in the city of Amsterdam.

At the turn of the century, following the expansion of regional knowledge and creative economies, the municipality had established programmes for reusing vacant spaces as areas of creative and artistic engagement and businesses. The 2001 *breeding places* policy (*broedplaatsen*) was a programme allowing the use of vacant estates as incubator spaces for all kinds of start-ups related to an idea of creativity, artistic expression, and knowledge production. These spaces included derelict buildings and vacant office towers turned into locations where artists, creatives, and entrepreneurs in the entertainment and knowledge sectors could settle. A few years later, in the aftermath of the housing bubble, the pressing vacancy of real-estate and land pushed the city to extend the same principles to other locations and fields of urban policy. The program *Free-state of Amsterdam* (*Vrijstaat Amsterdam*), initiated in 2012, was aimed at providing rent-free, vacant space for creative purposes based on a call for projects that were imaginative, experimental, and innovative. At the time, the catchphrase of the "self-organizing" city provided an umbrella narrative for different projects to mobilize inhabitants and their businesses in imagining new uses for urban space. Among the projects financed at that time were initiatives that became benchmark of futuristic imagination, such as *De Ceuvel*, a project of circular economy in the North of Amsterdam. The success of these practices, in combination with the rising relevance of the high-tech industry and smart-city discourses in the city, stimulated the proliferation of similar practices. The *Amsterdam StadsLab* (literally City-Lab), funded in 2013 by the national fund for creative industries (*Stimuleringsfonds Creatieve Industrie*), designed by the *Pakhuis de Zwijger*, represented one of the earliest attempts to systematically integrate a multiplicity of creative practices in the city under an overarching "movement" of experimentalism, where best practices of imagination could be showcased. Not yet formalized into policy, these practices set the early foundation of a large coalition of actors in the field of creative urbanism, experimental urban intervention, architecture, and technology (see above).

Throughout the last five years, living labs have proliferated in the city, and many of those take place in the locations that once were targets of creative use of vacant space. The Buiksloterham Living Lab is perhaps one of the best-known examples, where local home-owners are taking the lead in establishing an online community dedicated to "hacking" the system and designing new forms of urban living. At the Waag, The Amsterdam Living Lab.nl is another platform to collect and methodologically systematize a whole range of urban experimental practices and includes research institutes such as University of Amsterdam and the Amsterdam Metropolitan Solution Institute. In a different fashion, yet via the same concept, living laboratories are also mushrooming in the less wealthy areas of the West and South-East social housing districts. In these areas, projects of technological innovation and energy transitions are coupled with initiatives of community building and social cohesion (i.e. Energiek Zuidoost and City-Zen Nieuw West). Here, corporate actors (Cisco, Philips, Amsterdam Arena), research institutes (Amsterdam Polytechnic School, HvA), public authorities (City Police Department and Spatial Planning), ICT designers, local inhabitants, and visionary architects are prototyping alternative designs and urban forms. The power of the laboratory concept is epitomized today by the City-Lab (*Stadslab*) projects by the Museum of the City of Amsterdam (Amsterdam Museum, 2017), which involves pupils in the imagination of Amsterdam's urban future.

Despite the variety of initiatives covered by the label of "lab", the increased diffusion of this term allows us to trace an emerging logic through which planning processes organize imagination in cities. Firstly, laboratories commit to a logic of *experimentalism*. Living Labs are often described by their proponents as "test-beds" for alternative solutions to urban problems. Labs are instruments through which a particular set of actors (see my second point later) wishes to organize a process of collective imagination *and* action, which may have unknown results in the future. Laboratories entail a start-up culture of entrepreneurialism built on an epistemology of pragmatism to urban problems (Evans and Karvonen, 2011). The participants in these initiatives aim to find prototypes of solutions to problems that are identified in context. They are driven by the assumption that urban action needs to engage with concrete problems in those locations where these problems are the most evident and possible to be addressed (Voytenko et al., 2016). Emphasizing pragmatism, they seek tailor-made solutions and deliberately set up processes that are protected against ideological, political, and institutional constraints. However, this pragmatism at the same time contradicts their long-term expectations towards experimentation and solution-finding: on the one hand, living laboratories are presented as experiments that target site-specific problems; on the other, the same laboratories are set up with the ambition to design solutions that can be up-scaled or replicated somewhere else.

A second distinctive logic of laboratories is their *instrumental design*. Experiments may fail. They can result in something that cannot be up-scaled, or in nothing at all. Yet, in the organization of living laboratories, actors deploy a repertoire of techniques aimed at reducing these risks. The laboratory is an "environment", an "ecosystem" a "milieu" (Ståhlbröst, 2012), which is constructed or crafted. It is the product of a careful, methodologically systematic, and expertise-ridden organization, and it is instrumentally "designed" to the enablement of a particular social dynamic. Its design is a necessary condition for reducing the economic, political and social uncertainties that are inherent in any type of socio-spatial experiments. The repertoire of lab designers includes contingency plans, rules of conduct, specification of the roles of governmental, public, private actors, and, most importantly, a particular set of instruments and tools used to stimulate interaction. The use of a particular technology actually becomes the central pillar of the social dynamic, the point of reference for the whole process. The participants themselves are often preselected, based on a careful determination of the target areas where they take place (see my third point below). Future-building agencies provide these types of expertise, often presented as transition management skills. There is thus a second contradiction in the use of a living laboratory: the search for a surprising outcome of the experiment is coupled with an expert-driven crafting of the interactions in the present. Broadly inspired by methodologies of industrial design and product development, urban living laboratories build on a perspective of techno-driven imagination oriented to a substantive idea (to which a process is just a tool) of a future state of the city. Understanding the future as the end-object of a process turns imagination into a mechanism to be designed, a framework that is artificially abstracted from and deployed onto the ongoing social processes embedded in a particular place.

Experimental rationalities and instrumental design are necessary conditions to define a planning practice as an urban living laboratory, but not sufficient in and of themselves. The third characteristic of living labs is their *spatio-temporal isolation* from broader institutional, political, and social dynamics that may constrain, frustrate, and suppress processes of future-building. In the field of transition theory and evolutionary economics, the term "niche" is used today to refer to the bounded space and time of a potential innovative social practice (Geels, 2010; Kemp, 1994). In the context of urban policy-making, "niche" is mobilized to stress the need for safeguarding and nurturing an emergent dynamic from its environment (or regime), providing a safe space where alternative practices and imaginaries can emerge (Bertolini, 2018). Urban living laboratories have been broadly inspired by this evolutionary terminology, particularly in the field of sustainable transition management (Karvonen and van Heur, 2014).

This framing of imagination sets up the production of new imaginaries in some sort of tension with institutions. Yet, while this tension may be exactly what constitutes planning, the combination of this particular perspective with the abovementioned characteristics turns out to be problematic when understood prescriptively or normatively for the design of imagination (Caprotti and Cowley, 2016). Within an apparent dichotomy of imagination vs. institutions, imagination is situated within a spatially and temporally defined event (or ecosystem). It is thus the artificial spatial and temporal boundedness of experimental imagination that distinguishes a laboratory from other possible forms of imagination. The imaginative process is insulated within a preselected location and time horizon of the interaction. The autonomy from the context of rules, regulations, and physical-infrastructural constraints appears to be the way to allow planning to be effective in experimenting with alternative imaginaries and in setting-up new interactive processes. The laboratory is built on a state of "exception" from the ordinary planning process, oriented to the *here* and *now* of urban problems.

In the course of history, imaginative planning practices have been generally characterized by their orientation to the future, and often carefully organized in specific institutional settings. Moreover, planning always has spatial and temporal dimensions, whether it is concerned with a neighbourhood, a housing block, a public park, or a city-region. The three logics described above are not exclusive per-se to living laboratories or necessarily problematic in light of the urgency to define and enable radically imaginative processes. Urban living laboratories, however, show a peculiar combination of those logics. In contrast to other forms of participatory planning, laboratories approach the future as "something" to be constructed and discovered as a solution to particular problems. While they stress that the future is generally unknown, full of unexpected possibilities, and often surprising, they build on a careful crafting of socio-technical processes. While they seek radical innovation in social and technical dynamics, they often follow a defined set of aims and results. Often, it is on this basis that they are assessed and eventually funded or replicated. Their instrumental design is at odds with their experimentalism. Laboratories include a specific indication of the technologies and instruments to be applied (e.g. a platform, an app, a set of databases), functional to their expected outcomes – whether a new rule, a new technology, a building, a piece of architecture, or a metabolic park. The objectification of the imaginative process, regarded as "something" to be crafted, distinguishes these planning processes from others, where, for example, the aim is to build the social, cultural, and legal conditions for a long-lasting imaginative process to occur. Because of their instrumentalism, laboratories are also place- and time-specific. They have a beginning and an end. They are created as insulated niches, as states of exception.

The limits of crafted futures

Can imagination be crafted? Can an urban future be built in a lab? To what extent can processes of imagination be set up? Throughout history, urban planning has developed a broad repertoire of techniques, instruments, set-ups, and approaches to address those questions. Ultimately, the answer to those questions depends on the particular understanding that planners have of their role and on the underlying assumptions that they mobilize in organizing processes of imagination. At the beginning of this chapter, I have conceptualized planning as the practice of organized imagination, namely the production of imaginaries, that takes place within institutionalized imaginaries of cities. It has been pointed out that, as an embedded process of imagination, planning often takes place in tension with existent imaginaries, institutionalized in physical spaces, material rules, legal codes, and economic frameworks of urban development. Based on this assumption, I have discussed and criticized one particular way in which contemporary planning policy addresses this tension: the creation of "safe spaces" for experimental design and the simultaneous professionalization of future-building processes. Laboratories are set up in order to address the apparent hostility between existing institutions and the quest for social imagination. They do this by circumscribing imaginary practices within bounded spaces and times, instrumentally designing processes of interaction. Future-building experts combine a methodological knowledge of decision-making and human-technology interaction with a charismatic attitude to the exploration of an unknown future. In this chapter, I have argued that the laboratory embodies the imagination of the future as a substantive goal, namely an end product for which a particular expertise, a set of tools and technologies, and a specific location are needed.

Imaginative planning practices in contemporary cities are not limited to the initiation of laboratories, and imagination is not the exclusive responsibility of future-building agencies. Practices of radical imagination, tactical urbanism, and counter-cultural movements still offer a broad range of imaginative practices that attempt to redefine contemporary imaginaries of cities and urban change. Urban laboratories are diffuse, yet not everywhere and they are often complementary to ongoing planning processes in cities. Nonetheless, this chapter suggests that the underlying logic of urban laboratories is symptomatic of a particular governmentality of cities in times of techno-driven planning. They stand for an emerging constellation of actors, agencies, and funding mechanisms, which are likely to determine the formulation of urban, environmental, social, and economic policies of cities in the future. The rapid proliferation of urban living laboratories throughout Europe shows the economic and political success of a widely recognized methodology of imagination. It consolidates a planning approach founded on a novel mixture of socio-technical design, business entrepreneurialism, and techno-solutionism of the future. These practices ultimately convey an objectified understanding of the future, as a product rather than a process, diverting attention from the underlying problems of contemporary urban democracies.

The risks of living laboratories may lie more in the way laboratories are designed, funded, and carried out than in the actual principles of experimentalism and contextualization. The challenge of contemporary planning practice and research is, first of all, to avoid adopting a narrow understanding of imagination as something that can be instrumentally organized or designed. Imagination may be already happening under the radar of governmental policy-making or corporatized industrial innovation. This is a necessary condition for planning to appreciate the value of all kinds of non-organized or counter-organized movements of imaginary production. It may require a much broader understanding of builders of the future beyond the future-building agencies. Secondly, it is vital that planners and governments are aware of the biases that the notion of the "laboratory" entails. Experimentalism builds on an understanding of imagination as something that can be tried out and tested. Such a view inevitably suffers the limit of context, which biases experiments to local and particular conditions, missing the broad need of new global economic and social imaginaries. Instrumental design may instead impose biases to interactive processes and be prone to cooptation by powerful financers of laboratories. Design is never neutral, and the network of future-building agencies shows how the legitimate capacity of narrating the future is not equally distributed. Design is a way to enable and simultaneously limit imagination, perpetuated by setting a language, a semiotic, a particular symbolism of the discourses and technologies that are employed in the process. Finally, and most importantly, the spatial and temporal boundedness that the notion of the "testbed" entails inevitably avoids understanding imagination as a continuous process, which starts in the present. It artificially suspends time and space to enable imagination, and it may underestimate the value of long-term processes of democratic inclusion. Paradoxically, the focus on the here and now of laboratories constrains the capacity of experiments to uncover the conflicts and contradictions between imagination and the present socio-economic order. This is a necessary condition for "truly new" imaginaries to emerge.

References

Albrechts, L., Balducci, A., and Hillier, J. (2017) *Situated Practices of Strategic Planning: An International Perspective.* New York: Routledge.

Amsterdam Museum. (2017) Online page Stadslab Available at: https://www.amsterdammuseum.nl/stadslab.

Appadurai, A. (1996) *Modernity at Large: Cultural Dimensions of Globalization.* Minneapolis: University of Minnesota Press.

Bertolini, L. (2018) What Can We Learn from Evolutionary Theory When Confronting the Deep Challenges of Our Times? In: Salet, W. (ed.), *Public Norms and Aspirations: The Turn to Institutions in Action.* New York: Routledge.

Bulkeley, H., Castán Broto, V., and Edwards, G. (2015) Climate Governance and Urban Experiments. In: Bulkeley, H., Castán, B.V., and Edwards, G. (eds.), *An Urban Politics of Climate Change:*

Experimentation and the Governing of Socio-Technical Transitions. London, New York: Routledge, 3–31.
Caprotti, F. and Cowley, R. (2016) Interrogating Urban Experiments. *Urban Geography*: 1–10.
Dembski, S. (2014) Structure and Imagination of Changing Cities: Manchester, Liverpool and the Spatial In-between. *Urban Studies* 52(9): 1647–64.
Dutch Future Society (2016) Event: Picturing the Future 2016, http://dutchfuturesociety.nl/event/picturing-future-class-2016/.
EC (2006) European Network of Living Labs: Human Dimension of Technology. Available at: https://ec.europa.eu/digital-single-market/en/news/european-network-living-labs-human-dimension-technology.
Evans, J. and Karvonen, A. (2011) Living Laboratories for Sustainability: Exploring the Politics and Epistemology of Urban Transition. In: *Cities and Low Carbon Transitions*. London: Routledge, 126–41.
Evans, J., Karvonen, A., and Raven, R. (2016) *The Experimental City*. London: Routledge.
Gaonkar, D.P. (2002) Toward New Imaginaries: An Introduction. *Public Culture* 14(1): 1–19.
Geels, F.W. (2010) Ontologies, Socio-Technical Transitions (To Sustainability), and the Multi-Level Perspective. *Research Policy* 39(4): 495–510.
Graham, S. and Healey, P. (1999) Relational Concepts of Space and Place: Issues for Planning Theory and Practice. *European Planning Studies* 7(5): 623–46.
Groves, C. (2016) Emptying the Future: On the Environmental Politics of Anticipation. *Futures*. DOI: 10.1016/j.futures.2016.06.003.
Healey, P. (2004) The Treatment of Space and Place in the New Strategic Spatial Planning in Europe. *International Journal of Urban and Regional Research* 28(1): 45–67.
Healey, P. (2009) The Pragmatic Tradition in Planning Thought. *Journal of Planning Education and Research* 28(3): 277–92.
Karvonen, A. and Van Heur, B. (2014) Urban Laboratories: Experiments in Reworking Cities. *International Journal of Urban and Regional Research* 38(2): 379–92.
Kemp, R. (1994) Technology and the Transition to Environmental Sustainability. *Futures* 26(10): 1023–46.
Kitchin, R., Coletta, C., Evans, L., Heaphy, L., and MacDonncha, D. (2017) *Smart Cities, Urban Technocrats, Epistemic Communities and Advocacy Coalitions*. Available at: osf.io/preprints/socarxiv/rxk4r.
Law, J. and Urry, J. (2004) Enacting the Social. *Economy and Society* 33(3): 390–410.
Madanipour, A. (2013) Researching Space, Transgressing Epistemic Boundaries. *International Planning Studies* 18(3–4): 372–88.
Municipality of Amsterdam (2017) European Capital of Innovation, https://www.amsterdam.nl/bestuur-organisatie/volg-beleid/innovatie-0/european-capital/.
Peck, J. (2016) Economic Rationality Meets Celebrity Urbanology: Exploring Edward Glaeser's City. *International Journal of Urban and Regional Research* 40(1): 1–30.
Peck, J., Theodore, N., and Brenner, N. (2013) Neoliberal Urbanism Redux? *International Journal of Urban and Regional Research* 37(3): 1091–99.
Poplin, A. (2012) Playful Public Participation in Urban Planning: A Case Study for Online Serious Games. *Computers, Environment and Urban Systems* 36(3): 195–206.
Purcell, M. (2016) For Democracy: Planning and Publics without the State. *Planning Theory* 15(4): 386–401.
Sandercock, L. (2004) Towards a Planning Imagination for the 21st Century. *Journal of the American Planning Association* 70(2): 133–41.
Savini, F. (2016) Self-Organization and Urban Development: Disaggregating the City-Region, Deconstructing Urbanity in Amsterdam. *International Journal of Urban and Regional Research* 40(6): 1152–69.
Savini, F., Salet, W., and Majoor, S. (2014) Dilemmas of Planning: Intervention, Regulation and Investment. *Planning Theory* 14: 296–315.

Söderström, O., Paasche, T., and Klauser, F. (2014) Smart Cities as Corporate Storytelling. *City* 18(3): 307–20.
Ståhlbröst, A. (2012) A Set of Key Principles to Assess the Impact of Living Labs. *International Journal of Product Development* 17(1–2): 60–75.
Uitermark, J. (2009) An in Memoriam for the Just City of Amsterdam. *City* 13(2–3): 347–61.
Urry, J. (2016) *What Is the Future?* London: John Wiley & Sons.
Voytenko, Y., McCormick, K., Evans, J., and Schliwa, G. (2016) Urban Living Labs for Sustainability and Low Carbon Cities in Europe: Towards a Research Agenda. *Journal of Cleaner Production* 123: 45–54.

32

Urban space and the posthuman imaginary

Debra Benita Shaw

Introduction

In 2005, the Italian philosopher Giorgio Agamben gave a lecture in which he suggested that the processes of subjectivation in the contemporary city were becoming increasingly opaque. Currently, he suggested, we do not know enough about what he calls the "the dispositif or group of dispositifs" which construct contemporary urban subjects (Agamben, 2005). Here, he is following Michel Foucault in proposing that we need to understand how the deployment of institutionalized knowledges operates to determine the parameters of self-understanding but with a specific orientation towards what he calls "metropolis": the city itself understood as a historically constituted apparatus of individuation. According to Agamben, we are lacking an analysis through which we may understand on what he calls the "ontological or Spinozian level" how the processes that structure an attachment to a subjective identity affect individual agency. As he puts it, "what, in the processes whereby a subject somehow becomes attached to a subjective identity, leads to a change, an increase or decrease in his/her power to act" (Agamben, 2005)? This chapter will address Agamben's question through an analysis of the historical conditions under which bodies and cities are produced as objects of study. What are the discourses that construct the *dispositifs* which determine the conditions for what or who has a right to the city? How does built space both reflect and perpetuate these ideas? How might this knowledge be applied to understanding urban subjectivities and their expression in the cartography of urban space?

At the dawn of late modernity, Le Corbusier was insisting that his contemporaries must "measure afresh the consequences of being bodies", believing that architects must necessarily respond to a fundamental ontological change ushered in by the machine age, which would require a substantial revision in concepts of inhabitation. As far as Le Corbusier was concerned, in fact, it was the job of the architect to respond to the changing conditions brought about by industrial technology and design an environment in which a "new kind of consciousness" (Le Corbusier, 1964/1933: 36) would flourish. In other words, he was proposing an architecture that would actively produce a citizenship adapted to the new relations of production. Thus, in what Donna J. Haraway calls "the social relations of science and technology" (Haraway, 1991: 163), the built environment, as an expression of how these relations determine bodies and their right to inhabit urban space, is of considerable importance. My proposal is that we should return to Le Corbusier's proposition that we should "measure afresh the consequences of being bodies" under the new conditions ushered in by the information revolution. These conditions have not only provided for "smart" cities and the interconnected culture of global capitalism but have forced us to re-evaluate what it means to "be a body". How then might we evaluate the effects of this at the level of everyday life? And, more importantly, how might this knowledge be applied to addressing the escalating inequalities of global cities in the third millennium?

Modern cartographies of urban space

The organic body produced by the discourses of nineteenth and twentieth-century biological sciences knew its place. The hierarchical arrangement of species bequeathed by the work of Carolus Linnaeus in the eighteenth-century informed the development of social Darwinism through which emerged, not only taxonomies of human "types" but a cartography of distribution which provided much of the justification for colonialism. The *flâneur*, Charles Baudelaire's Parisian stroller and observer of the everyday life of the city, "botanizing on the asphalt" (Benjamin, 1999: J82a,3, 372), was a figure whose characteristic activity enabled this cartography to be brought home to the metropolis. He offered the assurance that the teeming life of the city could be mapped, classified, and brought under calculative control (Frisby, 1994: 86). As Walter Benjamin astutely observed, he was both a detective and a journalist, able to read the text of the city, identify its characters, and, through selling his findings to the newspapers, provide assurance to the bourgeois consumer of their superior position in the ranks of urban dwellers (Benjamin, 1999: M13a,2, 443 & M16,4, 447).

A similarly hierarchical cartography can be seen to have informed the urban sociology of the Chicago School in the early twentieth century. In general, the Chicago researchers understood the city as a "pseudo-biological organism" (Soja, 2000: 86). In this model, the processes of urban life were conceived of as "natural" and thus as amenable to scientific analysis as a rock pool. Many of the Chicago luminaries were influenced by the social Darwinism of William Graham Sumner (Parker, 2004: 40), whose famous dictum "root, hog or die" (Bowler, 1983: 302) expressed his understanding of social competition as the process through which human improvement was to be realized. The Chicago method mapped the city as a series of hierarchically arranged zones distinguished by significant patterns of behavior, levels of income, and population density, with movement through the zones conditioned by successful negotiation of the capitalist economy. Nineteenth-century organic metaphors applied to urban space provided the conditions under which this kind of scheme would make sense. The zones associated with the institutions of the state and the living arrangements of the bourgeoisie were correlated to the head or "mind" of the city, while the sewers and slums were associated with the organs of defecation and reproduction (Stallybrass and White, 1986: 145). Thus, the city produced as an idea through the conjoined discourses of bodily health and evolutionary biology could not only be diseased or infected but could be subjected to pseudo-medical techniques of quarantine, purging, and invasive surgery, with the bodies that inhabit it assimilated to models of viral invasion and deformity. Schemes for "cleansing" the city were thus developed on the basis of identifying, making visible, and eliminating forms of life that were deemed not to meet the conditions for human flourishing.

Indeed, European and American cities in the late nineteenth and early twentieth centuries were the stages for exhibitions of colonized and enslaved peoples as exemplary not-quite humans; object lessons in the identification of deviant physicalities. The most famous cases were Saartje Baartman, known as the "Hottentot Venus", who was paraded around Europe as an example of a supposed degenerate femininity (Gilman, 1985: 85) and Oto Benga, a Congolese pygmy who, in New York in 1906, was displayed in a cage with an orangutan (Sewell, 2009: 1). Late-nineteenth-century zoological gardens often exhibited whole villages of supposed "primitives" whose behavior was studied as examples of a less advanced stage of evolutionary development (Gilman, 1985: 110). These exhibitions performed segregation disguised as education, ensuring that people would monitor their fellow citizens for any taint of the "primitive". And, in the US, the so-called "ugly laws" explicitly prohibited persons whose performance of self did not conform to what was deemed normal from appearing on the streets until well into the late twentieth century. "[U]gly laws", writes Susan Schweik, "are part of the story of segregation and profiling in the United States, part of the body of laws that specified who could be where, who would be isolated and excluded, who had to be watched, whose comfort mattered" (Schweik, 2009: 184). Thus a medical model in which a race- and class-derived pathology is marked on the body and backed by the authority of science can be seen to produce modern social space at the same time as it determines paradigms of subjectivity.

In view of this, it is worth considering the location of institutions concerned with the body and its health and education in urban space. If the display of colonized and enslaved peoples was to be effective in demonstrating racial distinctions, then the dominant population required standards through which they could measure degrees of deviation. The role of hospitals and museums, which are to be found in all cities, generally housed in architecturally significant buildings, is particularly interesting. These, as I will demonstrate, are not only institutions for the production and dissemination of knowledge, but can be seen to deploy that knowledge as a mechanism for disciplining urban bodies. In the case of the modern hospital, this means both caring for the health of the population and educating them in the performance of what Zygmunt Bauman has called "separating operations" (Bauman, 1992: 155). These are techniques of segregation and control designed to protect bodies from contamination at the same time as they institute standards of normality and abnormality, health and disease and acceptable and unacceptable behaviors.

Of course, Michel Foucault (2003/1963) has written at length about classificatory medicine and the role of the hospital in the production and dissemination of medical discourse. But what I am interested in here is the specific architecture of hospitals, their siting within urban space and the way they function to impart a classificatory paradigm to the surrounding space. The separating operations, which make sense in the case of infectious diseases or accumulations of dirt that may harbor harmful bacteria and in which, in modern cities, bodies are trained from birth, have their origin in the topography of the hospital that was developed to both segregate hospital patients and subject them to regimes of medication and infection control. Similar techniques then contribute to the segregation of domestic space, "localizing families (one to a house) and individuals (one to a room)". As Foucault points out elsewhere, in nineteenth-century working-class housing estates, these arrangements ensured "a sort of spontaneous policing or control" (Foucault, 2004: 251). At the same time, separating operations are practiced diligently in the home where they become the measure of "good" and "bad" parenting and, in the case of women, are associated with sexual hygiene and "correct" performances of gender. For Bauman, these ideas find their symbolic expression in the consumption of cleaning technologies: "brooms and brushes, soaps, cleaning sprays, washing powders" (Bauman, 1992: 155). I would suggest they are further symbolized by the centrality of the kitchen in the design of the modern suburban home, which functions to condition family life through food preparation and consumption. Indeed, the function of the "ethnological" villages in nineteenth-century zoological gardens was precisely to demonstrate a deviant domesticity to which bourgeois practices could be compared. More recently, kitchens have become the focus for anxieties about correct nutrition and the micromanagement of separating operations aimed at policing the behavior of children. It is not difficult to see then how the hospital, as a representation of space (Lefebvre, 1991/1974: 45), is involved in the structuring of urban subjects through a discourse of pathology that separates according to racialized norms of health expressed through performances of sexuality, gender, and domesticity.

Tony Bennett (1995) has made a study of the natural history museum which reveals with startling clarity how it, like the hospital, functions to structure both the surrounding space and the identification of bodies within it. Indeed, as he demonstrates, the architectural arrangements of nineteenth-century museums were specifically oriented towards making visible both the artefacts on display and the visiting public. While the artefacts were arranged sequentially to impart knowledge of biological evolution as progressive and racial hierarchies as natural, the visiting public, and in particular the newly-admitted working class, were encouraged to conform to norms of deportment and behavior commensurate with an educated, middle-class sensibility and to police both themselves and others accordingly. What this effected was an identification of self according to the established authority of the biological sciences, alongside the training of the body in line with both a disciplinary architecture and an order of classificatory knowledge. This then can be seen to inculcate a form of performance in which the compensatory rituals of museum viewing are conflated with the determination of self in a taxonomic order. These rituals are then carried beyond the walls of the museum, serving to further establish the relationship between the organization of space and the identification of natural kinds. This is what Barbara Kirshenblatt-Gimblett calls "the museum effect". "[U]rban dwellers such as James Boswell", she writes, "reported that walking in the streets of London 1775 was 'a high entertainment of itself. I see a vast museum of all objects, and I think with a kind of wonder that I see it for nothing'" (Kirshenblatt-Gimblett, 1991: 410). Poor neighborhoods were favored by amateur ethnographers. Here, "[a]ny stranger could see openly on the streets what in better neighborhoods was hidden in an inaccessible domestic interior, in a closed carriage, or under layers of clothing" (Kirshenblatt-Gimblett, 1991: 411). The museum effect thus promotes a form of voyeurism in which the poorer neighborhoods function to establish the success of separating operations in the production of bourgeois subjects. The "natural history" of the city is established, made visible, and ordered accordingly.

What is significant here, then, is the way in which the discourse of the biological sciences can be seen to construct the cartography of urban space and to order the city in terms of a progressive ideology. Natural history museums did not necessarily intend to promote social Darwinism, but the effect of being exposed to a sequentially ordered reading of natural evolution in what amounts to a cathedral of knowledge should not be underestimated. The ugly laws, alongside the museum effect, can be seen to have established not only a hierarchy of acceptable and unacceptable bodies, but to manage bourgeois concerns about (particularly racial) degeneration alongside what were deemed inappropriate expressions of sexuality. Indeed, sexual degeneracy was often equated with physiological distinctions that were "discovered" to be prevalent among races thought to be less advanced in the "great chain of being". In this way, physical marks of degeneracy were established and the expulsion of the bodies bearing them from full participation in urban life could be justified by the authority of science and achieved through the kinds of mutual surveillance promoted in museum space.

What this amounts to is a constant sifting of bodies according to arbitrary and shifting standards of appearance, behavior, and deportment measured against an ideal that, fundamentally, does not exist. It is an axiom of feminist and post-colonial theory that the standards for full inclusion in the species category "human" are determined by comparison with the white European male who remains both the arbiter of difference and the unmarked body from which all other bodies receive their marks of deviation. While this, in terms of cultural representations of race and gender, is undoubtedly true, what is also clear is that a central concern and perhaps the driving force of scientific modernity has been to define the qualifications for human belonging. The Holocaust, one of the most pernicious crimes of the last century, remains the quintessential example of the violence inherent in this project and the escalation of anxieties that have attended it.

When, at the end of *The Order of Things* (1994), Foucault hopefully foresaw the death of "Man", he was signaling the demise of an ontological concept mired in the economic imperialism of the capitalist world order that had been legitimated by the scientific method applied to populations and social structures. "Man", however, was and always has been too indeterminate to pin under the objective lens, resulting in a series of approximations defined largely through invented categories of exclusion. "Man" then, is not only an unmarked body but an empty category that successive descriptive iterations have struggled to fill. He is probably best represented by Vitruvian Man, the Roman ideal of the perfectly proportioned male body that, as imaged by Leonardo, is an annoyingly persistent figure, often used to represent a vaguely defined universal humanity but, in fact, representing nothing at all. Vitruvius's "perfect" proportion is just that; an ideal that no body approximates, developed originally to illustrate the principles to be applied in the building of temples in ancient Rome. Vitruvian Man thus stands for both the transcendent promise of universal humanism and the idea that has conditioned the biopolitics of architecture. Hence Sven-Olov Wallenstein can claim that "the trajectory of architectural modernity can be interpreted as . . . a biopolitical instrument" (Wallenstein, 2009: 4) the goal of which is to "*produce subjectivity*" (Wallenstein, 2009: 20, emphasis in original).

I have referred to hospitals and museums as examples of how this process is caught up with the discourse of the biological sciences. There are, of course, many others, including, as I have suggested, the family home, a site of intense separating operations and a space that is still nominally private but is now thoroughly penetrated by surveillance in the interests of the child. As Roddey Reid has pointed out, "'family' and its liberal body have remained the very measure of the 'human'" (Reid, 1995: 190). However, contemporary moral panics about inadequate parenting, particularly among the poor, can be seen to have escalated in concert with more recent uncertainties about the integrity of the bourgeois body so necessary to the maintenance of the family ideal. The irony here is that the biotechnologies that are the product of convergences between information technology and the biological sciences, designed originally to finally determine the essence of what it means to be human through techniques like genome sequencing, are not only casting doubt on where the human ends and other forms of life begin. Under the pressure of markets for, for instance, sophisticated anti-ageing technologies, bodies and machines are being brought into ever closer entanglement. At the same time, as my further discussion will demonstrate, the new social relations of science and technology are significantly impacting how urban space is understood.

Posturban production

If the group of engineers and scientists collectively referred to as "transhumanists" are to be believed, human bodies are a problem to be solved by technology. For them, the advent of the digital age and the drive towards general artificial intelligence implies only one thing – that we are in the process of producing our successor species. Their plan, articulated by organizations like Humanity+, is to speed the process towards the point where we become immaterial; where what we thought was human becomes something other – a discorporate mind which, in full realization of the promise of enlightenment, need no longer fear the decay of the body and its end in death. There are various versions of the trajectory towards what Ray Kurzweil has called "the singularity" (Roden, 2015: 22; Thorpe, 2016: 96), most involving harnessing human intelligence to vast digital networks, employing the so-called NBIC technologies (nanotechnology, biotechnology, information technologies, and cognitive science) to finally transcend corporeality and become something that we can currently not even imagine. The singularity describes the point where "human" no longer applies to what we have become; where a new mode of being emerges, untethered from planet Earth.

This is not science fiction but a significant and well-funded project, supported by both governments and major corporations in several (mostly Western) countries (Fuller, 2011). The questions that it raises, in the context of a volume devoted to urban imaginaries, are many. For transhumanists, presumably, the city, along with the body, will no longer be a problem. Urban living, in this case, is to be understood as a necessary historical development where the concentration of capital in industrial and, later, information technologies during the twentieth and twenty-first centuries has provided for the promise of techno-salvation. The injustices of advanced capitalism and their expression in urban deprivation can thus be understood as sacrifices on the road to enlightenment. Leaving aside the as yet unsolved problem of isolating the physical properties of mind in the hope that it can be replicated in a silicon environment, the fact remains that the ideology of transhumanism is concomitant with the anti-urbanism of capitalist elites who commute by helicopter to high-rise apartments to avoid contact with the contaminating street (Graham, 2016: 99).

Fear of the body here is paramount. If the unpredictability of the body can be dispensed with, then fear of death – the one thing that wealth can never wholly guard against – is alleviated. At the same time, the promise is that the threat of other bodies, particularly the increasingly mobile and uncontrolled bodies of the poor and stateless, will no longer be a concern. Transhumanism aside, this is made explicit in the marketing of luxury housing developments which derive added value from their stringent security arrangements and self-contained facilities (Minton, 2017: 20–4). In a UK study completed in 2004, Rowland Atkinson and John Flint identified a series of "corridors" in urban space through which urban elites commuted in regular journeys between luxury apartments, shopping and entertainment facilities, and exclusive schools, immured in utility vehicles more suited to off-road competition than the school run (Atkinson and Flint, 2004: 888). The "danger" that is implied here and that must be avoided at all costs is the lived urban reality of the majority of the population. Like the world in 2019 as imagined in Ridley

Scott's celebrated *Blade Runner* (1982/1992/2007), cities are experienced as polluted by both the environmental effects of capitalism and its human and technological discards. Migration to the "off-world colonies" may not yet be possible but the Transhumanist Declaration (Humanity+, 2016–2017) includes a commitment to extraterrestrial migration alongside the technological "enhancements" that will usher in a new form of life that transcends both the urban and human condition.

This then is one type of urban imaginary for the twenty-first century and it is one that owes much to the discourse of urban lawlessness inaugurated during the early years of the industrial revolution, based in racial- and class-based anxieties and perpetuated throughout the growth of industrial and, later, consumer capitalism. Indeed, it is a discourse that has accompanied and largely structured the growth of the modern city and its institutions. It is a discourse, therefore, that achieves its effects through the dissemination of specific knowledges in urban space, marked by the architectonics of urban design. In other words, what I am suggesting is that urban architecture and design do not respond to a need for protection against the inherent violence of urban life but actively perpetuate it.

Furthermore, cognitive cartographies of urban space are increasingly communicated through screen-based media that reproduce the voyeurism of the museum effect through "gritty" drama and "reality" television, often juxtaposed with breathtaking views of the cityscape, making clear the separation between the city as an "experience" and as actually lived. This contributes significantly to what Sarah Chaplin and Eric Holding describe as the "posturban" city, where urban space is produced as a facsimile of itself as developers respond to the fiscal demands of tourism and commerce rather than the needs of its inhabitants (Chaplin and Holding, 2002). The marketing of cities through their representations on film and video, in digital gaming and advertising, as well as the staging of mega-events like the Olympics, contributes to a culture of urban cleansing in a drive to force cities, and the bodies that inhabit them, to correspond to their imaginary equivalents. In fact, it is probably not too much of a stretch to claim that cities *become* their imaginary equivalents to the extent that the removal of bodies that are out of place in the scheme can not only be justified but effected through architectural design as well as the kinds of draconian policing that has led to the disproportionate number of deaths of young black men at the hands of law enforcement officers in both the US and the UK (Townsend, 2017). In London, the politics of austerity combined with the sale of social housing sites to private developers has resulted in the displacement of people on low incomes to satellite towns, often miles from their work and their extended families (Minton, 2017: 81–94).

This then is part of the process that Mike Davis calls "imagineered urbanism" (Mike, 2007: 51) in which cities are engineered to appeal to what Anna Minton calls "Big Capital" (2017). This, in turn, provides for a sorting of bodies according to how closely they conform to the codes of behavior and appearance that the new imagineered spaces demand. In twenty-first century global cities, private spaces that look like public spaces have proliferated (parks and squares for instance). Here property law determines rules of access, rather than local ordinances. At the same time, new forms of heterotopia have emerged in the networked spaces of social media where everyday life is codified in terms of the spectacle and urban space is experienced solely through the cartographies of tourism and consumption. To give just one example, popular tourist hotels now provide "selfie spots" where tourists are able to locate themselves strategically in the frame with a notable landmark (Dinhopl and Gretzel, 2016: 136). "Tourists", write Anja Dinhopl and Ulrike Gretzel, "are . . . not looking through the screen at the destination, but at the screen to see themselves" (2016: 132). Shared and networked in real time, these are visual affirmations of the posturban subject, confirming emplaced prosumer identities at the same time as they enhance the spectacle.

This then goes some way towards answering Agamben's question. The group of dispositifs that constructed the modern industrial subject now emerge as techniques to contain the anxieties unleashed by the effects of urban living in the context of colonialism. The "botanizing" project of the *flâneur*, the museum effect, and the ugly laws are examples of how urban culture has evolved as an expression of these anxieties enacted through the control of bodies and their movements, informed by the discourses of the biological sciences. In the contemporary city, advanced capitalism has worked on the desire for knowledge of others as a means to define the self and thus has constructed the posturban in a realm *beyond* the street and its connection to lived reality. Imagineered urbanism is built for consumption and sold on the basis of disconnection from what threatens the integrity of the self, the body, the family, and, ultimately, the imagined community (Anderson, 1983) of the nation state. Transhumanism is this disconnection fully realized; a conflation of the discourses of the medical sciences with the "exiling of ... powers in a 'world beyond'" (Debord, 1995/1967: 18) identified by Guy Debord as the "religion" of the capitalist spectacle. Agency, in the contemporary metropolis, is thus conferred by successful separation from the life of the city, from the corporeality of other bodies and thus from anything that might be recognized as community.

The posturban, in fact, confers a dual ontology. Considering the late twentieth century migration of the *flâneur* to cyberspace, Kristin Veel (2003) offers a helpful distinction between labyrinth viewers and labyrinth walkers where the ideal is to be both; to both inhabit the city and take advantage of the overview perspective offered by information and communications technologies, such as Google Earth, drone mounted cameras, and video games like *Grand Theft Auto* (Rockstar Games) which offers the *frisson* of engagement with the street but with the safety controls firmly *on*. The reality is that imagineered urbanism offers monied elites a permanent exile in a pay-per-view utopia where the promise of these technologies is realized while labyrinth walkers remain as unpaid extras in a movie of their own lives. Witness the vast displacement of urban dwellers in cities where the Olympics have been staged. Olympic stadia, in fact, are the new architecture of separation, staging a spectacle of elite bodies whose characteristic modes of performance are emulated in everyday life. Monitoring here is through, again, informational devices that measure something called "health" as a quantity of steps, heartbeats, and calories. The function of the hospital as a "curing machine" which not only attends to sick bodies but inculcates disciplinary regimes has been digitized, miniaturized, and returned to the body as wearable technology.

Hardly surprising then that Matthew Gandy refers to "cyborg urbanization" to describe the way in which "the distinction between 'city' and 'non-city' becomes extensively blurred ... to produce a tendential landscape exhibiting different forms of integration between the body, technology and social practices" (2005: 41). Cyborg urbanization is an effect of postindustrial capitalism at work on the interrelation between urban infrastructures and urban subjects but, equally, as I have suggested and as Gandy also points out, it does not represent a break with the trajectory of nineteenth-century modernity, but is its realization in a new space which may be virtual but leaves nothing of the material of the city untouched. Cyborg urbanization, then, is one result of the profound changes in the social relations of science and technology brought about by the information revolution. It suggests that we might want to revise how we understand from what and into what we might be evolving and embrace a hybrid ontology in which the perfection of a putative ideal makes no sense. Nevertheless, despite this, as Rosi Braidotti puts it, "Vitruvian Man rises over and over again from his ashes, continues to uphold universal standards and to exercise a fatal attraction" (Braidotti, 2013: 29). In other words, the conditions exist through which new forms of subjectivity might be realized but the only future on offer seems to be based in the destruction of what we are for the sake of something we can never be. This was what Haraway was referring to when she offered the cyborg as a figure in which socialist-feminists of the late twentieth century might recognize themselves, in opposition to the nature-identified feminine subject, derived from the gendered suppositions of the seventeenth-century natural philosophers (Easlea, 1981) of much 1980s feminist discourse. Hence, I want to conclude here by offering some brief suggestions for how actively rejecting the human as a circumscribed category, and embracing a cyborg ontology, fully at home in the posturban and attuned to its as yet unrecognized potential, might free us to act effectively for a more egalitarian city.

Posthuman urbanism

That Haraway's argument has had major effects beyond its original constituency is testament to its potency in identifying the conditions for subject formation in general under the terms of informational capitalism and its attendant social structures. The cyborg is a figure for a posthuman world in which the biological sciences are increasingly focused on *making* bodies rather than describing them (Cooper, 2008) and the discourse of the informational sciences is becoming increasingly pervasive in the codes that describe life and the processes through which it is realized. Equally, it is a figure that makes sense of the everyday under the terms of cyborg urbanization and codes strategies for political organizing that take advantage of the way in which it is violent to the fixed determinations of space guaranteed by Vitruvian Man. Le Corbusier could dream an architecture that would both shape and accommodate bodies fully adapted to the necessary social conditions to realize the promise of modernity, but what might it mean to dream an architecture for bodies that, as Haraway says, do not "end at the skin" (Haraway, 1991: 178)? Or, perhaps more pertinently, what might be the consequences of accepting ourselves as always already posthuman; as having evolved *with* the technologies which have shaped our worlds and conditioned our ontology (Stiegler 1994, 2009)? At the very least it instructs us that the divide between nature and culture is meaningless and that, where there are no natural beings, there are equally no pre-determined forms or hierarchical taxonomies that structure the relation between bodies and space. Accepted divisions of gender, race, class, sexuality, and, more specifically, species emerge as arbitrary and open to challenge. Beyond this, the posthuman idea violates the sanctity of the institutions which have traditionally served the human ideal and have been largely responsible for its perpetuation. The isolated domesticity of the modern family and its association with private property, inherited wealth, and sanctioned reproduction, already under threat, begins to look distinctly alien to beings no longer invested in policing biological boundaries.

I do not mean to propose here a new architecture for posthumans or indeed a new form of urban planning. Tearing down and re-building, as accomplished by Baron von Haussmann in nineteenth-century Paris and, more recently, the "neo-Haussmanization" (Merrifield, 2014: xii) of global cities which makes imagineered urbanism a reality is, after all, not only labor- and resource-intensive but what continues to create and maintain urban hierarchies and divisive living conditions. My proposition is for a form of urban imaginary through which a posthuman politics can be realized that re-imagines and re-purposes what exists; which employs the kinds of aesthetics suggested by sampling and the ethics of code hacking. That is, it makes use of the morphological plasticity imparted to urban space by global information flows and the churn of market demand to, as Braidotti puts it "mobiliz[e] resources and visions that have been left untapped and . . . actualiz[e] them in daily practices of interconnection with others" (Braidotti, 2013: 191). Some of these practices already exist; others are yet to be invented.

Squatting, for instance, is a practice as old as private property and lends itself to protean re-invention under changing conditions (Vasudevan, 2017). The "movement of the squares" was essentially a series of squatter camps which, beginning with the 2011 occupation of Tahrir Square in Cairo, spread globally over the following three to five years (Gerbaudo, 2017). This was not squatting to alleviate homelessness (although it drew attention to the displacements effected by neoliberal capitalism) but an imposition of bodies in the spaces of the city where visibility is guaranteed which was co-ordinated, for the most part, by social media. It may, with hindsight, be understood as an expression of a nascent posthuman politics and not only because it depended on the co-presence of individuals simultaneously in urban space and the space of digital networks.

The fact that many cities have squares at all is a legacy of the Roman camp with its grid pattern determining a strict hierarchy of position and place (Betsky, 1995: 46). The central square was at the crossing of two major roads and was left deliberately empty. It thus provided an absent center that conditioned all the surrounding space and guaranteed its order. This, then, can be understood as the space of Vitruvian Man, representing the human ideal which real bodies can only ever approximate. Making home in this space, even if only temporarily, establishes the everyday in the heart of institutionalized space, challenges the human paradigm which keeps bodies in their ideologically allotted place, and confounds the separating operations effected by social architecture. Similarly, practices like parkour, which originated in the Paris *banlieues* and makes use of the verticality of city architecture to defy the restrictions of travel through the city at street level, and urban exploration, in which groups of practitioners delve into the world beneath the street and party in the spaces abandoned by urban "renewal" (Garrett, 2013) are glimpses of a nascent re-imagining of the city that defies the shepherding of bodies according to the dictates of imposed posturban cartographies. These practices have in common the deployment of advanced visual technologies and the use of social media through which the experience is communicated to a wider community. They can, in fact, be understood as new forms of psychogeography, which utilize the media of imagineered urbanism to defy the conditioning of bodies according to its spatial determinants.

Psychogeography, as employed by the Situationist International in the mid-twentieth century, was a mode of navigating the city intended to disrupt the imposed cognitive cartography of the worker under capital through creative meanders; to discover "a new city via a calculated drifting (*dérive*) through the old" (Wark, 2015: 17). This much misunderstood practice was pitted against the "rational" city of planners and architects like Le Corbusier. Crucially, the *dérive* was a conscious interposition of the unruly body in the managed spaces of the modern city and a deliberate attempt to demonstrate how "a new form of consciousness" could indeed flourish under the conditions of consumer capitalism. This was a revolutionary consciousness attuned to the potential for creative play in the interstices of the metropolis rather than, as Le Corbusier had intended, one suited to the progressive ideals of modernity.

For Stevphen Shukaitis, psychogeography gains new potential under postindustrial conditions where "the shaping of the city is embedded directly within the changing circuits of capital accumulation" (Shukaitis, 2016: 56). This is the posturban as constructed through immaterial labor; where the tourists employing selfie spots to accumulate images to post on social media produce the imagineered city even as they consume it. In light of this, practices that defy the strategic locating of bodies in urban space to optimize revenue flows under the guise of providing "experiences" and "consumer choice" have the potential to re-shape the city, particularly if they simultaneously expose how the maintenance of the posturban surface also relies on what Keller Easterling calls "infrastructure space" (Easterling, 2014: 13). This is the space that comprises the technologies reproducing the posturban in diverse global conditions; networked computing facilitated by fiber-optic cables deep under the city and images fed to the networks by satellites high above, which also facilitate surveillance and remote war. Ironically perhaps, they are also the technologies that have enabled doubt to be cast on the integrity of the organic body and have ushered in the possibility of a posthuman orientation towards the materiality of bodies and their effects in urban space.

Bradley Garrett describes urban explorers as "documenting archaeologies of the future" (2013: 129) when they unearth and photograph the remains of bankrupt businesses, failed building projects, and other decaying pockets of urban life which have escaped, however briefly, the overarching dominance of the spectacle. Here he is echoing Fredric Jameson's re-working of utopia to emphasize how its futurity can be built out of the "imaginary enclave[s]" which are "aberrant by-product[s]" of the continual differentiations that historically produce social space (Jameson, 2005: 15). But I want to suggest here that it could equally apply to the sense in which a posthuman politics does not pre-suppose either the body or the space appropriate to it. Urban exploration, in common with parkour, is a mis-use of urban monuments (Lamb, 2014); a psychogeography that deliberately defies the biopolitics inherent in the Vitruvian paradigm. We may not yet know what posthuman bodies can achieve; but we can make use of the new imaginaries the idea of posthuman selves makes possible, to subvert the architectonics of the city and re-invent the conditions of everyday life.

References

Agamben, G. (2005) *Metropolis*. Transcribed and translated by Arianna Bove. Available at: http://www.generation-online.org/p/fpagamben4.htm. Accessed 25 September 2017.

Anderson, B. (1983) *Imagined Communities: Reflections on the Origin and Spread of Nationalism*. London: Verso.

Atkinson, R. and Flint, J. (2004) Fortress UK? Gated Communities, the Spatial Revolt of the Elites and Time-Space Trajectories of Segregation. *Housing Studies* 19(6): 875–92.

Bauman, Z. (1992) *Mortality, Immortality & Other Life Strategies*. Cambridge: Polity.

Benjamin, W. (1999) *The Arcades Project*. Transl. H. Eiland and K. McLaughlin Cambridge, MA: The Belknap Press of Harvard University Press.

Bennett, T. (1995) *The Birth of the Museum: History, Theory, Politics*. London: Routledge.

Betsky, A. (1995) *Building Sex: Men, Women, Architecture and the Construction of Sexuality*. New York: William Morrow & Co.

Bowler, P.J. (1983) *Evolution: The History of an Idea*. Berkeley: University of California Press.

Braidotti, R. (2013) *The Posthuman*. Cambridge: Polity Press.

Chaplin, S. and Holding, E. (2002) Addressing the Post-Urban: Los Angeles, Las Vegas, New York. In: Leach, N. (ed.), *The Hieroglyphics of Space: Reading & Experiencing the Modern Metropolis*. London: Routledge, 185–99.

Cooper, M. (2008) *Life as Surplus: Biotechnology and Capitalism in the Neoliberal Era*. Seattle: University of Washington Press.
Corbusier, L. (1964/1933) *The Radiant City*. London: Faber & Faber.
Debord, G. (1995/1967) *The Society of the Spectacle*. Transl. D. Nicholson-Smith New York: Zone Books.
Dinhopl, A. and Gretzel, U. (2016) Selfie-Taking as Touristic Looking. *Annals of Tourism Research* 57: 126–39.
Easlea, B. (1981) *Science and Sexual Oppression*. London: George Weidenfeld & Nicholson.
Easterling, K. (2014) *Extrastatecraft: The Power of Infrastructure Space*. London: Verso.
Foucault, M. (1994/1967) *The Order of Things: An Archaeology of the Human Sciences*. New York: Vintage Books.
Foucault, M. (2003/1963) *The Birth of the Clinic*. London: Routledge.
Foucault, M. (2004) *Society Must Be Defended: Lectures at the Collège De France 1975–76*. Transl. D. Macey. Harmondsworth, UK: Penguin Books.
Frisby, D. (1994) The *Flâneur* in Social Theory. In: Tester, K. (ed.), *The Flâneur*. London: Routledge, 81–110.
Fuller, S. (2011) *Humanity 2.0: What It Means to Be Human Past, Present and Future*. Basingstoke: Palgrave.
Gandy, M. (2005) Cyborg Urbanization: Complexity and Monstrosity in the Contemporary City. *International Journal of Urban and Regional Research* 29(1): 26–49.
Garrett, B.L. (2013) *Explore Everything: Place-Hacking the City*. London: Verso.
Gerbaudo, P. (2017) *The Mask and the Flag: Populism, Citizenism and Global Protest*. London: Hurst & Co.
Gilman, S.L. (1985) *Difference and Pathology: Stereotypes of Sexuality, Race, and Madness*. Ithaca: Cornell University Press.
Graham, S. (2016) *Vertical: The City from Satellites to Bunkers*. London: Verso.
Haraway, D.J. (1991) *Simians, Cyborgs, and Women: The Reinvention of Nature*. London: Free Association Books.
Humanity+ (2016–2017). *Transhumanist Declaration*. Available at: http://humanityplus.org/philosophy/transhumanist-declaration/. Accessed 20 September 2017.
Jameson, F. (2005) *Archaeologies of the Future: The Desire Called Utopia and Other Science Fictions*. London: Verso.
Kirshenblatt-Gimblett, B. (1991) Objects of Ethnography. In: Karp, I. and Lavine, S.D. (eds.), *Exhibiting. Cultures: The Poetics and Politics of Museum Display*. Washington, DC: Smithsonian Institution Press, 386–443.
Lamb, M.D. (2014) Misuse of the Monument: The Art of Parkour and the Discursive Limits of a Disciplinary Architecture. *Journal of Urban Cultural Studies* 1(1): 107–26. DOI: 10.1386/jucs.1.1.107_1. Accessed 22 November 2017.
Lefebvre, H. (1991/1974) *The Production of Space*. Transl. D.O. Nicholson-Smith. Oxford: Blackwell.
Merrifield, A. (2014) *The New Urban Question*. London: Pluto.
Mike, D. (2007) Sand, Fear and Money in Dubai. In: Davis, M. and Monk, D.B. (eds.), *Evil Paradises: Dreamworlds of Neoliberalism*. New York: The New Press, 48–68.
Minton, A. (2017) *Big Capital: Who Is London For?* London: Penguin Random House.
Parker, S. (2004) *Urban Theory and the Urban Experience: Encountering the City*. London: Routledge.
Reid, R. (1995) "Death of the Family": Or, Keeping Human Beings Human. In: Halberstam, J. and Livingston, I. (eds.), *Posthuman Bodies*. Bloomington: Indiana University Press, 177–99.
Roden, D. (2015) *Posthuman Life: Philosophy at the Edge of the Human*. London: Routledge.
Schweik, S.M. (2009) *The Ugly Laws: Disability in Public*. New York: New York University Press.
Scott, R. Dir. (1982/1992/2007) *Blade Runner*. The Ladd Company, Shaw Bros., Warner Bros.
Sewell, D. (2009) *The Political Gene: How Darwin's Ideas Changed Politics*. London: Picador.

Shukaitis, S. (2016) *The Composition of Movements to Come: Aesthetics and Cultural Labor after the Avant-Garde*. London: Rowman & Littlefield International.
Soja, E.W. (2000) *Postmetropolis: Critical Studies of Cities and Regions*. Oxford: Blackwell.
Stallybrass, P. and White, A. (1986) *The Politics and Poetics of Transgression*. London: Methuen.
Stiegler, B. (1994) *Technics and Time, 1: The Fault of Epimetheus*. Transl. R. Beardsworth and Collins, G. Stanford: Stanford University Press.
Stiegler, B. (2009) *Technics and Time, 2: Disorientation*. Transl. S. Barker Stanford: Stanford University Press.
Swaine, J. and McCarthy, K. (2017) Young Black Men Again Faced Highest Rate of US Police Killings in 2016. *The Guardian*, 8 January 2017. Available at: https://www.theguardian.com/us-news/2017/jan/08/the-counted-police-killings-2016-young-black-men. Accessed 20 September 2017.
Thorpe, C. (2016) *Necroculture*. New York: Palgrave Macmillan.
Townsend, M. (2017) Deaths of Black Men in Custody Pose Challenge on 'Stephen Lawrence Level'. *The Guardian*, 3 September 2017. Available at: https://www.theguardian.com/uk-news/2017/sep/03/police-custody-deaths-black-men. Accessed 20 September 2017.
Vasudevan, A. (2017) *The Autonomous City: A History of Urban Squatting*. London: Verso.
Veel, K. (2003) The Irreducibility of Space: Labyrinths, Cities, Cyberspace. *Diacritics* 33(3–4): 151–72.
Wallenstein, S.-O. (2009) *Bio-Politics and the Emergence of Modern Architecture*. New York: Columbia University and Princeton Architectural Press.
Wark, M. (2015) *The Beach beneath the Street: The Everyday Life and Glorious Times of the Situationist International*. London: Verso.

Index

For Product Safety Concerns and Information please contact our EU representative GPSR@taylorandfrancis.com Taylor & Francis Verlag GmbH, Kaufingerstraße 24, 80331 München, Germany

Batch number: 10726294

Printed by Printforce, the Netherlands